The Principles of Clinical Cytogenetics

The Principles of Clinical Cytogenetics

Edited by

Steven L. Gersen, PhD

DIANON Systems, Inc., Stratford, CT

and

Martha B. Keagle, MEd

School of Allied Health Professions, University of Connecticut, Storrs, CT

Humana Press ✳ **Totowa, New Jersey**

Cover illustration: Figure 16 from Chapter 15, "Fluorescence *In Situ* Hybridization," by Jan K. Blancato.

Cover design by Patricia F. Cleary.

For additional copies, pricing for bulk purchases, and/or information about other Humana titles, contact Humana at the above address or at any of the following numbers: Tel.: 973-256-1699; Fax: 973-256-8341; E-mail: humana@humanapr.com; Website: http://humanapress.com

Printed in the United States of America. 10 9 8 7 6 5 4 3 2 1

Library of Congress Cataloging in Publication Data

The principles of clinical cytogenetics / [edited by] Steven L. Gersen and Martha B. Keagle.
       p. cm.
    Includes index.
    ISBN 0-89603-553-0 (alk. paper)
    1. Human chromosome abnormallities—Diagnosis. 2. Human cytogenetics. I. Gersen, Steven L. II. Keagle, Martha B.
    [DNLM: 1. Chromosomes—genetics. QH 600P957 1999]
    RB44.P75   1999
    616'.042—dc21
    DNLM/DLC   98-21743
    for Library of Congress   CIP

Preface

The study of human chromosomes plays a role in the diagnosis, prognosis, and monitoring of treatment involving conditions seen not only by medical geneticists and genetic counselors, but also by pediatricians, obstetrician/gynecologists, perinatologists, hematologists, oncologists, endocrinologists, pathologists, urologists, internists, and family practice physicians. In addition, cytogenetic testing is often an issue for hospital laboratory personnel and managed care organizations.

Few esoteric clinical laboratory disciplines have the potential to affect such a broad range of medical specialists, yet cytogenetics is often less well understood than most "specialized" testing.

One can attribute this to several causes:

- The cytogenetics laboratory is essentially the only setting in which living cells are required for traditional testing (fluorescence *in situ* hybridization [FISH] provides an exception to this rule). This unusal sample requirement is a potential source of confusion.
- Cytogenetics is still perceived, and rightly so, to be as much "art" as it is science in an era when most clinical testing is becoming more and more automated or "high tech."
- Genetics in general still does not receive sufficient emphasis in the training of medical personnel.

This issue has been complicated in recent years because, in an era of molecular medicine, chromosome analysis has become somewhat less of a stand alone discipline; as genes are mapped to chromosomes, traditional cytogenetics is often augmented with DNA analysis and/or FISH. The latter, often referred to as "molecular cytogenetics," represents the single most significant advance in this field in decades, and has become such an integral part of the typical cytogenetics laboratory, with such a wide variety of applications, that it warrants its own chapter in *The Principles of Clinical Cytogenetics.*

It is impossible to completely separate the relationships that exist today between the cytogenetics and the molecular genetics laboratories, from cases involving fragile-X-syndrome to those dealing with cancer patients, and for this reason, relevant molecular concepts are discussed in several chapters.

Entire volumes have been devoted to some of the topics covered in *The Principles of Clinical Cytogenetics*; these often serve as references or how-to manuals for those involved in providing genetics services, and in most cases provide a greater level of detail than is needed here. The purpose of the present book is to provide a comprehensive description of the basic concepts involved in chromosome analysis in a single volume, while at the same time producing a summary of sufficient depth to be of value to the practicing genetics professional. We hope that it will serve as a valuable reference to any health care provider, from the individual who utilizes cytogenetics routinely to someone who has need of it on rare occasions.

The Principles of Clinical Cytogenetics is divided into four sections. The first section provides an historical perspective and explanation of the concepts involved, including a detailed desciption of cytogenetic nomenclature and examples of its use. The second section is an overview of the processes involved. The purpose of this section is to provide a fundamental understanding of the labor-intensive nature of chromosome analysis. It is not, however, a "laboratory manual"; detailed protocols for laboratory use are available elsewhere and are not appropriate in this setting. The third section comprises the main focus of this book, namely, the various applications of chromosome analysis in clinical settings and the significance of abnormal results. The final section connects cytogenetics to the broader field of clinical genetics, with discussions of synergistic technologies and genetic counseling.

We gratefully acknowledge the hard work and attention to detail provided by the individuals who authored each chapter of *The Principles of Clinical Cytogenetics*, and thank our publisher for supporting this effort.

Steven L. Gersen, PhD
Martha B. Keagle, MEd

Contents

Contributors

JAN K. BLANCATO, PhD • *Institute of Human and Molecular Genetics, Georgetown University Medical Center, Washington, DC*

ANNEMARIE W. BLOCK, PhD • *Clinical Cytogenetics Laboratory, Division of Pathology and Laboratory Medicine, Roswell Park Cancer Institute, Buffalo, NY*

JUDITH D. BROWN, MS • *Diagnostic Genetic Sciences Program, School of Allied Health, University of Connecticut, Storrs, CT*

YAEL FURMAN, MS • *Genzyme Genetics, Tarrytown, NY*

STEVEN L. GERSEN, PhD • *DIANON Systems, Inc., Stratford, CT*

JACQUES GINESTET, MS • *Applied Imaging, Santa Clara, CA*

PATRICIA N. HOWARD-PEEBLES, PhD • *Genetics and IVF Institute, Fairfax, VA, and Department of Human Genetics, Medical College of Virginia, Richmond, VA*

KATHLEEN KAISER-ROGERS, PhD • *Department of Pediatrics, University of North Carolina at Chapel Hill, Chapel Hill, NC*

MARTHA B. KEAGLE, MEd • *Diagnostic Genetic Sciences Program, School of Allied Health, University of Connecticut, Storrs, CT*

CHRISTOPHER MCALEER • *Applied Imaging, Santa Clara, CA*

SOLVEIG M. V. PFLUEGER, PhD, MD • *Department of Pathology, Bay State Medical Center, Tufts University School of Medicine, Springfield, MA*

CYNTHIA M. POWELL, MD • *Division of Genetics and Metabolism, Department of Pediatrics, The University of North Carolina at Chapel Hill, Chapel Hill, NC*

LINDA MARIE RANDOLPH, MD • *Alfigen/The Genetics Institute, Pasadena, CA*

KATHLEEN RAO, PhD • *Department of Pediatrics, The University of North Carolina at Chapel Hill, Chapel Hill, NC*

AVIRACHAN THARAPEL, PhD • *Department of Pediatric and Ob/Gyn Genetics, University of Tennessee, Memphis, TN*

JIN-CHEN C. WANG, MD • *Alfigen/The Genetics Institute, Pasadena, CA*

I
Basic Concepts and Background

1

History of Clinical Cytogenetics

Steven L. Gersen, Ph.D.

The beginning of the science of human cytogenetics is generally attributed to Walther Flemming, an Austrian cytologist and professor of anatomy, who published the first illustrations of human chromosomes in 1882. Flemming also referred to the stainable portion of the nucleus as *chromatin* and first used the term *mitosis (1)*. In 1888, Waldeyer introduced the word *chromosome*, from the Greek words for "colored body" *(2)*, and several prominent scientists of the day began to formulate the idea that determinants of heredity were carried on chromosomes. After the "rediscovery" of Mendelian inheritance in 1900, Sutton (and, independently at around the same time, Boveri) formally developed a "chromosome theory of inheritance" *(3,4)*. Sutton combined the disciplines of cytology and genetics when he referred to the study of chromosomes as *cytogenetics*.

Owing in part to improvements in optical lenses, stains and tissue manipulation techniques during the late 19th and early 20th centuries, the study of cytogenetics continued, with an emphasis placed by some on determining the correct number of chromosomes, as well as the sex chromosome configuration, in humans. Several reports appeared, with differing estimates of these. For example, in 1912, von Winiwarter concluded that men have 47 chromosomes and women 48 *(5)*. Then, in 1923, T.S. Painter studied (meiotic) chromosomes derived from the testicles of several men who had been incarcerated, castrated, and ultimately hanged in the Texas State Insane Asylum. Based on this work, Painter definitively reported the human diploid chromosome number to be 48 (double the 24 bivalents he saw), even though, 2 years earlier, he had preliminarily reported that some of his better samples produced a diploid number of 46 *(6)*. At this time, Painter also proposed the X and Y sex chromosome mechanism in humans. One year later, Levitsky formulated the term *karyotype* to refer to the ordered arrangement of chromosomes *(7)*.

Despite continued technical improvements, there was clearly some difficulty in properly visualizing or discriminating between individual chromosomes. Even though Painter's number of 48 human chromosomes was reported somewhat conservatively, it was increasingly treated as fact with the passage of time, and was "confirmed" several times over the next few decades. For example, in 1952, T.C. Hsu reported that, rather than depending upon histologic sections, examination of chromosomes could be facilitated if one studied cells grown with tissue culture techniques published by Fisher *(8)*. Hsu then demonstrated the value of this method by using it to examine human embry-

From: *The Principles of Clinical Cytogenetics*
Edited by: S. Gersen and M. Keagle © Humana Press Inc., Totowa, NJ

onic cell cultures, from which he produced both mitotic metaphase drawings and an ideogram *(9)* of all 48 human chromosomes!

As with other significant discoveries, correcting this inaccuracy required an unplanned event—a laboratory error. Its origin can be found in the addendum that appears at the end of Hsu's article:

> It was found after this article had been sent to press that the well-spread metaphases were the result of an accident. Instead of being washed in isotonic saline, the cultures had been washed in hypotonic solution before fixation.

The hypotonic solution caused water to enter the cells via osmosis, which swelled the cell membranes and separated the chromosomes, making them easier to visualize. This accident was the key that unlocked the future of clinical human cytogenetics. Within one year, Hsu, realizing the potential of this fortuitous event, reported a "hypotonic shock" procedure *(10)*. By 1955, Ford and Hamerton had modified this technique, and had also worked out a method for pretreating cells grown in culture with colchicine so as to destroy the mitotic spindle apparatus and thus accumulate dividing cells in metaphase *(11)*. Joe Hin Tjio, an American-born Indonesian, learned about these procedures and worked with Hamerton and Ford to improve upon them further.

In November of 1955, Tjio was invited to Lund, Sweden to work on human embryonic lung fibroblast cultures in the laboratory of his colleague, Albert Levan, a Spaniard who had learned the colchicine and hypotonic method in Hsu's laboratory at the Sloan-Kettering Institute in New York. Tjio and Levan optimized the colchicine/hypotonic method for these cells, and in January of 1956 (after carefully reviewing images from decades of previously reported work) diplomatically reported that the human diploid chromosome number appeared to be 46, not 48 *(12)*. They referenced anecdotal data from a colleague who had been studying liver mitoses from aborted human embryos in the spring of 1955, but temporarily abandoned the research "because the workers were unable to find all the 48 human chromosomes in their material; as a matter of fact, the number 46 was repeatedly counted in their slides." Tjio and Levan concluded their article:

> . . . we do not wish to generalize our present findings into a statement that the chromosome number of man is $2n = 46$, but it is hard to avoid the conclusion that this would be the most natural explanation of our observations.

What was dogma for over 30 years had been overturned in one now classic paper. Ford and Hamerton soon confirmed Tjio and Levan's finding *(13)*. The era of clinical cytogenetics was at hand. It would take three more years to arrive, however, and it would begin with the identification of four chromosomal syndromes.

The concept that an abnormality involving the chromosomes could have a phenotypic effect was not original. In 1932, Waardenburg suggested that Down syndrome could perhaps be the result of a chromosomal aberration *(14)*, but the science of the time could neither prove nor disprove his idea; this would take almost three decades. In 1958, Lejeune studied the chromosomes of fibroblast cultures from patients with Down syndrome, and in 1959 described an extra chromosome in each these cells *(15)*. The trisomy was reported to involve one of the smallest pairs of chromosomes, and would eventually be referred to as trisomy 21. Lejeune had proved Waardenburg's hypothesis by reporting the first example of a chromosomal syndrome in humans, and in Decem-

Fig. 1. Jérôme Lejeune receives a Joseph P. Kennedy Jr. Foundation International Award for demonstrating that Down syndrome results from an extra chromosome. (Photo provided by the John F. Kennedy Library, Boston, MA.)

ber of 1962, he received one of the first Joseph Kennedy Jr. Foundation International Awards for his work (see **Fig. 1**).

Three more chromosomal syndromes, all believed to involve the sex chromosomes, were also described in 1959. Ford reported that females with Turner syndrome have 45 chromosomes, apparently with a single X chromosome and no Y *(16),* and Jacobs and Strong demonstrated that men with Klinefelter syndrome have 47 chromosomes, with the additional chromosome belonging to the group that contained the X chromosome *(17).* A female with sexual dysfunction was also shown by Jacobs to have 47 chromosomes and was believed to have an XXX sex chromosome complement *(18).*

The sex chromosome designation of these syndromes was supported by (and helped explain) a phenomenon that had been observed 10 years earlier. In 1949, Murray Barr was studying fatigue in repeatedly stimulated neural cells of the cat *(19).* Barr observed a small stained body on the periphery of some interphase nuclei, and his records were detailed enough for him to realize that this was present only in the nuclei of female cats. This object, referred to as sex chromatin (now known as X chromatin or the Barr body), is actually the inactivated X chromosome present in nucleated cells of all normal female mammals but absent in normal males. The observation that the Turner syndrome, Klinefelter syndrome, and putative XXX patients had zero, one, and two Barr bodies, respectively, elucidated the mechanism of sex determination in humans, confirming for the first time that it is the presence or absence of the Y chromosome that determines maleness, not merely the number of X chromosomes present, as in *Drosophila.* In 1961, the single active X chromosome mechanism of X-dosage compensation in mammals was developed by Mary Lyon *(20)* and has been since known as the Lyon hypothesis.

It was not long after Lejeune's report of the chromosomal basis of Down syndrome that other autosomal abnormalities were discovered. In the April 9, 1960 edition of The Lancet, Patau et al. described two similar infants with an extra "D group" chromosome who had multiple anomalies quite different from those seen in Down syndrome *(21)*. In the same journal, Edwards et al. described "a new trisomic syndrome" in an infant girl with yet another constellation of phenotypic abnormalities and a different autosomal trisomy *(22)*. The former became known as Patau syndrome or "D trisomy" and the latter as Edward syndrome or "E trisomy." Patau's article incredibly contains a typographical error and announces that the extra chromosome "belongs to the E group," and Edwards reported that "the patient was . . . trisomic for the no. 17 chromosome," but we now know these syndromes to be trisomies 13 and 18, respectively.

Also in 1960, Nowell and Hungerford reported the presence of the "Philadelphia chromosome" in chronic myelogenous leukemia, demonstrating, for the first time, an association between chromosomes and cancer *(23)*.

In 1963 and 1964, Lejeune et al. reported that three infants with the *cri du chat* ("cat cry") syndrome of phenotypic anomalies, which includes severe mental retardation and a characteristic kittenlike mewing cry, had a deletion of the short arm of a B-group chromosome, designated as chromosome 5 *(24,25)*. Within 2 years, Jacobs et al. described "aggressive behavior, mental subnormality, and the XYY male" *(26)*, and the chromosomal instabilities associated with Bloom syndrome and Fanconi anemia were reported *(27,28)*.

Additional technical advancements had facilitated the routine study of patient karyotypes. In 1960, Peter Nowell observed that the kidney bean extract phytohemagglutinin, used to separate red and white blood cells, stimulated lymphocytes to divide. He introduced its use as a mitogen *(23,29)*, permitting a peripheral blood sample to be used for chromosome analysis. This eliminated the need for bone marrow aspiration, which had previously been the best way to obtain a sufficient number of spontaneously dividing cells. It was now feasible to produce mitotic cells suitable for chromosome analysis from virtually any patient.

Yet, within nine years of the discovery of the number of chromosomes in humans, only three autosomal trisomies, four sex chromosome aneuploidies, a structural abnormality (a deletion), an acquired chromosomal abnormality associated with cancer, and two chromosome breakage disorders had been described as recognizable "chromosomal syndromes." A new clinical laboratory discipline had been created; was it destined to be restricted to the diagnosis of a few abnormalities?

This seemed likely. Even though certain pairs were distinguishable by size and centromere position, individual chromosomes could not be identified, and as a result, patient-specific chromosome abnormalities could be observed but not defined. Furthermore, the existence of certain abnormalities, such as inversions involving a single chromosome arm (so-called *para*centric inversions) could be hypothesized but not proven, because they could not be visualized. Indeed, it seemed that without a way to definitively identify each chromosome (and more importantly, regions of each chromosome), this new field of medicine would be limited in scope to the study of a few disorders.

For three years, clinical cytogenetics was so relegated. Then, in 1968, Torbjörn Caspersson observed that, when plant chromosomes were stained with fluorescent quinacrine compounds, they did not fluoresce uniformly, but rather produced a series of bright and dull areas across the length of each chromosome. Furthermore, each pair fluoresced with a different pattern, so that previously indistinguishable chromosomes could now be recognized *(30)*.

Caspersson then turned his attention from plants to the study of human chromosomes. He hypothesized that the quinacrine derivative quinacrine mustard (QM) would preferentially bind to guanine residues, and that C–G-rich regions of chromosomes should therefore produce brighter "striations," as he initially referred to them, whereas A–T-rich regions would be dull. Although it ultimately turned out that it is the A–T-rich regions that fluoresce brightly, and that ordinary quinacrine dihydrochloride works as well as QM, by 1971 Caspersson had successfully produced and reported a unique "banding" pattern for each human chromosome pair *(31,32)*.

For the first time, each human chromosome could be positively identified. The method, however, was cumbersome. It required a relatively expensive fluorescence microscope and a room that could be darkened, and the fluorescence tended to fade or "quench" after a few minutes, making real-time microscopic analysis difficult.

These difficulties were overcome a year later, when Drets and Shaw described a method of producing similar chromosomal banding patterns using an alkali and saline pretreatment followed by staining with Giemsa, a compound developed for identification, in blood smears, of the protozoan that causes malaria *(33)*. Even though some of the chromosome designations proposed by Drets and Shaw have been changed (essentially in favor of those advocated by Caspersson), this method, and successive variations of it, facilitated widespread application of clinical cytogenetic techniques. Although the availability of individuals with the appropriate training and expertise limited the number and capacity of laboratories that could perform these procedures (in some ways still true today), the technology itself was now within the grasp of any facility.

What followed was a cascade of defined chromosomal abnormalities and syndromes: aneuploidies, deletions, microdeletions, translocations, inversions (including the paracentric variety), insertions, and mosaicisms, plus an ever-increasing collection of rearrangements and other cytogenetic anomalies associated with neoplasia, and a seemingly infinite number of patient- and family-specific rearrangements.

Thanks to the host of research applications made possible by the precise identification of smaller and smaller regions of the karyotype, genes began to be mapped to chromosomes at a furious pace. The probes that resulted from such research have given rise to the discipline of molecular cytogenetics, which utilizes the techniques of fluorescence *in situ* hybridization (FISH). In recent years, this exciting development and the many innovative procedures derived from it have created even more interest in the human karyotype.

This brings us to the present. Some 900,000 cytogenetic analyses are now performed annually in approximately 500 laboratories worldwide *(34)*, and this testing is now often the standard of care. Pregnant women over the age of 35, or those with certain serum-screening results, are routinely offered prenatal cytogenetic analysis. For children with phenotypic and/or mental difficulties, and for couples experiencing reproductive problems, chromosome analysis has become a routine part of their clinical work-up. Cytogenetics also provides information vital to the diagnosis, prognosis, and monitoring of treatment for a variety of cancers.

It was really not so long ago that we had 48 chromosomes. One has to wonder whether Flemming, Waldeyer, Tjio, Levan, Hsu, or Lejeune could have predicted the modern widespread clinical use of chromosome analysis. But perhaps it is even more exciting to wonder what lies ahead for medical and molecular cytogenetics as we enter the 21st century.

REFERENCES

1. Flemming, W. (1882) In *Zellsubstanz, Kern und Zellteilung.* Vogel, Leipzig.
2. Waldeyer, W. (1888) Über Karyokineze und ihre Beziehung zu den Befruchtung-svorgängen. *Arch. Mikr. Anat.* **32,** 1.
3. Sutton, W.S. (1903) The chromosomes in heredity. *Biol. Bull. Wood's Hole* **4,** 231.
4. Boveri, T. (1902) Über mehrpolige Mitosen als Mittel zur Analyse des Zellkerns. *Verh. Physmed. Ges. Würzburg, N.F.* **35,** 67–90.
5. Von Winiwarter, H. (1912) Études sur la spermatogenèse humaine. I. Cellule de Sertoli. II. Hétérochromosome et mitoses de l'epitheleum seminal. *Arch. Biol. (Liege)* **27,** 91–189.
6. Painter, T.S. (1923) Studies in mammalian spermatogenesis. II. The spermatogenesis of man. *J. Exp. Zool.* **37,** 291–336.
7. Levitsky, G.A. (1924) *Materielle Grundlagen der Vererbung.* Kiew, Staatsverlag.
8. Fisher, A. (1946) *Biology of Tissue Cells.* Cambridge University Press.
9. Hsu, T.C. (1952) Mammalian chromosomes in vitro. I. The karyotype of man. *J. Hered.* **43,** 167–172.
10. Hsu, T.C. and Pomerat, C.M. (1953) Mammalian chromosomes in vitro. II. A method for spreading the chromosomes of cells in tissue culture. *J. Hered.* **44,** 23–29.
11. Ford, C.E. and Hamerton, J.L. (1956) A colchicine, hypotonic citrate, squash sequence for mammalian chromosomes. *Stain Technol.* **31,** 247.
12. Tjio, H.J. and Levan, A. (1956) The chromosome numbers of man. *Hereditas* **42,** 1–6.
13. Ford, C.E. and Hamerton, J.L. (1956) The chromosomes of man. *Nature* **178,** 1020–1023.
14. Waardenburg, P.J. (1932) Das menschliche Auge und seine Erbanlagen. *Bibliogr. Genet.* **7.**
15. Lejeune, J. Gautier, M. and Turpin, R. (1959) Étude des chromosomes somatiques de neuf enfants mongoliens. *Compt. Rend. Acad. Sci.* **248,** 1721–1722.
16. Ford, C.E., Miller, O.J., Polani, P.E., Almeida, J.C. de, and Briggs, J.H. (1959) A sex-chromosome anomaly in a case of gonadal dysgenesis (Turner's syndrome). *Lancet* **i,** 711–713.
17. Jacobs, P.A. and Strong, J.A. (1959) A case of human intersexuality having a possible XXY sex-determining mechanism. *Nature* **183,** 302–303.
18. Jacobs, P.A., Baikie, A.G., MacGregor, T.N., and Harnden, D.G. (1959) Evidence for the existence of the human "superfemale." *Lancet* **ii,** 423–425.
19. Barr, M.L. and Bertram, L.F. (1949) A morphological distinction between neurones of the male and the female and the behavior of the nucleolar satellite during accelerated nucleoprotein synthesis. *Nature* **163,** 676–677.
20. Lyon, M.F. (1961) Gene action in the X-chromosome of the mouse. *Nature* **190,** 372–373.
21. Patau, K., Smith, D.W., Therman, E., and Inhorn, S.L. (1960) Multiple congenital anomaly caused by an extra chromosome. *Lancet* **i,** 79-0–793.
22. Edwards, J.H., Harnden, D.G., Cameron, A.H., Cross, V.M., and Wolff, O.H. (1960) A new trisomic syndrome. *Lancet* **i,** 711–713.
23. Nowell, P.C. and Hungerford, D.A. (1960) A minute chromosome in human chronic granulocytic leukemia. *Science* **132,** 1497.
24. Lejeune, J., Lafourcade, J., Berger, R., Vialatte, J., Boeswillwald, M., Seringe, P., and Turpin, R. (1963) Trois cas de délétion partielle du bras court d'un chromosome 5. *C.R. Acad. Sci. (Paris)* **257,** 3098–3102.
25. Lejeune, J., Lafourcade, J., Grouchy, J. de, Berger, R., Gautier, M., Salmon, C., and Turpin,

R. (1964) Délétion partielle du bras court du chromosome 5. Individualisation d'un nouvel état morbide. *Sem. Hôp. Paris* **18,** 1069–1079.

26. Jacobs, P.A., Brunton, M., Melville, M.M., Brittain, R.P., and McClermont, W.F. (1965) Aggressive behavior, mental subnormality and the XYY male. *Nature* **208,** 1351–1352.

27. Schroeder, T.M., Anschütz, F., and Knopp, F. (1964) Spontane Chromosomenaberrationen bei familiärer Panmyelopathie. *Hum. Genet.* **I,** 194–196.

28. German, J., Archibald, R., and Bloom, D. (1965) Chromosomal breakage in a rare and probably genetically determined syndrome of man. *Science* **148,** 506.

29. Nowell, P.C. (1960) Phytohaemagglutinin: an initiator of mitosis in cultures of normal human leukocytes. *Cancer Res.* **20,** 462–466.

30. Caspersson, T., Farber, S., Foley, G.E., Kudynowski, J., Modest, E.J., Simonssen, E., Waugh, U., and Zech, L. (1968) Chemical differentiation along metaphase chromosomes. *Exp. Cell Res.* **49,** 219–222.

31. Caspersson, T., Zech, L., and Johansson, C. (1970) Differential binding of alkylating fluorochromes in human chromosomes. *Exp. Cell Res.* **60,** 315–319.

32. Caspersson, T., Lomakka, G., and Zech, L. (1971) The 24 fluorescence patterns of the human metaphase chromosomes—distinguishing characters and variability. *Hereditas* **67,** 89–102.

33. Drets, M.E. and Shaw, M.W. (1971) Specific banding patterns in human chromosomes. *Proc, Nat. Acad. Sci, USA* **68,** 2073–2077.

34. Knudsen, T., ed. (1997) 1996–1997 *AGT International Laboratory Directory.* Association of Genetic Technologist, Lenexa, KS.

2

DNA, Chromosomes, and Cell Division

Martha Keagle, M.Ed. and Judith D. Brown, M.S.

INTRODUCTION

The molecule deoxyribonucleic acid (DNA) is the raw material of inheritance, and ultimately influences all aspects of the structure and functioning of the human body. A single molecule of DNA, along with associated proteins, comprises a chromosome. Chromosomes are located in the nuclei of all human cells (with the exception of mature red blood cells), and each human cell contains 23 different pairs of chromosomes.

Genes are functional units of genetic information that reside on each of the 23 pairs of chromosomes. These units are linear sequences of nitrogenous bases that code for protein molecules necessary for the proper functioning of the body. The genetic information contained within the chromosomes is copied and distributed to newly created cells during cell division. The structure of DNA provides the answer to how it is precisely copied with each cell division, and to how proteins are synthesized.

DNA STRUCTURE

James Watson and Francis Crick elucidated the molecular structure of DNA in 1953 using X-ray diffraction data collected by Rosalind Franklin and Maurice Wilkins, and model building techniques advocated by Linus Pauling (1,2). Watson and Crick proposed the double helix—a twisted, spiral ladder structure consisting of two long chains wound around each other and held together by hydrogen bonds. DNA is composed of repeating units, the nucleotides. Each nucleotide consists of a deoxyribose sugar, a phosphate group, and one of four nitrogen-containing bases: adenine (A), guanine (G), cytosine (C), or thymine (T). Adenine and guanine are purines with a double-ring structure, whereas cytosine and thymine are smaller pyrimidine molecules with a single-ring structure. Two nitrogenous bases positioned side by side on the inside of the double helix form one rung of the molecular ladder. The sugar and phosphate groups form the backbone, or outer structure of the helix. The fifth (5′) carbon of one deoxyribose molecule and the third (3′) carbon of the next deoxyribose are joined by a covalent phosphate linkage. This gives each strand of the helix a chemical orientation with the two strands running opposite or antiparallel to one another.

Biochemical analyses performed by Erwin Chargaff showed that the nitrogenous bases of DNA were not present in equal proportions and that the proportion of these bases varied from one species to another (3). Chargaff noticed, however, that concen-

From: *The Principles of Clinical Cytogenetics*
Edited by: S. Gersen and M. Keagle © Humana Press Inc., Totowa, NJ

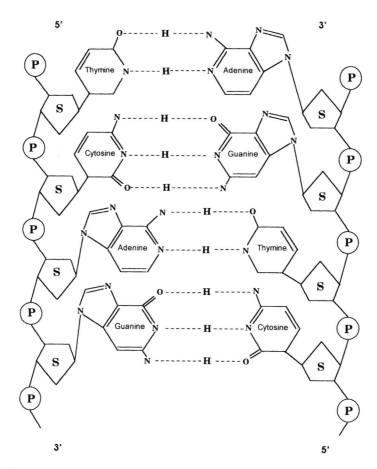

Fig. 1. DNA structure. Schematic representation of a DNA double helix, unwound to show the complementarity of bases and the antiparallel structure of the phosphate (P) and sugar (S) backbone strands.

trations of guanine and cytosine were always equal, as were the concentrations of adenine and thymine. This finding became known as Chargaff's rule. Watson and Crick postulated that to fulfill Chargaff's rule and to maintain a uniform shape to the DNA molecule, there must be a specific complementary pairing of the bases: adenine must always pair with thymine and guanine must always pair with cytosine. Each strand of DNA therefore contains a nucleotide sequence that is complementary to its partner. The linkage of these complementary nitrogenous base pairs holds the antiparallel strands of DNA together. Two hydrogen bonds link the adenine and thymine pairs, while three hydrogen bonds link the guanine and cytosine pairs (**Fig. 1**). The complementarity of DNA strands is what allows the molecule to replicate faithfully. The sequence of bases is critical for DNA function because genetic information is determined by the order of the bases along the DNA molecule.

DNA SYNTHESIS

The synthesis of a new molecule of DNA is called replication. This process requires many enzymes and cofactors. The first step of the process involves breakage of hydrogen bonds that hold the DNA strands together. DNA helicases and single-strand binding proteins separate the strands and keep the DNA exposed at many points along the length of the helix during replication. The area of DNA at the active region of separation is a Y-shaped structure referred to as a replication fork. Replication forks originate at structures called replication bubbles, which in turn are at DNA sequences called replication origins. The molecular sequence of the replication origins has not been completely characterized. Replication takes place on both strands, but nucleotides can be added only to the 3′ end of an existing strand. The separated strands of DNA serve as templates for production of complementary strands of DNA following Chargaff's rule of base pairing.

The process of DNA synthesis differs for the two strands of DNA owing to its anti-parallel structure. Replication is straightforward on the leading strand. The enzyme DNA polymerase I facilitates the addition of complementary nucleotides to the 3′ end of a newly forming strand of DNA. To add further nucleotides, DNA polymerase I requires the 3′ hydroxyl end of a base-paired strand.

DNA synthesis on the lagging strand is accomplished by the formation of small segments of nucleotides called Okazaki fragments *(4)*. After separation of the strands, the enzyme DNA primase uses ribonucleotides to form a ribonucleic acid primer.

The structure of ribonucleic acid (RNA) is similar to that of DNA, except that each nucleotide in RNA has a ribose sugar instead of deoxyribose and the pyrimidine thymine is replaced by another pyrimidine, uracil (U). RNA also differs from DNA in that it is a single-stranded molecule. An RNA primer is found at the beginning of each Okazaki segment to be copied. This primer provides a 3′ hydroxyl group, and is important for the efficiency of the replication process. The ribonucleic acid primer then attracts DNA polymerase I. DNA polymerase I brings in the nucleotides and also removes the RNA primer and any mismatches that occur during the replication process. Okazaki fragments are later joined by the enzyme DNA ligase. The process of replication is semiconservative, as the net result is creation of two identical DNA molecules, each consisting of a parent DNA strand and a newly synthesized DNA strand. The new DNA molecule grows as hydrogen bonds form between the complementary bases (**Fig. 2**).

PROTEIN SYNTHESIS

The genetic information of DNA is stored as a code, a linear sequence of nitrogenous bases. Every 3 base triplet codes for an amino acid; these amino acids are subsequently linked together to form a protein molecule. The process of protein synthesis involves several types of ribonucleic acids.

The first step in protein synthesis is transcription. During this process, DNA is copied into a complementary piece of messenger RNA (mRNA). Transcription is controlled by the enzyme RNA polymerase, which links ribonucleotides together in a sequence complementary to the DNA template strand. The attachment of RNA polymerase to a promoter region, a specific sequence of bases that varies from gene to gene, starts transcription. RNA polymerase moves off the template strand at a termination sequence to complete the synthesis of an mRNA molecule (**Fig. 3**).

Messenger RNA (mRNA) is modified at this point by the removal of introns, segments of DNA that do not code for a mRNA product. In addition, some nucleotides are removed from the 3′ end of the molecule, and a string of adenine nucleotides are added.

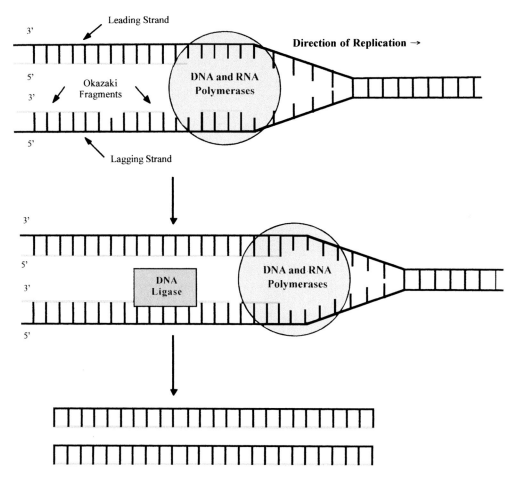

Fig. 2. Semiconservative replication. Complementary nucleotides are added directly to the 3′ end of the leading strand, while the lagging strand is copied by the formation of Okazaki fragments.

This poly(A) tail helps in the transport of mRNA molecules to the cytoplasm. Another modification is the addition of a cap to the 5′ end of the mRNA, which serves to aid in attachment of the mRNA to the ribosome during translation. These alterations to mRNA are referred to as mRNA processing (**Fig. 4**). At this point, mRNA, carrying the information necessary to synthesize a specific protein, is transferred from the nucleus into the cytoplasm of the cell where it then associates with ribosomes. Ribosomes, composed of ribosomal RNA (rRNA) and protein, are the site of protein synthesis. Ribosomes consist of two subunits that come together with mRNA to read the coded instructions on the mRNA molecule.

The next step in protein synthesis is translation. A chain of amino acids is synthesized during translation by using the newly transcribed mRNA molecule as a template, with the help of a third ribonucleic acid, transfer RNA (tRNA). Marshall Nirenberg and Har Gobind Khorana determined that three nitrogen bases on an mRNA molecule constitute a codon *(5,6)*. With four nitrogenous bases there are 64 possible three-base

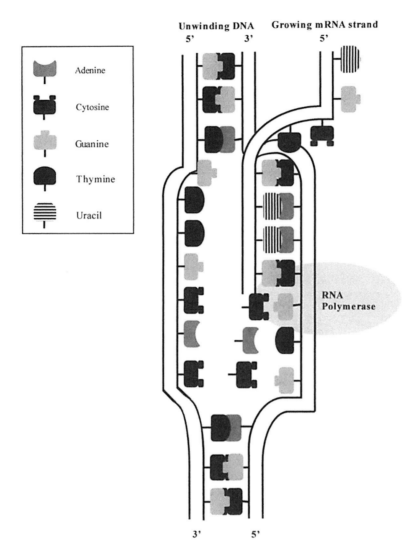

Unwinding DNA   Growing mRNA strand
5'   3'   5'

Adenine

Cytosine

Guanine

Thymine

Uracil

RNA
Polymerase

3'   5'

Fig. 3. Transcription. A DNA molecule is copied into messenger RNA with the help of RNA polymerase.

codons. Sixty-one of these code for specific amino acids, whereas the other three are "stop" codons that signal the termination of protein synthesis. There are only 20 amino acids but 61 codons, therefore, most amino acids are coded for by more than one mRNA codon. This redundancy in the genetic code is referred to as degeneracy.

Transfer RNA (tRNA) molecules contain "anticodons," nucleotide triplets that are complementary to the codons on mRNA. Each tRNA molecule has attached to it the specific amino acid it codes for. Ribosomes read mRNA one codon at a time. Transfer RNA molecules transfer the specific amino acids to the growing protein chain (**Fig. 5**). The amino acids are joined to this chain by peptide bonds. This process is continued until a stop codon is reached. The new protein molecule is then released into the cell milieu and the ribosomes split apart (**Fig. 6**).

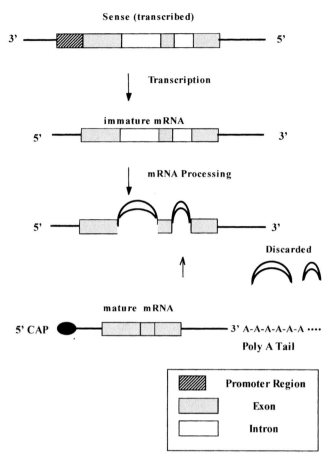

Fig. 4. Messenger RNA processing. The transcribed strand of DNA is modified to produce a mature messenger RNA transcript.

DNA ORGANIZATION

Human chromatin consists of a single continuous molecule of DNA complexed with histone and nonhistone proteins. The DNA in a single human diploid cell, if stretched out, would be approximately 2 m in length *(7)* and therefore must be condensed considerably to fit within the cell nucleus. There are several levels of DNA organization that allow for this.

The DNA helix itself is the first level of condensation. Next, two molecules of each of the histones H2A, H2B, H3, and H4 form a protein core, the octamer. The DNA double helix winds twice around the octamer to form a 10 nm nucleosome, the basic structural unit of chromatin. Adjacent nucleosomes are pulled together by a linker segment of the histone H1. Repeated, this gives the chromatin the appearance of "beads on a string." Nucleosomes are further coiled into a 30 nm solenoid, with each turn of the solenoid containing about six nucleosomes. The solenoids are packed into DNA looped domains attached to a nonhistone protein matrix. Attachment points of each loop are fixed along the DNA. The looped domains coil further to give rise to highly compacted units, the chromosomes, which are visible with the light microscope only during cell

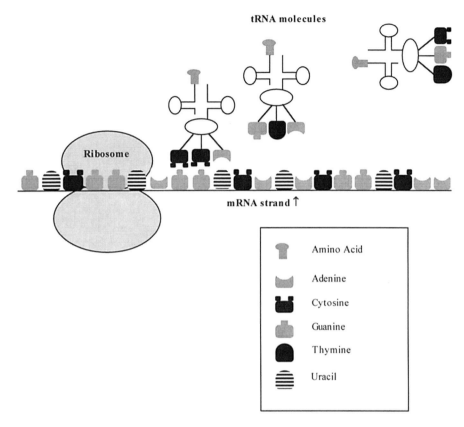

tRNA molecules

Ribosome

mRNA strand ↑

Amino Acid	
Adenine	
Cytosine	
Guanine	
Thymine	
Uracil	

Fig. 5. Translation. Transfer RNA molecules bring in specific amino acids according to the triplet codon instructions of messenger RNA that are read at the ribosomes.

division. Chromosomes reach their greatest extent of condensation during mitotic metaphase (**Fig. 7**).

CHROMOSOME STRUCTURE

A metaphase chromosome consists of two sister chromatids, each of which is comprised of a contracted and compacted double helix of DNA. The centromere, telomere, and nucleolar organizing regions are functionally differentiated areas of the chromosomes (**Fig. 8**).

The Centromere

The centromere is a constriction visible on metaphase chromosomes where the two sister chromatids are joined together. The centromere is essential to the survival of a chromosome during cell division. Interaction with the mitotic spindle during cell division occurs at the centromeric region. Mitotic spindle fibers are the functional elements that separate the sister chromatids during cell division.

Human chromosomes are classified based on the position of the centromere on the chromosome. The centromere is located near the middle in metacentric chromosomes, near one end in acrocentric chromosomes, and it is between the middle and end in submetacentric chromosomes. The kinetochore apparatus functions at the molecular

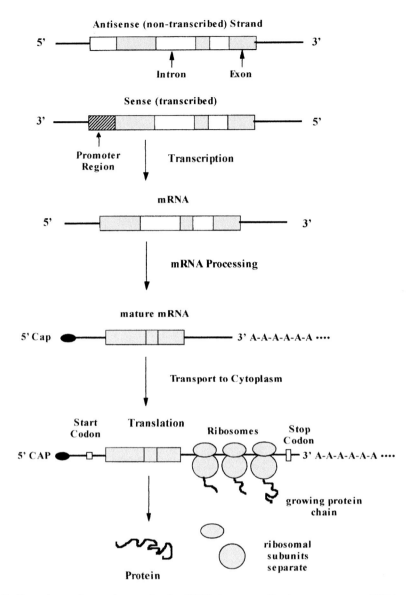

Fig. 6. Overview of protein synthesis. DNA is transcribed to messenger RNA, which is modified to a mature transcript and then transferred to the cytoplasm of the cell. The codons are read at the ribosomes and translated with the help of transfer RNA. The chain of amino acids produced during translation is joined by peptide bonds to form a protein molecule.

level to attach the chromosomes to the spindle fibers during cell division. Although the kinetochore is located in the region of the centromere, it should not be confused with the centromere.

The Nucleolar Organizer Regions

The satellite stalks of human acrocentric chromosomes contain the nucleolar organizer regions (NORs), so called because this is where nucleoli form in interphase cells.

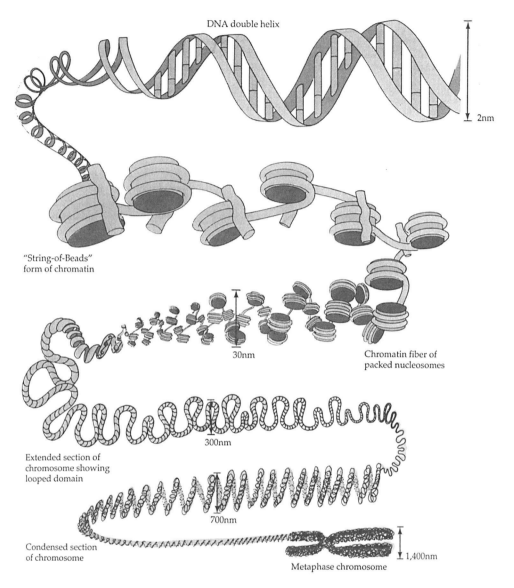

DNA double helix

2nm

"String-of-Beads"
form of chromatin

30nm

Chromatin fiber of
packed nucleosomes

300nm

Extended section of
chromosome showing
looped domain

700nm

Condensed section
of chromosome

1,400nm

Metaphase chromosome

Fig. 7. The levels of DNA organization. (From Mange and Mange [20]. Reprinted with permission).

NORs are also the site of ribosomal RNA genes and production of rRNA. In humans there are theoretically 10 nucleolar organizer regions, although all may not be active during any given cell cycle.

The Telomeres

The telomeres are the physical end of chromosomes. Telomeres contain repeats of a short TTAGGG sequence of nitrogenous bases (8). Telomeres act as protective caps to chromosome ends, preventing end-to-end fusion of chromosomes and DNA degradation resulting after chromosome breakage. A nonhistone protein complexed with

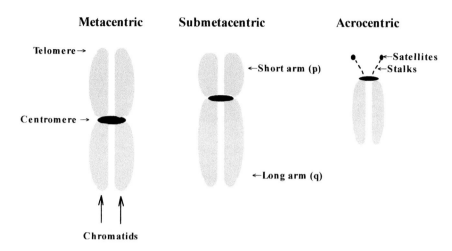

Fig. 8. The functional and structural components of metaphase chromosomes.

telomeric DNA serves to protect the ends of chromosomes from nucleases located within the cell *(9)*.

Telomeres are replicated differently than other types of linear DNA. The enzyme telomerase synthesizes new copies of the telomere TTAGGG repeat using an RNA template that is a component of the telomerase enzyme. Another role of telomerase is to counteract the progressive shortening of chromosomes that results from many cycles of normal DNA replication. Telomere length gradually decreases with the aging process and with increased number of cell divisions in culture. The progressive shortening of human telomeres appears to be a tumor suppressor mechanism *(10)*. The maintenance of telomeric DNA permits the binding of telomeric proteins that form the protective cap at chromosome ends and regulate telomere length *(10)*. Cells that have defective or unstable telomerase will exhibit shortening of chromosomes, leading to instability and cell death.

TYPES OF DNA

DNA is classified into three general categories: unique sequence, highly repetitive sequence DNA ($>10^5$ copies), and middle repetitive sequence DNA (10^2–10^4 copies). Unique sequence or single-copy DNA is the most common class of DNA, comprising about 75% of the human genome *(11)*. This DNA consists of nucleotide sequences that are represented only once in a haploid set. Genes that code for proteins are single-copy DNA. Repetitive or repeated sequence DNA makes up the remaining 25% of the genome *(11)* and is classified according to the number of repeats and whether the repeats are tandem or interspersed among unique sequence DNA.

Repetitive, tandemly arranged DNA was first discovered with a cesium chloride density gradient. Repetitive, tandem sequences were visualized as separate bands in the gradient. This DNA was termed satellite DNA *(12)*. Satellite DNA is categorized based on length of sequences that make up the tandem array and the total length of the array, as α-satellite, minisatellite, and microsatellite DNA.

Alpha-satellite DNA is a repeat of a 171-base-pair sequence organized in a tandem array of up to a million base pairs or more in total length. Alpha-satellite DNA is generally not transcribed, and is located in the heterochromatin associated with the centromeres of chromosomes (see later). The size and number of repeats of satellite DNA is chromosome specific *(13)*. Although α-satellite DNA is associated with centromeres, its role in centromere function has not been determined. A centromeric protein, CENP-B, has been shown to bind to a 17-base-pair portion of some α-satellite DNA, but the functional significance of this has not been determined *(14)*.

Minisatellites have repeats that are 20–70 base pairs in length, with a total length of a few thousand base pairs. Microsatellites have repeat units of two, three or four base pairs and the total length is usually less than a few hundred base pairs. Minisatellites and microsatellites vary in length among individuals and as such are useful markers for gene mapping and identity testing.

The genes for 18S and 28S ribosomal RNAs are middle repetitive sequences. Several hundred copies of these genes are tandemly arranged on the short arms of the acrocentric chromosomes.

Dispersed repetitive DNA is classified as either short or long. The terms SINES (short interspersed elements) and LINES (long interspersed elements) were introduced by Singer *(15)*. SINES range in size from 90 to 500 base pairs. One class of SINES is the Alu sequence. Many Alu sequences are transcribed and are present in nuclear pre-mRNA and in some noncoding regions of mRNA. Alu sequences have high G–C content, and are found predominantly in the Giemsa-light bands of chromosomes *(16)*. LINES can be as large as 7000 bases. The predominant member of the LINEs family is a sequence called L1. L1 sequences have high A–T content and are predominantly found in the Giemsa-dark bands of chromosomes *(16)*. See Chapters 3 and 4.

CHROMATIN

There are two fundamental types of chromatin in eukaryotic cells: euchromatin and heterochromatin. Euchromatin is loosely organized, extended, and uncoiled. This chromatin contains active, early replicating genes, and stains lightly with GTG banding techniques.

There are two special types of heterochromatin that warrant special mention: constitutive heterochromatin and facultative heterochromatin. Both are genetically inactive, late replicating during the synthesis (S) phase of mitosis, and are highly contracted.

Constitutive Heterochromatin

Constitutive heterochromatin consists of simple repeats of nitrogenous bases that are generally located around the centromeres of all chromosomes and at the distal end of the Y chromosome. There are no transcribed genes located in constitutive heterochromatin, which explains that fact that variations in constitutive heterochromatic chromosome regions apparently have no effect on the phenotype. Chromosomes 1, 9, 16, and Y have variably sized constitutive heterochromatic regions. The heterochromatic regions of these chromosomes stain differentially with various special staining techniques, revealing that the DNA structure of these regions is not the same as the structure of the euchromatic regions on the same chromosomes. The only established

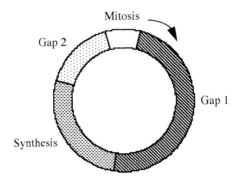

Fig. 9. The cell cycle: gap 1, synthesis, gap 2, and mitosis.

function of constitutive heterochromatin is the regulation of crossing over, the exchange of genes from one sister chromatid to the other during cell division *(17)*.

Facultative Heterochromatin

One X chromosome of every female cell is randomly inactivated. The inactivated X is condensed during interphase and replicates late during synthesis stage of the cell cycle. It is termed facultative heterochromatin. Because these regions are inactivated, it has been proposed that facultative heterochromatin regulates gene function *(18)*.

CELL DIVISION

An understanding of cell division is basic to an understanding of cytogenetics. Dividing cells are needed to study chromosomes using traditional cytogenetic techniques, and many cytogenetic abnormalities result from errors in cell division.

There are two types of cell division: mitosis and meiosis. Mitosis is the division of somatic cells, whereas meiosis is a special type of division that occurs only in gametic cells.

The Cell Cycle

The average mammalian cell cycle lasts about 17–18 h and is the transition of a cell from one interphase through cell division and back to interphase *(19)*. The cell cycle is divided into four major stages. The first three stages, gap 1 (G1), synthesis (S), and gap 2 (G2), comprise interphase. The fourth and final stage of the cell cycle is mitosis (M) (**Fig. 9**).

The first stage, G1, is the longest and typically lasts about 9 h *(19)*. Chromosomes exist as single chromatids during this stage. Cells are metabolically active during G1, and this is when protein synthesis takes place. A cell may be permanently arrested at this stage if its does not undergo further division. This arrested phase is referred to as gap zero (G0).

Gap 1 is followed by the synthesis phase, which lasts about 5 h in mammalian cells *(19)*. This is when DNA synthesis occurs. The DNA replicates itself and the chromosomes then consist of two identical sister chromatids.

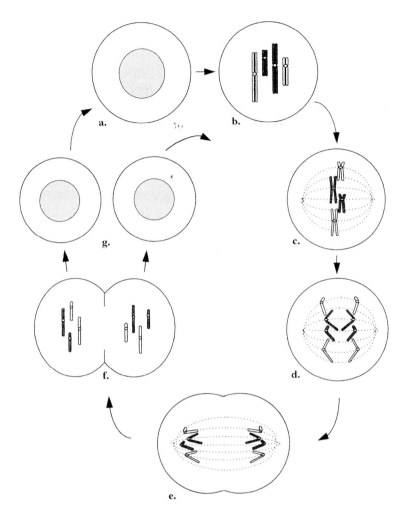

Fig. 10. Mitosis. Schematic representation of two pairs of chromosomes undergoing cell division. **a**: interphase, **b**: prophase, **c**: metaphase, **d**: anaphase, **e**: telophase, **f**: cytokinesis, **g**: interphase of next cell cycle.

Some DNA replicates early in the S phase, and some replicates later. Early replicating DNA contains a higher portion of active genes than late replicating DNA. By standard G-banding techniques, the light staining bands usually replicate early, whereas the dark staining bands and the inactive X chromosome in females replicate late in the S phase.

Gap 2 lasts about 3 h *(19)*. During this phase, the cell prepares to undergo cell division. The completion of G2 represents the end of interphase.

The final step in the cell cycle is mitosis. This stage lasts only 1–2 h in most mammalian cells. Mitosis is the process by which cells reproduce themselves, creating two daughter cells that are genetically identical to one another and to the original parent cell. Mitosis is itself divided into stages (**Fig. 10**).

MITOSIS

Prophase

Chromosomes are at their greatest elongation and are not visible as discrete structures under the light microscope during interphase. During prophase, chromosomes begin to coil, become more condensed, and begin to become visible as discrete structures. Nucleoli are visible early in prophase, but disappear as the stage progresses.

Prometaphase

Prometaphase is a short period between prophase and metaphase during which the nuclear membrane disappears and the spindle fibers begin to appear. Chromosomes attach to the spindle fibers at their kinetochores.

Metaphase

During metaphase, the mitotic spindle is completed, the centrioles divide and move to opposite poles, and chromosomes line up on the equatorial plate. Chromosomes reach their maximum state of contraction during this phase. It is metaphase chromosomes that are traditionally studied in cytogenetics.

Anaphase

Centromeres divide longitudinally and the chromatids separate during this stage. Sister chromatids migrate to opposite poles as anaphase progresses.

Telophase

The final stage of mitosis is telophase. The chromosomes uncoil and become indistinguishable again, the nucleoli reform, and the nuclear membrane is reconstructed. Telophase is usually followed by cytokinesis, or cytoplasmic division. Barring errors in DNA synthesis or cell division, the products of mitosis are two genetically identical daughter cells, each of which contains the complete set of genetic material that was present in the parent cell. The two daughter cells enter interphase and the cycle is repeated.

MEIOSIS

Meiosis takes place only in the ovaries and testes. A process involving one duplication of the DNA and two cell divisions (meiosis I and meiosis II) reduces the number of chromosomes from the diploid number ($2n = 46$) to the haploid number ($n = 23$). Each gamete produced contains only one copy of each chromosome. Fertilization restores the diploid number in the zygote.

Meiosis I

Meiosis I is comprised of several substages: prophase I, metaphase I, anaphase I, and telophase I (**Fig. 11**).

Prophase I

Prophase I is a complex stage that is further subdivided.

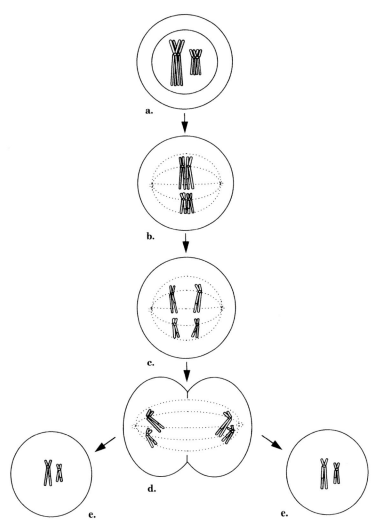

Fig. 11. Schematic representation of two chromosome pairs undergoing meiosis I. **a**: prophase I, **b**: metaphase I, **c**: anaphase I, **d**: telophase I, **e**: products of meiosis I.

LEPTOTENE

In leptotene there are 46 chromosomes, each comprised of two chromatids. The chromosomes begin to condense, but are not yet visible by light microscopy. Once leptotene takes place, the cell is committed to meiosis.

ZYGOTENE

Zygotene follows leptotene. Homologous chromosomes, which in zygotene appear as long threadlike structures, pair locus for locus. This pairing is called synapsis. A tripartite structure, the synaptonemal complex, can be seen with electron microscopy. The synaptonemal complex is necessary for the phenomenon of crossing-over that will take place later in prophase I.

Synapsis of the X and Y chromosomes in males occurs only at the pseudoautosomal regions. These regions are located at the distal short arms and are the only segments of the X and Y chromosome containing homologous loci. The nonhomologous portions of these chromosomes condense to form the sex vesicle.

PACHYTENE

Synapsis is complete during pachytene. Chromosomes continue to condense and now appear as thicker threads. The paired homologs form structures called bivalents, sometimes referred to as tetrads because they are composed of four chromatids.

The phenomenon of crossing-over takes place during pachytene. Homologous or like segments of DNA are exchanged between nonsister chromatids of the bivalents. The result of crossing-over is a reshuffling or recombination of genetic material between homologs, creating new combinations of genes in the daughter cells.

DIPLOTENE

In diplotene, chromosomes continue to shorten and thicken and the homologous chromosomes begin to repel each other. Repelling continues until the homologous chromosomes are held together only at points where crossing-over took place. These points are referred to as chiasmata. In males, the sex vesicle disappears and the X and Y chromosomes associate end-to-end.

DIAKINESIS

Chromosomes reach their greatest contraction during this last stage of prophase.

Metaphase I

Metaphase I is characterized by disappearance of the nuclear membrane and formation of the meiotic spindle. The bivalents line up on the equatorial plate with their centromeres randomly oriented toward opposite poles.

Anaphase I

During anaphase I the centromeres of each bivalent separate and migrate to opposite poles.

Telophase I

In telophase the two haploid sets of chromosomes reach opposite poles, and the cytoplasm divides. The result is two cells containing 23 chromosomes, each composed of two chromatids.

Meiosis II

The cells move directly from telophase I to metaphase II with no intervening interphase or prophase. Meiosis II proceeds much like mitotic cell division except that each cell contains only 23 chromosomes (**Fig. 12**).

The 23 chromosomes line up on the equatorial plate in metaphase II, the chromatids separate and move to opposite poles in anaphase II, and cytokinesis occurs in telophase II. The net result is four cells, each of which contains 23 chromosomes, each consisting of a single chromatid. The effects crossing-over and random assortment of homologs results in each of the new cells differ genetically from one another and from the original cell.

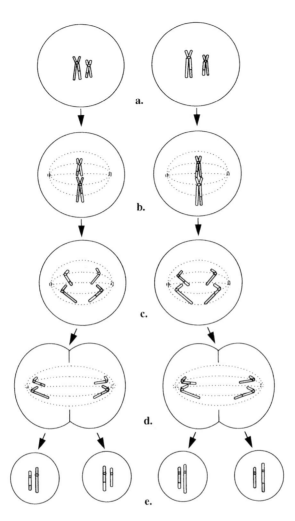

Fig. 12. Schematic representation of two chromosome pairs undergoing meiosis II. **a**: products of meiosis I, **b**: metaphase II, **c**: anaphase II, **d**: telophase II, **e**: products of meiosis.

Spermatogenesis and Oögenesis

The steps of spermatogenesis and oögenesis are the same in human males and females; however the timing is very different (**Fig. 13**).

Spermatogenesis

Spermatogenesis takes place in the seminiferous tubules of the male testes. The process is continuous and each meiotic cycle of a primary spermatocyte results in the formation of four nonidentical spermatozoa. Spermatogenesis begins with sexual maturity and occurs throughout the postpubertal life of a man.

The spermatogonia contain 46 chromosomes. Through mitotic cell division they give rise to primary spermatocytes. The primary spermatocytes enter meiosis I and give rise to the secondary spermatocytes, which contain 23 chromosomes, each consisting of two chromatids. The secondary spermatocytes undergo meiosis II and give

Spermatogenesis   **Oögenesis**

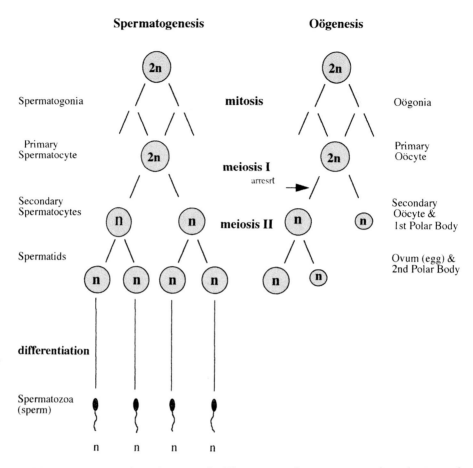

Fig. 13. Spermatogenesis and oögenesis. The events of spermatogenesis and oögenesis are the same, but the timing and net results are different. Oögenesis begins prenatally and is arrested in meiosis I until the postpubertal life of a woman.

rise to spermatids. Spermatids contain 23 chromosomes, each consisting of a single chromatid. The spermatids differentiate to become spermatozoa, or mature sperm.

Oögenesis

Oögenesis in human females begins in prenatal life. Ova develop from oögonia within the follicles in the ovarian cortex. At about the third month of fetal development, the oögonia, through mitotic cell division, begin to develop into diploid primary oöcytes. Meiosis I continues to diplotene, where it is arrested until sometime in the postpubertal reproductive life of a woman. This suspended diplotene is referred to as dictyotene.

Subsequent to puberty, several follicles begin to mature with each menstrual cycle. Meiosis I rapidly proceeds with an uneven distribution of the cytoplasm in cytokinesis of meiosis I, resulting in a secondary oöcyte containing most of the cytoplasm, and a first polar body. The secondary oöcyte, which has been ovulated, begins meiosis II. Meiosis II continues only if fertilization takes place. The completion of meiosis II results in a haploid ovum and a second polar body. The first polar body may undergo meiosis II or it may degenerate. Only one of the potential four gametes produced each menstrual cycle is theoretically viable.

Fertilization

The chromosomes of the egg and sperm produced in meiosis II are each surrounded by a nuclear membrane within the cytoplasm of the ovum. These pronuclei fuse to form the diploid nucleus of the zygote and the first mitotic division begins.

REFERENCES

1. Watson, J.D. and Crick, F.H.C. (1953) A structure for deoxyribose nucleic acid. *Nature.* **171,** April 25.
2. Watson, J.D. and Crick, F.H.C. (1953) The structure of DNA. *Cold Spring Harbor Symp Quant Biol.* **18,** 123–131.
3. Chargaff, E. (1951) Structure and function of nucleic acids as cell constituents. *Fed. Proc.* **10,** 654–659.
4. Okazaki, R., Okazaki, T., Sakabe, K., Sugimoto, K., and Sugino, A. (1968) Mechanism of DNA chain growth, I. Possible discontinuity and unusual secondary structure of newly synthesized chains. *Proc. Natl. Acad. Sci. USA* **59,** 598–605.
5. Leder, P. and Nirenberg, M. (1964) RNA codewords and protein synthesis. *Science* **145,** 1399–1407.
6. Khorana, H.G. (1968) Synthesis in the study of nucleic acids. *Biochem. J.* **109,** 709–725.
7. Sharma, T., ed. (1990). *Trends in Chromosome Research.* Narosa, New Delhi.
8. Moyzis, R.K., Buckingham, J.M., Cram, L.S., Dani, M., Deaven, L.L., Jones, M.D., Meyne, J., Ratcliffe, R.L., and Wu, J. (1988) A highly conserved repetitive DNA sequence, (TTAGGG)n present at the telomeres of human chromosomes. *Proc. Natl. Acad. Sci. USA* **85,** 6622–6626.
9. Zakian, V.A. (1989) Structure and function of telomeres. *Annu. Rev. Genet.* **23,** 579–604.
10. Smith, S. and De Lange, T. (1997) TRF1, a mammalian telomeric protein. *Trends Genet.* **13,** 21–26.
11. Spradling, A., Penman, S., Campo, M.S., and Bishop, J.O. (1974) Repetitious and unique sequences in the heterogeneous nuclear and cytoplasmic messenger RNA of mammalian and insect cells. *Cell* **3,** 23–30.
12. Hsu, T.C. (1979) *Human and Mammalian Cytogenetics: An Historical Perspective.* Springer-Verlag, New York.
13. Willard, H.F. and Waye, J.S. (1987) Hierarchal order in chromosome-specific human alpha satellite DNA. *Trends Genet.* **3,** 192–198.
14. Willard, H.F. (1990) Centromeres of mammalian chromosomes. *Trends Genet.* **6,** 410–416.
15. Singer, M.F. (1982) SINEs and LINEs: highly repeated short and long interspersed sequences in mammalian genomes. *Cell* **28,** 433–434.
16. Korenberg, J.R. and Rykowski, M.C. (1988) Human genome organization: Alu, LINEs and the molecular structure of metaphase chromosome bands. *Cell* **53,** 391–400.
17. Miklos, G. and John, B. (1979) Heterochromatin and satellite DNA in man: properties and prospects. *Am. J. Human Genet.* **31,** 264–280.
18. Therman, E. and Susman, M. (1993) *Human Chromosomes: Structure, Behavior, and Effects.* Springer-Verlag, New York.
19. Barch M.J., Knutsen T., and Spurbeck J.L., eds. (1997) *The AGT Cytogenetic Laboratory Manual.* Raven-Lippincott, Philadelphia.
20. Mange, E.J. and Mange, A.P. (1994) *Basic Human Genetics.* Sinauer Associates, Sunderland, MA, p. 37.

3

Human Chromosome Nomenclature

An Overview and Definition of Terms

Avirachan Tharapel, Ph.D.

INTRODUCTION

Advancements in methodology and discovery of the diploid human chromosome number invigorated further research in human cytogenetics *(1,2)*. The eventful years that followed witnessed the birth of a new specialty—"human cytogenetics"—which provided answers to many intriguing phenomena in medicine. Little was it known at the time that human cytogenetics would form the backbone of present-day "human genetics," providing answers to questions regarding human reproduction, behavior, aging, and disease while generating knowledge that could be applied to the treatment and prevention of many disorders.

The discovery of the chromosomal etiology of Down, Turner, Klinefelter, Edward, and Patau syndromes further added to the knowledge that variations from the normal diploid chromosome number and structure can render severe phenotypic malformations and mental impairment. The investigators responsible for these early discoveries hailed from both sides of the Atlantic. Working independently of each other, they devised their own terminology and nomenclature to describe chromosome abnormalities. Confusion in the scientific literature was the result. The need for guidelines and standardization of terminology thus became imperative. At a conference held in Denver, 14 attendees from different countries argued for 3 days. In the end, they agreed upon guidelines for describing human chromosomes and chromosome abnormalities. This historic document is called the "Denver Conference (1960): A Proposed Standard System of Nomenclature of Human Mitotic Chromosomes" *(3)*.

Although the basic principles adopted in Denver have prevailed to date, new technologies and ever-increasing knowledge in human cytogenetics necessitated periodic revision and update of the nomenclature document (**Table 1**). The Chicago Conference nomenclature *(4)* was widely used from 1966–1971 during the prebanding era (**Fig. 1**). At the Paris Conference *(5,6)* the document was expanded to include nomenclature for banded chromosomes (**Fig. 2**). With the Stockholm Conference in 1977, the proceedings came to be known as the International System for Human Cytogenetic Nomenclature or "ISCN" *(7)*. Each ISCN is identified by the year of its publication *(8–10)*. The document currently in use is "An International System for Human Cytogenetic

From: *The Principles of Clinical Cytogenetics*
Edited by: S. Gersen and M. Keagle © Humana Press Inc., Totowa, NJ

Table 1
International Conferences on Human Chromosome Nomenclature

Conference/Document		Year of Publication
Denver Conference		1960
London Conference		1963
Chicago Conference		1966
Paris Conference		1971
Paris Conference (Supplement)		1975
Stockholm-1977	ISCN	1978
Paris-1980	ISCN	1981
	ISCN	1985
Cancer Supplement	ISCN	1991
Memphis—1994	ISCN	1995

Nomenclature (1995)," abbreviated as "ISCN 1995," as agreed upon by the conferees in Memphis, Tennessee in October, 1994 *(11)*. ISCN 1995 established a uniform code for designating both constitutional (congenital) and acquired chromosome abnormalities as well as one for describing and reporting results obtained from *in situ* hybridization methodologies.

ISCN 1995 has a certain uniqueness. It has provided a new 850-band-level ideogram based on actual measurements of bands. For comparative purposes, it includes G- and R-banded composite photographs of chromosomes at band resolutions ranging from about 400 to 850 bands. It has introduced specific ways to accurately describe Robertsonian translocations, whole-arm translocations, and uniparental disomy. Keeping up with technical developments, this document for the first time has established nomenclature guidelines for the description of fluorescence *in situ* hybridization (FISH). In the following pages an attempt has been made to simplify the use and point out the highlights of ISCN 1995. The examples that appear in this chapter are based on the dictates of this nomenclature document. However, for a detailed understanding of ISCN 1995 the reader is requested to refer to the original document *(11)*.

HUMAN CHROMOSOMES

Of the 46 chromosomes in a normal human somatic cell, 44 are autosomes and 2 are sex chromosomes. The autosomes are designated as pairs 1–22. The numbers are assigned in descending order of the length, size, and centromere position of each chromosome pair. In a normal female the sex chromosomes are XX and in a normal male they are XY.

Until the advent of certain specialized staining techniques, arbitrary identification of individual chromosome pairs was based on the size and position of the centromere *(4)*. Variability in the centromere position of different chromosomes allowed them to be classified into three basic categories. A chromosome with its centromere in the middle is *metacentric,* one with the centromere closer to one end is *submetacentric,* and one with the centromere almost at one end is *acrocentric* (**Fig. 3**). Based on decreasing relative size and centromere position, a karyotype comprised of seven groups labeled A through G was devised. The X chromosome belonged to the third or "C" group,

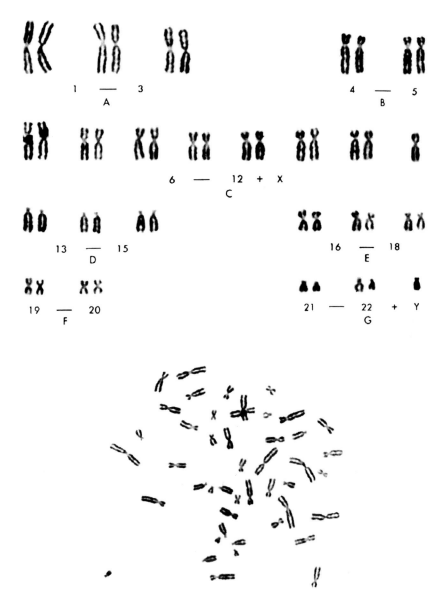

Fig. 1. Unbanded metaphase spread (bottom) and corresponding karyotype (top) per the Chicago conference.

while the Y was often placed separately. Although still used occasionally, these letter group names are now considered obsolete.

Chromosome Banding and Identification

Unequivocal identification of individual chromosomes and chromosome regions became possible with the technical developments of the late 1960s (refer to Chapters 1 and 4). When chromosome preparations are treated with dilute solutions of proteolytic enzymes (trypsin, pepsin, etc.) or salt solutions (2× SSC) and treated with a chromatin stain such as Giemsa, alternating dark and light stained demarcations called *bands* ap-

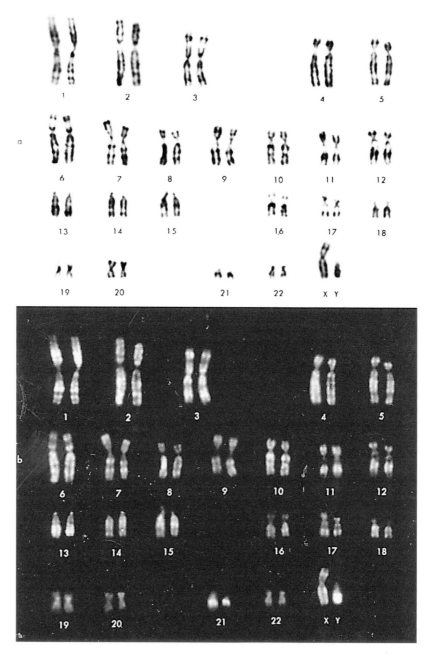

Fig. 2. Normal 46,XY male karyotype. Characteristic G-band pattern (a) and fluorescent Q-banding (b). The same cell was used for both methodologies to demonstrate the complementary banding patterns.

pear along the length of each chromosome. The banding patterns produced are specific for each chromosome pair, thus enabling the identification not only of individual chromosomes, but of regions within each chromosome as well. Methods commonly used to produce these discriminative banding patterns include Giemsa or G-banding, quina-

metacentric   submetacentric   arocentric

Fig. 3. Examples of metacentric, submetacentric and acrocentric chromosomes.

Table 2
Frequently Used Banding Methods and Their Abbreviations

Banding Method	Abbreviation
Q-bands	Q
Q-bands by quinacrine derivatives and fluorescence microscopy	QFQ
G-bands	G
G-bands by trypsin and Giemsa	GTG
C-bands	C
C-bands by barium hydroxide and Giemsa	CBG
R-bands	R
R-bands by acridine orange and fluorescence microscopy	RFA
R-bands by BrdU and Giemsa	RBG
Telomere bands or T-bands	T

crine mustard or Q-banding, and reverse or R-banding, each with its own uniqueness. In the United States and Canada the most frequently used methods for routine cytogenetic analysis are G- and Q-bands (**Fig. 2**), whereas in other countries (France, for example) R-banding is more common. Additional banding methods are occasionally employed to exemplify specific abnormalities or chromosome regions. Abbreviations commonly used to denote the various banding techniques appear in **Table 2**.

Chromosome Regions and Band Designations

The chromosomal details revealed by the new banding techniques necessitated the introduction of additional terminology and modifications of certain existing ones. This task was accomplished by a standing committee appointed at the Fourth International Congress of Human Genetics in Paris. The recommendations of the committee were published as "Paris Conference (1971): Standardization in Human Cytogenetics." Through a diagrammatic representation of banding pattern the document elucidated the typical band morphology for each chromosome *(5)* (see **Fig. 4**). The Paris Conference (1971) introduced a numbering system helpful in designating specific bands and regions. New terminology and abbreviations were introduced to help explain chromo-

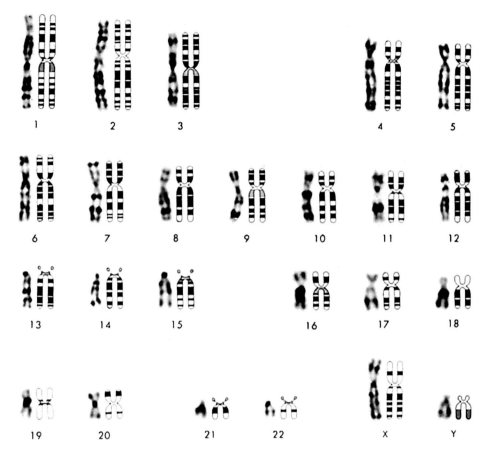

Fig. 4. A composite karyotype of G-banded chromosomes (left) along with the corresponding 1971 Paris Conference ideograms (right).

some abnormalities in a more meaningful way. Other conferences then followed, with the latest held in Memphis in 1994. Descriptions of human chromosomes and their abnormalities utilize a series of symbols and abbreviations. A partial list of recommended symbols and abbreviations in ISCN 1995 appears in **Table 3**.

The centromere, "cen," divides a chromosome into a short or "p" arm (from the French *petit*) and a long or "q" arm. For descriptive purposes the centromere is composed of two portions. The portion of the centromere lying between its middle and the first band on the short arm is designated as "p10." Similarly the portion of the centromere lying between its middle and the first band on the long arm is designated as "q10." The designations p10 and q10 allow us to describe accurately the nature and organization of centromeres in isochromosomes, whole-arm translocations, and Robertsonian translocations (see later). Each arm ends in a terminus ("ter", thus "pter" and "qter"), where telomeres are present to prevent the chromosomes from having "sticky ends."

Each chromosome arm is divided into *regions*. This division is based on certain *landmarks* present on each chromosome. By definition a landmark is "a consistent and

Table 3
Selected List of Symbols and Abbreviations Used in Karyotype Designations[a]

Abbreviation or Symbol	Description
add	additional material, origin unknown
arrow ($\rightarrow$ or $\rightarrow$)	from – to, when using long form
[ ] square brackets	number of cells in each clone
cen	centromere
chi	chimera
single colon (:)	break
double colon (:)	break and reunion
comma (,)	separates chromosome number, sex chromosomes, and abnormalities
del	deletion
der	derivative chromosome
dic	dicentric
dmin	double minute(s)
dup	duplication
fis	fission
fra	fragile site
h	heterochromatin
i	isochromosome
ins	insertion
inv	inversion
mar	marker chromosome
mat	maternal origin
minus sign (−)	loss
mos	mosaic
multiplication sign (×)	multiple copies; also designates copy number with ish
p	short arm of chromosome
pat	paternal origin
Ph or Ph[1]	Philadelphia chromosome
plus sign (+)	gain
q	long arm of chromosome
question mark (?)	uncertainty of chromosome identification or abnormality
r	ring chromosome
rcp	reciprocal
rec	recombinant chromosome
rob	Robertsonian translocation
s	satellite
slash (/)	separates cell lines or clones
semicolon (;)	separates chromosomes and breakpoints in rearrangements involving more than one chromosome
stk	satellite stalk
t	translocation
upd	uniparental disomy

[a]For a complete listing of symbols and abbreviations, refer to ISCN 1995 *(11)*.

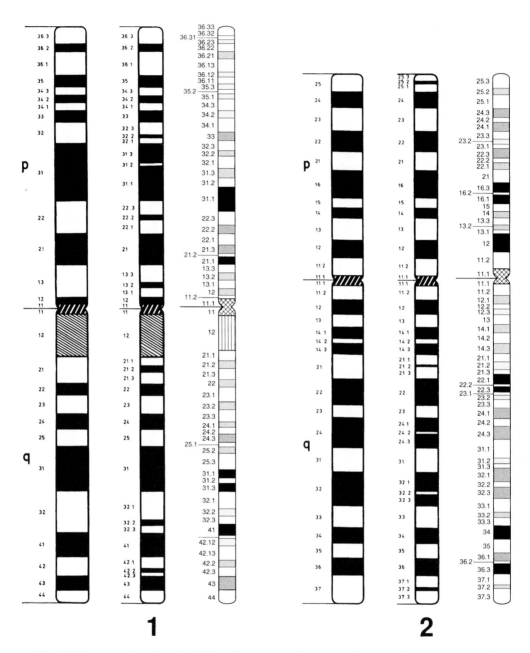

Fig. 5. Ideogram showing the G-banding pattern for normal human chromosomes at three different band resolutions. The left chromosome represents a haploid karyotype (one of each chromosome) of approximately 400 bands. The middle chromosome is at an approximately 550 band level, and the right chromosome represents about 850 bands. Reproduced from ISCN 1995 with permission of S. Karger AG, Basel.

distinct morphologic area of a chromosome that aids in the identification of that chromosome." A *region* is an area that lies between two landmarks. The two regions immediately adjacent to the centromere are designated as "1" (p1 and q1), the next distal as "2," and so on. Regions are divided into *bands* and the bands into *sub-bands* (see **Fig. 5**). A

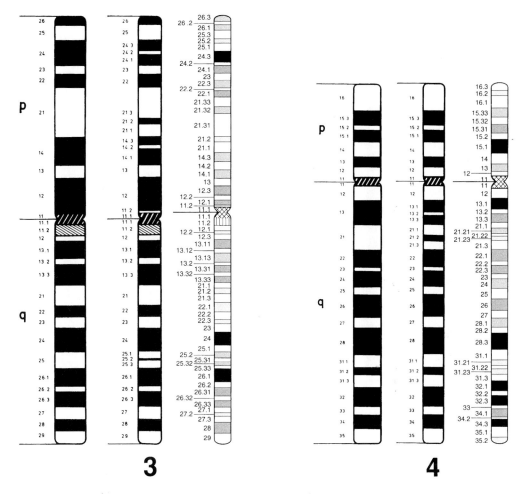

Fig. 5. (*continued*)

band is that part of a chromosome that is distinctly different from the adjacent area by virtue of being lighter or darker in staining intensity. Sequential numbering of chromosome arms and bands helps make the designation of specific bands easy. For example, the terminal band on the long arm of chromosome 2 is written as 2q37 and means: chromosome 2, long arm, region 3, band 7 and is referred to as "two q three-seven, NOT "two q thirty-seven."

Karyotype Descriptions

Karyotype descriptions follow certain basic rules. When designating a karyotype, the first item specified is the total number of chromosomes including the sex chromosomes present in that cell, followed by a comma, and the sex chromosomes, in that order. Thus a normal female karyotype is written as 46,XX and a normal male karyotype as 46,XY. The characters are contiguous, without spaces between items. Chromosome abnormalities, when present, follow the sex chromosome designation using abbreviations or symbols denoting each abnormality (see **Table 3**). These are listed in a specific order: sex chromosome abnormalities are described first, followed by auto-

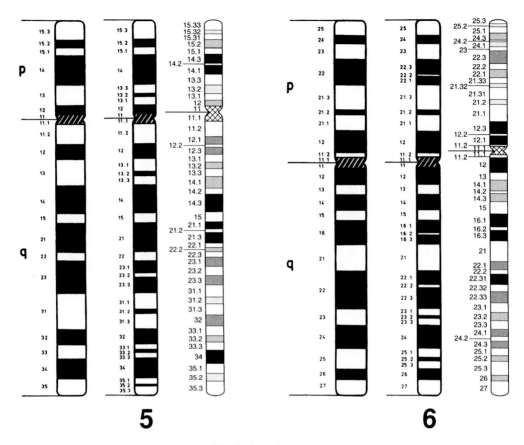

Fig. 5. (*continued*)

somal changes in numerical order. For each chromosome described, numerical changes are listed before structural abnormalities.

Most karyotypes can be described using the "short form" of the nomenclature, which is used in this chapter. However, it should be noted that, for certain complex rearrangements, this can produce ambiguity. ISCN therefore provides for a "long form," in which abnormal chromosomes can be described from end to end, with all structural changes "spelled out" in detail. Some examples are provided throughout the chapter; the reader is encouraged to refer to the original document (*11*) for additional information.

This remainder of this chapter discusses the current method of using the ISCN nomenclature to describe chromosome abnormalities. A section on interpretation of karyotype descriptions follows.

NUMERICAL ABNORMALITIES OF CHROMOSOMES

The term "numerical abnormality" refers to gain or loss of chromosomes. As outlined previously, all such abnormalities are presented in numerical order with the exception of the X and Y, which are always listed first. To designate an additional or a missing chromosome, plus (+) and minus (–) signs are placed before the specific chromosome number. Thus, –7,+18 would mean a missing chromosome 7 and having an extra chromosome 18, respectively. Note that these abnormalities are presented in nu-

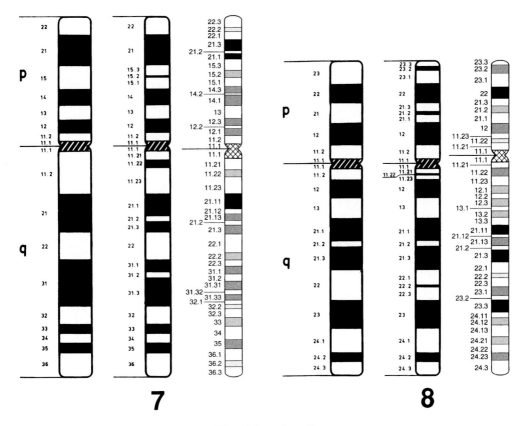

Fig. 5. (*continued*)

merical order, regardless of whether they involve gain or loss of a chromosome. A + sign may also be used to denote additional copies of derivative chromosomes or accessory marker chromosomes, for example, +der(6) or +mar (see later).

Numerical Abnormalities Involving the Sex Chromosomes

These can be constitutional (congenital) or acquired. ISCN 1995 provides special ways to distinguish between the two. As shown in the examples below, the + and – signs are not needed to designate constitutional sex chromosome aneuploidies.

Constitutional Sex Chromosome Aneuploidies

45,X	Classical monosomy X or Turner syndrome
47,XXY	Classical Klinefelter syndrome
47,XXX	A female with three X chromosomes
48,XXYY	Variant of Klinefelter syndrome with two X and two Y chromosomes

Acquired Sex Chromosome Aneuploidies

These involve chromosome changes seen in certain leukemias and solid tumors, and are restricted to the affected tissues.

45,X,–X

This describes a normal female with two X chromosomes but with the loss of one X chromosome in her tumor cells.

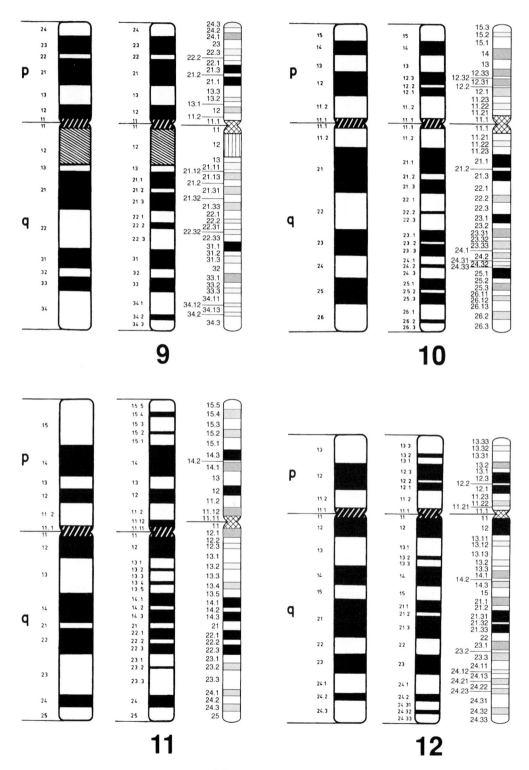

Fig. 5. (*continued*)

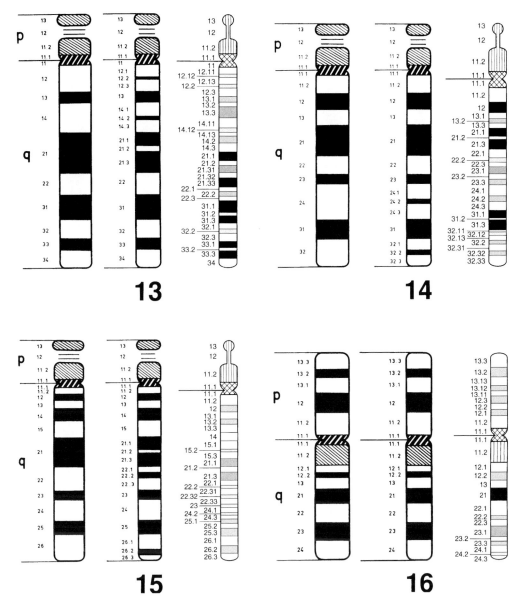

Fig. 5. (*continued*)

47,XX,+X

This is a normal female with two X chromosomes and gain of an extra X chromosome in her tumor cells.

45,X,–Y

This is a normal male with XY chromosomes and loss of the Y chromosome in his tumor cells.

48,XY,+X,+Y

This describes a male with acquired X and Y chromosomes in his tumor cells.

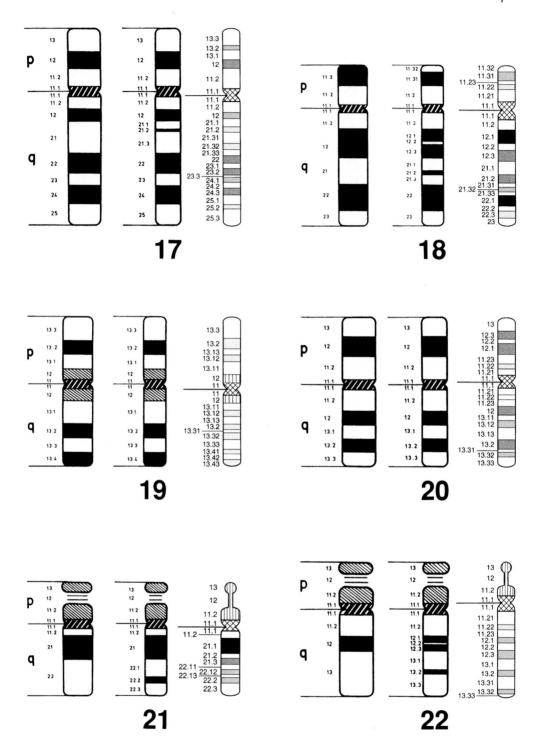

Fig. 5. (*continued*)

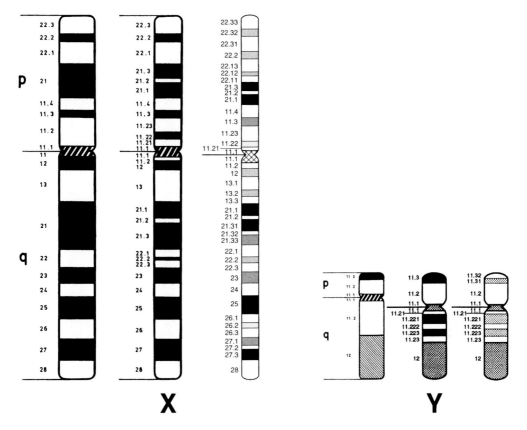

Fig. 5. (*continued*)

48,XXYc,+X

Here is a patient with Klinefelter syndrome who has an acquired X chromosome in his tumor cells. The letter "c" is placed next to XXY to show that the patient's sex chromosome complement is XXY and *not* XY or XXXY.

46,Xc,+X

This is a Turner syndrome patient (45,X) with gain of an X chromosome in her tumor cells.

Numerical Abnormalities of the Autosomes

The situation here is similar to that involving the sex chromosomes, with the exception that (+) and (–) signs are used to designate the constitutional loss or gain of autosomes:

47,XY,+18	Male with trisomy 18
48,XX,+18,+21	Female with both trisomy 18 and trisomy 21
45,XY,–21	Male with monosomy 21
46,XY,+21c,–21	Male trisomy 21 patient with loss of one chromosome 21 in his tumor cells
48,XX,+21c,+21	Female with trisomy 21 and gain of an additional chromosome 21 in her tumor cells

Mosaics and Chimeras

An individual with two or more cell types, differing in chromosome number or structure, is either a mosaic or a chimera. If the two cell types originated from a single zygote, the individual is a mosaic (mos). If the cell types originated from two or more zygotes that subsequently fused, the individual is a chimera (chi). In designating mosaic or chimeric karyotypes a slash (/) is used to separate the cell lines. The actual number of cells detected in each clone may be given within square brackets [ ]. The largest clone is recorded first, then the next largest, and so on. Whenever a normal cell line is present, it is always recorded last irrespective of the number of normal cells detected.

Examples:

mos 45,X[4]/46,XX[16]

> This is a Turner mosaic with two cell lines. Analysis of 20 cells showed that this individual has four cells that are 45,X and 16 cells that are 46,XX.

mos 45,X[5]/47,XYY[5]/46,XY[10]

> This represents a mosaic with three cell lines.

mos 47,XX,+13[15]/46,XX[5]

> This is a mosaic with trisomy 13 and normal cell lines.

In a chimera where the two cell lines are normal (46,XX and 46,XY) and both are present in equal proportions, either one of them may be listed first. If one cell line is larger than the other, the larger clone is listed first.

chi 46,XX[10]/46,XY[10]

> This describes a chimera with female and male cells in equal number.

chi 47,XX+21[15]/46,XY[5]

> This is a chimera with both female and male cell lines. The female cell line shows trisomy 21, while the male cell line is normal.

chi 69,XXX[20]/46,XY[5]

> This represents a chimera with triploid and diploid cell lines. The triploid line is XXX while the diploid line is XY.

> Use of the abbreviations chi and mos is optional, as the presence of chimerism or mosaicism is usually evident from the karyotype.

STRUCTURAL CHROMOSOME ABNORMALITIES

This category of abnormalities includes several subclasses that are discussed under separate headings. All chromosomes involved in abnormalities are designated in numerical order, except for the X and Y, which are listed first.

When designating an abnormality that is limited to a single chromosome, the abbreviation for that abnormality is used followed by the chromosome number in parentheses, for example, r(X), del(2), ins(4), dup(5). If two or more chromosomes are involved in a rearrangement, as with translocations, a semicolon (;) is used to separate chromosome numbers within parentheses, for example, t(3;4), t(2;5;10;) or t(15;17). Again, chromosomes are listed in numerical order unless a sex chromosome is involved, for example, t(X;1) or t(Y;15). If, in the same cell, a specific chromosome is involved in both a numerical and a structural rearrangement, the numerical abnormality is designated first, for example, +13,t(13;14).

For ease of reference, the abnormalities covered are presented in alphabetical order. For a thorough description of the mechanisms and clinical significance of structural chromosome abnormalities, see Chapter 9.

Additional Material, Origin Unknown (add)

When a chromosome has additional material attached to it, the origin of this material may not be identifiable with conventional banding methods. This is especially likely if the abnormality is subtle and originated *de novo* or is acquired. The abbreviation "add" (from the Latin *additio*) is used to designate additional material of unknown origin.

46,XX,add(17)(p13)
>Additional material of unknown origin is attached to chromosome 17 at band p13 (the terminal band of 17p).

46,XX,add(9)(q22)
>Additional material of unknown origin is attached to chromosome 9 at band q22. The region 9q22→qter is missing and has been replaced by this material.

Deletions (del)

A deletion is an aberration in which a part of a chromosome is lost. Deletions can be either terminal, where all chromosomal material from the breakpoint to the chromosome terminus is lost, or interstitial, in which an internal section of one arm is missing. To introduce the reader to the long form of the nomenclature a few of the following abnormalities are presented using both the short and long forms.

Terminal Deletions

46,XY,del(1)(q32)   (Short form)
46,XY,del(1)(pter→q32:)   (Long form)
>This karyotype describes a terminal deletion involving the long arm of chromosome 1. The colon present in the long form indicates a break at band 1q32. The region distal to the breakpoint is deleted. The rest of the chromosome, from 1qter to 1q32, is present.

Interstitial Deletions

46,XY,del(1)(p21p32)   (Short form)
46,XY,del(1)(pter→p21::p32→qter)   (Long form)
>Breakage and reunion are represented in the long form by a double colon (::). Here, this occurred involving bands 1p21 and 1p32. The segment between them has been deleted.

Derivative and Recombinant Chromosomes

Derivative Chromosomes (der)

A structurally rearranged chromosome generated by events involving two or more chromosomes or due to multiple events within a single chromosome is a derivative chromosome. Thus, each unbalanced product of a translocation event is a derivative chromosome. The identity of a derivative chromosome is determined by its centromere.

Examples:

46,XY,der(3)t(3;6)(p21;q23)

> The derivative chromosome 3 in this example is the result of a translocation between the short arm of chromosome 3 at band p21 and the long arm of chromosome 6 at band q23. The der(3) replaces one normal chromosome 3. Both chromosomes 6 and the other chromosome 3 are normal. This unbalanced karyotype results in monosomy (loss) of region 3p21→pter and trisomy (gain) of 6q23→qter. This karyotype is the product of adjacent-1 segregation (see Chapter 9).

45,XY,der(3)t(3;6)(p21;q23),–6

> The der(3) is the same as in the above example and again replaces one of the normal chromosomes 3. However, there is only one normal chromosome 6 in this case, resulting in monosomy for both 3p21→pter and 6pter→q23. This is the result of 3:1 segregation (see Chapter 9).

47,XY,+der(3)t(3;6)(p21;q23)mat

> The der(3) is the same as in the above examples. A 3:1 segregation in the mother resulted in a normal 3 and the derivative 3 being retained in the ovum. The father contributed a normal 3 as well. The patient is therefore trisomic for both 3p21→qter and 6q23→qter.

Recombinant Chromosomes (rec)

Recombinant chromosomes are also structurally rearranged chromosomes. They arise *de novo* from meiotic crossing-over between homologous chromosomes when one is structurally abnormal, often in an inversion heterozygote.

Take, for example, an individual with the karyotype 46,XY,inv(3)(p21q27). As described later, this man has one chromosome 3 with a pericentric inversion involving the segment between bands p21 and q27. During meiosis, crossing over within the inverted segment could result in two recombinant chromosomes, each of which has a duplication of one part of the chromosome and deletion of another part; this is described in detail in Chapter 9.

46,XY,rec(3)dup(3p)inv(3)(p21q27)

> One normal chromosome 3 has been replaced by a recombinant chromosome 3. The segment 3p21→pter is duplicated, and the segment from 3q27→qter is deleted. The key to interpreting this karyotype is "dup(3p)."

46,XY,rec(3)dup(3q)inv(3)(p21q27)

> Here, the other possible recombinant chromosome is present, resulting in duplication of the segment 3q27→qter and loss of the segment 3p21→pter. In this case, note "dup(3q)."

Insertions (ins)

An insertion is a structural rearrangement in which a part of a chromosome is typically interstitially repositioned into a different area of the karyotype. Insertions can occur within a chromosome or between two chromosomes. They can be direct, in which the inserted segment retains its orientation relative to the centromere, or inverted, where the inserted segment has been "flipped over" (rotated 180°). Although the symbols "dir" and "inv" can be used to distinguish between the two, they are optional, as the orientation of the inserted material is typically evident from the nomenclature.

Insertion Within a Chromosome

In these cases, only one chromosome need be described. The first band listed is the break at the point of insertion, followed by the breakpoints that define the inserted segment itself. No punctuation is used.

46,XX,ins(3)(p21q27q32)

> This represents a direct insertion. The long arm segment between bands 3q27 and 3q32 has broken away and has been inserted into the short arm of the same chromosome at band p21. The orientation of the inverted segment has not changed, that is, band q27 is still proximal to the centromere relative to band q32.

46,XX,ins(3)(p21q32q27)

> In this case the inserted segment is inverted; band q32 is now closer to the centromere than band q27.

Insertion Between Two Chromosomes

Here, both chromosomes are listed, with the recipient chromosome presented first, irrespective of numerical order. As with other rearrangements, a semicolon separates the chromosome numbers.

46,XX,ins(4;9)(q31;q12q13)

> The long arm segment between bands 9q12 and 9q13 has been inserted, in its original orientation, into the long arm of chromosome 4 at band q31.

Inversions (inv)

A chromosomal aberration in which a segment of a chromosome is reversed in orientation but not relocated is called an inversion. There are two types of inversions. Paracentric inversions involve only one arm of a chromosome, while pericentric inversions involve both arms of a chromosome and therefore include the centromere. The type of inversion does not have to be specified, as this will be evident from the breakpoints.

Paracentric Inversions

46,XY,inv(3)(q21q27)

   Break and reunion occurred at bands q21 and q27 in the long arm of chromosome 3. The segment lying between these breakpoints has been reattached with its bands in reverse (inverted) order.

Pericentric Inversions

46,XY,inv(2)(p21q31)

   Break and reunion occurred at bands p21 (short arm) and q31 (long arm) of chromosome 2. The segment between these bands, including the centromere, was reattached with its bands in inverted order.

Isochromosomes (i)

   An abnormal chromosome in which one arm is duplicated (and the other lost) is an isochromosome, abbreviated as "i" in the nomenclature. The breakpoint in an isochromosome is assigned to the centromere, at band p10 or q10, depending upon which arm is duplicated:

46,XX,i(18)(p10)

   This describes an isochromosome for the short arm of chromosome 18, as evident by assigning the breakpoint to band p10.

46,XX,i(18)(q10)

   This describes an isochromosome for the long arm of a chromosome 18; the breakpoint is assigned to q10.

Isodicentric Chromosomes (idic)

   Unlike isochromosomes, isodicentric chromosomes contain two copies of the same centromere. One of the two centromeres may be inactive, in which case the chromosome is pseudodicentric (psu dic). The breakpoints in isodicentric chromosomes are usually on the band adjacent to the centromere *on the opposite arm.*

46,XX,idic(18)(q11.2)

   This is an isodicentric chromosome comprised of two copies of the entire short arm of chromosome 18, two copies of the centromere, and two copies of the small portion of the long arm between the centromere and band q11.2.

Marker Chromosomes (mar)

   Marker chromosomes (mar) are supernumerary, structurally abnormal chromosomes of which no part can be identified. If any part of such a chromosome is identifiable it is not a marker but a derivative chromosome. The presence of a "mar" in a karyotype is always recorded by a plus (+) sign.

47,XY,+mar

   This is a male karyotype with a marker chromosome.

48,XY,+2mar

   This is a male karyotype with two marker chromosomes.

48,XY,t(5;12)(q13;p12),+21,+mar
>   This describes a male karyotype with a translocation involving chromosomes 5 and 12, an extra chromosome 21, and a marker chromosome.

Ring Chromosomes (r)

A ring chromosome is a structurally abnormal chromosome with two breaks, one on the short arm and one on the long arm, in which the broken ends are attached to form a circular configuration. The net result is deletion of at least the terminal ends of both arms, and potentially more of either or both arms.

46,X,r(X)
>   This is a female karyotype with only one normal X chromosome and a ring X chromosome with no information on breakpoints.

46,X,r(X)(p22q24)
>   This describes a female karyotype with one normal X chromosome and a ring X chromosome with breakage and reunion at bands p22 and q24. The material distal to both breakpoints is lost.

Translocations (t)

The interchange or transfer of chromosomal segments between two nonhomologous chromosomes is defined as a translocation.

Reciprocal Translocations

If the translocation involves a mutual exchange of segments between two chromosomes it is referred to as a reciprocal translocation. To describe a reciprocal translocation the abbreviation "rcp" can be substituted for the "t," but this is generally not done, as all translocations are, in one sense, theoretically reciprocal, even if this is not readily apparent visually. As always, sex chromosomes are listed first, with autosomes presented in numerical order. If a translocation involves three or more chromosomes the same rule applies to the first chromosome listed; however, in these rearrangements the second chromosome specified will be the one that received the segment from the first and so on.

46,XX,t(7;10)(q22;q24)
>   Breakage and reunion occurred at bands 7q22 and 10q24. The segments distal to these bands were interchanged. The translocation event has not altered the total DNA content of this cell. Therefore, the translocation is microscopically (cytogenetically) balanced.

46,X,t(X;1)(p21;q32)
>   Breakage and reunion occurred at bands Xp21 and 1q32. The segments distal to these bands were interchanged. The translocation is balanced. Note that the X chromosome is specified first.

46,X,t(Y;15)(q11.23;q21.2)
>   Breakage and reunion occurred at sub-bands Yq11.23 and 15q21.2. The segments distal to these bands were interchanged. This translocation is cytogenetically balanced. Here again the sex chromosome is specified first.

46,XY,t(9;22)(q34;q11.2)

> Breakage and reunion has occurred at bands 9q34 and 22q11.2. The segments distal to these bands have been interchanged. This represents the typical "Philadelphia" rearrangement associated with CML and also seen in ALL and AML (see Chapter 13).

46,XX,t(1;7;4)(q32;p15;q21)

> This is an example of a complex translocation involving three chromosomes. The segment on chromosome 1 distal to band q32 has been translocated onto chromosome 7 at band p15, the segment on chromosome 7 distal to band p15 has been translocated onto chromosome 4 at band q21, and the segment on chromosome 4 distal to band q21 has been translocated onto chromosome 1 at band q32. The translocation is cytogenetically balanced.

These same general principles also apply to describing translocations involving more than three chromosomes.

Whole-Arm Translocations

Whole-arm translocation is a type of reciprocal translocation in which the entire arms of two nonacrocentric chromosomes are interchanged. Such rearrangements are described by assigning the breakpoints to the arbitrary centromeric regions designated as p10 or q10, as the actual ultimate composition of the centromeres is not known. If both chromosomes have exchanged the same arms, so that the resultant rearranged chromosomes are still comprised of one short arm and one long arm, the breakpoint p10 is assigned to the chromosome with the lowest number (or the sex chromosome, if applicable). Consequently, the other chromosome will have the breakpoint at q10.

If the chromosomes have exchanged opposite arms, so that one resultant chromosome consists of two short arms, and the other consists of two long arms, the breakpoint p10 is assigned to both chromosomes.

46,XX,t(3;8)(p10;q10)

> This represents a balanced whole arm translocation between chromosomes 3 and 8. In this example, the short arm of chromosome 3 and the long arm of chromosome 8 have been fused. Reciprocally, the long arm of chromosome 3 has fused with the short arm of chromosome 8, but only one combination need be written. The composition of the resultant centromeres is not known.

46,XX,t(3;8)(p10;p10)

> This is a balanced whole-arm translocation in which the short arms of chromosomes 3 and 8 have been fused, as have the long arms of these chromosomes. Note that the breakpoints designate the short arms of both chromosomes. The reciprocal product [t(3;8)(q10;q10)] need not be written, as its presence is obvious from the chromosome number of 46.

Whole-arm translocations are not always balanced, as in the following examples:

45,X,der(X;3)(p10;q10)

> This derivative chromosome consists of the short arm of an X and the long arm of chromosome 3. The reciprocal product consisting of the long arm of the X and the

short arm of 3 is missing. Note: the total chromosome number is 45, indicating the loss of the reciprocal product; no (–) sign is used. The net result is monosomy for both the long arm of X and short arm of 3.

47,XX,+der(X;3)(p10;q10)

This karyotype has an extra derivative chromosome consisting of the short arm of an X and the long arm of chromosome 3, the same derivative chromosome as in the previous example. However, in this case, two normal X chromosomes and two normal chromosomes 3 are also present, and so the derivative chromosome is extra (note that the total number of chromosomes is 47). The net result is trisomy for both the short arm of the X and the long arm of chromosome 3.

Robertsonian Translocations

These originate through centric fusion (recent data suggests this may not always be so; see Chapter 9) of the long arms of acrocentric chromosomes (pairs 13, 14, 15, 21, and 22) and were first described by Robertson, whose name they have been given. The short arms, which all contain redundant copies of ribosomal RNA genes, are lost in these rearrangements; this is of no clinical significance. Because Robertsonian translocations are still treated as a type of whole-arm translocation, they can be adequately described using the same nomenclature, but using the abbreviation "der" instead of "t."

45,XX,der(13;14)(q10;q10)

This describes a Robertsonian translocation between chromosomes 13 and 14. The origin of the centromere is unknown, so the breakpoints are designated as 13q10 and 14q10 to indicate that both long arms are involved. This derivative chromosome has replaced one chromosome 13 and one chromosome 14; there is no need to indicate the missing chromosomes. The karyotype contains one normal 13, one normal 14, and the der(13;14). The short arms of the 13 and 14 are lost, which is why the abbreviation "der" is used instead of "t" to describe the translocation. The loss of these short arms is not clinically significant, and therefore this description represents a balanced Robertsonian translocation (an individual with this karyotype is referred to as a balanced carrier) even though only 45 chromosomes are present. The abbreviation "rob" can also be used to describe Robertsonian translocations.

46,XX,+13,der(13;14)(q10;q10)

The derivative chromosome consists of the long arms of chromosomes 13 and 14, as in the above example. However, in this karyotype there are two normal 13s and one normal 14, plus the der(13;14). The net result is trisomy for the long arm of chromosome 13, clinically identical to trisomy 13. The additional chromosome 13 is shown by the designation +13. In this example there is a numerical and structural abnormality involving the same chromosome, and therefore the numerical abnormality is designated first.

Uniparental Disomy (upd)

Representation of both maternally and paternally inherited genes is required in many areas of the genome for normal development to occur. This phenomenon is referred to as genomic imprinting, and involves selective inactivation of certain genes by methylation. Uniparental disomy is a situation in which both homologs of a specific chromosome pair are inherited from the same parent, and in some cases is associated with an abnormal phenotype. Uniparental disomy can occur, for example, in an embryo that starts out trisomic for a given chromosome and then loses one copy of this chromosome early enough in development to "rescue" what would have been a pregnancy doomed to abort spontaneously. If, by chance, the two remaining copies were inherited from one parent, the individual is said to have uniparental disomy (upd) for that chromosome. For example, some patients with Prader–Willi/Angelman syndrome and no deletion of chromosome 15 have been shown to have upd for this chromosome. Inheriting two paternal chromosomes 15 results in Angelman syndrome, while receiving two maternal 15s results in Prader–Willi syndrome (see Chapter 16).

Nomenclature examples:

46,XY,upd(15)pat

This is a male patient with uniparental disomy for paternally derived chromosomes 15.

46,XY,upd(22)pat[10]/47,XY,+22[6]

This represents a mosaic male karyotype involving one cell line that contains two paternally derived chromosomes 22, and the other with trisomy 22. Here, both cell lines are abnormal and therefore the larger clone is recorded first.

46,XX,upd pat

This describes a complete hydatidiform mole with XX sex chromosomes (very rare). All 46 chromosomes are paternally derived.

46,XY,upd pat

This describes a complete hydatidiform mole with XY sex chromosomes. All 46 chromosomes are paternally derived.

46,XX,upd mat

This is an ovarian teratoma. All 46 chromosomes are maternally derived.

NEOPLASIA

The basic rules for using the nomenclature apply when describing the karyotypes associated with cancer. However, special situations, requiring additional guidelines, may arise in these cases. Therefore special ISCN definitions and rules have been devised for use with neoplasia.

Clones

A clone is defined as two cells that share the same abnormality or abnormalities, unless the change involves loss of a chromosome, in which case three such cells are required (due to the possibility of coincidental random chromosome loss). During tumor progression, related subclones may evolve; related or unrelated clones are separated by slashes [/] and the number of cells observed for each is given in square brackets [ ].

Mainline, Stemline, Sideline, and Clonal Evolution

These terms can be confusing, and are often misunderstood. The mainline (ml) is the term used to describe the most common clone, that is, the one represented by the most cells. This is a quantitative issue only. It does not necessarily indicate the most basic clone in tumor progression, which is referred to as the stemline (sl). Clones that evolve from the stemline are referred to as sidelines (sdl).

46,XY,t(9;22)(9q34;q11.2)[5]/47,XY,+8,t(9;22)(q34;q11.2)[11]/46,XY,t(9;22)(q34;q11.2),i(17)(q10)[4]

    ↑   ↑   ↖ ↗
 stemline   mainline   sidelines

When more than one clone is present but no clear clonal progression is evident, the mainline is listed first, followed by each clone in order of relative size. When clonal evolution is present, the stemline is listed first, with sidelines listed in order of increasing complexity whenever possible, or by clone size when more than one sideline evolves independently from the stemline, as in the previous example.

Composite Karyotype (cp)

When a clone contains multiple abnormalities, a frequent occurrence is that not all changes are present in every cell, yet the interpretation may be made that these cells do in fact represent a single abnormal clone rather than an evolving process. To report such a phenomenon, the clone is described as a *composite,* listing all of the abnormalities observed, using the abbreviation "cp" before the number of cells evaluated in brackets. It should be noted that this can occasionally produce seemingly contradictory data, as some cells will contain additional copies of a chromosome that is missing in others.

INTERPRETING A KARYOTYPE DESCRIPTION

Receiving a cytogenetic report that contains the description of a patient's karyotype can create confusion, particularly if complex rearrangements or multiple clones are present. Interpretation of the description of a karyotype can be facilitated by breaking this description into its component parts.

First, determine whether more than one cell line is present. This will happen if constitutionally the patient is a mosaic or a chimera, and is often the case with acquired cytogenetic abnormalities, particularly in patients whose neoplasm is progressing. Because the first item described is always the number of chromosomes present, each clone or cell line present will start with this number, and each is separated by a slash (/). Each cell line can then be examined individually. If abnormalities present in the first clone listed are also present in others, the description may be simplified by using the abbreviation "idem" to indicate this; note that idem always refers back to the first cell line described, which in these cases will be the stemline.

As discussed previously, the sex chromosome complement follows the chromosome count. Sex chromosome abnormalities are listed first, followed by autosomal abnormalities in numerical order. When abnormalities involve the same chromosome number, the numerical changes are presented first, followed by the structural ones listed in alphabetical order, using the abbreviations in Table 3.

Commas separate each abnormality listed, and so by examining the karyotype from comma to comma, the abnormalities involved can be interpreted.

Consider the following example from a patient with AML:

47,XY,del(5)(q13q33),+8,t(9;22)(q34;q11.2)[4]/48,idem,+9,i(17)(q10)[12]/46,XY[4]

At first blush, receiving a report with this karyotype might be enough to scare away even the most confident clinician! By breaking this karyotype down into its component parts, its interpretation will be simplified.

The slashes, brackets, and listings of number of chromosomes tell us that three different clones are present.

47,XY,del(5)(q13q33),+8,t(9;22)(q34;q11.2) **[4]**

/48,idem,+9,i(17)(q10)**[12]**

/46,XY**[4]**

Of the 20 cells examined, the first clone has 47 chromosomes, and is represented by four cells. The second clone has 48 chromosomes; 12 of these cells were observed. Finally, four normal 46,XY cells are present.

The first cell line is the stem line and has an XY sex chromosome complement. It also has three cytogenetic abnormalities. One chromosome 5 has an interstitial deletion of the material between bands q13 and q33 (on the long arm):

47,XY,**del(5)(q13q33)**,+8,t(9;22)(q34;q11.2)[4]/48,idem,+9,i(17)(q10)[12]/46,XY[4]
        ↗

There is an extra copy of chromosome 8,

47,XY,del(5)(q13q33),**+8**,t(9;22)(q34;q11.2)[4]/48,idem,+9,i(17)(q10)[12]/46,XY[4]
                     ↑

and there is a translocation involving the long arms of chromosomes 9 and 22, at band q34 of chromosome 9 and band q11.2 of chromosome 22:

47,XY,del(5)(q13q33),+8,**t(9;22)(q34;q11.2)**[4]/ 48,idem,+9,i(17)(q10)[12]/46,XY[4]
                        ↗

This is the "Philadelphia" rearrangement, which is sometimes also seen in patients with AML.

The second cell line contains the sex chromosomes and all of the abnormalities present in the first, therefore the abbreviation "idem" can be used to represent these:

47,XY,del(5)(q13q33),+8,t(9;22)(q34;q11.2)[4]/48,**idem**,+9,i(17)(q10)[12]/46,XY[4]
                                                 ↑

There is also an additional copy of chromosome 9:

47,XY,del(5)(ql3q33),+8,t(9;22)(q34;q11.2)[4]/48,idem,**+9**,i(17)(q10)[12]/46,XY[4]
                                                      ↑

and an isochromosome for the long arm of chromosome 17:

47,XY,del(5)(ql3q33),+8,t(9;22)(q34;q11.2)[4]/48,idem,+9,**i(17)(q10)**[12]/46,XY[4].
               ↑

Because this is the largest clone present (with 12 cells), it represents the mainline.

Finally, the third cell line represents cells with a normal male karyotype and is therefore listed last:

47,XY,del(5)(q13q33),+8,t(9;22)(q34;q11.2)[4]/ 48,idem,+9,i(17)(q10)[12]/**46,XY**[4]
                     ↑

By examining the components of a reported karyotype using the rules outlined above, together with the abbreviations listed in **Table 3** or in the nomenclature document itself *(11),* what initially may have appeared as an indecipherable compilation of numbers and symbols becomes a concise, universal method of describing the results of a patient's chromosome analysis.

FLUORESCENCE AND OTHER *IN SITU* HYBRIDIZATION

Recent advances in human cytogenetics include the development and application of *in situ* hybridization (ish) protocols to incorporate and bind labeled, cloned DNA or RNA sequences to cytological preparations. These techniques facilitate the localization of specific genes and DNA segments onto specific chromosomes, ordering the position and orientation of adjacent genes along a specific chromosome, the identification of microduplications or microdeletions of loci that lie beyond the resolution of conventional cytogenetics but manifest themselves as abnormal clinical phenotypes, and the detection of aneuploidies involving whole chromosomes or chromosomal regions (see Chapter 15). For these reasons, nomenclature to designate various ish applications was introduced in ISCN 1995. The symbols and abbreviations used in ish nomenclature are listed in **Table 4**.

Prophase or Metaphase Chromosome In Situ Hybridization (ish)

Even though fluorescence microscopy is most commonly used to view *in situ* hybridization signals, the abbreviation ish, not FISH, is used in the karyotype description. If chromosome analysis was done prior to ish, the karyotype is first designated using conventional rules. A period (.) is then placed to record the end of the cytogenetic findings. This is then followed by the ish results. If a standard cytogenetic analysis was not done and only ish studies were done, the ish results are presented directly.

When presenting an abnormal ish result, the abbreviation for that specific abnormality is recorded (e.g., ish del) followed by the chromosome number, the breakpoints, and a designation for the probe used, all listed in separate parentheses (e.g., ish del(4)(p16p16)(D4S96-)). Whenever possible, Genome Data Base (GDB) designations for loci are used. These consist of the letter "D" (for DNA), the chromosome of origin, the letter "S" (segment) and the GDB number of the probe. The example above uses the 96th DNA segment assigned to chromosome 4, D4S96. The locus designation must be

Table 4
Selected List of Symbols and Abbreviations Used for *in situ*
Hybridization (ish) Nomenclature

Abbreviation or Symbol	Description
–	absent on a specific chromosome
+	present on a specific chromosome
++	duplication on a specific chromosome
x	precedes the number of signals seen
.	period, separates cytogenetic results from ish results
con	connected or adjacent signals
ish	refers to *in situ* hybridization. When used by itself, ish refers to hybridization to chromosomes.
nuc ish	nuclear or interphase *in situ* hybridization
pcp	partial chromosome paint
sep	separated signals (which are usually adjacent)
wcp	whole chromosome paint

 [a] For a complete listing of symbols and abbreviations refer to ISCN 1995 *(11)*.

given using capital letters only. When a GDB designation is not available, a probe name can be used. If more than one probe from the same chromosome is used, these are listed in order from pter to qter. If probes from two different chromosomes are used, they are separated by a semicolon.

Given below are a series of examples that illustrate the ish karyotype designations, using patients suspected of having various disorders.

Patients with Possible DiGeorge/VCF Syndrome
46,XX.ish 22q11.2(D22S75x2)

This example illustrates the basic rules in describing an ish karyotype when both chromosome and ish results are normal. The test was performed on prophase/metaphase chromosomes. The probe used, D22S75, detects about 80% of deletions leading to DiGeorge/velocardiofacial (VCF) syndrome. First, the cytogenetics result 46,XX is recorded, followed by a period and then the abbreviation "ish." A single space is left, after which the chromosome and region numbers are given together without parentheses, 22q11.2, followed by the GDB locus designation of the probe used, within parentheses (D22S75) and the number of times the probe signal is observed (x2, as in a normal cell neither chromosome 22 would have a deletion).

46,XX.ish del(22)(q11.2q11.2)(D22S75–)

This patient has a normal karyotype resulting from standard chromosome analysis, but a deletion in the DiGeorge region of chromosome 22, at band q11.2, was detected by ish using a probe for that locus, D22S75. Note: the chromosome 22 and the region tested are now placed within parentheses because an abnormality (no signal, indicated by a minus sign) is being described.

46,XX,del(22)(q11.2q11.2).ish del(22)(q11.2q11.2) (D22S75–)

Here is a karyotype in which a deletion was identified with standard chromosome analysis and confirmed with ish using a probe for locus D22S75.

Patients with Possible Prader–Willi/Angelman Syndrome

46,XX.ish 15q11.2q13(D15S11x2,GABRB3x2)

This patient has normal chromosomes, but was suspected of having a microdeletion in the Prader–Willi/Angelman syndrome region. She was studied by ish using probes for loci D15S11 and GABRB3, both of which map to the region 15q11.2→q13. Hybridization showed two copies each of the two probes, suggesting no deletion for either locus.

46,XX.ish 15q11.2q13(D15S11x2,GABRB3x2,D15S10x2,SNRPNx2,)

Because of the negative ish results but continued clinical suspicion, the above patient was retested using the additional probes for loci D15S10 and SNRPN. There is still no deletion detected. Because of the high degree of suspicion, this patient is a candidate for uniparental disomy analysis.

46,XX,del(15)(q11.2q13).ish del(15)(q11.2q13)(D15S10–,SNRPN–)

Here a deletion of 15q11.2→q13 was detected with cytogenetic analysis and was confirmed with ish. The probes used were for loci D15S10 and SNRPN. Both were absent from one chromosome 15. The ish "deletion" is denoted by a minus sign.

Patients with Possible Williams Syndrome

46,XX.ish 7q11.23(ELNx2)

This patient had a diagnosis of Williams syndrome and normal chromosomes. *In situ* hybridization with a probe for the Elastin–Williams syndrome (ELN) locus produced hybridization at band 7q11.23 on both chromosomes 7. There is no deletion.

46,XX.ish del(7)(q11.23q11.23)(ELN–)

As above, this is a patient with Williams syndrome and normal cytogenetic results. *In situ* hybridization with a probe for the *ELN* locus showed a deletion on one chromosome 7.

46,XX.ish del(7)(q11.23q11.23)(ELN–x2)

Again, this is a patient with Williams syndrome and normal cytogenetic results. However, in this case, ish with a probe for the *ELN* locus showed deletions on both chromosomes 7.

Patients with Possible Charcot–Marie–Tooth Syndrome

46,XX.ish 17p11.2(CMT1Ax2)

This is a patient with Charcot–Marie–Tooth syndrome and normal chromosomes. *In situ* hybridization with a probe for the *CMT1A* locus showed normal hybridization on both chromosomes 17 and thus no deletion or duplication of the locus.

46,XX,ish dup(17)(p11.2p11.2)(CMT1A++)

This Charcot–Marie–Tooth syndrome patient also has normal chromosomes. ish with a probe for the *CMT1A* locus showed duplication of the locus on one chromosome 17(++).

Chromosome Abnormalities Identified with Whole Chromosome Paints

In situ hybridization can be performed using a cocktail of chromosome-specific probes that will hybridize along the entire length of that chromosome pair, effectively "painting" them. Because the procedures used ensure that no other chromosomes are "painted," these probes provide a way to identify or confirm the identity of chromosomal material.

46,XX,add(20)(p13).ish dup(5)(p13p15.3)(wcp5+)

In this patient, one chromosome 20 has extra material attached to it at band p13. By using a whole chromosome paint for chromosome 5 (wcp5), the extra segment was identified as a partial duplication of chromosome 5. Subsequent analysis of the band morphology using Giemsa banded preparations allowed the duplicated material to be identified as the segment 5p13→p15.3. The diagnosis is essentially made by going from G-banding to ish and then back to G-banding.

Both whole-chromosome paints and locus-specific ish probes can be used in combination in order to determine the composition of an abnormal chromosome:

46,X,r(X).ish r(X)(p22.3q13.2)(wcpX+,DXS1140+,DXZ1+,XIST+,DXZ4–)

This is an example of a ring X that was identified with G-banding and then further defined by ish. First, a whole chromosome paint for the X confirmed the origin of the ring. Next, probes localized to Xp22.3 (DXS1140), the X centromere (DXZ1), Xq13.2 (XIST), and Xq24 (DXZ4) were used. The last probe, DXZ4, produced no hybridization signal (–), narrowing down the portion of the X that was lost during formation of the ring.

46,X,r(?).ish r(X)(DYZ3–,wcpX+)

In this case, a small ring chromosome of indeterminate origin was detected with G-banding. Hybridization with the Y probe DYZ3 showed that the ring was not derived from the Y. Follow-up hybridization with a whole chromosome paint for the X showed that it originated from an X.

47,XX,+mar.ish der(12)(wcp12+,D12Z1+)[10]/46,XX[10]

This patient is a mosaic, with normal cells and cells with an extra marker chromosome. The marker was identified with ish as being derived from chromosome 12, using a whole chromosome paint for chromosome 12 and chromosome 12 centromere probe D12Z1.

Painting probes that hybridize to specific parts of chromosomes have also been developed, and are referred to as partial chromosome paints (pcp). Consider the following example:

46,XX.ish inv(16)(p13.1q22)(pcp16q sp)

> In this case, what appeared to be a normal female karyotype with routine G-banding was found by ish to have a pericentric inversion of chromosome 16. When a partial chromosome paint for band 16q22 was used, this band was shown to be split (sp) between the long and short arms.

Identification of Cryptic Translocations Using ish

Some translocations are beyond the limits of microscopic resolution. Take, for example, an individual who has a child with Miller–Dieker syndrome. Although routine chromosome analysis produced a normal karyotype the child was shown by ish to have a microdeletion for this locus on chromosome 17.

46,XX.ish del(17)(p13.3pl13.3)(D17S379–)

> The patient and her husband wish to know if this condition could have been inherited as the result of a microscopically undetectable (cryptic) translocation carried by one of them. Both of their karyotypes are normal with standard chromosome analysis, but ish analysis demonstrates that the mother carries a cryptic translocation.

46,XX.ish t(16;17)(q24;p13.3)(D17S379+;wcp16+)

> After hybridizing the Miller–Dieker locus probe D17S379 to previously banded cells, one signal was observed in its proper location on the short arm of chromosome 17, but the other appeared on the long arm of chromosome 16. Subsequent hybridization with a whole chromosome paint for chromosome 16 showed that part of this chromosome is now on chromosome 17, confirming the presence of a reciprocal cryptic rearrangement.

> Sometimes, a translocation appears to be present with standard cytogenetics, but is so subtle that this is not certain, and must be confirmed with ish.

46,XX,?t(4;7)(p16;q36).ish (wcp7+,D7S427+,D4S96–;wcp4+,D4S96+,D7S427–)

> Here, a cryptic translocation between the short arm of chromosome 4 and the long arm of chromosome 7 was suspected with G-banding, but was not a certainty via cytogenetics alone, hence the "?." The presence of this rearrangement was confirmed with ish using probe D7S247 localized to 7qter, D4S96 localized to 4pter, and whole chromosome paints for both chromosomes 4 and 7. The distal short arm of the der(4) was wcp7+, D7S247+ and D4S96–. The distal long arm of the der(7), on the other hand, was wcp4+, D4S96+, and D7S247–.

Interphase or Nuclear In Situ Hybridization (nuc ish)

In situ hybridization can be performed on interphase nuclei to provide information concerning the number and/or relative positions of the probes (and therefore the loci) involved. Thus it can be used as a screening method for the rapid detection of aneuploidies and gene rearrangements. Typically performed prior to or in the absence of standard chromosome analysis, interphase ish results are abbreviated nuc ish.

Designation of the Number of Signals

When designating interphase ish results the abbreviation nuc ish is followed by a space, the chromosome band to which the probe is mapped, and then, in parentheses, by the GDB locus designation, a multiplication sign, and the number of signals detected.

nuc ish Xcen(DXZ1x1,DYZ3x1)

● = probe for DXZ1

○ = probe for DYZ3

One copy of DXZ1, a probe for the X centromere, is detected, as is one copy of the Y chromosome probe DYZ3. This implies the presence of one X and one Y chromosome, suggesting an XY sex chromosome complement. No other information is presented, and so this report cannot specify whether these sex chromosomes are normal. This type of data is therefore generally used when the sex chromosome information itself is of value, for example, when monitoring the progress of an mixed-gender bone marrow transplant.

nuc ish Xcen(DXZ1x2)

● = probe for DXZ1

Here, two copies of the DXZ1 locus were detected. This implies the presence of two X chromosomes.

nuc ish Xcen(DXZ1x3)

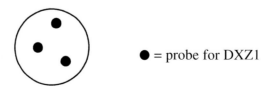

● = probe for DXZ1

Three copies of the DXZ1 locus are detected, implying the presence of three X chromosomes.

nuc ish 13q14(Rb1x1)

○ = probe for Rb1

Only one copy of a probe for the retinoblastoma locus Rb1 is detected. This implies a deletion of the *Rb1* gene from one chromosome 13.

nuc ish Yp11.2(DYZ3x2)

○ = probe for DYZ3

In this case, two copies of DYZ3 are detected, implying that an extra copy of this locus is present. It is not clear whether the extra copy is due to the presence of two Y chromosomes or to an isochromosome involving the short arm of the Y.

nuc ish 4cen(D4Z1x2),4p16.3(D4S96x1)

○ = probe for D4Z1
● = probe for D4S96

This is an example of a structural abnormality identified with ish. Two copies of chromosome 4 are implied by two D4Z1 signals. However, only one copy of D4S96 is detected, implying a deletion of this locus from one chromosome 4. In the nomenclature, two or more probes for the same chromosome are separated by commas.

nuc ish Xp22.3(STSx2),13q14(Rb1x3)

○ = probe for STS
● = probe for Rb1

This is an example of ploidy detection. Two copies of the steroid sulfatase locus on the X chromosome and three copies of the Rb1 locus are detected. This implies the presence of two X chromosomes and trisomy 13. Probes from different chromosomes are separated by commas.

Designation of Relative Positions of the Signals

Under normal conditions, if probes from two chromosomes are tested simultaneously, the signals are expected to appear separated. However, chromosome rearrangements, such as the BCR/ABL gene fusion, can bring signals together.

nuc ish 9q34(ABLx2),22q11.2(BCRx2)

Two ABL and two BCR loci seen and they are well separated. No gene rearrangement is evident.

nuc ish 9q34(ABLx2),22q11.2(BCRx2)(ABL con BCRx1)

Here, signals from two ABL and two BCR loci are seen. However, one ABL signal and one BCR signal are juxtaposed (or connected, "con"), suggesting that they now reside on the same chromosome. This is the pattern observed when a t(9;22) or "Philadelphia rearrangement" is present.

nuc ish 9q34(ABLx3),22q11.2(BCRx3)(ABL con BCRx2)

Here three ABL and three BCR signals are present. However, two pairs of BCR/ABL signals are juxtaposed. This is the pattern observed when both a t(9;22) and an additional der(22) ["Philadelphia chromosome"] are present.

Sometimes a rearrangement can be detected when signals that are normally juxtaposed become separated.

nuc ish Xp22.3(STSx2,KALx2)

○ = probe for STS
● = probe for KAL

STS and KAL are two loci on Xp22.3 that are adjacent to each other. Because they map to the same band, they are reported within the same parentheses. Under normal circumstances, ish signals will appear side by side, as they do here. However, the signals may appear to be independent of each other.

nuc ish Xp22.3(STSx2,KALx2)(STS sep KALx1)

○ = probe for STS
● = probe for KAL
○● = STS sep KAL

In this last example, one STS and one KAL locus are separated, most likely as the result of a rearrangement involving this area of the X chromosome. The nomenclature term "sep" is used to designate this change.

ISCN 1995 has indeed made a good beginning concerning nomenclature for various *in situ* hybridization scenarios. However, it is the author's opinion that this document will require revision to accommodate the technical explosion we are witnessing in this arena (see Chapter 15). Primed *in situ* labeling (PRINS), comparative genomic hybridization (CGH), use of five or more probes simultaneously for rapid aneuploidy screening, and arm-specific subtelomeric detection screening, to name a few, will need to be included in future versions.

REFERENCES

1. Tjio, J.H. and Leven, A. (1956) The chromosome number in man. *Hereditas* **42,** 1–16.
2. Ford, C.E. and Hamerton, J.L. (1956) The chromosomes of man. *Nature* **178,** 1010–1023.
3. Denver Conference (1960) A proposed standard system of nomenclature of human mitotic chromosomes. *Lancet* **i,** 1063–1065; also reprinted in (4).
4. Chicago Conference (1966) Standardization in human cytogenetics. *Birth Defects Orig. Article Ser.* VIII, 2. The National Foundation, New York.
5. Paris Conference (1971) Standardization in human cytogenetics. *Birth Defects Orig. Article Ser.* VII, 7; The National Foundation, New York, 1972; also in *Cytogenetics* **11,** 313–362.
6. Paris Conference (1971), Supplement (1975) Standardization in human cytogenetics. *Birth*

Defects Orig. Article Ser. XI; The National Foundation, New York, 1975; also in *Cytogenet. Cell Genet.* **15,** 201–238.

7. ISCN (1978) An international system for human cytogenetic nomenclature. *Birth Defects Orig. Article Ser.* XIV, The National Foundation, New York; 1978; also in *Cytogenet. Cell Genet.* **21,** 309–404.

8. ISCN (1981) An international system for human cytogenetic nomenclature. *Birth Defects Orig. Article Ser.* XVII, March of Dimes Birth Defects Foundation, New York; also in *Cytogenet. Cell Genet.* **31,** 1–32.

9. Harnden, D.G. and Klinger, H.P. eds. (1985) *ISCN (1985): An International System for Human Cytogenetic Nomenclature.* S. Karger, Basel, in collaboration with *Cytogenet Cell Genet.;* also in *Birth Defects Orig. Article Ser.* Vol XXI, No. 1. March of Dimes Birth Defects Foundation, New York.

10. Mitelman, F. ed. (1991) *ISCN (1991): Guidelines for Cancer Cytogenetics, Supplement to An International System for Human Cytogenetic Nomenclature.* S. Karger, Basel.

11. Mitelman, F., ed. (1995) *ISCN (1995): An International System for Human Cytogenetic Nomenclature.* S. Karger, Basel.

II
Examining and Analyzing Chromosomes

Editor's Foreword to Section II

Despite a prolific increase in technological equipment available to the clinical laboratorian, involving both the actual number of devices themselves as well as an ever-increasing variety of tasks they can perform, cytogenetics is still a relatively labor-intensive discipline. The basic tool of the cytogenetics technologist remains the microscope, and the number of steps involved in generating chromosomes for analysis has essentially not changed for decades.

Although it is sometimes difficult to make a distinction between "how it's done" and "how to do it," the former is our goal in presenting the following four chapters. It is not our intention to prepare the reader for a career in the cytogenetics laboratory, or to provide the cytogeneticist with yet another lab manual. Our aim, rather, is to impress upon the reader the amount of effort and attention to detail required to accurately diagnose a myriad of cytogenetic conditions from the variety of tissue types routinely submitted to the laboratory.

However, given the diverse background of the individuals likely to be referring to this book, a certain amount of technical data has been retained for those who might find it interesting and/or informative.

It is our hope that this section will leave the reader with a newfound respect for the effort required to perform this often critical aspect of patient care.

4

Basic Laboratory Procedures

Martha Keagle, M.Ed and Steven L. Gersen, Ph.D.

INTRODUCTION

The study of chromosomes using traditional cytogenetic techniques requires cells that are actively dividing. Chromosomes are individually distinguishable under the light microscope only during cell division and are best examined during metaphase. Metaphase chromosomes can be obtained from specimens that contain spontaneously dividing cells or ones that are cultured and chemically induced to divide *in vitro*.

Specimens that contain spontaneously proliferating cells include bone marrow, lymph nodes, solid tumors, and chorionic villi. If there are not enough naturally dividing cells for a chromosome analysis, these specimen types may also be cultured in the laboratory. Peripheral blood lymphocytes, tissue biopsies, and amniotic fluid samples are routinely cultured to obtain dividing cells; lymphocytes usually require the addition of a mitotic stimulant. The choice of specimen for chromosome analysis depends on clinical indications and whether the diagnosis is prenatal or postnatal.

The individual details of culture initiation, maintenance, and cell harvest vary somewhat for the different sample types; however, the general steps and requirements are similar. These are summarized below.

OVERVIEW OF CELL CULTURE AND HARVEST

Culture Initiation	→	Culture Maintenance	→	Cell Harvest
• Living cells		• Sterility		• Arrest division
• Sterility		• Optimal temperature		• Swell cells
• Proper growth medium		• Optimal pH		• Fix cells
• ± Mitotic stimulant		• Optimal humidity		• Prepare slide
• Microbial inhibitors		• Optimal time interval		• Stain/band

The most critical requirement is that *living cells* capable of cell division be received by the laboratory. The manner in which the sample is collected and subsequently handled will greatly influence whether or not the cells will grow and divide, as well as the quality of the resulting metaphases. Specimen containers must be sterile and must be labeled with the patient's name. The laboratory may reject specimens that are unlabeled or improperly labeled.

From: *The Principles of Clinical Cytogenetics*
Edited by: S. Gersen and M. Keagle © Humana Press Inc., Totowa, NJ

SPECIMEN COLLECTION AND HANDLING

Requirements: Peripheral Blood Specimens

Peripheral blood samples should be collected in sterile syringes or vacuum tubes containing preservative-free *sodium* heparin. Vacuum tubes should be discarded if outdated. Peripheral blood cultures can be initiated several days after the blood is drawn; however, for best results blood samples should be set up within 24 h of collection. Temperature extremes must be avoided if samples are transported or stored. Specimens should be kept at *room temperature* or refrigerated above 4°C until they can be processed. Culture medium is sometimes added to small blood samples, as these have a tendency to dry up, especially if collected in large containers.

A repeat sample should be requested if these requirements are not met (e.g., the sample is received clotted, on ice, more than 24 h old, etc.). It is not always practicable or possible to obtain a new sample and in such cases the laboratory should attempt to salvage the original specimen. There may be enough viable cells for a cytogenetic analysis, although the number and quality of cells may be compromised.

Requirements: Bone Marrow Specimens

The collection requirements for bone marrow samples are essentially the same as for peripheral blood. Bone marrow specimens should be collected in sterile syringes or vacuum tubes containing preservative-free, *sodium* heparin and transported at room temperature. The first few milliliters of the bone marrow tap contain the highest proportion of cells and are the best sample for the cytogenetics laboratory. Blood dilutes the bone marrow sample in later taps and reduces the number of actively dividing cells present in the sample. The success of bone marrow culture is dependent on the number of actively dividing cells. Bone marrow specimens should be processed without delay upon receipt to avoid cell death.

Requirements: Amniotic Fluid Specimens

Amniocentesis can be performed from as early as 10 weeks gestation until term (see Chapter 11). From 15 to 30 mL of amniotic fluid should be obtained under sterile conditions and collected in a sterile container approved for cell culture. For amniocenteses performed earlier than 15 weeks, 1 mL of fluid is generally drawn for each week of gestation. The first few milliliters of an amniotic tap are the most likely to be contaminated with maternal cells and should not be submitted to the cytogenetics laboratory. Samples should be transported at room temperature. Temperature extremes and long transport times should be avoided.

There is an inherent, albeit small, risk of miscarriage from the amniocentesis procedure, and it should not be repeated unless absolutely necessary. Every effort should be made to salvage samples improperly collected or handled to diminish the need for a repeat tap.

Requirements: Solid Tissue Biopsies

Solid tissue sources include skin biopsies, chorionic villi, products of conception, and stillbirth biopsies. Products of conception and stillbirths are one-of-a-kind specimens that cannot be recollected, and repeat collection of chorionic villi increases the risk of abortion, although subsequent amniocentesis is an option here. Microbial contamination is a common problem associated with many types of solid tissue samples. Unlike amniotic fluid, blood, bone marrow, and chorionic villi, most solid tissue specimens are not sterile prior to collection. In addition, viable cells may be few or even nonexistent. These factors threaten the integrity of the sample and pose problems for the laboratory.

Small samples should be collected and transported in sterile culture vessels containing growth or tissue culture medium (not formalin). Sterile saline is not optimal for this purpose, but should be used if no other option is available. If distance and timing permit the laboratory to receive and process the sample at once, it may be delivered with no liquid added at all. Larger samples may be sent to the laboratory *in toto* for dissection. Solid tissue samples should be transported and stored on ice until culture is established. Storing tissue specimens on ice slows the action of enzymes that degrade the tissue and slows microbial growth in the event of contamination.

CULTURE INITIATION
Growth Media

All specimens for chromosome preparation are grown and maintained in an aqueous growth medium. Some media are formulated for specific cell types (e.g., AmnioMax® or CHANG® for amniocytes), whereas others are appropriate for a broad spectrum of cell types (e.g., RPMI 1640, MEM). All culture media are balanced salt solutions with a variety of additives including salts, glucose, and a buffering system to maintain the proper pH. Phenol red is often used as a pH indicator in many media. If the medium becomes too acidic it will turn yellow, whereas medium that is too basic becomes pink or purple.

Commercial media are available either in powder forms that must be rehydrated or as ready-to-use aqueous solutions. Both complete and incomplete types are commercially available, but most commercial media are incomplete. Incomplete media do not contain all of the nutrients and additives necessary for cell growth. Incomplete culture media must be supplemented with one or more additives before being used for cell culture, as described in the following paragraphs.

L-*Glutamine*

L-Glutamine is an amino acid essential for cell growth. L-Glutamine is unstable and breaks down on storage to D-glutamine, a form that cannot be used by cells. L-Glutamine must therefore be stored frozen to retain its stability, and it is optimal to add it to the culture medium just prior to use. Some commercially available complete media contain L-glutamine.

Serum

Serum is essential for good cell growth. Too little does not allow for maximum cell growth, but too much can have a detrimental effect. Fetal bovine serum (FBS) is preferred; culture medium is generally supplemented with 10–30% FBS.

Antibiotics

Microbial inhibitors are added to culture media to retard the growth of microorganisms. This is a stopgap measure at best and should never be relied upon to compensate for sloppy technique. *Good sterile technique is always the best defense against contamination.*

Penicillin/streptomycin, kanamycin, and gentamicin are bacterial inhibitors commonly used in tissue culture. Fungicides routinely used include nystatin and amphotericin B. Fungicides can adversely affect cell growth and are generally used only when the potential for contamination outweighs this potentially negative effect.

Bacterial contamination of cultures imparts a cloudy appearance to the culture medium. Fungal contamination presents to the unaided eye as "woolly" masses in the medium, or when observed under an inverted microscope as branching hyphae. Mycoplasma and viral contamination can be hard to detect and treat. Mycoplasma should be suspected if the background level of chromosome breaks and rearrangements is higher than usual.

Mitotic Stimulants (Mitogens)

Some cells, particularly mature lymphocytes, do not spontaneously undergo cell division and must be stimulated to divide by the addition of an appropriate mitogen to the cell culture. Phytohemagglutinin (PHA) is an extract of red kidney beans that stimulates division primarily of T lymphocytes. Cell division starts 48 h after the addition of PHA, with additional waves of division at 24-h intervals. The culture period for blood specimens is based on this knowledge. For routine peripheral blood cultures, 72 h is usually optimal. Blood specimens from newborns may require a shorter culture period.

Some leukemia and lymphoma studies require stimulation of B lymphocytes. A number of B-cell mitogens are available, including Epstein–Barr virus, LPS (lipopolysaccharide from *E. coli*), protein A, TPA (12-*O*-tetradecanoyl-phorbol-13-acetate), and pokeweed.

Growth Factors

A variety of additional growth factors are commercially available and are used by some laboratories to achieve optimal cell growth for different sample types.

Culture Vessels

Choice of culture vessel depends in part on the growth needs of the sample, and in part on the individual preference of the laboratory. Blood and bone marrow samples consist of single free-floating cells. For such suspension cultures, sterile centrifuge tubes or tissue culture flasks (T-flasks) may be used. The cells from samples such as amniotic fluid, chorionic villi, skin biopsies, and other solid tissues need to attach to a surface to grow. Such samples may be grown in T-flasks, or with an *in situ* method.

Flask Method

Cells are grown on the inner surface of T-flasks until adequate numbers of dividing cells are present. Cell growth is monitored using an inverted phase contrast microscope. To remove the cells from the surface of the culture flask where they have been growing, the cultures are treated with an enzyme such as trypsin. This enzymatic treatment releases the individual cells into the fluid environment and permits their collection, harvest, or subculture, as needed.

In Situ *Method*

Amniotic fluid, chorionic villus (CVS), and other tissue samples can be grown directly on coverslips in small petri dishes, in "flaskettes," or in slide chambers. Growth of these cultures is also monitored with an inverted phase contrast microscope. They are harvested as "primary" cultures (those that have not been subcultured) when adequate numbers of dividing cells are present, and cells do not have to be enzymatically removed prior to harvest. The cells can therefore be analyzed as they grew *in situ*.

Advantages of the In Situ *Method over the Flask Method*

The primary advantage of using the *in situ* method is that it provides information about the colony of origin of a cell. This is important when deciding whether an abnormality seen in some, but not all cells represents true mosaicism (constitutional mosaicism) or an artifact of tissue culture (pseudomosaicism). True mosaicism is said to be present when there are multiple colonies from more than one culture with the same chromosomal abnormality. Pseudomosaicism is suggested if a single colony with all or some cells exhibiting a chromosomal abnormality is found. In such cases, all available colonies should be studied to rule out the possibility of true mosaicism. If only a single colony with a potentially viable abnormality is found, it may result in an equivocal diagnosis. Low-level mosaicism cannot be completely ruled out in such cases. Clinical correlation may help clarify the picture. A repeat amniocentesis may confirm the presence of true mosaicism, but cannot of course eliminate the results of the first study.

No inference can be made about the origin of cells when using the flask method, as cells from all colonies are mixed together after they are released from the growing surface. It is impossible to tell if multiple cells exhibiting the same chromosomal abnormality arose from one or multiple colonies. Thus, two or more cells exhibiting the same structural abnormality or having the same extra chromosome, or three or more cells lacking the same chromosome must be treated as potential true mosaics if the flask method is used. However, it should be noted that the presence of multiple abnormal colonies **on the same *in situ* coverslip** might also represent artifact. Guidelines for interpretation of mosaicism are available for both flask and *in situ* methods.

Another advantage of the *in situ* method is that there is usually a shorter turnaround time (TAT) since only primary cultures are harvested. Flask cultures are often subcultured, adding days to the culture time.

Preparation of Specimens for Culture

Amniotic fluid specimens, whole blood, and bone marrow samples arrive in the laboratory as single cells in a fluid environment. Whole blood or bone marrow can be added directly to the culture medium or the white blood cells can be separated from the other blood elements and used to inoculate the culture medium. Separation of the white blood cells is easily accomplished by centrifuging the sample or allowing it to rest undisturbed until the blood settles into three distinct layers. The lowest layer consists of the heavier red blood cells; the top layer consists of plasma; and the narrow middle layer, the buffy coat, consists of the desired white blood cells. The buffy coat can be removed and used to establish the suspension culture.

Amniotic fluid contains a variety of cells that arise from the fetal skin, urinary and gastrointestinal tract, and the amnion. These are collectively referred to as amniocytes. Most of the cells in an amniotic fluid sample are dead or dying and are not suitable for cytogenetic analysis. Amniotic fluids are centrifuged at low speed (800–1,000 rpm) to retrieve the small number of viable cells. The cell pellet is then used to establish the cultures. The supernatant may be used for a variety of biochemical tests including α-fetoprotein (AFP) and acetylcholinesterase (AChE) assays for open fetal defects.

Solid tissue samples received in the cytogenetics laboratory are usually too large to culture directly and must be disaggregated before use. To obtain single cells the sample must be finely minced using sterile scissors or scalpels, or alternately, cell dispersion can be achieved by enzymatic digestion of the sample using collagenase or trypsin.

CULTURE MAINTENANCE

After cultures have been initiated, they are allowed to grow under specific conditions of temperature, humidity, and pH until adequate numbers of dividing cells are present. The optimal temperature for human cell growth is 37°C and it is essential that incubators be maintained at this temperature. Cultures are maintained in either "open" or "closed" systems, depending upon the type of incubator used.

Open systems are those that allow the free exchange of gases between the atmosphere inside the culture vessel and the surrounding enviornment of the incubator. To facilitate the exchange of gases, the tops or caps of tissue culture vessels are loosely applied. A CO_2 incubator is required for open systems to maintain the 5% CO_2 level necessary to sustain the ideal pH of 7.2–7.4. A humidity level of 97% should be maintained to prevent cell death due to cultures drying out. This can be accomplished by placing pans of sterile water in the bottom of the incubator. A major disadvantage of open systems is that they are susceptible to microbial contamination, especially fungi, owing to the moist warm surfaces in the incubator. An open system is required for samples grown on coverslips by the *in situ* method.

Closed systems are those in which the culture vessels are tightly capped to prevent exchange of gases. Humidification is self maintained, and CO_2 incubators are not required. Commercial media are buffered to the appropriate pH necessary to sustain short-term cultures such as those from blood and bone marrow samples. Long-term cultures from amniotic fluid and solid tissue specimens require the use of additional buffering systems to maintain the proper pH over the longer culture period. Microbial contamination is not as great a risk with closed systems.

In the final analysis, the decision to use an open or closed system, or a combination of both, involves the type of sample being processed and the preference of the laboratory.

Culture Maintenance and Growth Interval

Once the culture requirements are met, the cells must be allowed time to grow and divide. The time in culture varies depending upon the cell type involved.

Peripheral blood cultures require little maintenance once the growth requirements have been met. The culture vessels are placed in an incubator for a specified period of time, usually 72 h.

Likewise, bone marrow cultures need little attention once the culture has been initiated. Bone marrow contains actively dividing cells and therefore can be harvested directly, without any time in culture, or a 24- to 48-h culture time may be used to increase the mitotic index. Longer culture periods are generally not advised because the abnormal cancerous cells may be lost over time or be diluted out by normal precursor cells that may be present. A short growth period usually provides a more accurate reflection of makeup of the tumor; however there are exceptions, as some tumor cells are slow growing.

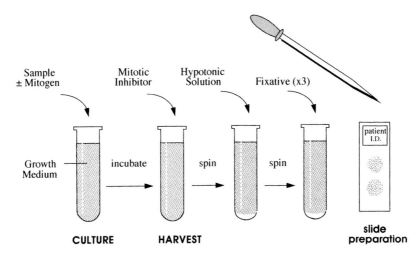

Fig. 1. Overview of peripheral blood cell culture and harvest for chromosome analysis. This procedure, with minor variations, is utilized for all specimen types.

Amniotic fluid and solid tissue specimens require longer culture periods and do not grow at predictable rates. Cell growth is monitored periodically until there are sufficient numbers of dividing cells present, indicating that the culture is ready for harvest. An inverted phase contrast microscope is used to visualize the mitotic cells, which appear as small, refractile spheres. *In situ* amniotic fluid cultures are generally harvested at 6–10 days, sometimes earlier. For amniotic fluid and solid tissue specimens grown using the flask method, the culture interval may be 2 weeks or more.

Amniotic fluid and solid tissue specimens cultured with either the *in situ* or flask method become depleted of required nutrients and additives during the culture period. Depleted medium must be removed and replenished with fresh medium. This process is called "feeding" the culture and is done on a regular basis throughout the culture maintenance period dependent upon the number of cells growing, the length of time in culture, and the protocol of the laboratory. Exhausted medium becomes acidic and will appear yellow if the medium contains a pH indicator such as phenol red.

CELL HARVEST

After the cell cultures have grown for the appropriate period of time and there is a sufficient number of dividing cells, the cells are harvested. Harvest is the procedure of collecting the dividing cells at metaphase, their subsequent hypotonic treatment and fixation, and the placement of the chromosomes on glass slides so they may be stained and microscopically examined. The basic steps of cell harvest are the same for all specimen types, with minor variation. An example is shown in **Fig. 1**.

Mitotic Inhibitor

A mitotic inhibitor must be used to obtain adequate numbers of cells in metaphase. Colcemid®, an analogue of colchicine, is used in most cytogenetics laboratories. Colcemid® binds to the protein tubulin, obstructing formation of the spindle fibers or destroying those already present. This prevents separation of the sister chromatids in anaphase, thus collecting the cells in metaphase. Exposure time to colcemid is a trade-off between quantity and quality. A longer exposure results in more metaphases being collected, but they will be shorter because chromosomes condense as they progress through metaphase. Longer chromosomes are generally preferred for cytogenetic studies. Exposure time to Colcemid® varies by specimen type.

Hypotonic Solution

A hypotonic solution is added to the cells after exposure to Colcemid®. The hypotonic solution has a lower salt concentration than the cell cytoplasm, allowing water to move into the cell by osmosis. This swells the cells and is critical for adequate spreading of the chromosomes on the microscope slide. Timing is crucial, as too long an exposure will cause the cells to burst. Too short an exposure to hypotonic solution will not swell the cells sufficiently, which results in poor spreading of the chromosomes.

There are a variety of acceptable hypotonic solutions including 0.075M potassium chloride (KCl), 0.8% sodium citrate, dilute balanced salt solutions, dilute serum, and mixtures of KCl and sodium citrate. Morphology of the chromosomes is affected by the hypotonic solution used. The choice of hypotonic solution is based on specimen type and laboratory protocol.

Fixative

A modified Carnoy's solution of three parts absolute methanol to one part glacial acetic acid is used to stop the action of the hypotonic solution and to fix the cells in the swollen state. This fixative also lyses any red blood cells present in the sample. The fixative must be prepared fresh before use because it readily absorbs water from the atmosphere, which adversely affects chromosome quality and staining.

Slide Preparation

The final step of the harvest procedure is slide preparation. Fixed cells from suspension cultures are dropped onto glass slides to allow for subsequent staining and analysis. The concentration of the cell suspension can be adjusted to achieve optimal results. Fixed cells from *in situ* cultures are not dropped because they are already attached to a coverslip or other solid surface. The prepared slides or coverslips are dried under conditions that favor optimal chromosome spreading, and are checked with a phase-contrast microscope for metaphase quality and number. A good slide preparation has sufficient numbers of metaphases that are not crowded on the slide, metaphases that are well spread with minimal overlapping of the chromosomes, and no visible cytoplasm.

A number of variables affect the rate of evaporation of fixative from the slide, the spreading of chromosomes, and the overall quality of the slide preparation. Ambient temperature and humidity and length of time in hypotonic treatment all affect spreading of chromosomes. Increased temperature and humidity enhance chromosome spreading, whereas cooler temperature and lower humidity decrease it. Longer exposure to hypotonic solution makes cells more fragile and increases spreading, but an inadequate exposure can result in cells that are difficult to burst. Every technologist must have an arsenal of techniques to effectively deal with these variables.

Other variables in slide preparation include the height from which the cells are dropped; the use of wet or dry slides; the use of cold, room temperature, or warm slides; the use of steam; air- or flame-drying the slides; and the angle at which the slide and/or pipette is held.

After slides are prepared they are "aged" overnight at 60°C or for 1 hour at 90°C to enhance chromosome banding. There are also techniques that allow chromosomes to be "aged" by brief exposure to UV light.

CHROMOSOME STAINING AND BANDING

Prior to the 1970s, human chromosomes were "solid" stained using orcein or other stains with an affinity for chromatin. The chromosomes were classified according to their overall length, centromere position, and the ratio of the short arm to the long arm. Solid stains provided limited information. Simple aneuploidies could be recognized, but structural aberrations were difficult to characterize, and in some cases impossible to detect. In addition, it was not possible to specifically identify individual chromosomes.

A large number of banding and staining techniques have since been developed. These can be divided into two broad categories: those that produce specific alternating bands along the length of each entire chromosome, and those that stain only a specific region of some or all chromosomes.

Methods that produce specific alternating bands along the length of the chromosomes create unique patterns for each individual chromosome pair. This property allows for the positive identification of the individual chromosome pairs and permits characterization of structural abnormalities. These banding techniques answer many questions by facilitating the numerical and structural examination of the entire karyotype.

Those techniques that selectively stain specific regions of chromosomes are used in special circumstances when a particular piece of information cannot be obtained using a routine banding method. These special stains are typically utilized to obtain such specific data.

Techniques that Create Bands Along the Length of the Chromosomes

An important measurement associated with these methods is the level of banding resolution obtained. As chromosomes condense during mitosis, subbands begin to merge into larger landmarks along the chromosome. Obviously, as this progresses, the ability to visualize subtle abnormalities is reduced. Chromosomes with a greater number of visible bands and subbands (higher resolution) are therefore more desirable. Laboratories accomplish this in two ways—by optimizing the banding and staining procedures themselves so that a maximum number of sharp, crisp bands is produced, and by choosing (and in some cases manipulating cultures to produce) cells with longer, less condensed chromosomes.

Banding resolution is an estimate of the number of light and dark bands in a haploid set, arrived at by counting these bands in one chromosome of each homologous pair. Minimum estimates usually begin at approximately 400 bands. Well-banded, moderately high-resolution metaphases are usually in the 500- to 550-band range, whereas prometaphase cells can achieve resolutions of 850 or more bands.

Fig. 2. G-Banding (Giemsa banding). Note the light and dark bands along the length of each chromosome.

G-Banding (Giemsa Banding)

G-Banding is the most widely used routine banding method in the United States. GTG banding (<u>G</u> bands produced with <u>t</u>rypsin and <u>G</u>iemsa) is one of several G-band techniques. With this method, prepared and "aged" slides are treated with the enzyme trypsin and then stained with Giemsa. This produces a series of light and dark bands that allow for the positive identification of each chromosome (**Fig. 2**). The dark bands are A–T-rich, late replicating, heterochromatic regions of the chromosomes, whereas the light bands are C–G-rich, early replicating, euchromatic regions. The G-light bands are biologically more significant, because they represent the most active regions of the chromosomes, whereas the G-dark bands contain relatively few active genes. There are also G-banding techniques that actually utilize stains other than Giemsa, such as Wright's and Leishman's stains.

Q-Banding (Quinacrine Banding)

Q-Banding is a fluorescent technique, and was the first banding method developed for human chromosomes. Certain fluorochromes, such as quinacrine dihydrochloride, will bind to DNA and produce distinct banding patterns of bright and dull fluorescence when excited with the proper wavelength of light. Because adjacent A–T pairs are necessary to create binding sites, the brightly fluorescing regions are A–T-rich. The Q-banding pattern is similar to the G-banding pattern with some notable exceptions. In particular,

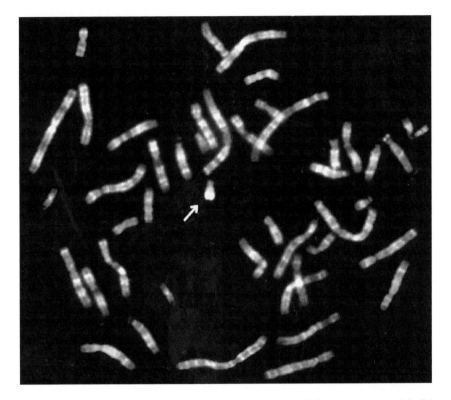

Fig. 3. Q-Banding. The fluorescence banding pattern is essentially the same as with G-banding. Note, however, the bright fluorescence on the long arm of the Y chromosome (*arrow*).

the large polymorphic pericentromeric regions of chromosomes 1 and 16, and the distal long arm of the Y fluoresce brightly; the distal long arm of the Y chromosome is the most fluorescent site in the human genome. Q-Banding is therefore useful to confirm the presence of Y material or when studying the cited polymorphic regions. (See **Fig. 3**.)

Most fluorescent stains are not permanent, and require the use of expensive fluorescence microscopes and a darkened room. Q-Banding is therefore not conducive to routine work in most laboratories.

R-Banding (Reverse Banding)

R-Banding techniques produce a banding pattern that is the opposite or reverse of the G-banding pattern. There are fluorescent and nonfluorescent methods. The C–G-rich, euchromatic regions stain darkly or fluoresce brightly whereas the A–T-rich heterochromatic regions stain lightly or fluoresce dully. The euchromatic, R-band positive regions are the more genetically active regions of the chromosomes. Many human chromosomes have euchromatic terminal ends that can be difficult to visualize with standard G-band techniques because the pale telomeres may fade into the background. R-Banding is a useful technique for the evaluation of these telomeres. R-Banding is typically used as an additional procedure in many countries, but is the standard method for routine banding in France (see **Fig. 4**).

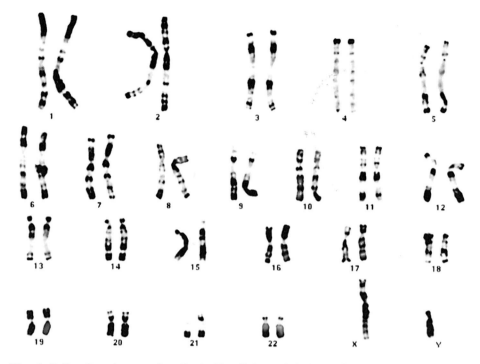

Fig. 4. R-Banding (reverse banding). The light and dark bands are the opposite of those obtained with G-banding. R-Banding can also be performed with fluorescent staining. (Image courtesy of Dr. Patricia Lewin, Laboratoire cerba, Val d'Oise, France.)

Techniques that Stain Selective Chromosome Regions

C-Banding (Constitutive Heterochromatin Banding)

C-Banding techniques selectively stain the constitutive heterochromatin around the centromeres, the areas of inherited polymorphisms present on chromosomes 1, 9, 16, and the distal long arm of the Y chromosome. C-band positive areas contain highly repetitive, late replicating sequences of α-satellite DNA. The function of constitutive heterochromatin is not understood, but it is stable and highly conserved evolutionarily.

With CBG banding (C-bands, by barium hydroxide, using Giemsa) the DNA is selectively depurinated and denatured by barium hydroxide, and the fragments are washed away by incubation in a warm salt solution. Constitutive heterochromatin resists degradation and is therefore the only material left to bind with the Giemsa stain. The result is pale, almost ghostlike chromosomes with darkly stained areas around the centromeres, at the pericentromeric polymorphic regions of chromosomes 1, 9, and 16, and at the distal Y long arm (**Fig. 5**). C-Banding is useful for determining the presence of dicentric and pseudodicentric chromosomes, and also for studying marker chromosomes.

T-Banding (Telomere Banding)

T-Banding is an offshoot of R-banding that results in only the terminal ends or telomeres of the chromosomes being stained. A more harsh treatment of the chromo-

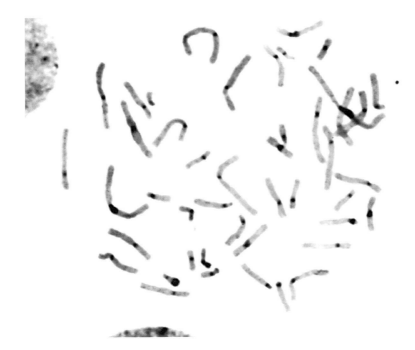

Fig. 5. C-Banding. This technique stains the constitutive heterochromatin found in each chromosome (hence the term C-banding), and is useful for clarification of polymorphisms. Note the large heterochromatic regions in some of the chromosomes.

somes diminishes staining except at the heat-resistant telomeres. There are fluorescent and nonfluorescent T-banding techniques.

Cd Staining (*C*entromeric *d*ot or Kinetochore Staining)

This technique produces a pair of dots at each centromere, one on each chromatid. These are believed to represent the kinetochores or the chromatin associated with them. The dots are specific to the centromeric region and are not the same as C-bands. Only active or functional centromeres will stain with Cd staining, in contrast to C-banding, which will stain inactive as well as active centromeric regions. Cd staining can be used to differentiate functional from nonfunctional centromeres, and to study Robertsonian translocations (centromere-to-centromere translocations of acrocentric chromosomes), ring chromosomes, and marker chromosomes.

G-11 Banding (*G*iemsa at pH *11*)

This technique specifically stains the pericentromeric regions of all chromosomes; the heterochromatin regions of chromosomes 1, 9, 16, and the distal Yq; and the satellites of the acrocentric chromosomes. An alkaline treatment of the chromosomes causes loss of the Giemsa binding sites. Optimal results are achieved at pH of 11.6. At this high alkaline pH only the azure component of Giemsa binds with the majority of the chromosomes, staining them light blue. The eosin component of Giemsa binds specifically to the heteromorphic regions cited previously, staining them magenta. G-11 banding is used to delineate these heterochromatin polymorphisms.

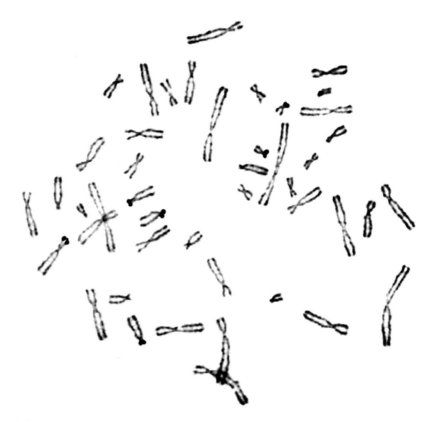

Fig. 6. NOR staining (silver staining). This procedure identifies active nucleolar organizer regions, found on the stalks of acrocentric chromosomes. Silver nitrate produces dark staining in these areas.

G-11 banding also has research applications. It is used to differentiate between human and rodent chromosomes in hybrid cells. The human chromosomes stain pale blue while the rodent chromosomes stain magenta.

NOR Staining (Silver Staining for N̲ucleolar O̲rganizer R̲egions)

This technique selectively stains the nucleolar organizer regions (NORs) located on the satellite stalks of the acrocentric chromosomes. These regions contain the genes for ribosomal RNA and can be stained with silver nitrate. Theoretically, there are 10 NORs per cell, one for each acrocentric chromosome. However, not all will usually stain at any one time because the silver stains the activity, not presence, of rRNA genes. NOR staining is useful for the identification of marker chromosomes and rearrangements or polymorphisms involving the acrocentric chromosomes. See **Fig. 6**.

DAPI/DA Staining (4,6-D̲iamino-2-P̲henole-I̲ndole/D̲istamycin A̲)

This stain combines DAPI, a fluorescent dye, with distamycin A, a nonfluorescent antibiotic. Both form stable bonds preferentially to similar, but not identical, A–T-rich, double-stranded regions of DNA. Used together, DAPI/DA fluoresces certain A–T-rich areas of constitutive heterochromatin in the C-band regions of chromosomes 1, 9, 16, the distal Yq, and the short arm of chromosome 15. Prior to the development of fluorescence *in situ* hybridization techniques, this was the stain that differentiated between satellite regions of any of the acrocentric chromosomes.

DAPI/DA is used to identify rearrangements of chromosome 15; to confirm variations in the polymorphic regions of chromosomes 1, 9, 16, and distal Yq; and to study marker chromosomes with satellites.

Fluorescence In Situ *Hybridization (FISH)*

The development of fluorescence *in situ* hybridization technology represents an important advancement in cytogenetics. FISH is a marriage of classic cytogenetics and new molecular technologies and has a large number of applications. Chapter 15 deals with this topic in depth.

HIGH-RESOLUTION STUDIES

Chromosomes are routinely examined during metaphase, when they are at their most contracted state. Although this is often sufficient for chromosomal analysis, small structural abnormalities may not be detected in chromosomes of metaphase length. In such cases, longer, less contracted prophase or prometaphase chromosomes are needed. To achieve longer chromosomes, the cells can be synchronized and harvested earlier in the cell cycle, or chemical elongation techniques can be used to prevent condensation of the chromosomes.

Cell Synchronization Techniques

Randomly dividing cells can be synchronized with knowledge of the average timing of the stages of the human cell cycle. The cells are blocked and then released at the appropriate time so that a large percentage of cells accumulate in prophase or prometaphase at the time of harvest. There are several protocols for generating such synchronization.

One method involves the addition of FUdR (5-fluorodeoxyuridine) to peripheral blood cultures prior to harvest. FUdR is an inhibitor of thymidylate synthetase, which plays an important role in the folic acid pathway. Folic acid is required for incorporation of thymidine during DNA synthesis. The addition of FUdR blocks cell division at the G1/S border. After 17 h, the accumulated cells are released from the block by the addition of a high level of thymidine. The peak prometaphase index occurs 5–6 h later, and this is when the harvest is performed.

Methotrexate (MTX), also called L-(+)- amethopterin, can also be used to achieve cell synchrony, and BrdU (5-bromodeoxyuridine), an analog of thymidine, can be used to release the block.

Chemical Elongation

Ethidium bromide (EB) can be added to cultures prior to harvest to achieve longer chromosomes. Ethidium bromide acts by intercalating between the bases of DNA, thus preventing or slowing its contraction. This results in the collection of long, if not truly prometaphase, chromosomes. The procedure is technically very simple and can be used routinely on blood and bone marrow cultures to obtain better length chromosomes. The major drawback to using EB is that it is highly mutagenic. Extreme care must therefore be taken when utilizing this reagent.

CULTURE FAILURE

All culture failures must be investigated and the circumstances of the failure should be recorded as a part of an ongoing Quality Assurance program. A record of failure rates for each specimen type in the laboratory must be kept as a baseline so that deviations from the norm can be detected. It is important to isolate the reason(s) for a culture failure so that steps can be taken to prevent future similar failures. Some culture failure is unavoidable, but adherence to strict standards and rigorous investigation of all failures should keep this number to a minimum.

There are many possible origins of culture failure. It can be due to improper specimen collection or transport, improper laboratory technique, or the condition of the sample. There are general sources of failure that apply to all sample types, and specific ones that pertain to one or more of the sample types.

Errors in sample collection and handling include failure to submit an adequate amount of sample, collection under nonsterile conditions resulting in microbial contamination, use of an inappropriate collection vessel or medium, failure to use an anticoagulant, use of an inappropriate or expired anticoagulant, delay in transport, and improper storage before and/or during transport of the sample.

In the laboratory, errors can occur at any step from culture initiation to staining. Failure to follow proper protocol can cause loss of a culture. This is one reason for establishing multiple cultures for all samples and harvesting them at different times. Faulty media, sera, or other reagents can also result in culture failure. It is therefore important to test all new lots of media and sera for sterility and ability to support cell growth before using these on patient samples. It is also important to maintain a log of lot numbers of all reagents used and the date each was put into use to help identify the source of any problem. During the culture period, improper temperature, CO_2 level, or pH of the culture can have deleterious results. The temperature and CO_2 levels of all incubators must therefore be monitored and recorded at least daily and samples should be split and grown in separate incubators in the event an incubator malfunction. In general, all equipment used in the laboratory must be monitored at regular intervals and maintained to prevent malfunction.

Lack of viable cells or unsuitable cell type can compromise amniotic fluid samples. Samples from patients with advanced gestational age (20 weeks or greater) may consist primarily of mature nondividing cells, or dead cells. Some samples consist principally of epithelial cells, which typically produce few metaphases of poorer quality than the desired fibroblasts.

Amniotic fluid samples are usually clear yellow in appearance. A brown fluid indicates prior bleeding into the amniotic cavity which may suggest fetal death or threatened miscarriage. In such samples there may be few if any viable cells present. Bloody taps containing large numbers of red blood cells can be problematic. The physical presence of large numbers of red blood cells can prevent the amniocytes from settling on and attaching to the growth surface of the culture vessel. In addition, the red cells utilize nutrients in the culture medium, thereby competing with the amniocytes.

Patient factors can influence the success of peripheral blood and bone marrow samples. Disease conditions, immunosuppression, and use of other drugs can affect both the number of lymphocytes present and their response to mitotic stimulants. The laboratory is not always made aware of these confounding factors. Bone marrow samples that have been contaminated with blood may not have adequate numbers of spontaneously dividing cells present. For this reason, it is important that the cytogenetics laboratory receive the first few milliliters of the bone marrow tap. Bone marrow samples are notorious for producing poor quality metaphases. There are sometimes adequate numbers of metaphases, but the chromosomes are so short and so poorly spread that analysis is difficult or impossible. In addition, metaphases of poor quality often represent an abnormal clone.

The failure rate of solid tissues may be quite high, and is often due to the samples themselves. In the case of products of conception or stillbirths, the sample may not contain viable cells or the wrong tissue type may have been collected. In addition, microbial contamination is a frequent contributing factor because many solid tissue samples are not sterile prior to collection.

PRESERVATION OF CELLS

Cells do not survive indefinitely in tissue culture. After a period of time they become senescent and eventually die. At times a sample may need to be saved for future testing, to look at retrospectively, or because it is unusual or interesting and might be of some value in the future. In such cases the cells need to be kept alive and capable of division long term or indefinitely.

Cultured cells can be kept alive by cryopreservation, the storage of cells in liquid nitrogen. The freezing process is critical to cell survival. Rapid freezing will cause cell death due to formation of ice crystals within the cells. Improper freezing can also denature proteins, alter the pH, and upset electrolyte concentrations. The cells must be cooled slowly so that water is lost before the cells freeze. The addition of 10% glycerol or dimethyl sulfoxide (DMSO) to the storage medium lowers the freezing point and aids in this process. One-milliliter aliquots of the sample in storage medium are placed in cryogenic freezing tubes. The samples are then slowly frozen under controlled conditions at a rate of 1°C per minute to a temperature of −40°C. The sample can then be rapidly frozen to about −80°C. Alternately, the samples may be placed in a −70°C freezer for 1–4 h. After this initial freezing has been accomplished, the cells are stored in the liquid phase at about −190°C.

Thawing of the sample is also critical. Rapid thawing is necessary to prevent the formation of ice crystals.

B Lymphocytes can be transformed so that they will proliferate indefinitely in tissue culture by exposing them to Epstein–Barr Virus (EBV). These immortalized lym–phoblastoid cell lines do not become senescent and can therefore be maintained indefinitely in culture.

CHROMOSOME ANALYSIS

Selection of the correct specimen for chromosome analysis and additional tests is not always straightforward, and the submission of an inappropriate sample to the laboratory can create frustration for both patient and clinician.

This was not always as complex an issue as it is today. In the 1970s, prenatal diagnosis involved an amniotic fluid specimen, often obtained at exactly 17 weeks of gestation, for chromosome analysis and α-fetoprotein testing. Other tests were available, but rarely ordered. The cytogenetic contribution to hemoncology essentially involved whether a bone marrow specimen was "positive or negative" for the "Philadelphia chromosome." Constitutional chromosome analysis from peripheral blood implied that the patient had to be an adult or a child.

Today's prenatal caregivers and their patients must choose between traditional amniocentesis, early amniocentesis, chorionic villus sampling or, sometimes, percutaneous umbilical blood sampling. A decision concerning whether ploidy analysis via FISH is warranted must be made, and acetylcholinesterase is often a factor in the diagnosis of certain open fetal lesions, but AFP and AChE analyses cannot be performed on all sample types. Many disorders can be also diagnosed by biochemical or molecular methods, and ethical dilemmas surround the potential to prenatally diagnose late-onset disorders such as Huntington's disease. Screening for increased potential or predisposition to develop certain cancers or other diseases will create new moral and ethical dilemmas. Each of these may ultimately affect the number of cells available for chromosome analysis, and all of these issues can play a role in the timing and choice of sampling procedure.

Today, the cytogenetics laboratory provides indispensable information for the diagnosis, prognosis, or monitoring of patients with a wide variety of hematological disorders and other neoplasms, using not only bone marrow, but in some cases blood, lymph node biopsies, or tumor tissue or aspirates. Treatment decisions often rest on the results of a chromosome analysis, but some tissue types are appropriate only under certain conditions, and an incorrect selection here can delay a vital diagnosis.

In addition to blood samples from patients suspected of having constitutional chromosomal abnormalities, samples can also be obtained prenatally from fetuses, and from patients with hematological disorders. These must all be handled differently, and the information they provide is unique in each circumstance.

Procedure

After all of the appropriate laboratory manipulations and staining procedures have been performed, there are several steps involved in the clinical analysis of chromosomes. These start with the microscope, where selection of appropriate metaphases begins the process. Although technologists are trained to recognize well-spread, high-quality cells under low-power magnification, they must also examine some poor quality metaphases when looking for acquired changes, as these often represent abnormal clones.

Under high power, the chromosome morphology and degree of banding (resolution) are evaluated. If these are appropriate, the number of chromosomes is counted, and the sex chromosome constitution is typically determined. The microscope stage coordinates of each metaphase are recorded, and in many laboratories an "identifier" of the cell is also noted. This is typically the position of one or more chromosomes at some reference point(s), and serves to verify, should there be a need to relocate a cell, that the correct metaphase has been found. Any other characteristics of the metaphase being examined, such as a chromosome abnormality or quality of the banding and chromosome morphology are also noted.

In the United States, certifying agencies such as the College of American Pathologists (CAP) require that a minimum number of metaphases be examined for each type of specimen, barring technical or clinical issues that can sometimes prevent this. There are also requirements for a more detailed analysis (typically band-by-band) of a certain number of cells, as well as standards for the number of metaphases from which karyotypes are prepared. Regulations notwithstanding, it is clearly good laboratory practice to analyze every chromosome completely in several cells, and even more important to check all chromosomes in certain situations, such as when analyzing cancer specimens. Depending upon the results obtained and/or initial diagnosis, additional cells may be examined to correctly identify all cell lines present.

Once the appropriate number of mitotic cells has been examined and analyzed, a representative sample must be selected for imaging and ultimate preparation of karyotypes. This will involve either traditional photography and manual arrangement of chromosomes, or computerized image capture and automated karyotyping (see Chapters 5 and 7). Many laboratories also image additional cells, to be included as references in the patient chart. Ultimately, summary information (patient karyotype, banding resolution, number of cells examined, analyzed, and photographed, etc.) is recorded in the patient's file and is used, either manually or via computer, in the clinical report.

The final steps of the process typically involve a clerical review of all relevant clinical, technical and clerical data, examination of the patient's chart and karyotypes by the laboratory director (often preceded by the supervisor and/or other senior laboratory personnel), and generation of the formal clinical report. In addition to the appropriate physician and patient demographic information, this should include the number of metaphases that were examined microscopically, the banding resolution obtained for the specimen, the number of cells that were analyzed in detail, the number of karyotypes prepared, the patient's karyotype, and a clinical interpretation of the results, including, where appropriate, recommendations for additional testing and/or genetic counseling.

SUMMARY

The purpose of this chapter is to provide a general overview of the many steps involved from receipt of a sample in the cytogenetics laboratory to the generation of a patient report, and to impress upon the reader the labor-intensive nature of this work. Although the basic procedure is always the same, there are culturing and processing variations that are sample type dependent, choices of methodology that are diagnosis dependent, and microscopic analysis decisions that are results dependent. All of these in turn depend upon individuals with the appropriate expertise and dedication to patient care.

REFERENCES UTILIZED

Due to the nature of this chapter, individual citations were not practical. In addition to the authors' personal experience, the following were used as supplemental sources of information:

- Barch, M.J., Knutsen, T., and Spurbeck, J.L., eds. (1997) *The AGT Cytogenetic Laboratory Manual.* Raven-Lippincott, Philadelphia.
- Rooney, D.E. and Czepulkowski, B.H. (1992). *Human Cytogenetics: A Practical Approach,* Vol. I: *Constitutional Analysis.* IRL Press, Oxford University Press, New York.
- Verma, R.S. and Babu, A. (1995). *Human Chromosomes.* McGraw-Hill, New York.

5

Microscopy and Photomicrography

Christopher McAleer

INTRODUCTION

Microscopy and photomicrography are essential techniques for clinical cytogenetic analysis, and have traditionally been an integral part of the cytogenetics laboratory. Computer imaging (see Chapter 7) is a much newer technology that has become a component of many cytogenetics laboratories over the last decade.

Many variables are involved in obtaining good microscope images and photomicrographs. By themselves, the individual processes may have little impact on a specimen image, but together they can have a significant influence on the final product. It is perhaps key to recognize that each step in microscopy and photomicrography tends to build upon those before it. Achievement of the best image quality thus requires a full understanding and application of many principles. This chapter will explore in some detail a process that many take for granted.

BRIGHTFIELD MICROSCOPY

The brightfield microscope is arguably the most important piece of equipment in the cytogenetics laboratory. Knowledge of its component parts and their proper use is fundamental.

The Transmitted Light Source and Power Supply

A transmitted light source is found in the base of a microscope, often in the rear, but occasionally in the front. The bulb housings of many microscopes will often automatically center the light bulb. For some microscopes, it may be necessary to center and focus the bulbs manually.

The power supply for transmitted light sources is typically located within the base of a microscope, but may be an external unit. In either instance, the intensity of the microscope light is usually controlled by regulating the bulb voltage through a rheostat.

Even illumination and precise control of the light supply are important for good microscope images, and are critical for the proper function of photomicrographic and computer imaging systems. Proper alignment of the bulb and use of dispersion filters are requirements for illumination that is free of shadows. If the quality and age of a light bulb, the microscope rheostat, or the quality of the power supplied to the microscope do not permit a stable supply of light, they can quickly degrade images. The path of light through a brightfield microscope is depicted in **Figure 1**.

From: *The Principles of Clinical Cytogenetics*
Edited by: S. Gersen and M. Keagle © Humana Press Inc., Totowa, NJ

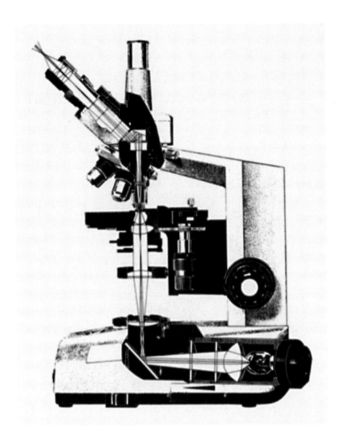

Fig. 1. Cut-away view depicting the lightpath of a brightfield microscope (reprinted with permission of Olympus America, Inc.)

Microscope Filters for Brightfield Microscopy

In general, the resolution of a microscope image improves with the quantity of focused light passing from a specimen into the objective lens. In addition, the color of light also plays an important role in resolving capacity of a microscope. *Resolving power* can be thought of as the ability to visually distinguish two separate objects that lie in close proximity. As the distance between two such objects approaches the wavelength of light, the ability to see separate points becomes highly dependent upon the wavelength used for viewing. For this reason, shorter light wavelengths (green–blue) have greater resolving capabilities than longer light wavelengths (yellow–red).

Resolving power is also influenced by the color difference that two objects share, as well as the color of the background on which the objects appear. When an image has a low range of contrast, it will be more difficult for the eye to detect structures resolved by the microscope components. Stains used for G-banding act to increase the contrast range of a chromosome by absorbing transmitted light where the stain has bound. This produces bands of varying intensity, as regions with more stain allow less light to be detected. Green light enhances this absorption effect, as G-banding stains strongly absorb green wavelengths. This increases the overall contrast range of the image, allowing the eye to detect more of the subtle details of chromosome morphology and banding. For these reasons, a green filter is recommended for routine chromosome analysis.

A variety of filters are available for improving the image contrast of cytogenetic specimens. "Green glass" filters are the least expensive option and act to improve the contrast range of G-banded chromosomes. Green interference filters, a more expensive option, absorb all but a single wavelength to a narrow band of green wavelengths. Because the optics of some microscopes rely upon the use of monochromatic green light to produce quality images, investment in a green interference filter can be a requirement. Interference filters are easily identified by their partial reflective quality and unusual tint (often orange) when viewed at an angle.

Field Diaphragm, Condenser, and Condenser (Aperture) Diaphragm

The field diaphragm, condenser, and condenser (aperture) diaphragm gather and focus the microscope light, passing it through the specimen and into the objective lens. These components play an important role in image contrast and resolution.

As light passes through a specimen, the light rays will bend or diffract from their original path. It is important to understand that the smaller structures of a specimen will diffract light to a greater degree relative to the diffraction of larger structures. To obtain well-resolved images a microscope must gather as many of these highly diffracted light rays as possible for viewing *(1)*.

The process of seeing an image begins with the field diaphragm, which is used to control the area of specimen illumination. From here light passes into the condenser, through the specimen and into the small opening of the objective lens. The numerical apertures (NA) of the condenser and objective lens are a measurement of their ability to gather highly diffracted light, and thus a measure of a microscope's resolving potential.

The condenser influences the number of light rays passing into an objective lens by how it is used to illuminate the specimen. A properly used condenser illuminates the small structures of a specimen from many different angles. This increases the likelihood of producing light rays with an angle of diffraction that will pass through the small opening of an objective, and thus be present for viewing.

Microscope condensers come in a variety of numerical apertures, ranging from 1.4 to less than 1.0. Because the operating numerical aperture of an objective lens cannot be greater than that of the condenser, each component should have a similar NA value.

Condensers require correction for two basic groups of optical imperfections or aberrations that are created as light passes through any lens. Monochromatic aberrations are those that can occur with any wavelength of light, whereas chromatic aberrations are problems unique to a specific wavelength. A microscope is equipped with one of three of condensers: abbe, aplanatic, or aplanatic/achromatic, each differing in its ability to correct optical aberrations. Condensers are generally labeled with the type of optical correction and the numerical aperture.

Abbe condensers are the most basic type of condenser and are not corrected for either type of optical aberration. They are not recommended for use in the cytogenetics laboratory.

Aplanatic condensers are corrected solely for monochromatic aberrations and rely upon green light for the greatest degree of correction *(2)*. The performance of aplanatic condensers is highest when a monochromatic green interference filter is used.

Aplanatic/Achromatic condensers are corrected for both types of optical aberration and do not require the use of green light for the correction *(2)*.

The aperture diaphragm of a condenser is adjusted to achieve a balance between the resolving power of the microscope and image contrast. When the aperture diaphragm is completely open, the small structures of a specimen are illuminated by light from the greatest number of angles and resolving capacity is at its highest. Unfortunately, the details of these structures are so well illuminated that they lose the "shadowing" or variations in color that give such structures perspective.

As the aperture diaphragm is closed, the structures of the specimen are illuminated from fewer angles, resulting in a loss of resolving power, but an improvement in the "shadowing" or contrast of the image. Considering this, the aperture diaphragm is set to produce a suitable balance between image contrast and resolution. Many microscope manufacturers recommend setting the condenser aperture between 66% and 75% open to achieve the best balance.

Köhler Illumination

Centering and focusing the condenser (Köhler illumination) is crucial for optimum image quality. The process begins by closing the field diaphragm so that light traveling through the condenser can be visually centered and focused. Once achieved, the field diaphragm is opened so that the light illuminates the specimen just beyond the field of view. Finally, the aperture diaphragm is adjusted to generate the desired image contrast.

The Phase-Contrast Condenser

Phase-contrast microscopes use a special condenser and objective lens to increase the contrast range of a specimen by darkening areas of greater density, and lightening areas of lesser density. Phase-contrast is often used for visualizing living or other unstained cells, but can be used successfully to increase the contrast range of G-banded specimens. A microscope equipped for phase-contrast will make use of a special condenser as well as phase objective lenses. Proper use of a phase-contrast microscope requires individuals to first achieve Köhler illumination, followed by an adjustment to align the phase components of the microscope. Of key importance is the selection of a phase condenser setting that matches the "Ph" number of the phase objective.

Immersion Oil

As light rays enter and exit a lens, they can reflect (refract) off of its outer and inner surfaces. This occurs because the lens and air each have a different refractive property (refractive index).

Most microscopes have been engineered to reduce the incidence refraction to that which can occur between the condenser and microscope slide, as well as between the specimen and objective lens. Immersion oil can be used to greatly reduce the remaining refraction by removing the air from these spaces, filling them with a substance that has a nearly identical refractive index. In short, more light rays will exit one surface and pass into the next than would occur if air occupied these spaces, thus increasing the resolving capacity of the microscope.

The question that is often raised is whether oiling a condenser will actually improve the image quality to the extent that it is worth having to clean an oily condenser. The highest operating numerical aperture of an "un-oiled" condenser and objective lens will be slightly less than 1.0, as the refractive index of the air prevents a higher operating numerical aperture *(3)*. If the numerical aperture of a condenser or objective lens is less than 1.0, no benefit will result from applying immersion oil.

Even with higher numerical apertures, the loss may not be of practical significance. Remembering that resolving capacity must be balanced with image contrast, the act of applying immersion oil increases the operating numerical aperture, but also lowers image contrast. In turn, closing the aperture diaphragm improves image contrast, but also lowers the resolving capacity of the microscope. Considering this, the application of immersion oil to a condenser may be beneficial only when a specimen can be viewed at a reasonable contrast while the aperture diaphragm of the condenser is nearly wide open.

Immersion oil comes in several types that vary in viscosity and fluorescent properties. Each is formulated to have a refractive index of 1.5150 ± 0.0002 at 23°C (essentially the refractive index of glass).

Noncoverslipped microscope slides should be cleaned of immersion oil at the completion of microscopic analysis. Immersion oil can cause fading of unprotected G-banded chromosomes if left on the slides for prolonged periods of time.

The Microscope Stage and Coordinate Location

The microscope stage provides a flat, level surface for the microscope slide, and a means of affixing the slide to the stage. Controls on a mechanical stage allow the microscope slide to be moved in *x*- and *y*-axes. Mechanical stages usually have a coordinate grid on each axis to precisely identify the location of an object on the slide. The microscope stage can also be moved in an up-and-down manner (*z*-axis) by using the coarse and fine focus controls.

Coordinate Location

Recording accurate coordinates is essential for documentation of cytogenetic findings. In most instances, notation of the *x* and *y* coordinates is used for this purpose.

VERNIER GRIDS AND ENGLAND FINDERS®

When a metaphase is to be relocated at a microscope other than that used for the original analysis, a system of coordinate conversion between the two microscopes needs to be employed.

The stages of microscopes of the same manufacturer and model can often be aligned so that the coordinates of one scope can be used at another. Vernier grids or England Finders® allow for easy conversion of coordinates between similar microscopes whose stages cannot be aligned, or when the microscopes are made by different manufacturers. This technique provides a printed grid whose value is read at one microscope, and then simply relocated at the second.

Microscope Slides, Coverslips, and Mounting Media

The microscope slides, coverslips, and mounting media play a significant role in the contrast and resolution of an image. Microscope slides and coverslips should be made from high-quality glass to allow light to pass with the least generation of optical aberrations. A microscope slide with a thickness of 1.0 mm is well suited for cytogenetics microscopy. Coverslip thickness can be 0.17 mm to 0.18 mm, depending upon the recommendation of the microscope manufacturer. It is important to note that high numerical aperture lenses have a very low tolerance to variance of slide, mounting medium, and coverslip thickness (± 0.05 mm for NAs greater than 0.7) *(3)*. Images that cannot be brought into good Köhler illumination are often a sign of a specimen whose thickness has exceeded the capacity of the microscope lenses.

Magnification and Objective Lenses

The objective lens plays the largest role in the magnification and resolution of the microscope image. Choice of objective lens often begins by selecting a lens that allows a specimen's detail to be comfortably analyzed at the microscope. For most cytogenetic analyses, this is either a 63× or 100× lens. Consideration of the image size as it is projected on to a recording medium is also an important criterion for lens selection. Images that are too small as they emit onto the electronic chip of a camera will suffer from empty magnification as they are later enlarged and printed for viewing. Considering this, 100× objective lenses have the capacity to allow the greatest degree of enlargement by a computer imaging system (see Chapter 7). Sixty three (63)× objective lenses can also be used with computer imaging systems, but require a secondary means of magnification to produce printed enlargements of equal size and quality.

There are a variety of objective lenses.

Plan Objective Lenses

"Plan" objective lenses are corrected for flatness of field, a term that describes an image that is in focus at its center and its periphery. A flat field is critical for photomicroscopy.

Achromat Objective Lenses

As light passes through any lens, each wavelength will come into focus at a different plane within the lens (red at the highest point, green in the middle, and blue at the lowest). This phenomenon is referred to as chromatic aberration. Achromatic objective lenses are engineered to bring the red and blue wavelengths into focus at nearly the same point, greatly reducing the size of the overall focal plane.

Achromat lenses also encounter a second group of optical aberrations referred to as monochromatic, or those that can occur with any wavelength of light. Achromat objective lenses are often engineered for correction of monochromatic aberrations, requiring green light for the greatest degree of correction *(4)*. Thus, the use of a monochromatic green interference filter will greatly reduce the incidence of both chromatic and monochromatic aberrations experienced with achromat objective lenses.

Apochromat Objective Lenses

Apochromat objective lenses are regarded as the highest quality lenses. Apochromatic objective lenses bring three or four colors into focus at nearly the same focal plane within the objective. Apochromatic objective lenses also correct for monochromatic aberrations and do not rely upon green light for the correction *(4)*.

Fluorite Objective Lenses

Fluorite objective lenses (also referred to as FL, Fluars, Neofluars, or semiapochromat objective lenses) are lenses of intermediate quality between that of achromatic and apocromatic objective lenses and are more commonly used for fluorescent microscopy.

Tube Length

The tube length is the distance from the top of the objective to the back lens of the eyepiece. It is the distance at which an objective lens will bring an image into focus within the microscope. Tube lengths have been standardized by some microscope manufacturers at 160 mm, and by others at 170 mm *(3)*. When appropriate, the tube length of an objective is engraved on its outer surface (see Infinity Correction). Objectives with identical tube lengths can sometimes be exchanged between different microscopes, but should be done only after first consulting with the microscope manufacturer.

Coverslip Correction

A predictable working distance between the condenser, specimen, and objective lens is required for microscopy. Microscope slides are made a specific thickness so as to place the specimen at the proper working distance between these components. When a slide is coated with mounting medium and a coverslip is attached, the objective lens must be corrected for this additional thickness.

The correction factor is engraved on the objective immediately following the tube length (often 0.17 or 0.18 mm). A coverslip and mounting medium of that thickness must be used in order for the objective to maintain its optimal resolving power. If the value is listed as "–", the objective may be used with or without a coverslip.

Non-Coverslip-Corrected Lenses

Non-coverslip-corrected lenses are occasionally found in cytogenetics laboratories and are used exclusively with microscope slides without coverslips.

Numerical Aperture

The numerical aperture (NA) of an objective lens is a statement of its resolving capacity. Objective lenses of differing magnification but identical NA values have similar resolving potential.

Many factors can act to reduce the operating numerical aperture of an objective lens, resulting in a decrease in the microscope's overall resolving capacity. These include the alignment of the light bulb, the wavelength of light, the quality of the condenser, the aperture diaphragm setting, the presence of immersion oil between the condenser and slide, and the characteristics of the microscope slide and mounting medium.

Oil Immersion Lenses

The designation "Oil" or "Oel" engraved on an objective denotes an oil immersion lens. These lenses use oil as a means of increasing the refractive index of the space between the specimen and objective lens, thus increasing the operating numerical aperture and resolving capacity of the microscope.

High-Dry Objective Lenses

"High-dry" objective lenses are not used with immersion oil. High-dry lenses substantially sacrifice the resolving potential of a microscope and are generally not recommended for diagnostic use in a cytogenetics laboratory. They are, however, commonly used to check the morphology of nonstained chromosomes. In addition, they can be used to check the banding characteristics of stained chromosomes before the application of immersion oil.

The spaces between the condenser, specimen, and objective lens are filled with air when a high-dry lens is used. Therefore, the operating numerical aperture of high-dry lenses is limited to that of air, or 1.0. In practice, the operating NA of a high-dry objective is typically 0.95 or less, compared with the NA of 1.3 to 1.4 found in oil immersion lenses.

High-dry objectives typically have a very low tolerance to slide thickness *(4)*. In addition, when a high-dry lens is used on a coverslipped specimen, a correction collar should also be present to allow compensation for the additional thickness generated by the mounting medium and coverslip.

Infinity Correction

The symbol "∞" engraved on an objective identifies it as an *infinity*-corrected lens. Infinity-corrected objectives project parallel light rays into the microscope. These rays do not come into focus until they pass through a special tube lens, where they are brought into focus at the back lens of the microscope eyepiece. Because infinity corrected light rays are parallel as they pass through the microscope, tube length is not an issue and a variety of features can be added to a microscope without generating a change in image magnification. Infinity-corrected lenses are typically engineered for a specific microscope and are not usually interchangeable between microscopes of the same or differing manufacture.

Objective Lens Diaphragms

A diaphragm is sometimes present in fluorescence objective lenses to control the intensity of fluorescent light. When used for brightfield imaging, closing this diaphragm even slightly will significantly decrease the resolving capacity of the objective, and greatly affect image quality.

Optivar Lenses and Magnification Changers

A microscope may have additional lenses that increase the total magnification of the image. A magnification changer or optivar lens can appear as a rotating control, or a sliding bar located between the objective lenses and the eyepieces. Optivar lenses allow microscopes using a mid-range objective lens (63×) with a high numerical aperture to increase the size of an image without the loss of resolution.

Use of optivar lenses to increase the magnification of an image beyond that provided by the objective lens should be done with care, as image magnification that exceeds the resolving capacity of a microscope will result in "empty magnification."

Eyepieces

Microscope eyepieces increase the magnification of the microscope image, as well as position an image so that is can be seen by each eye. Eyepieces may also be engineered with a variety of features, including those that correct chromatic aberrations (C, K), those that provide a wide field of view (WF), and those that allow individuals to view images from a greater distance (H) (thus allowing the microscopist to wear corrective eyeglasses!) *(5)*. Most eyepieces are adjustable so that the focus characteristics of each eye can be optimized for the individual viewing an image. This allows the

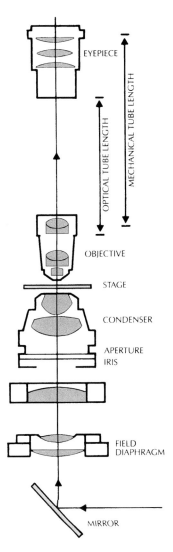

Fig. 2. Schematic diagram of a brightfield microscope illustrating the basic components (reprinted with permission of Olympus America, Inc.).

images at each eyepiece to be brought into simultaneous focus, permitting individuals with vision deficits to be able to use the microscope without the need for wearing corrective lenses. Finally, cross hairs may also be present in an eyepiece to provide an indication of the image focus at the microscope compared to the image focus at a photographic camera.

Beam Splitter

A beam splitter is present on microscopes capable of photomicrography or electronic image capture. A beam splitter allows the light to be diverted between the eyes and the photographic port at various intensities.

The basic components of a brightfield microscope are illustrated in **Figure 2**.

FLUORESCENCE MICROSCOPY

Microscopes used for brightfield microscopy may also be equipped for fluorescence microscopy. The additional components for fluorescence microscopy include an epifluorescence lamp housing, a horizontal attachment for the fluorescent light path, fluorescence filters, and fluorescence objective lenses.

Essentials of the Fluorescence Microscope

Epifluorescence Lamp Housing and Microscope Attachment

The housing for the epifluorescent light source is mounted on the rear of many microscopes and is located just above the housing for the brightfield light source. The fluorescence housing is mounted to an epifluorescent microscope attachment, which is used to direct fluorescent light into the microscope, and also to house the fluorescence filters. Most fluorescent lamp housings include bulb alignment controls, as well as an adjustable collector lens to control the dispersion of the light across the microscope field of view. The epifluorescent microscope attachment also includes a light shutter to block the passage of light into the microscope, an aperture diaphragm to control the area of illumination, and a means of inserting infrared or neutral density filters.

Filters

Infrared (IR) filters can be useful for electronic cameras when a fluorescence filter emits in the infrared spectrum. IR filters can reduce the overall intensity of the fluorescent light and should be used on an "as-needed" basis. In addition, IR filters should not be used with fluorescent dyes that rely upon IR, or near-IR wavelengths for excitation.

Neutral density (ND) filters allow the intensity of the fluorescent light to be reduced without affecting the wavelengths of the light. Neutral density filters come in several intensities and are labeled with the percentage of light that will be removed from the image (e.g., ND 6 represents a 6% reduction in light intensity). Neutral density filters are often used with fluorescent preparations that have unusually bright fluorescence intensity, or that produce a great deal of fluorescent flare. Neutral density filters can dramatically reduce the intensity of a limited fluorescent light supply and should be used only when necessary.

Fluorescence Filters and the Fluorescence Filter Housing

The basic principal of fluorescence microscopy involves the excitation of a fluorochrome (fluorescent stain) with one wavelength of light, causing the emission of a second wavelength of light. The wavelength used for excitation will vary for each fluorochrome, but will be of a higher energy than the emission wavelengths they produce (e.g., green excitation wavelengths may produce red emission wavelengths).

Fluorescent filters include three basic components: the excitation filter, the dichroic mirror, and the emission filter (see **Fig. 3**). The excitation filter and dichroic mirror work together to produce precise excitation wavelengths, and to direct the light down into the objective lens and onto the specimen. The resulting emission wavelengths then travel up through the objective and pass through the mirror and emission filter so that precise wavelengths reach the eyes of the microscopist or microscope camera.

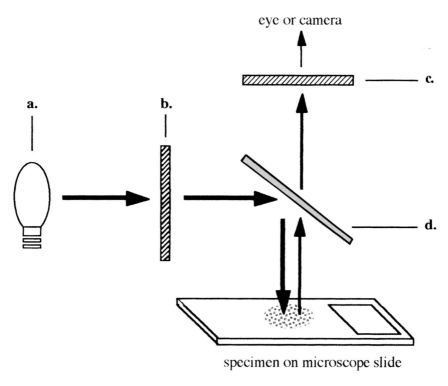

Fig. 3. Schematic diagram of the basic components of a fluorescent microscope: (a) light source, (b) excitation filter, (c) dichroic mirror, and (d) emission or barrier filter.

Protocols for the various staining techniques specify the fluorochromes and filters required for their specific use.

Viewing Fluorochromes at the Microscope

The best fluorescent stain and filter combination for specimens stained with more than one fluorochrome is one in which the color of each fluorochrome contrasts strongly against all other fluorochromes present.

Single-emission fluorescence filters limit the wavelengths of light so that only one fluorochrome is visible at a time. Dual and triple emission filters allow multiple fluorochromes to be seen at the same time.

Single-emission filters offer several advantages over dual- or triple-emission filters, including the use of peak excitation wavelengths for particular fluorochromes, resulting in stronger emission intensities. In addition, single-emission filters usually allow a wider band of wavelengths to emit to the eyes or camera, thus increasing the overall intensity of the fluorescent image. Finally, viewing only one fluorochrome at a time allows for the visualization of very low intensity fluorescence, without becoming lost in the fluorescence of other fluorochromes. The disadvantage of single-emission filters is that the fluorescence filters must be frequently changed to allow all fluorochromes to be seen when multiple fluorochromes are used simultaneously. Scanning and analysis can be less tedious when dual- or triple-emission filters are used in such situations.

Fluorescence Objective Lenses

Fluorescence objective lenses are made of fluorite or quartz, not glass. This extends the useful range of the objective into the ultraviolet spectrum, as UV wavelengths (a requirement for DAPI microscopy) are absorbed by glass.

The fluorescence objective lens diaphragm controls the fluorescent light intensity. Although this diaphragm can be adjusted to reduce the intensity of fluorescence and image flare, closing it will result in a significant loss of resolution, and is therefore not recommended. A slight adjustment of the collector lens, use of an ND filter, or other controls present on the epifluorescent microscope attachment are better solutions to control image intensity in these instances.

Immersion Oil

Immersion oil is very important for fluorescent imaging, as the fluorescent light source is very limited. Low fluorescing oil is recommended to allow the contrast of the fluorescing objects to stand out against a dark background.

Beam Splitters

As fluorescent specimens produce low-intensity images, it is very important to equip a microscope with a beam splitter that allows all light to be directed either to the eyes for microscopic analysis, or to photographic ports for photography or electronic image capture.

The Brightfield Condenser

The brightfield condenser is not used for eipfluorescent microscopy, and can allow ambient light to interfere with the fluorescent image quality. For this reason, it is recommended that the condenser be blocked off, or defocused so that ambient light will not be passed into the objective lens.

PHOTOMICROGRAPHY

Photomicrography reproduces microscope images in physical form. Photomicrographs should reproduce microscopic images with consistent resolution, contrast, and, when appropriate, color. Photographic film has the capacity to resolve more detail than will appear within the best resolved images from microscopes used in the cytogenetics laboratory; however, photomicrography is not an enhancement tool for poor specimens or microscope image quality. If photomicrographs do not faithfully reproduce what is seen under the microscope, changes in the photomicrographic technique are probably in order.

Two types of photomicrography are common in cytogenetics laboratories: black and white and color fluorescence photography. Photomicrography involves four basic processes: capturing the microscope image, developing the photographic film, exposing the photographic paper, and developing the photographic paper.

Components of a Photomicrographic System

The basic components of a photomicrographic system include a photo eyepiece, photomicrographic port and beam splitter, camera mounting device, camera, and camera control box.

Photo Eyepiece

A photo eyepiece is used to increase the magnification of an intermediate microscope image in much the same way that eyepieces of the microscope increase the magnification for the eye.

As the center of a microscope image is likely to have the best image quality (under most conditions), photomicrography components are engineered to project this area onto the photographic film *(1)*. A reticle within the microscope eyepiece or beam splitter delineates the photographic area of the field of view. Several reticles may be present, each defining the photographic area at a specific magnification, or with a particular film format.

Beam Splitter and Photographic Port

The beam splitter distributes the microscope light between the eyepieces and the photo port of the camera. When the light source is limited, or when low light level fluorescence techniques are used, it is important to project all light from the microscope onto the photographic film. Light levels that are too low may require longer exposure times than are recommended for the photographic film (see Reciprocity Failure). Long exposures increase the potential for camera vibration and quenching of fluorescent specimens.

Camera Mounting Device

A camera mounting device is the interface between camera and microscope. It attaches the camera to the microscope and may include a number of accessories such as an adjustable bellows and glass display, a light meter, a focusing telescope, and settings to identify film format, speed, and exposure duration.

LIGHT METER

The light meter detects the intensity of light projecting into the camera. Once the intensity has been measured, the exposure duration is calculated for the photographic film being used.

FOCUSING TELESCOPE

Microscopes that do not have an adjustable reticle to identify the focal plane of the camera require the use of a focusing telescope. A diopter ring in the focusing telescope is rotated to bring the crosshairs of its reticle into sharp focus. The fine focus control of the microscope is then adjusted until an image of the specimen (viewed through the telescope) also appears in sharp focus. Finally each microscope eyepiece is adjusted so that the microscope image appears in focus with that of the focusing telescope.

FILM SETTINGS

Some camera mounts have controls allowing the microscopist to indicate film format (e.g., 35 mm) and film speed (e.g., ASA 100) and provide an indication of exposure duration. In some instances these settings reside on an external control box for the camera.

Cameras

Photographic cameras can utilize a variety of film formats. Some microscopes can be equipped with more than one style of camera, allowing more than one format to be used.

Camera Control Box

An external camera control box is often present to supply a camera with power and to control many of its features. These units can vary widely with model of microscope.

Many control boxes allow the film speed and reciprocity settings to be identified. These are then used with the light meter data to calculate the duration of an automatic exposure. The camera control box is also used to initiate the exposure, allowing automatic or manual exposures.

Focus of the Photomicrographic Image

When determining image focus at the microscope, individuals may note a range of adjustment that will produce acceptable image focus (depth of field). It is important to note that the range of image focus in a photographic camera (depth of focus) may not correspond to the depth of field of a particular objective lens. In fact, objectives that favor a broad depth of field (63×) actually favor a low depth of focus, whereas objectives with a narrow depth of field (100×) actually favor a broad depth of focus *(3)*. Considering this, objectives that appear to require great precision to achieve focus at the microscope should actually produce more consistently focussed photomicrographs. When objectives with a broad depth of field are used, extra care should be exercised when calibrating the image focus at the camera to the focus at the microscope eyepieces.

Film Exposure and Development

Precise control of image focus, microscope light intensity, the characteristics of the film, the exposure duration, and variables in the development process each play an important role in achieving high-quality photomicrographs.

Film Composition and Reaction to Light

Photographic film is composed of several layers including a base, a scratch-resistant coating, and a backing to prevent light from passing completely through the film. The critical layer in black and white film is the light-sensitive emulsion of silver halide crystals. When this layer is exposed to light the silver halide crystals react to form an image on the film. The development process converts the exposed crystals into metallic silver, dissolves unreacted silver crystals, and cleanses the film of processing chemicals *(6)*.

Color film is composed of several layers, including emulsion layers of silver halide crystals sensitive to one of the three primary colors of light. As the emulsion layers are developed, color dyes are released by the film in proportion to amount of silver developed. The final stages vary with the type of film, but include dissolving the silver from the negative, and fixing and stabilizing the color dyes in the film *(6)*.

Black and White Photography

SENSITIVITY TO COLOR

Black and white films are available in a variety of sensitivities to color. Panchromatic film is generally recommended for black and white photomicrography in cytogenetics because it is sensitive to a range of color that might be produced by a variety of specimen stains, as well as the green filters used in the laboratory.

SENSITIVITY TO LIGHT AND FILM SPEED

Film speed, or the sensitivity of film to light, is directly related to the size of the silver halide crystals in the film emulsion. Larger crystals are more sensitive to light, allowing the film to react at lower light intensities, and with shorter exposure duration *(6)*.

Film speed is expressed in ASA (American Standards Association), DIN (Deutsche Indistrie Norm), or ISO (International Standards Organization) values, with higher values indicating higher film speeds. An ISO value of 100/20° would translate to an ASA value of 100, or a DIN value of 20° *(6)*. As ASA of 100 or lower is generally recommended for black and white photomicrography of brightfield images for its speed, resolution, and definition characteristics.

GRAIN AND GRANULARITY

The grain of a film refers to the size of the crystals within the film emulsion and is therefore related to the film's speed *(10)*. High-speed films tend to produce photographs with a grainy quality because of their large crystal size. Slower films have a much finer grain and are thus better suited for photo enlargements, or for resolving detail.

Granularity is a term that describes the metallic silver particles within a negative image following the development process. The large crystal size of a fast film tends to produce large silver particles. Selection of developer and development time can also greatly affect the granularity of a photographic negative *(6)*.

RESOLVING CAPACITY

The resolving capacity of film is a measure of its ability to recreate small, detectable structures. It is directly related to the contrast range of the specimen being photographed, the contrast characteristics of the developed film, the silver halide crystal size in the film emulsion, and the silver grain size following development.

Slower film speeds generally have a higher resolving capacity, as they begin with a finer grain. The development process can significantly affect the size of the silver grains, and is generally regarded as having the greatest impact on a film's resolving capacity.

DEFINITION

Definition describes the clarity of the resolved detail. It can be thought of as how "sharp" the transition is between one object and the next *(6)*. The small crystal size and distribution of silver grains in slower films favor high definition, but are influenced by the development process.

DENSITY AND CONTRAST

Density is a measure of the concentration of silver grains in one area of the negative. Contrast is a measure of the range of silver concentration throughout the negative. Areas where silver crystals are exposed to higher light intensities have greater concentrations of metallic silver, and thus greater density *(6)*.

The contrast of a negative image is significantly affected by the development process. Underdeveloped film will not allow the exposed silver crystals to react to their fullest capacity, reducing the difference between the darkest and lightest areas of the image (lower contrast). Overdevelopment will allow more silver to build up within exposed areas of the negative. This favors the presence of very dark colors within the exposed areas, and very light colors within the nonexposed areas (high contrast).

Color Fluorescence Photomicrography

FILM TYPE

Color print or slide film can be used for the photomicrography of fluorescent images. Color print film produces a negative image that is subsequently converted to a positive color image on photographic paper. Color slide film creates a positive image that can be viewed directly on the film or projected onto a large screen.

Color films are available with tungsten or daylight corrections. Fluorescent light more closely approximates a daylight source, and daylight films are therefore recommended to attain the best color rendition.

The choice of whether to use positive or negative film is often a matter of color reproduction. Accuracy of color can be a significant issue with color negative films, where one color dominates the entire image *(3)*. Negative films are also very unforgiving to under- or overexposure, making precise exposure control a requirement *(3)*. Positive and negative films can often be recognized by the suffix attached to the brand of film. Color-positive films generally end with the suffix "chrome," while color-negative films end with the suffix "color."

Reciprocity Failure

Reciprocity is a term that describes the predictable fashion in which photographic film reacts to light. When the light intensity is such that the exposure duration is longer or shorter than the recommended range, the result is a decrease in image density, contrast, and resolution. This is reciprocity failure.

In most instances, reciprocity failure in cytogenetics microscopy is due to a low light intensity. Low-wattage light sources, phase-contrast microscopy, and fluorescence imaging may all generate low light levels that can create exposure durations longer than are recommended for many film types. Each photographic film requires a different correction for a particular exposure setting.

The Film Development Process

The development process includes loading the film onto processing reels; immersing it in developer; stopping development; and fixing, washing, and drying the film. Most cytogenetics laboratories process only black and white film, and therefore the development process for color photography is not described here. It should also be noted that machines that automate the following processes are available.

FILM LOADING

The film is removed from its cassette in total darkness, carefully loaded onto a processing reel, and sealed in a development canister. Care should be taken when removing film from its container to prevent scratching. In addition, care should be taken to prevent the film from making contact with itself as it spirals into its reel in order to ensure proper development.

FILM DEVELOPING

Developer acts on the film to convert the exposed silver halide crystals to metallic silver grains. Many factors can affect the final image contrast, definition, and resolution, including type of developer, temperature, dilution, and time in developer.

The choice of developer can impact the grain size and contrast qualities of a negative image. A developer that produces a fine grain and offers good contrast should be used. The temperature of the developer will influence the speed of the development process. Higher temperatures shorten the development time, but may not yield consistent results.

STOPPING THE DEVELOPMENT PROCESS

A stop bath halts the action of the developer. The development process will continue until the film comes in contact with the stopping solution.

FIXING

A fixing reagent dissolves the undeveloped silver halide from the film emulsion, allowing it to be rinsed from the photographic negative.

WASHING

At the end of the fixing process, negatives are washed with running tap water. A clearing agent can be used to speed up the washing process.

DRYING

After the negatives are washed they are removed from the film reels. A drying agent such as Kodak Photo-Flo™ can be used to prevent streaking of the negatives while they are drying, or special wiping products can be used to remove remaining water droplets.

Making Photographic Prints

After the film has been developed, photographic paper is exposed to the negative's image, and the exposed paper is processed. A photographic enlarger is used to project the negative image at the desired magnification onto the photographic paper, creating a positive image. The choice of paper, f-stop of the enlarger, exposure duration, and the development process all contribute to the final resolution and contrast of the printed image.

The Photographic Enlarger

A wide variety of photographic enlargers are commercially available. The basic features of most enlargers are a light source, condenser, negative carrier, enlarger lens, aperture or f-stop control, stage, and a control box for timing and performing exposure.

LIGHT SOURCE AND CONDENSER

The light source and condenser are very much like the light source and condenser of a microscope in that their primary function is to create even illumination.

NEGATIVE CARRIER

The negative carrier is used to flatten the photographic negative and position it for optimal illumination. Care should be taken when placing a negative into a carrier, as scratches can result when the carrier is not used in the correct fashion.

ENLARGER LENSES

These lenses enlarge the negative image to the desired magnification.

CONTRAST FILTERS

Photographic paper that can produce varying degrees of contrast (polycontrast paper, see later) requires special filters to be inserted between the light source and the paper. Most enlargers can accommodate these, and some perform this function automatically via a control box.

f-Stop

The term f-stop refers to the aperture opening of the enlarger lens. The values of the f-stop are an indication of the amount of light that passes through the lens, and the amount of exposure that will occur on the photographic paper. The larger the f-stop number, the less light is allowed to pass through the lens. As an f-stop value increases by one step, the amount of light passing through the lens is reduced by half, and the exposure duration must be increased by a factor of two to produce the same degree of exposure *(6)*. For example, 10 seconds at f-8 will result in the same exposure as 5 seconds at f-5.6, or 20 seconds at f-11.

Focusing Telescope

The focusing telescope is a device that allows the silver grains of film to be visualized. When the silver grains are in sharp focus, the image will be in focus on the paper.

Safe Lights

Safe lights produce light at an intensity and wavelength that allow objects to be seen in a darkroom without exposing light-sensitive materials. An amber light of less than 15 W, located at least 4 ft from light-sensitive materials is often recommended *(6)*. Safe lights that are too bright or too close to light-sensitive materials can reduce image quality of the prints.

Photographic Paper

Black and white photographic paper, like film, is composed of a light-sensitive silver halide emulsion layer that is processed to form metallic silver in the final image. Also like film, photographic paper has characteristics of speed, grain, and contrast. There are several characteristics unique to paper, including weight, base, tone, texture, and brilliance.

Texture and Brilliance

Texture and brilliance refer to the characteristics of a paper's surface. It may be smooth or have one of a variety of surface textures. Brilliance is a measure of a paper's "shininess" and can range from glossy to varying degrees of a "matted" finish *(6)*.

Speed

Speed is a measure of a paper's sensitivity to light. Sensitivity to light increases with the size of silver halide crystals.

Contrast Grades

Photographic papers have the capacity to produce a wide or narrow range of image contrast. The contrast grade of a paper can be used to improve the image produced from negatives with high or low image contrast, as well as from specimens with a poor contrast range. Polycontrast paper can produce a range of contrast grades in conjunction with special filters housed in the enlarger.

Paper Weights, Tones, and Base Composition

Paper is available in a variety of weights. The weight designation is different for each for each manufacturer, but often includes LW (light weight), SW (single weight), MW (medium weight), and DW (heavy weight) *(6)*.

Paper tone refers to the color of the silver within the image. Paper tones are available in cold (blue), neutral (black), and warm (brown) tints *(6)*. Tint refers to the color of the paper itself and is available in various shades of white. Because good contrast is important for rendering image detail, a white paper with a neutral tone (black) is best for use in a cytogenetics laboratory.

A paper's base refers to its composition and varies in permeability to water and processing chemicals. Fiber-based papers absorb water and processing chemicals, lengthening the fixing and washing times. RC (resin coated) papers are resistant to processing fluids, and therefore have shorter fixing and washing times.

Developing the Print

The development process for paper is similar to the development process for film. The steps include developing, stopping, fixing, clearing, washing, and drying the print. Paper processing is often automated in cytogenetics laboratories.

DEVELOPER

The choice of developer is highly dependent upon the paper used for printing. Developers are available in warm (brown), neutral (black), and cold (blue) grades, each affecting the tint of the metallic silver formed in the development process *(6)*.

STOP BATH

An acid stop bath is used to quickly stop the development process and to increase the effectiveness of the fixer.

FIXING

Fixing is important in producing stable prints and must be carefully controlled to fully remove the unexposed silver halide from the paper. Proper agitation, temperature, time in the fixer, and freshness of the fixer are critical for the production of quality prints.

CLEARING AGENTS

A clearing agent can be used to speed the washing process of non-resin-coated papers, allowing the fixer to be removed with greater speed and with less exposure to water. Clearing agents should not be used on resin-coated (RC) papers *(6)*.

WASHING AND DRYING

Thorough washing with frequent changes of water rids the paper of processing chemicals, and prevents the prints from discoloring or fading.

STABILIZATION PROCESSING

This is a rapid method of print making used by some laboratories. Instead of standard developing and printing, the paper is "activated" and "stabilized." No other process, besides brief air drying, is necessary; hence increased speed and therefore greater lab efficiency are possible. The drawback to this process is that prints are not permanently fixed and will discolor over time.

ACKNOWLEDGMENTS

I would like to thank Andrew J. Hunt of B&B Microscopes, Pittsburgh, Pennsylvania for his contribution to the information in this chapter. I would also like to thank the following individuals:

Gerald Hints and Dorene Cantelmo of Olympus America
Sandra Doleman of Carl Zeiss Instruments

Lee Shuette of Nikon Instruments
John P. Vetter RBP, FBPA
Paul Milman of Chroma Technologies

REFERENCES

1. (1994) *Microscopy from the Very Beginning.* Carl Zeiss, Oberkochen, Germany
2. Abramowitz, M. (1985) *Microscope: Basics and Beyond,* Vol. 1. Olympus Corporation, Lake Success, NY.
3. Delly, J.G. (1988) *Photography Through the Microscope.* Eastman Kodak, Rochester, NY.
4. Abramowitz, M. *Optics: A Primer.* Olympus Corporation, Lake Success, NY.
5. *Basics of the Optical Microscope.* Olympus Corporation, Lake Success, NY.
6. Grimm, T. (1986) *The Basic Darkroom Book,* Revised Edition. Plume Books, published by the Penguin Group.

Quality Control and Quality Assurance

Christopher McAleer and Steven L. Gersen, Ph.D.

INTRODUCTION

Upon receiving news that results of a chromosome analysis are abnormal (and even sometimes that they are normal), a patient will frequently ask, "How do I know that the lab didn't make a mistake? How do I know that the sample they reported on was really mine? How can I be certain that this is all correct?" Most would be surprised to learn of the myriad of checks and balances that exist, some because of regulatory agencies and some a function of common sense, to prevent clinical and clerical errors. These comprise the area of laboratory medicine known as quality assurance and quality control (QA/QC).

A proper QA/QC program requires that policies for validation of protocols and reagents, training and credentials of individuals performing chromosome analysis, sample identification, safety for laboratory staff, and other compliance issues must all be in place. Laboratories are periodically inspected by various state and federal entities, and most have internal regulations and guidelines as well.

Many steps occur between obtaining a specimen for chromosome analysis and the generation of a final clinical report. After collection of the specimen itself, accessioning, culturing, harvesting, slide preparation and staining, microscopic analysis, photography or electronic imaging, karyotype production, creation of a final report, and actual reporting of results are the path that specimens follow as they progress into and out of the cytogenetics laboratory. During this process, many variables can subject a specimen or data to a variety of conditions that must be managed to reach a proper diagnosis.

Central to any QA/QC program is the laboratory's Standard Operating Procedure Manual. This often formidable document contains requirements that must be met for the laboratory to perform chromosome analysis: physical space and mechanical requirements, sample collection, transport and amount requirements, personnel experience and credential requirements, and safety and protection requirements for those personnel. It contains a detailed protocol for every procedure the laboratory performs, often including the clinical and technical rationale behind each one. It includes sections on training and compliance with the various regulatory agencies that monitor and inspect laboratories, and, finally, it may contain a section pertaining to quality assurance and quality control.

From: *The Principles of Clinical Cytogenetics*
Edited by: S. Gersen and M. Keagle © Humana Press Inc., Totowa, NJ

Books could be written that address each of these issues in detail; entire chapters could be devoted to labels alone! Such detail is beyond the scope of this book, however; this chapter provides an overview of the ways in which laboratories deal with many of these steps to ensure proper patient care.

SUBMITTING A SPECIMEN

Specimens are almost always collected by individuals who rely upon the laboratory to provide a requisition form and instructions for specimen collection and transport. Considering this, QA/QC begin through an interaction with the health care providers who collect and submit specimens for chromosome analysis.

Collection Protocol

A collection protocol from the cytogenetics laboratory is of critical importance, as it establishes the collection guidelines for individuals who are not intimately familiar with the operating procedures of the laboratory. A collection protocol should include:

- Ideal volume of specimen for collection
- Suitable transport containers, anticoagulants, and media
- Transport temperature and the maximum permissible transport time to ensure optimum specimen growth
- Confirmation of the identification of the patient from whom the specimen was collected
- Specimen container labeling and requisition form requirements
- Laboratory hours, phone numbers, contact individuals, and after-hours procedures

Once established, it is important to keep copies of this protocol anywhere a specimen might be collected, including a hospital's general laboratory, departmental clinics, and operating room suites, as well as outpatient clinics and referring physician's offices. It is also a good idea to routinely discuss collection protocols with the appropriate individuals, especially those who submit samples infrequently to the laboratory. Regular interaction will promote a complete understanding of collection requirements, as well as general expectations for a sample submitted for cytogenetic analysis. It will also provide an opportunity to discuss questions, concerns, or suggested improvements of collection or submission procedures.

Specimen Labeling and Requisition Forms

Accurate specimen identification is perhaps one of the most important policies to implement. Specimen labels should include at least two sources of identification, such as patient name, date of birth, date of collection, etc. for proper identification in the event of a labeling error.

The requisition form is equally important, as it supplies the laboratory with the patient and clinical data associated with the specimen. For obvious reasons, it is desirable to have a properly completed requisition accompany each specimen submitted to the laboratory, but it is also important for the laboratory to develop a policy for dealing with specimens that are not accompanied by a requisition, or for requisitions that have not been filled out completely. Of special importance are those requests for chromosome analysis that are made verbally with the laboratory. In these instances, it is important for the laboratory to obtain written or electronic authorization for the study, unless a requisition that was submitted to the facility for other testing can be appropriately annotated.

Rejection Criteria

It is very important for individuals to clearly understand the minimum requirements for submission of a specimen for chromosome analysis, and what circumstances would prevent a laboratory from performing analysis. The collection protocol and requisition forms should clearly state these requirements. Although extremely rare, circumstances can arise that prevent a laboratory from accepting a specimen for analysis.

In the event of a problem with a sample, the laboratory should make immediate contact with the individual submitting the specimen, either to obtain clarification of the specimen identity or to discuss potential difficulties in obtaining a result. In most instances, both parties will elect to proceed, knowing that the success of the analysis may be impacted. In some instances, the problems are insurmountable and a repeat sample is needed. When this occurs, it is a requirement for the lab to carefully document the reason for rejection, as well disposition of the specimen in the patient report and appropriate log.

ACCESSIONING A SPECIMEN

Once a specimen has been received, an accession process is used to log it into the laboratory and to prepare it for analysis. During this time an accession and/or laboratory number is assigned to a specimen, relevant patient and clinical data are entered into a log book and/or database, and the culture and analysis requirements for the studies requested are identified.

Assessing the Condition of the Specimen and Requisition

Once a specimen is received, the accessioning individual must check the sample and requisition for the appropriate labels, transport reagents (media, anticoagulants, etc.), specimen condition (clotted, adequate sample size, transport temperature, etc.), and date of collection. When a problem is detected, the individual should follow the laboratory procedure for informing the "submitter" of the specimen, and take appropriate actions. Problems with the specimen and action taken might also be documented.

Accession Numbers and Patient Database

It is important to assign a unique identifier to a specimen as it enters a laboratory, distinguishing it from the other lab specimens, as well as distinguishing it from a patient's previous studies. The lab number, patient data, and clinical information are then often transferred into a logbook or electronic database, creating a patient record that can be tracked and crossreferenced against previous or future studies. In addition, other data can be entered into a database record as a study progresses, allowing the laboratory to track:

- Culture conditions
- Results
- Turnaround times
- Dates and individuals issuing reports
- Cytogenetic results vs. the findings of patients with similar histories or abnormalities
- Culture failures, labeling errors, and misdiagnoses
- Incidence of submission problems

Electronic databases need to be managed within a laboratory to ensure the accuracy of the data as well as patient confidentiality.

Once a specimen has been logged into the cytogenetics laboratory, the individual who has done so should follow the procedure for preparing it for tissue culture. This may include storing a specimen under appropriate conditions; creation of culture records and container labels; notification of the appropriate individuals, if others, that the samples have been received; and creation of a patient folder for paper records.

It is also important to have a system for identifying specimens that require accelerated study, a preliminary report, or completion by a certain date to meet anticipated turnaround times. These requirements should be clearly indicated on all appropriate forms and/or computer fields, and all individuals involved with the study should be notified. Although specimens requiring rapid results are frequently marked "STAT," this term is not appropriate in the cytogenetics laboratory, and the individual who has submitted such a sample should be contacted and informed that results will be available "ASAP."

Specimen Labels

The accuracy of any laboratory result requires correct specimen labeling. Up to this point, labels have been assigned to the specimen container and requisition form. After the accessioning process, specimen labels may take on many forms, involving a culture worksheet, culture flasks or petri dishes, microscope slides, a microscope analysis worksheet, photographic film, photographic or electronically produced prints, karyotypes, and reports. With these things in mind, it is important for a laboratory to establish a labeling policy so that patient identification can be crosschecked in the event of a labeling error.

SPECIMEN CULTURE, HARVESTING, SLIDE PREPARATION, AND STAINING

All equipment and supplies used for culture and harvesting of cells, preparation of slides, and banding and staining of chromosomes should be monitored to provide high-quality analyses.

Cell Culture

When preparing a specimen for culture it is important, when possible, to generate duplicate or independently established cultures for all samples. It is also important to place duplicate cultures in separate incubators, each equipped with its own power, CO_2 source, and emergency alarm. When this is not possible, a backup procedure must be created that specifies how cultures will be maintained in the event of a power, gas, or incubator failure.

As a specimen is added to culture medium, a culture is transferred between containers, or reagents are added to a specimen culture, it is important to take precautions to prevent contamination.

Working with specimens within the area of a laboratory designated for biological hazardous materials, and using sterile technique in laminar flow hoods will greatly reduce the risk of bacterial contamination. In addition, using rubber gloves, cleaning work surfaces with alcohol before and after use, and exposing container openings, pipettes, or other measuring devices to a flame will also reduce the likelihood of contamination.

Working with <u>one specimen at a time</u>, and disposing of all used pipettes or containers that come into contact with a specimen (before moving onto the next) will greatly reduce the likelihood of cross-contamination or improper identification. It is also important to note that the transfer of reagents into a culture should be done using a fresh pipette when there is any risk of contact with a specimen or specimen aerosol.

Culture Protocols

Tissue culture begins with a culture protocol that outlines tested and reproducible steps to produce metaphase chromosomes for analysis.

Whenever a new protocol or a procedural modification is to be used for tissue culture, or a new medium or medium additive will be introduced, it is important to validate protocol changes under controlled conditions. The method of validation should be one that is appropriate for the reagent or technique being tested, and may include parallel testing of current methodology vs. new, testing on nonclinical control specimens, or direct analysis using reference materials. It is also important to track the history of protocol modifications, allowing a comparison of past culture techniques and successes. The format of a culture protocol is an important consideration and should comply with the requirements of the agency used for laboratory accreditation.

Equipment Maintenance

Consistency and reliability of laboratory procedures cannot be accomplished without well-maintained equipment, and there are many regulations that reflect this.

Refrigerators, freezers, and water baths should be closely monitored daily for temperature and follow regular cleaning schedules. Centrifuges should be monitored for accurate speed semiannually. Laminar flow hoods should be cleaned before and after use, and be equipped with an antibacterial light, or cover, to prevent contamination during periods of non-use. Laminar flow hoods also should be checked and certified annually for airflow and bacterial contamination. pH meters should be cleaned and calibrated regularly. Balances should be kept clean of laboratory reagents and calibrated regularly to ensure proper weight measurements. Ovens need to be monitored daily for temperature. Trays for slide preparation and storage should be kept clean to reduce chemical contamination of staining reagents.

Incubator temperature and gas concentration should be monitored continuously and documented daily. Incubators should be on a regular cleaning schedule and, as discussed previously, should also be equipped with separate power and gas sources, as well as emergency alarms. Incubator gas and power supplies should also have a backup in the event of a failure, and the laboratory should maintain an emergency plan in the event of complete incubator failure. Records of equipment monitoring and maintenance should be documented in an equipment log.

Automation of the harvesting procedure is used by many cytogenetics laboratories as a way of increasing laboratory productivity and improving consistency. However, automation does not imply "care-free"! Laboratories that utilize such technology must closely follow the manufacturer's recommended operational guidelines, and also closely monitor the equipment for acceptable performance. A procedure for the use of automated equipment must be prepared that details the procedural steps for operation, appropriate reagents, calibration and cleaning requirements, and preventive maintenance. It is also important for individuals operating the equipment to receive proper training before using the harvester on clinical specimens.

Harvesting, Slide Making, and Staining

The transition from tissue culture to microscopically analyzable chromosomes is achieved by harvesting the dividing cells (mitotic arrest, osmotic swelling of cell membranes, and fixation), spreading of chromosomes on microscope slides, and staining chromosomes with one of various methods that produce an appropriate banding pattern (see Chapter 4). Each of these steps must be optimized to facilitate correct diagnoses.

Protocols

Harvesting, slide making, and banding/staining determine the quality of metaphase chromosomes made available by successful tissue culture. Following validated protocols is very important for these procedures, but frequent modifications may be required to address changing laboratory conditions. It is important to note that these procedures can be especially sensitive to individual technique, and that mastery of these skills requires individuals to observe and document minor variations in procedure or laboratory conditions that improve or detract from chromosome morphology.

New protocols, procedural changes, and introduction of new reagents, reagent concentrations, or microscope slides, etc. must be validated under controlled conditions. The method of validation should be one that is appropriate for the reagent or technique being tested and again may include parallel testing of current vs. new, testing on nonclinical control specimens, or direct analysis using reference materials. It is also important to track the history of harvesting, slide preparation, and staining protocol modifications to allow a comparison of past techniques to present successes. Documentation of proactive and reactive factors from these procedures is important to ensure quality metaphase chromosomes as well as to identify and track problems that reduce specimen quality.

Slide Preparation

The chromosomes present in harvested metaphases must be spread apart so that they can be microscopically analyzed. They must lie flat, so that staining will be uniform and plane of focus will not be affected, and they must be aged (literally or artificially) for most banding and staining procedures to work properly.

When all else has gone well with the tissue culture and harvesting procedures, poor slide preparation can result in scarce, poorly spread, or improperly aged metaphase spreads for staining and microscope analysis. Consider the following variables:

- Harvesting method (centrifuge tubes vs. *in situ* processing)
- The humidity and temperature of the laboratory or drying chamber utilized
- The number of fixations and the method of fixing the specimen
- The slide temperature
- Wet or dry slide? How much water?
- The angle of the slide during specimen application
- The method of applying the specimen
- The method of drying the slide
- The slide aging technique

Each of these factors significantly contributes to the success of slide preparation. As these can be variable from day to day or between individuals, close observation and documentation of technique may allow the highest proficiency of these skills.

Banding and Staining

Slide preparations and aging may play a deciding factor in the ability to successfully stain a specimen, requiring adjustments to solution concentration, the time a slide is left in a staining solution, etc. Careful reagent preparation and documentation of adjustments made to staining procedures helps laboratories refine techniques.

The shelf life and storage conditions of banding and staining reagents are important considerations and should also be documented in a staining log. As reagents arrive in the laboratory, lot numbers should be recorded and compared with previous lots used. Reagent containers should be labeled with the reagent name, quantity, concentration, storage requirements, the date received, and an expiration date. Reagents that require refrigeration should have minimum and maximum permissible temperatures documented, and these should not be exceeded. Existing supplies of reagents should be rotated so that they are depleted before using new supplies.

Although good specimen staining is important for microscope analysis, it is also necessary to consider the microscope on which a specimen will be analyzed, and the staining requirements of the recording media (photography or electronic image capture). When a laboratory has a variety of microscopes, each may have a light source, green filter, objectives, or other lenses that produce images with a unique set of visual characteristics. Individual taste will also play an important factor in identifying a staining intensity that is well suited for microscope analysis.

For either photography or electronic capture it is important to identify staining intensities that produce the following:

- Chromosome pale ends that contrast well against background areas
- A wide range in mid-gray intensity
- Dark bands in close proximity appear as distinct bands

It is not unusual to find that a staining protocol may not be well suited for both microscope analysis, as well as photography/electronic capture.

Comparing the requirements of the individual performing the microscope analysis against the requirements of the recording media, and documentation of ideal conditions in a staining log will help laboratories gain control of the many variables of a staining procedure.

SPECIMEN ANALYSIS

Any chromosome analysis must begin by identifying specific requirements for the specimen type being examined. Following this, the basic steps involved are microscopic analysis itself (location of metaphase spreads suitable for analysis, counting chromosomes and determining sex chromosome complement, and band-pattern analysis), photographic or computerized imaging, preparation of karyotypes, and documentation and reporting of results. The procedure begins with a protocol that must be accessible and thoroughly understood by all individuals performing chromosome analysis.

Analysis Protocols

An analysis protocol must identify the general requirements for each specimen type. The protocol should identify normal parameters and normal variants, and should distinguish between true abnormality and artifact. The number of cells from which chromosomes are to be counted, sex chromosomes examined, and bands analyzed in detail must be clearly stated, including whether each type of examination is to occur at the microscope, on a photograph, or via a karyotype. A protocol should set standards for the selection of suitable metaphase spreads, as well as the number of cultures (and colonies, when applicable) from which cells should be examined. When an abnormality is detected, the appropriate steps to take should be specified. Other items, such as an appropriate banding resolution level, maximum allowable number of overlapping chromosomes, random chromosome loss, and dealing with metaphases in close proximity might also be included.

A protocol should identify the procedures used to document each metaphase, as well as the data to be recorded on a microscope analysis worksheet, requirements for imaging, the number of cells to be karyotyped, the number of individuals who should take part in performing the analysis, and the individual who should verify the results.

Finally a protocol should establish the policies for the storage of microscope slides, both during analysis and once analysis has been completed.

Personnel Requirements

Besides understanding specimen analysis requirements, it is important to identify personal criteria to be met for and by individuals performing the analysis. The experience level, credentials, and work load of each technologist are important considerations, and the laboratory must be appropriately staffed to allow for complete, accurate, and timely results of all samples received. When possible, it is often recommended to split the analysis of a specimen between two individuals in some way, increasing the potential for detection of a subtle abnormality.

Establishing goals for individuals or groups to meet, such as turnaround time and the number of cases to be completed in a week, are important aspects of effective laboratory management. The quality of analysis should not, however, be sacrificed in the attainment of these goals, and performance monitors should include frequent statistical analysis of failure rates and percentage of abnormal cases.

Microscopy

A significant part of quality microscopy lies within the training an individual receives on the components of a microscope and their proper use. One of first subjects detailed in any protocol for microscopy should therefore be training, including quality checks to identify equipment in need of service, or individuals in need of additional training.

The selection of a microscope for analysis and documentation of results (image production) is also a very important consideration. It is not unusual for a laboratory to have microscopes of various quality grades, and individuals need to understand the limiting factors of any given scope. "Newer" does not necessarily imply "better" in this instance, as many "veteran" microscopes can produce excellent images, and it is often the resolution of the objective (lens), not extraneous accessories, that is the key to image clarity. Also, keep in mind that good images are more likely to come from well-prepared microscope slides. Controlling the slide preparation process and using a microscope with the appropriate lenses and features will promote quality cytogenetic analysis and image documentation.

General Analysis Requirements

Regulatory bodies typically specify minimum requirements for each sample type processed for chromosome analysis. Laboratories themselves and individual States of the United States frequently augment these. It should be noted that minimum requirements are just those; the standard of care frequently requires more rigid guidelines. It must also be remembered that most listed standards apply to chromosomally normal samples. Once an abnormality has been discovered, it is important to confirm its presence or absence in each cell examined, and to identify additional procedures that may be necessary for correct diagnosis.

Some general guidelines follow.

Sample Type	Basic Analysis Requirements
PHA-stimulated blood (non-neoplastic disorders)	Chromosome count and sex chromosome complement determination of at least 20 cells, or at least 30 cells when a mosaic condition is suspected or detected. At least five metaphases should be completely analyzed, and at least two karyotypes should be prepared. If more than one cell line is present, at least one karyotype must be prepared from each. A minimum resolution of 400–450 bands (500–550 is preferable) should be attained, greater for high-resolution or focused studies of a specific chromosome pair.
Amniotic fluid, *in situ* method	Chromosome count and sex chromosome complement determination of one cell from each of at least 15 colonies; as many colonies as possible should be examined when a true mosaic condition is detected, or in some cases to confirm pseudomosaicism. Cells must originate from at least two independent cultures (initiated from more than one sample syringe or tube, when possible). At least five metaphases should be completely analyzed, and at least two karyotypes should be prepared. If more than one cell line is present, at least one karyotype must be prepared from each. A minimum resolution of 400–450 bands is recommended.
Amniotic fluid, flask method	Chromosome count and sex chromosome complement determination from at least 20 cells from at least two independent cultures as above. Other requirements are the same as for the *in situ* method.
Chorionic villus samples	Chromosome count and sex chromosome complement determination of least 20 cells from at least two independent cultures. Although opinions differ, most laboratories examine both "direct" preparations and cultured cells when possible. Additional cells should be examined when

	mosaicism is detected, particularly when there are discrepancies between the direct and cultured preparations, often an indication of confined placental mosaicism (see Chapter 11). At least five metaphases should be completely analyzed, and at least two karyotypes should be prepared. If more than one cell line is present, at least one karyotype must be prepared from each. A minimum resolution of 400 bands should be achieved.
Solid tissues (non-neoplastic studies)	Chromosome count and sex chromosome complement determination of least 20 cells from at least two independent cultures. At least five metaphases should be completely analyzed, and at least two karyotypes should be prepared. If more than one cell line is present, at least one karyotype must be prepared from each. A minimum resolution of 400 bands should be achieved.
Neoplastic studies (bone marrow, tumor biopsy or aspirate, unstimulated peripheral blood)	Thorough examination of 20 cells when possible. All metaphases should be analyzed, and at least two karyotypes should be prepared. If more than one cell line is present, at least one karyotype must be prepared from each. The resolution should be at least 400 bands. When fewer than 20 cells can be analyzed and an abnormality has been detected, the number of abnormal and normal (if any) cells is reported. When fewer than 20 cells can be examined and an abnormality is not detected, the number of cells studied is reported and additional procedures [fluorescence *in situ* hybridization (FISH), molecular analysis] or a repeat study, when clinically appropriate, may be recommended. For studies of minimal residual disease or engraftment studies, additional metaphases may be examined.
Fragile X syndrome	Although guidelines were created for the diagnosis of this disorder via cytogenetic analysis, current standard of care now involves analysis via molecular methods. See Chapter 14.

Analysis Worksheets

Laboratories routinely use some form of worksheet to document microscopic analysis data. This is the technologist's working document, but becomes part of the patient's permanent laboratory chart, and as such serves as an additional clinical and clerical crosscheck.

The analysis worksheet typically includes patient data (patient name, laboratory accession and case numbers), indication for study, and specimen type. The identification of each slide examined should be verified, and previous studies might be noted. The technologist performing the analysis and the date should be recorded. The microscope being used is often indicated, and microscopic coordinates are recorded for each metaphase examined, along with other data (slide number, culture of origin, banding method, and identifiers for relocating the cell). The number of chromosomes and sex chromosome complement is typically noted, along with other relevant data such as quality of banding, abnormalities, polymorphisms or chromosome breakage observed, whether the cell was analyzed and/or imaged, which cells should be considered for karyotype, etc. Finally, a summary of the results, including the patient's karyotype, can be included, along with indications of clerical review.

PHOTOGRAPHY, IMAGING SYSTEMS, AND KARYOTYPE PRODUCTION

During or upon completion of the microscopic analysis, a specimen is ready to be photographed or imaged electronically, printed, and karyotyped. Photography or electronic capture are the tools used to record the microscope image, allowing the chromosomes to be documented and reanalyzed as necessary. Understanding how to operate, optimize, and maintain the equipment used in these processes is necessary to achieve optimum results from any sample.

Photography

Like microscopy, good photography is not a difficult technique, but is one that relies heavily upon proper training. Technologists must understand the factors that go into successful transfer of a microscopic image to film, and must be thoroughly familiar with their photomicrographic equipment, as well as with the chemicals, equipment, and procedures for developing film and preparing photographic prints. (See Chapter 5.)

One of the most important photographic processes is accurate labeling of film and developed prints. A protocol for the laboratory and its darkroom should include a system for labeling these items, as well as information that will allow a label to be crosschecked against the microscope analysis worksheet.

Computerized Imaging Systems

Computer-driven imaging systems are essentially the digital equivalent of photography; otherwise, the steps involved are similar. Instead of photographing a cell, it is electronically captured in digital form. Instead of developing film and using filters to produce prints with the appropriate contrast and background, the image is electronically enhanced to achieve a similar appearance, and a laser or other type of printer provides a hard copy. Finally, images are stored not as photographic negatives, but as digital files on optical disks, tape drives, or other digital storage media. (See Chapter 7.)

Otherwise, as with photography, an understanding of theory and hardware, generated by the appropriate amount of training, are required so that laboratory staff can utilize an imaging system properly and efficiently.

Karyotype Production

Although not used in all countries, the final laboratory manipulation required for chromosome analysis is typically generation of the ordered arrangement of chromosomes known as a *karyotype*.

If there was ever a perfect example of the value of training in laboratory medicine, it is this process. A bright individual with a modest comprehension of the theory behind cytogenetics and essentially normal pattern-recognition and motor skills can be taught the human karyotype well enough to perform this task in about a week. Yet the comment most often made by visitors to a cytogenetics lab is typically, "These chromosomes all look alike. How do you tell them apart? I'd never be able to do that." In reality, all that is required is a sufficient number of images for repeated attempts, plus sufficient patience on the part of the individuals doing the training. By making attempt after attempt (and receiving the appropriate corrections each time), one eventually begins to recognize certain pairs, and then eventually all pairs. Mastery of the subtleties, sufficient to perform actual microscopic analysis, of course requires much more training, but in many laboratories interns or other students are often employed to generate karyotypes. A good rule to follow when this occurs is that no student be permitted to karyotype an entire case without supervision or review by a trained technologist.

Karyotype production is one method laboratories use to divide analyses between two or more technologists. This can be accomplished with a guideline that specifies that the technologist who performed the microscopic analysis cannot prepare or review the karyotypes for that patient. When one adds to this a review by the laboratory supervisor or another senior individual, followed by final review by the laboratory director, it can be seen that a well-designed protocol can ensure that at least four trained "pairs of eyes" examine chromosomes from every patient, increasing the likelihood of detecting a subtle abnormality or clerical error.

A special consideration in this area involves the use of the computerized imaging system to prepare patient karyotypes. Although in the past this essentially involved "cutting and pasting" the chromosomes with a trackball or mouse, pattern-recognition software has improved to the point that many sophisticated systems can now arrange the chromosomes with little or no human input (see Chapter 7). This, of course, creates a quality concern. Laboratories deal with this by putting in place protocols that require appropriate review of all computer-generated karyotypes. When properly monitored, such systems can increase laboratory efficiency by markedly reducing the time required for karyotype production.

PRELIMINARY AND FINAL REPORTS

Reporting the results of chromosome analysis will have a direct impact on the diagnosis and treatment of a patient. Considering this, it is important to establish a reporting procedure that will:

- Summarize the findings of the laboratory
- Crosscheck the findings against the various specimen labels for labeling errors
- Establish a reporting process to outside individuals so that the data, individual issuing the report, individual receiving the data, and the report date are properly documented

Preliminary Reports

Although potentially risky, preliminary results are sometimes released by a laboratory before the full chromosome analysis has been completed. Preliminary reports are often issued verbally once enough data have been collected to formulate a likely indication of the final result, or once the data already available are clinically critical and must be communicated to a physician. Once verified, it is important to follow an established procedure for reporting preliminary results. Individuals reporting the data should be qualified to interpret the preliminary findings and to give an indication of the possible outcome once a complete study has been conducted. It is important for this individual to document the microscope analysis data, the patient and cytogenetic data reported, date of the report, and individual receiving preliminary data. It is also vital to impress upon the person receiving the report what may change once the study is completed.

Final Reports

A final report will summarize and interpret the results of the study. A procedure for the creation of final reports should include a checklist to ensure that all appropriate procedures have been completed and that all data are clerically correct. Once completed, final reports can be generated electronically or on paper.

Once the final report has been completed, a record should be kept of the individuals to whom a report was issued, as well as the date(s) of issue. In most instances a report is typed or printed electronically by a computer database and filed in a patient folder. Patient folders are retained in the laboratory or filed in an outside facility. Whether stored within a laboratory or at an off-site facility, it is important to have access that allows prompt data retrieval.

QUALITY ASSURANCE

Laboratories can experience a variety of difficulties with samples themselves; some of these are inevitable and therefore are not preventable (insufficient volume, wrong sample type, no living cells present, etc.). Others may be due to collection or transport of the specimen; incorrect labeling; or other human error at sample collection, transport, or in the laboratory. There are also a number of difficulties that can arise in the laboratory after an appropriate specimen has been received without incident.

Any of these can result either in an incorrect diagnosis, or in failure to reach one at all. It is therefore very important for a laboratory to document all problems that arise and, by determining ways to prevent similar occurrences, improve overall quality.

Specimen Failures

The inability of a laboratory to provide a diagnostic result is typically due to one of two basic reasons: cells from the sample do not grow in culture, and therefore no mitotic cells are produced, or a problem occurs in one of the many postculture steps, rendering the processed material useless. The purpose of this section is not to convince the reader that problems are inevitable, but rather to impress upon him or her the amount of care and attention to detail required, and the critical role quality assurance plays in the cytogenetics laboratory.

Culture Failure

As described in Chapter 4, the basic procedure for producing chromosomes for analysis from any tissue type requires living cells that can somehow be coaxed into active division. Without mitosis, there can be no chromosomes to process and examine.

There are several possible reasons for cell culture failure:

- *The sample did not contain any living cells.* In some cases, this is clinically not surprising; it is frequently the case with products of conception obtained from fetal demise, or in necrotic bone marrow samples. Other times, one can deduce the cause (such as a delay in sample transport). In still other instances, no explanation is readily available. In these cases, the entire path the specimen followed between the point of collection and delivery to the laboratory is suspect, and must be investigated.

- *An inappropriate specimen is submitted to the laboratory.* This may involve peripheral blood with no circulating blasts being collected instead of bone marrow. (Without blasts in the periphery, there are no spontaneously dividing cells present and the unstimulated cultures used for hematopoietic disorders will not produce metaphases.) It might be due to the wrong collection tube being used, or to products of conception being placed in formalin and then sent to the lab. The specimen and the way it is collected must match the intended application of chromosome analysis.
- *An insufficient specimen is submitted to the laboratory.* For example, "2 mL of extremely bloody amniotic fluid" or "0.5 mL of watery bone marrow" is the type of description that frequently accompanies a culture failure record. It should be pointed out, however, that *all such samples should be submitted to the laboratory,* which will do everything it can to generate a result, no matter how unlikely this may seem.
- *The laboratory suffers a catastrophic equipment failure.* With proper precautions in place, this is unlikely. Specimens should be divided and multiple cultures, placed in separate incubators, should be initiated whenever possible. Appropriate backup power, redundant CO_2, and alarm/warning systems should also be in place. Nevertheless, unusual hardware problems do occur.
- *Reagent failure.* There are rare but unfortunate examples of supplies that are supposedly quality-controlled by the manufacturer being released (unknowingly) for purchase by laboratories without actually meeting the appropriate criteria. Improperly cleaned water storage tanks have poisoned entire lots of culture medium, and syringes made with natural rubber stoppers have periodically resulted in amniotic cell death on contact. Again, with proper precautions in place (testing all supplies before use and dividing all cultures between two lots of everything) this risk can be minimized.
- *Human error.* Although also unlikely, it is always possible for a technologist to inadvertently prepare culture medium incorrectly, forget to add the appropriate mitogen, or utilize equipment improperly.

Every culture failure must be documented and the cause investigated to the extent possible. The laboratory should keep records of these, along with periodic measurements of culture failures *for each specimen type,* as a way of detecting an increasing trend before it becomes a serious problem.

Postculturing Errors

There are few things as frustrating to the cytogenetics laboratory as having beautiful cell cultures or routine blood cultures produce no usable metaphases. Although these are admittedly rare events, they do occur and, as with culture failures, must be fully investigated and documented. Some examples are:

- *Harvesting errors.* As outlined in Chapter 4, there are a variety of steps in the harvest procedure, and each provides the potential for error. If Colcemid® is not added, an insufficient of mitotic cells can be the result. If fixative is added before the hypotonic solution (unfortunately an easy thing to do, but a mistake a good technologist makes only once), cells will not swell and chromosome separation is impossible. If a centrifugation step is omitted, all cells except those that have settled via gravity will be removed via aspiration or pouroff. Other errors, such as adding the wrong hypotonic solution, making any of the reagents incorrectly, or using incorrect timing can also render a harvest unusable. Finally a catastrophic event that results in the loss of all material (e.g., spillage or breakage of a rack of tubes) will of course result in loss of usable material.
- *Problems with a robotic harvester.* As described in Chapter 7, cultures that are processed with the *in situ* method are amenable to harvesting on a robotic fluid handling device. Although the concept when using such a machine is freeing up technologist time for other

vital functions, it is not good practice merely to load the cultures onto the harvester, press the start button, and walk away. Solution bottles must be filled with the proper reagents, lines must be free of clogs, and the computer program must be functioning correctly. All of these must be verified before a technologist leaves the machine alone, and periodic checks up until the cultures are in fixative is a good policy for the laboratory to adopt.

- *Slide making/culture drying errors.* It has often been said that clinical cytogenetics is part art and part science. Producing high-quality metaphases during the slide making process is one example. This procedure is described in Chapter 4 and also discussed in Chapter 7; suffice it to say that if not done properly, the laboratory's ability to correctly analyze a patient's chromosomes can be compromised.
- *Banding/staining errors.* This is another example of the art of cytogenetics. Correct "aging" and timing of each step in this process is critical to producing well-banded chromosomes (see Chapter 4), and a failure to interpret results and adjust parameters accordingly can ruin even the best of preparations.
- *Miscellaneous accidents or human error.* Although each of the basic postculture steps has been covered, strange things can still occur at any point in the process, from wiping the wrong side of a slide to breaking it completely.

Labeling Errors

The result of a labeling error can range from an incorrect laboratory number appearing on a report to the misdiagnosis of a specimen. Collection containers, requisition forms, computer databases, culture flasks, culture worksheets, microscope slides, and microscope analysis worksheets are all places where specimen labeling errors can occur. Regardless of the outcome, labeling errors lead to improper identification of or assign inaccurate information to a specimen and are therefore a significant concern of any laboratory. Processing specimens one at a time using controlled, standardized procedures serves to greatly reduce the likelihood of labeling errors. Nevertheless, it is important to remember that people make mistakes, and the laboratory must therefore implement a system that crosschecks the accuracy of the labels assigned to a patient as well as the data collected from a cytogenetic study. Each step that creates the possibility for misidentification should have a crosscheck built into it, and some form of overall clerical review of a patient chart is frequently carried out before results are released.

IMPROVING LABORATORY QUALITY

In addition to the numerous steps already described, cytogenetics labs, like all other clinical laboratories, are subject to many external guidelines, inspections, and tests that ensure and improve quality. These vary from country to country and even from State to State in the United States.

United States

Accreditation, Inspections, and External Proficiency Testing

Under the Clinical Laboratory Improvement Amendment of 1988 (CLIA '88), every laboratory performing moderate- to high-complexity testing (i.e., every cytogenetics laboratory) must enroll in external proficiency testing programs. In fact, virtually all clinical laboratories in the United States do so under the auspices of the College of American Pathologists (CAP). This accrediting organization inspects laboratories and provides proficiency testing, according to CLIA requirements, several times a year. A lab's ability to perform and be reimbursed for testing depends upon successful participation in each aspect of this process, as failure can lead to loss of accreditation.

CAP sends a team, typically from another laboratory, to inspect each facility every other year; during off-years the laboratory must conduct and report the results of a self-inspection. Proficiency testing varies according to specialty; in cytogenetics, the proficiency tests generally consist of four unknowns in the form of banded metaphase preparations plus sufficient clinical information for the lab to make a diagnosis. A fifth unknown, in the form of a peripheral blood sample, is also frequently submitted, but the reader will appreciate the logistical and medical challenges of this procedure; there are enough cytogenetics laboratories in the United States that care must be taken not to exsanguinate the individual (typically a carrier of some rearrangement) who has volunteered to be the test subject!

State requirements can be quite variable. One of the more rigorous programs is administered by the New York State Department of Health, which conducts its own inspections and proficiency tests of all labs in the United States that process specimens from New York State residents. This body also has its own certification process (see later).

Laboratory Staff Qualifications

Many states in the United States require, either formally or informally, that the individual who signs chromosome analysis reports (typically the director of the cytogenetics lab) is board certified in Clinical Cytogenetics by the American Board of Medical Genetics (ABMG), a body that is recognized by both the American Board of Medical Specialties and the American Medical Association. Such certification is awarded to a doctor (M.D. or Ph.D.) who passes a comprehensive examination in General Genetics as well as a specialty exam (in this case in Clinical Cytogenetics). Both exams must be passed for an individual to be board certified. Current diplomates must recertify after 10 years, whereas those who passed previous rounds of examinations are considered "grandfathered" and need not recertify (although this will be encouraged).

Many technologists, supervisors, and even directors in clinical cytogenetics labs across the United States learned how to perform chromosome analysis on the job, and such experience was all that was needed to find employment. Today, degreed programs in cytogenetics exist in several colleges and universities, and a technologist can now be certified as Clinical Laboratory Specialist in Cytogenetics [CLSp(CG)] by the National Credentialing Agency for Laboratory Personnel (NCA). Initial certification results from passing an examination, and lasts for 5 years. Recertification can be accomplished either via continuing education (every 2 years) or reexamination (every 4 years).

The State of New York requires documentation of sufficient postdoctoral experience in order for an individual to receive a Certificate of Qualification (COQ) in various laboratory disciplines, including cytogenetics. Individuals, even ABMG board certified ones, who do not have such a COQ in cytogenetics, are not permitted to sign chromosome analysis reports for samples from New York State residents.

International QA/QC

There are cytogenetics laboratories located in North America, South America, Central America, eastern and western Europe, Africa, Australia, and Asia. Although a comprehensive listing would not be feasible here, the following are some examples of the way quality issues are handled around the world.

Canada

Cytogenetics laboratories in Canada fall under provincial jurisdiction, and regulations vary from province to province. In some, quality control is legislated whereas in others it is optional. However, many laboratories voluntarily follow CAP guidelines and/or participate in CAP proficiency testing. There is also no uniformity in the credentials needed to be a cytogenetics laboratory director, although many are certified by the Canadian College of Medical Genetics (CCMG).

Europe

For many years, only the United Kingdom had a comprehensive, formal quality control program in place. Criteria are similar to those found in the United States. However, because the British do not prepare hard-copy karyotypes, greater emphasis is given to slide quality.

In November, 1996, 24 geneticists representing 15 European countries met in Leuven, Belgium to discuss quality control and quality assurance guidelines for prenatal genetic diagnosis in Europe. This committee endorsed the formation of "pan-European external quality assessment (EQA) networks," and produced "Quality Guidelines and Standards for Genetic Laboratories/Clinics in Prenatal Diagnosis on Foetal Samples Obtained by Invasive Procedures—an attempt to establish a common European framework for quality assessment."

Since this meeting, the French have already modified their accreditation procedures under the auspices of the Ministry of Health, and formal national guidelines are under development. At this point in time, the major difference between France and the United States is that only an M.D. may interpret and report cytogenetic results in France, whereas in the United States many laboratory directors are Ph.D.s.

Requirements and guidelines for other European countries vary widely at the present time. Some have detailed policies, some use those of other countries, and some have no formal guidelines in place, although individual laboratories may.

Japan

Clinical cytogenetics laboratories are not widespread in Japan. National quality guidelines do not exist, and therefore neither do inspections or proficiency tests. The Human Genetics Society of Japan has developed recommendations that most labs adhere to voluntarily, and some labs use the College of American Pathologists guidelines. Samples are also often referred to labs outside the country. This creates a unique quality issue, as reference laboratories in Japan are not permitted to make clinical interpretations or diagnoses; this responsibility rests with the physician. Therefore, even though formal training and board certification for cytogenetics do not exist, it is the referring physician and not the laboratory who must decide whether a chromosome, karyotype, or polymorphism is normal or abnormal! This scenario therefore requires a special relationship between doctor and laboratory.

RELATED TOPICS

We have covered most issues involved in the generation of clinical results in the cytogenetics laboratory, which was the goal of this chapter. However, no such work would be complete without making mention of the ancillary QA/QC issues that must also be dealt with on an ongoing basis.

Safety

In decades past, laboratory design and protocols put the specimen first and the technologist second. Mouth pipetting was common, even with potentially toxic reagents (Giemsa stain, for example, is frequently dissolved in methanol). Gloves were not used, and "medical waste" was any garbage can that had come in contact with a specimen. Cytogenetics labs often reeked of acetic acid (used in fixing samples; see Chapter 4). Laminar flow hoods ("sterile hoods") were constructed with no separation between the specimen and the tech, and utilized a back-to-front horizontal flow of filtered air. The sample was protected from microbial contamination as air blew over it directly into the technologist's face! The reader is reminded that hepatitis existed long before AIDS.

Today's hoods feature split vertical air flow and protective glass windows. Pipetting devices are typically required, and, in the United States MSDS (material safety data sheets) for every reagent used in the lab must be available to all employees. Acceptable concentrations of all volatile reagents are maintained via ventilation systems and are monitored, and universal precautions govern every process that involves contact with patient samples. Most laboratory inspections include a safety component.

FISH

Many of the multitudinous and ever-expanding uses of this versatile technology are employed in the clinical cytogenetics laboratory (and many, we should remember, are used in other settings). Guidelines are still under development, but those already in place typically involve probe validation, controls, and training.

Reference Laboratories

Not every cytogenetics lab performs every type of test on every type of sample. Some specimens require additional non-cytogenetic testing. Some laboratories experience backlogs or other similar difficulties which require that some samples be sent to another lab to enable them to "catch up." For these reasons, proper record keeping and other regulations exist to ensure proper handling and timely reporting of results for such specimens.

Ethics Policies

While most laboratories that perform prenatal testing consider themselves to be "pro-choice" concerning a patient's right to make informed decisions, many feel compelled to contribute only clinically relevant data to such a process. Prenatal analysis for "gender selection" does not fall into this category, and such studies are therefore often refused by laboratories with this type of policy. Because of the obvious difficulties faced by all involved with such issues, a written policy, created by an internal ethics committee, can be extraordinarily helpful.

Compliance Training

Many labs, particularly those in commercial settings, find themselves subjected to an increasing number of restrictions designed to prevent "kickbacks" or other potentially fraudulent finance-related practices. While the average technologist is unlikely to be faced with decisions that may involve such regulations, training in this area is becoming common as a precaution.

ACKNOWLEDGMENTS AND CONTRIBUTORS

The authors would like to sincerely thank the following for their direction and generous contributions to this chapter:

Renee R. Fordyce-Boyer, M.S., CLSp(CG) of the University of Nebraska Medical Center, Omaha NE.
Sandra J. Bomgaars, CLSp (CG), Iowa Methodist Medical Center, Des Moines, IA.
Sally J. Kochmar, M.S., CLSp(CG), Magee Women's Hospital, Pittsburgh PA.
Dr. Jay W. Moore, Ph.D., Columbus Children's Hospital, Columbus, OH.

In addition, we would also like to thank the following for their contributions to this chapter:

Suzanne Merrick, CLSp(CG), Reproductive Genetics Laboratory, Denver CO
Dennis Hulseberg, CLSp(CG), Denise Wilson, CLSp(CG), and Carol Johnson, CLSp (CG), University of Iowa Hospitals and Clinics, Iowa City, IA.
Matt Williams, CLSp(CG), Stanford University Hospital, Stanford, CA.
Patricia Lewin, M.D., Laboratoire cerba, Val d'Oise, France
Raleigh Hankins, Ph.D., Health Sciences Reasearch Institute, Yokohama, Japan
Lola Cartier, McGill University, Montreal, Canada

7

Automation in the Cytogenetics Laboratory

Steven L. Gersen, Ph.D. and Jacques Ginestet

INTRODUCTION

Ask anyone to envision a typical clinical laboratory, and a host of blinking, whir-ring, computer-controlled machines that analyze samples and spit out results usually comes to mind. Even traditionally labor-intensive settings such as the cytology labora-tory are frequently populated by automatic stainers, and machines that prepare and automatically analyze Pap smears are becoming ever more popular.

This image does not hold up when one takes a closer look at the modern cytogenet-ics laboratory. Although certain procedures have been automated in recent years, most processes are still performed manually. One message that the reader should take away from this chapter, therefore, is that while technology can be utilized in any setting, the world of cytogenetics is still essentially one of manual manipulation and diagnosis.

Nevertheless, no description of the steps involved in producing a cytogenetic diag-nosis would be complete without mention of the instrumentation that has been devel-oped to assist the chromosome laboratory. Such instrumentation can assist with both sample preparation and chromosome analysis, and falls into several basic categories: robotic harvesters, environmentally controlled drying chambers, and computerized imaging systems. There have also been devices developed to eliminate some of the manual steps involved in performing fluorescence (FISH) *in situ* hybridization analy-sis. It should be pointed out that some cytogenetics laboratories use all of these devices, most use one or two, and some do not use any.

ROBOTIC HARVESTERS

As described in Chapter 4, harvesting of mitotic cells for cytogenetic analysis involves exposing the cells to a series of reagents that separate the chromosomes, fix them, and prepare them for the banding and staining process. This traditionally involves pelleting the cells by centrifugation between steps, to aspirate one reagent and add another, a process that, by its very nature, is not amenable to any form of automation. However, the *in situ* method of culture and harvest of amniotic fluid (and other) speci-mens requires that the cells remain undisturbed in the vessel in which they were cul-tured. Reagents are therefore removed and added without the need to collect the cells in a tube that can be centrifuged. If the culture vessel is a petri dish with a removable cover or a similar type of "chamber slide," the harvest process does lend itself to automation.

From: *The Principles of Clinical Cytogenetics*
Edited by: S. Gersen and M. Keagle © Humana Press Inc., Totowa, NJ

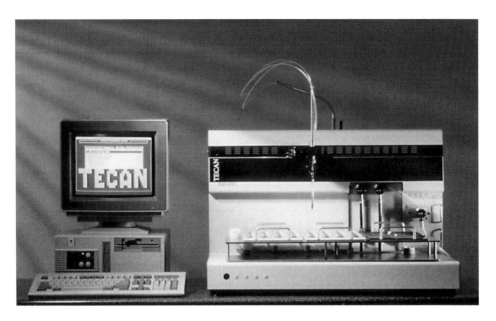

Fig. 1. Robotic Harvester for processing *in situ* cultures. (Courtesy of Tecan U.S., Inc.)

Webster defines a robot as " . . . an automatic apparatus or device that performs functions ordinarily ascribed to human beings . . .". In this context those functions are aspiration of the growth medium from the culture dish, addition of a hypotonic solution, and, after an appropriate incubation time, removal of the hypotonic solution and addition of several changes of fixative, each with its own duration. What is required, then, is a device that can both aspirate and dispense liquids, monitor the timing of each step, and control these steps correctly regardless of the number of cultures being processed at any one time, i.e, some form of computer control that can be "told" how many dishes there are and where on the device they are.

Such automatic liquid handling devices have been available for many years, and only a minor modification was required for petri dishes to be accommodated. An example is shown in **Fig. 1**. Two of the racks that sit on the base of the robot are designed to accept 35-mm petri dishes. The arm moves along both the *x*- and *y*-axes, and contains, in this case, three plastic tubes, one for aspiration, one for dispensing the hypotonic solution, and one for dispensing the fixative. The incubation times for each step are programmed into a computer that controls the robot, as is the number of dishes and their locations on the racks. After Colcemid® treatment, the dishes are placed on the harvester. The robot will then aspirate the culture medium, add hypotonic solution, and proceed to the next dish. The process continues until all dishes are filled with hypotonic solution. After the first dish has incubated for the proper amount of time, the hypotonic solution is aspirated and fixative is added; some protocols call for addition of a small amount of a "pre-fix" first. It should be pointed out that the computer program will not accept more dishes than it can process without perturbing the timing of these steps. The end result is culture dishes that contain fixative, which must be removed to properly spread the chromosomes.

DRYING CHAMBERS

Again, as described in Chapter 4, the typical end product of the cytogenetic harvest is a centrifuge tube with fixed cells, both mitotic and nonmitotic. Spreading of chromosomes is achieved by placing one or more drops of this suspension on a number of microscope slides, and is controlled by the height from which the suspension is dropped, the temperature and condition of the slide, and any number of manipulations while the slide is drying (including the ambient conditions in the laboratory). Any slide that is not satisfactory can be discarded and replaced, and trained individuals can determine the adjustments necessary to improve drying and spreading. Provided that such adjustments are made properly and quickly, running out of cell suspension is generally not a problem.

This is not the case with *in situ* harvesting. Most cytogenetics laboratories initiate four to six cultures from each sample, depending on the condition of the specimen upon receipt. Regulations and good clinical sense require that cells from at least two of these are examined, and in many cases three cultures are required. When one considers that at least one culture or some other form of backup should be retained against an unexpected need for additional testing, it becomes evident that every culture dish must produce usable metaphases. The concept of discarding one and trying again, as is so often done when making slides from cell suspensions, does not apply. Further complication is introduced by the fact that the physical force generated by dropping the cells onto a glass slide is not available when *in situ* processing is used, and so spreading of chromosomes is accomplished *solely by the manner in which the cultures are dried.*

As the 3:1 methanol/acetic acid fixative used in cytogenetics laboratories dries, it "pulls" the cell membrane across the slide or coverslip with it, allowing the chromosomes of mitotic cells to separate. If this process is viewed with a phase-contrast microscope, the metaphases appear to open much like a flower blossom. Clearly, the ambient temperature and humidity, as well as air flow over the cells (and possibly, as suggested by some studies, the barometric pressure) all affect the rate of drying, and therefore, when utilizing *in situ* processing, controlling these parameters is the only way to control chromosome spreading *(1)*.

In fact, of greatest importance is not merely controlling conditions, but maintaining them with a high degree of consistency. With each change in any one parameter, drying and spreading of chromosomes changes; once the correct combination is achieved, it is of paramount importance that it be maintained throughout the entire harvest.

There are probably as many solutions to this situation as there are cytogenetics laboratories. Some have constructed enclosed chambers in which air flow, humidity, and temperature can be varied, although these are typically prone to failure whenever the air conditioning breaks, as it is easy to warm the air inside the chamber but extremely difficult to cool it. Some labs have designed climate-controlled rooms; these frequently function well, but the drawbacks here are the need to maintain conditions while properly venting out fixative fumes (an engineering challenge, but certainly possible) and the potential to expose the technologist to uncomfortable conditions. Such rooms are also often costly to build.

Recently, several companies have developed self-contained chambers specifically for the purpose of drying *in situ* cultures; an example is shown in **Fig. 2**. Initially

Fig. 2. Bench top drying chamber. (Courtesy of Percival Scientific, Inc.)

developed for the culture of insect cells (which are grown at room temperature, and so the incubator must be capable of cooling as well as heating), this chamber has been modified to control humidity as well, and fans have been installed to allow for control of air flow over the coverslips. The advantages to this type of hardware are its ability to maintain conditions, quick recovery time after opening the chamber to insert or remove dishes, and potential for external venting if necessary. The disadvantage is the necessity to remove the fixative prior to placing the dishes in the chamber, creating the potential for drying to begin under noncontrolled conditions if there is any delay in getting the dishes into the chamber.

A variation on this theme is shown in **Fig. 3**. Here, the entire drying process, including aspiration of fixative, can take place inside the chamber. The technologist sits at the unit and manipulates the processing with a glove-box approach. The drawback to this concept is the large size of the unit, and a somewhat more cumbersome and limiting setup; removing one or more cultures for examination (an absolute requirement) can be more intrusive to the workflow.

These condition-controlled chambers are gaining in popularity in cytogenetics laboratories, and some use them not only for *in situ* processing, but for routine slide making as well, because of the consistency they provide.

Fig. 3. Floor model drying chamber. (Courtesy of Thermotron Industries.)

COMPUTERIZED IMAGING

The photograph is still considered by many to be the standard when it comes to a hard copy representation of the image of chromosomes produced by a microscope. The long-standing traditional relationship between microscope, camera, and dark room is described in detail in Chapter 5, and the final product it generates are photographs of metaphase chromosomes which are then cut out and arranged to form karyotypes. Although some will argue that this remains the best way to generate such karyotypes, it must be conceded that this process contributes to the already labor-intensive nature of clinical cytogenetics.

Because the right combination of hardware and software can now do in seconds what it takes 15–20 min to do manually, imaging systems clearly represent the single most significant example of automation in the cytogenetics laboratory. Indeed, today, computerized imaging systems are commonplace in many cytogenetics laboratories, which have grown to depend on them for their day-to-day operations, whether to produce karyotypes, generate printed images, analyze the various fluorescence techniques that have become so widespread, locate metaphase spreads, or detect rare cells. The impact of computerized imaging systems has been so significant that they are covered here in much greater detail.

Introduction and History

While the fundamental optical systems of microscopes have changed very little over the last quarter of a century, the same period has seen tremendous progress in computers, software design, and camera technologies. The impact of computerized imaging systems upon cytogenetics has been significant and they have become an accepted component of practice in many cytogenetics laboratories.

In 1964, Ledley first reported that computerized imaging systems could be useful in clinical cytogenetics laboratories *(2)*. The first prototype extensively evaluated within a clinical setting was built at the Jet Propulsion Laboratory (JPL, Pasadena, California, USA) *(3)*. The mid-1970s saw the introduction of a number of commercial automated cytogenetics systems *(4)*. All were based on proprietary image processing hardware or minicomputers. Although the user interfaces of these systems seem archaic and cumbersome by today's standards, they did make significant use of automation (e.g., automatic chromosome separation and identification) and set the standard for all future systems. These first systems were, however, quite expensive (> $100K per workstation) and had limited capabilities (insufficient image quality, poor resolution, difficult to use, etc.). In particular, the printers of this period had low reliability and produced unstable printed images. Further development was needed before imaging systems would gain widespread clinical acceptance *(5–13)*.

The explosive growth of the personal computer (PC) marketplace in the late 1980s allowed the next generation of imaging systems to be based upon low-cost PCs, providing a practical and more productive means of automation *(14–15)*. These systems turned the promise of the previous decade into a practical, clinical reality and became widely used in cytogenetics laboratories. They offered improved user interfaces, increased image quality, improved reliability, and a lower cost (< $50K per workstation). An important evolutionary step was the development of networks in which many workstations could be linked to a common archival device, patient and image database, and group of printers. This offered a practical way for laboratories to grow and increase productivity. Printers also improved during this period both in terms of reliability and print quality.

Today, every automated cytogenetics imaging system is based on a PC platform. All are essentially built from off-the-shelf components (cameras, computers, printers), use standard software operating systems, offer automated or interactive karyotyping capabilities, and support a variety of imaging modalities (brightfield, fluorescence, FISH, comparative genomic hybridization (CGH), etc.). Although the cost per station has not changed much since the late 1980s, performance, image resolution, degree of automation, speed, ease of use, and reliability have all continued to improve.

The first and foremost application of an imaging system in a cytogenetics laboratory is the production of karyotypes, either from brightfield (G-band) or fluorescence (Q-band or R-band) images. The next major application domain is fluorescence *in situ* hybridization (painting probes, single-locus probes, multicolor FISH, CGH, etc.). Rare event detection (e.g., automated metaphase or fetal cell finding, or fluorescence spot counting) represents a growing application for imaging systems in cytogenetics.

The major benefit of a computer imaging system is a reduction in the amount of time required to complete each analysis. Whether automatically locating suitable metaphases, automating the karyotyping process, and/or eliminating the need for photomicrography and darkroom processing, computerized imaging systems can save valuable operator time. Imaging systems also provide convenient image and patient data management features, telecommunications capabilities, hard copy options, and long-term archival tools. They further offer the ability to view microscope images instantaneously, to dramatically enhance images, to visualize chromosomes in different ways, to reveal unseen details, and to analyze images quantitatively, options not possible with traditional photomicrography. These features provide additional information and open new possibilities to the cytogenetics laboratory. Certain of the most recent advances in cytogenetics are possible only with a suitably equipped computerized imaging system (e.g., CGH, multicolor FISH).

Bits and Bytes

A bit (*binary digit*) is the most basic form of data that can be stored by a computer. It can only take two values, 0 or 1. A two-bit number would then be able to represent 2^2 or four values (0, 1, 2, 3). An eight-bit number, also called a byte, can therefore represent 2^8 or 256 discrete values.

Computers represent the various intensities in an image by using such digital numbers, made of bits. A one-bit image would have only two shades (2^1, black and white), a 6-bit image has 64 shades (2^6), an 8-bit image has 256 shades (2^8), a 10-bit image has 1024 shades (2^{10}), and a 24-bit image has 16,777,216 shades (2^{24}). Monochrome images are often represented as 8-bit images, while color images often use 24-bit values (8 bits for each of the primary colors, red, green, and blue).

Overview

The basic components of any cytogenetics imaging system are essentially the same, and include:

- ➤ An image acquisition subsystem, consisting of:
  - A microscope camera adapter (photo-eyepiece)
  - A camera
  - A frame grabber (digitizer), located inside the computer
- ➤ A computer subsystem, consisting of:
  - A personal computer with its operating system software
  - One or more devices that allow the operator to interact with the computer
  - A monitor for image display
  - Optional networking software and hardware
- ➤ Highly specialized, often proprietary cytogenetics image analysis software
- ➤ A hard copy printer
- ➤ An archival device

Systems used for the imaging of FISH preparations may use a motorized excitation filter wheel, and metaphase finding systems require a motorized stage and focus drive. **Figure 4** is a schematic block diagram of a fully equipped cytogenetics computerized imaging system. **Figure 5** shows an example of a commercially available instrument.

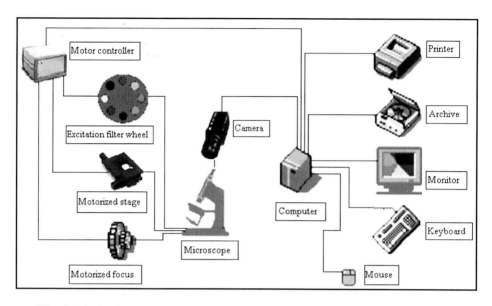

Fig. 4. Block diagram of a typical computerized imaging system for cytogenetics.

Fig. 5. Quips computerized imaging system. (Courtesy of Vysis.)

Components of a Computerized Cytogenetics Imaging System

Image Acquisition Subsystem

Of all the components of a computerized imaging system, those used to convert the optical image into electronic data are perhaps the most critical. Any degradation in this conversion process will affect the ultimate quality of the system's output. Poor image acquisition ultimately produces poor printed karyotypes. The image acquisition subsystem consists of four basic components: the microscope camera adapter, the camera, the frame grabber and the image capture software.

Microscope Camera Adapter

The microscope camera adapter is simply an interface device that is designed to attach a camera to the microscope, to project the primary microscope image onto the image sensing plane of this camera, and to change the magnification of the primary image so it fills this sensing area.

Cameras

Historically, cameras based on image tubes were the backbone of electronic imaging. In their simplest form, these devices have a target (photocathode) made of a photoconductive material. This material exhibits different electrical characteristics (e.g., resistance) when exposed to light or kept in the dark. An electron beam scans the target, as in a cathode-ray tube (CRT). This in turns generates an electrical signal in the photocathode circuit, which varies in proportion to the light intensity. The output of an image tube camera is therefore an electrical signal, which represents the image impinging on the target *(16)*.

Although image tube cameras offer excellent resolution and very good sensitivity, they face issues of linearity (a response that is not truly proportional to the light intensity), geometry (distortions), fragility, and high cost.

The limitations of image tube cameras, and the revolution in semiconductor manufacturing techniques, spurred the development of solid-stage imaging devices. From their commercial introduction in the 1980s, they quickly overran the older image-tube cameras and have actually begun to supplant photographic media in many applications, from astronomy to photomicrography and even portable "photographic" cameras *(17–19)*. Offering more flexibility, superior performance, and lower cost, solid-state cameras today are the choice for all cytogenetics applications.

Several types of silicon-based imaging technologies have been introduced over the years; of these, the CCD (charge coupled device) has received the most attention because of its superior performance.

Silicon can be grown into crystals, forming a three-dimensional lattice, where each silicon atom is covalently bonded to its three neighboring atoms. The fact that such crystals are sensitive to light is the principle behind CCD cameras. Light photons can break the covalent bonds, freeing electrons in the process. A positively charged, thin metal layer is deposited on the surface of the silicon, creating a potential well where these freed electrons can be stored. As more photons reach the silicon, more free electrons are accumulated in the potential wells. Each such potential well can be thought as a single picture element (pixel). The photosensitive area of the CCD is then composed of many such potential wells, laid out in a series of rows and columns. Images are therefore created by exposing the sensor to light, and then electronically reading the potential wells in a specific sequence. As the wells are emptied in this way, a series of electrical signals passes out of the silicon chip, each with an intensity proportional to the quantity of light received by a given well. Typically, CCD chips are two-dimensional arrays of between 512×512 and 1024×1024 potential wells.

Cooled vs. Uncooled CCD Cameras

Thermal agitation of the electrons in the silicon crystals can also cause random bond breakage, resulting in the production of "thermal" electrons. These are indistinguishable from photoelectrons and also accumulate in the potential wells, adding thermal noise to the image. This effect is dependent on the temperature of the CCD chip, and doubles for every 6°C rise. If long exposure times are required (e.g., low light imaging in fluorescence), this thermal noise will adversely affect image quality. CCD cameras are therefore available with cooling mechanisms to lower the CCD chip's temperature, reducing this effect and enabling longer exposure times. However, cooled cameras are obviously more complex and, as a result, more expensive than conventional CCD cameras.

With current technology, uncooled CCDs can be used without significant degradation of the image quality with exposure times of up to about 4 or 5 s. This makes them suitable for most cytogenetics applications, including the vast majority of FISH procedures *(20)*. When cooled down to about 5°C (–15°C from ambient), exposures of 15–20 s become possible. Such cameras are still reasonably priced, as they avoid problems linked with operating at temperatures below freezing. When cooled down to about –20°C (–40°C from ambient) exposures of up to 20 min are possible. The cost of such cameras, however, makes their choice justifiable only for the most demanding low light applications (e.g., multicolor FISH).

MONOCHROME AND COLOR CCD CAMERAS

Silicon is sensitive to photons with wavelengths between 300 mm and 1100 mm. CCD chips, however, are sensitive only to the intensity of light, and not to its color (wavelength). As a consequence, a standard CCD camera is inherently monochromatic (i.e., produces a gray scale image). Thus, additional techniques have to be used to design color cameras.

In a basic *one-chip* design, a small color filter with red, green, and blue segments is placed in front of the CCD, making every third pixel sensitive to red, green, or blue light. A trio of adjacent pixels is thus needed to represent a single point in the final color image. This stripped filter approach therefore essentially reduces the resolution of the camera to one third of its potential.

In a *three-chip* design, the light is chromatically divided by a prism or equivalent, and sent to three different chips. Each one therefore responds only to a portion of the full-color spectrum, the blue, green, and red regions respectively. The output from the three chips is then merged to generate the final color image. Such cameras do not compromise resolution, as a full array is used to generate each color component. They do, however, exhibit lower sensitivity, because the color splitting is less than 100% efficient. Finally, the added complexity obviously makes them more expensive than single-chip designs.

Although both types of color cameras can be found in cytogenetics imaging systems, actually most systems today use monochromatic cameras. This holds true even for systems designed for "color" applications such as FISH, and particularly for quantitative applications such as CGH or multicolor FISH.

VIDEO AND SLOW-SCAN CAMERAS

Electronic cameras were first developed for broadcast applications. To simplify interfacing between cameras, TV monitors, etc., the broadcast industry specified a number of standards that such equipment had to meet. Cameras that follow these standards, first developed by the broadcast industry, are called video cameras, and their outputs are called video signals.

Contrary to tube-based cameras, the basic output signal from a CCD camera is digital by nature. To operate as a video camera, this signal thus has to be converted to an analog waveform, meeting the above mentioned standards. Video cameras are therefore sometimes referred to as analog cameras. Such a video signal can then be, for example, directly connected to a TV monitor for viewing. However, to input it to a computer, it needs to be converted back to digital format. This is the role of the video frame grabber (see later).

Rather than going through these conversions, a number of CCD cameras directly output the signal from the chip to the computer in digital form. These cameras are called "slow scan" cameras (to contrast them with video cameras), or digital cameras (to contrast them with analog cameras). These cameras have not yet been standardized to the degree video cameras have. Their output cannot be directly connected to a TV monitor for viewing. For interfacing to a computer, they also require a frame grabber (although of a different, and in a way simpler, design than video frame grabber). Slow-scan cameras are most common in cytogenetics imaging systems used for low-light level fluorescence applications.

CAMERA SELECTION

The choice of a particular camera depends on its application. In cytogenetics, while there are no absolute rules, the following general guidelines, based on an analysis of currently available technology, can help in the selection of a camera:

- The camera of choice is a monochrome CCD camera, capable of extended exposure times.
- Uncooled cameras are sufficient for imaging objects visible with the naked eye.
- Video cameras are sufficient for all but the most demanding applications.

FRAME GRABBERS (DIGITIZERS)

Once the image has been converted into an electronic signal by the camera, that signal needs to be inputted to the computer. The electronic sampling of this signal is called image acquisition, or image capture, and is performed by devices known as frame grabbers (also referred to as digitizers or capture boards). The frame grabber converts the camera signal into an array of data points (the image pixels), each containing a digital value representative of the image intensity at a given point. This digitization process is necessary for the image to be stored by the computer and subsequently analyzed, processed, printed, archived, or displayed. Digitization is also required by video printers for the purpose of directly printing a camera image. Frame grabbers come in two types: those for video cameras (analog signals) and those for slow-scan cameras (digital signals).

In the case of video signals, the frame grabber must first sample the analog video waveform as a function of time to obtain the image pixels, and then must sample these selected time points as a function of voltage to obtain the image intensities (amplitudes) for these pixels.

In the case of slow-scan cameras the, spatial digitization is done entirely on the CCD chip (i.e., the locations of the various pixels from the chip directly map to the pixels of the digital image). In most instances, the amplitude sampling is also done directly by the camera electronics. Thus the frame grabber, while still necessary, becomes no more than an interface card, making slow-scan frame grabbers much simpler than their video counterparts.

IMAGE CAPTURE SOFTWARE

The last component of the image acquisition subsystem is the software that controls the camera and the frame grabber, and saves the acquired image. Beyond the basic interface to the hardware (camera, frame grabber, computer memory, and/or hard disk), and the control of their key characteristics (integration time, gain, offset, etc.), the software performs essential processing functions that directly relate to the acquisition of high-quality images.

Contrast Enhancement. In this operation the gray scale range of the image is remapped to better fit the range available for digital processing and display. For example, if the pixels of the acquired image ranged in intensity from 10 to 240, the software remaps them into the 0–255 range, improving contrast of the image. Such remapping of a captured image is often referred to as "normalization." Although it is best to always acquire images that already use the full range (through appropriate adjustment of the frame grabber parameters) normalization can be used as a matter of convenience.

Localized normalization or contrast enhancement is a more sophisticated implementation of this approach. Instead of examining the whole image, the software looks at small, localized neighborhoods around each pixel (or object) in the image, and adjusts the contrast independently in each of these regions. This can provide substantial improvements in image quality, but is not recommended for further quantitative work (as intensity relations between different objects are lost).

Background Subtraction. In this operation, the background of the image is somehow estimated or measured, and subtracted from the original image. Many different image processing techniques can be used for this purpose. In all cases, background artifacts are removed, and the contrast between objects of interest and background is increased.

Shading Correction. Shading correction is an image processing technique that can remove shadowing problems. This shadowing usually comes from nonuniform illumination of the scene by the microscope, further optical distortions, or artifacts of the camera. Shading correction compensates for all these sources of nonuniformity.

Frame Averaging. Another useful feature of image acquisition software is frame averaging. Because cytogenetic images are stationary, it is possible to capture multiple images of the same field and average them into a single image. Some types of noise are essentially random over time, and averaging therefore reduces the fluctuations due to such sources. Noise sources that are stationary over time, such as pixel defects, cannot be eliminated by image averaging, and require different image processing techniques. Vibrations can also render frame averaging ineffective.

Saving the Captured Image. Image capture software also provides several features for the storage of image data. Most system manufacturers store their images using proprietary image formats. Because of this, images from one system may not be automatically compatible with other technologies. Hence, these systems also offer data conversion into universally recognized image formats (e.g., TIF, GIF, JPEG, etc.), allowing images to be more easily interchanged.

Imaging systems also capture images to be linked to a patient database that provides case information (e.g., patient name, ID number, etc.) as well as image-specific information (e.g., slide ID, coordinates, image type, etc.).

Computer Subsystem

The function of the computer is to control the camera and frame grabber, operate the automated microscope parts, interface to the archival and hard copy devices, perform image analysis and display, and process inputs from the user. As such, it links together all the elements of the imaging system.

It is important at this point to make a distinction between dedicated, turnkey systems, and self-made systems. Turnkey systems are commercially available from manufacturers designing systems dedicated to the cytogenetics market. All components and subsystems are designed and supplied by the same vendor, who also supplies training, service, etc. This makes turnkey systems easier to use, better integrated, and better supported. Alternatively, laboratories may decide to build their own imaging systems. The growing compatibility between hardware components, the standardization of computer software, the increased sophistication of computer operating systems, as well as the growing individual mastery of computer skills, make many of the tools one would need to construct a powerful imaging station readily available. However, such self-made systems can face issues ranging from regulatory compliance to troubleshooting of failed equipment.

COMPUTER

Whether evaluated in terms of processing power, memory size, hard disk size, available peripherals, operating systems capabilities, user-friendly interfaces, etc., the performance of current personal computers is well up to the challenge of cytogenetics applications. Fueled by the demands of a growing business and consumer market, this performance continues to improve at a furious pace. Therefore, given today's available technology and foreseeable market trends, the platform of choice for computerized imaging systems is clearly the PC.

Given the rapid evolution of these products, one very important consideration is that of upgradability, that is the ability to update the system as hardware and software evolves. Although no one can predict the future, the best guaranty of upgradability is for the computer and its software to conform to official or *de facto* industry standards.

Similarly, compatibility with older, preexisting equipment can be equally important. As a computerized cytogenetics imaging system is used over the years, a considerable archive can be amassed that may not be compatible with future generation products. Many turnkey imaging systems currently available are in some form backward compatible with their predecessor products, as well as with the archives from competitive imaging systems.

OPERATING SYSTEM

The operating system is the basic, low-level piece of software, upon which are built all application-specific programs (such as karyotyping software). It essentially fills three roles: a control program, a resource manager, and a task coordinator. As a control program, it is an interface between other software and the hardware, a monitor over a set of conventions for accessing data. As a resource manager, it allocates CPU time, system memory, data, I/O devices, etc. among the various programs. As a task coordinator, it provides services allowing programs to cooperate, share data, and coexist.

Today's operating systems available for personal computers, whether Mac OS® or Windows 95/NT®, have all the necessary functionality to support the various software packages required for cytogenetics applications. Such functionality includes broad support of peripheral devices, networking ability, graphical user interface, conformance to industry standards, multitasking capability, wide availability of application software, etc.

A key aspect of an operating system is the concept of a multitasking environment, or one that allows multiple operations to occur simultaneously. As an example, a system that has a powerful multitasking environment can allow images to be captured, while other images are being archived or printed, while networked systems are requesting and sending data, etc.

INPUT DEVICES

Input devices are used to give commands to the imaging system and can range from keyboards or mice (with 1, 2, 3 . . . 25+ buttons), to light pens or voice automation. These input devices are those typically found on personal computers.

MONITOR

Today's high-resolution graphic monitors are perfectly suited to the task of displaying cytogenetics images. Display resolutions of at least 1024×768 by 32,768 colors, with monitor size of at least 17 inches are probably minimum requirements. Such monitors are still based on the bulky, but well-established CRT (cathode-ray tube) technology.

NETWORK

One very important feature of a computerized imaging system is its ability to support local and wide area networking (LAN and WAN). A LAN will network multiple workstations together, while a WAN might support connections to laboratory information systems (LIMS), hospital computing systems, remote laboratories, or even to the Internet. Local area networks allow users to share common printers, archiving devices, telecommunications facilities, patient database, or images. WANs allow them to exchange information with other computer systems. While reducing the overall cost per station, and allowing more users to access the imaging system, this sharing of resources and data also improves workflow and increases productivity. This is so true that networking is now a standard component available on all turnkey computerized imaging systems.

Cytogenetics Software

In terms of the software required for processing captured images, a fundamental distinction between dedicated cytogenetics software and general-purpose image analysis software needs to be made.

DEDICATED CYTOGENETICS SOFTWARE

Dedicated cytogenetics imaging software was developed specifically with the cytogenetics laboratory in mind. It focuses upon the applications relevant to these laboratories, tailors the user interface to these users, and offers the convenience and functions useful in such environments. It is engineered to be intuitive, to increase productivity, and to be forgiving. Dedicated functions for karyotyping (either automated or semi-automated), banding analysis, and ideogram display are standard. Optional components for functions such as metaphase finding and FISH analysis are available.

Because dedicated cytogenetics software is commercially marketed and sold for diagnostic applications, in the United States manufacturers fall under the jurisdiction of the Food and Drug Administration (FDA). This requires them to obtain and maintain marketing clearance from the FDA. They must follow fairly strict quality guidelines for the design, manufacture, documentation, validation, and service of their products. This in turn guaranties a certain quality level to the laboratories purchasing this software.

Dedicated cytogenetics software is currently available only as part of turnkey imaging systems.

GENERAL PURPOSE IMAGING SOFTWARE

Alternatively, general purpose image analysis software can be used. This is software that is designed to capture, process, and analyze images in a generic manner, regardless of the specific application (astronomy, microscopy, industrial inspection, etc.). It is available as software only packages, and can be obtained from a variety of sources.

Although the cost of these packages is much lower, they are not likely to offer the level of automation, or seamless integration, found on turnkey products. They will probably be less convenient for routine clinical usage (e.g., karyotyping) and may lack functionality in certain areas (e.g., ideograms). In addition, they require the users to integrate the other components of the imaging system, and to deal with all the associated compatibility or service issues. Finally, to date, manufacturers of general purpose imaging software have chosen not to submit their products for clearance by the FDA for use in diagnostic applications. Such software packages may on the other hand have advantages in research applications, as they are likely to offer greater flexibility, as well as access to very sophisticated image analysis features.

Hard Copy

Hard copy output of high-quality images is probably more important for cytogenetics than it is for many other biological imaging applications. Often, photorealistic prints are needed for diagnostic reasons and for record keeping requirements, and a hard copy image of a karyotype is frequently sent to the referring physician. Given this, print quality is arguably the second most significant feature of an imaging system, preceded only by the image acquisition subsystem.

HALFTONE VS. CONTINUOUS TONE

All printers use one of two basic processes to generate an image on paper. *Halftone* printers, like Impressionist painters, create images by building up patterns of tiny dots. At each addressable point on the paper, a printer can either lay down a dot, or leave the space blank. Each dot has the same intensity. For all shades beyond black and white, the printer spaces the dots closer together or farther apart, creating a series of half tones, or dithered dots, which the eye blends to form the desired shade. Halftones are what one sees on newspaper prints.

Color printers, instead of just black ink, use the subtractive primary colors—cyan, magenta, and yellow—with sometimes a separate black (CMY or CMYK). The printer can overlay two subtractive primaries to produce red, green, and blue, and can overlay all three to make black. To go beyond the eight fundamental colors (cyan, magenta, yellow, red, green, blue, black, and white), halftone (dithered) patterns are used.

However, some printers are capable of producing *continuous tone* images by directly varying the intensity (and/or size) of the printed dots. Just like their halftone cousins, continuous tone printers can produce either monochrome or color images. Continuous tone is what one sees on photographic prints.

PRINTER TECHNOLOGIES

Cytogenetics imaging systems continue to benefit from the tremendous technological advances of the printer industry. Indeed, many different technologies could be used to print images of chromosomes. Historically, *dry silver paper printers* were the mainstay technology for printing karyotypes. However, because of their high cost, inability to print in color, paper instability, and poor reliability, they have been largely replaced by more modern technologies. High-resolution *laser* printers (> 1200 dpi) are halftone printers that offer a very affordable solution. While the output quality is not quite comparable to photography, they are nevertheless very adequate for most cytogenetics applications. *Dye sublimation* printers, on the other hand, are the only true continuous tone printers currently available that can meet the most demanding requirements of the cytogenetics laboratory. For true photorealistic output, they have no equal. However, this quality comes at a significant, and sometimes prohibitive, cost. *Liquid ink jet* or *thermal paper* printers are generally considered marginal for use in cytogenetics, while *solid ink jet, dot matrix, thermal wax,* and *photographic* printers are considered inadequate for various reasons.

Which particular printer is best for a particular cytogenetics laboratory depends on the tasks at hand (e.g., karyotypes vs. FISH), the expected volume of prints, the required print quality, cost, and personal preference. Selecting more than one printer in some cases may represent a rational, cost-effective choice (e.g., a laser or ink jet printer for proofing or documentation and a dye sublimation printer for diagnostic quality prints).

Archiving

Data generated by the computerized imaging system can be stored for a short period of time on the computer's hard disk, but must eventually be moved to an archive medium for long-term storage. The requirements for long-term data storage can vary with the cytogenetics application, but in some instances are legally imposed and can involve extended periods of time. Archiving can also be used for information interchange, between users or between imaging systems.

Even more so than with printers, there is a wide array of available technologies for the storage of digital data, each with its own advantages and disadvantages *(21)*. Such technologies fall into three basic categories: tape, optical disk, and magnetic disk storage. They are rapidly evolving and improving technologies, mostly due to the growth of the personal computer market. As with the choice of camera or printer, the archival device that is best for a given cytogenetics laboratory depends on many factors. These include the specific application (backup, archive, data interchange, etc.), the expected volume of data to store (a few megabytes per week, or several hundred megabytes), and the required duration of the data storage (months or years). It may be that no single solution addresses all the issues satisfactorily and that the best choice will be to select several archival devices (e.g., a tape drive for backups, a removable magnetic drive for data interchange, and an optical disk for long-term archival).

Motorized Microscopes

In a number of applications, it is useful, and sometimes necessary, to have the computer directly control one or more of the components of a microscope. This is particularly true for digital fluorescence microscopy (FISH, CGH, multicolor FISH, etc.), where the fluorescence filters need to be controlled, and for slide scanning (metaphase finding), where the stage and focus need to be controlled.

EXCITATION FILTER WHEELS

When using an electronic imaging system for multifluorochrome applications (e.g., FISH), it is often much more convenient to have the computer control when the optical filters are switched in and out of the light path. The method of choice is probably to use a motorized filter wheel with multiple excitation filters, while all the other optical components remain fixed within a single epifluorescence cube. To capture an image, the software sequentially moves each excitation filter into position, and acquires a series of monochrome images for all fluorochromes studied. The same software then combines all these images to form a final color image.

MOTORIZED STAGE AND FOCUS

Another application of electronic imaging in cytogenetics is for computerized scanning of slides (e.g., metaphase finders). For such systems, it is necessary to equip the microscope with a motorized stage (for motions along the x or y directions, for the actual scanning), as well as a motorized focus drive (for motions along the z direction, for maintaining proper focus as the slide is scanned). See later.

Cytogenetics Applications of Computerized Imaging Systems

Photo Documentation

Photo documentation is a basic application of electronic imaging in cytogenetics. The computer simply replaces the photographic camera and darkroom. Monochrome and color images can be acquired and printed from brightfield and fluorescence images.

Karyotyping

Karyotyping remains the primary activity of most imaging systems. Such systems capture images of metaphases, provide software to help the user create a karyotype (electronic cut and paste), and generate photorealistic prints.

There are three basic ways to create a karyotype using computerized imaging systems. In a manual method, the user "cuts" each chromosome, outlining it with mouse, and classifies it. With a semi-automated method the computer "cuts" the chromosomes automatically, but the user still classifies them. With fully automated methods the computer automatically cuts and identifies the chromosomes and creates the complete karyotype, which the user can review and correct. Dedicated cytogenetics software will offer at least manual and semi-automated approaches and often a fully automated approach as well. General purpose image processing software usually can support only manual approaches.

Fig. 6. CytoVision Ultima metaphase finding system. (Courtesy of Applied Imaging Corporation.)

Metaphase Finding

Another common use of computerized systems in cytogenetics is for the actual location of metaphases suitable for analysis. For most samples, only cells of appropriate quality are analyzed, while in cancer patients it is often the poor quality cells that represent abnormal clones, and metaphases of any quality can often be few and far between. Computerized imaging can help cytogenetics laboratories by eliminating the manual search for the appropriate type and/or number of cells.

Metaphase finding requires a dedicated system, with software specifically tailored to this application. In addition to the basic components, described earlier in this chapter, such a system includes a microscope equipped with a motorized stage and a focus drive. While this can save technologist time, scanning one slide might, in the real world, be of little value. An additional component to these systems therefore involves a stage capable of hosting multiple slides or an automatic slide loader; these offer unattended operation for extended periods of time. Modern systems are able to scan an entire slide, in brightfield, in less than 10 min. An example of a complete metaphase finder system is shown in **Fig. 6**.

Criteria are entered into the program so that the computer can discriminate between the correct target (a mitotic cell) and artifact, and most programs allow for a "rating" system, so that, if desired, cells can be presented to the operator in order of quality. Images of the metaphases that are identified are captured by the software for eventual review by a human being.

In this way, the laboratory can prepare multiple slides from multiple patients, load them into a stacker, and have each of them scanned completely overnight. The following morning, a technologist can query the computer and it will display, one by one, images of the metaphases it has located, with coordinates for relocation of each cell in the microscope.

Key factors for metaphase finders include scanning speed (throughput), quality of the metaphases found, ease of review of the located metaphases, sensitivity of the scanning (i.e., the percentage of metaphases actually being found), and the accuracy of metaphase relocation. The ability to locate, with equal ease, absorption-stained as well as fluorescence-stained metaphases is also very important.

Automated Scanning

While metaphase finders can be invaluable time savers in certain situations, they are still rarely found in the typical cytogenetics laboratory. However, new investigative applications are making this type of automated scanning instrumentation increasingly more valuable. For example, the isolation of fetal cells from maternal blood relies on the detection and examination of extremely rare objects *(22)*. Similarly, interphase FISH, which uses DNA probes hybridized to interphase cells, requires, for proper statistical significance, the examination of a large number of cells. In both cases, automated scanning instrumentation, very similar to a metaphase finder, can relieve the tedium of the manual scanning process and save valuable time.

Digital Fluorescence Microscopy

Within the last 10 years, fluorescence microscopy has developed into one of the most important tools in cell biology. Cytogenetics is no exception, with interest in FISH, CGH, multicolor FISH, etc. This is due to a combination of several factors including the advent of new fluorescent probes and markers and new fluorescent dyes (see Chapter 15). The availability of electronic imaging systems to visualize these sometimes very faint probes and markers, and to generate quantitative data, has created new possibilities for this increasingly important technology.

FLUORESCENCE *IN SITU* HYBRIDIZATION (FISH)

While karyotyping and metaphase finding can be performed visually, and/or with a photographic camera, fluorescence *in situ* hybridization (FISH) can be performed more rapidly, less expensively, and with less tedium using a computerized imaging system *(23,24)*.

Such systems should be able to perform a variety of tasks including acquiring very low light level images, counting fluorescent dots, isolating and measuring cells, chromosomes or dots, and estimating intensity proportions between fluorescent objects.

The simplest applications, such as image capture and dot counting, can be very easily served by general purpose image processing software. More advanced, quantitative applications, such as intensity ratios, can be handled only by dedicated software and systems.

COMPARATIVE GENOMIC HYBRIDIZATION

An application of FISH technology, comparative genomic hybridization (CGH), allows assessment of relative changes in copy number of DNA sequences over the whole genome. This technique has proved to be very useful in locating possible regions of genetic imbalance in abnormal cells, which can then be analyzed by higher resolution techniques *(25)*. See Chapter 15.

For this application, a computerized imaging system is an absolute necessity. The system should first meet all requirements for a high-quality, quantitative, FISH imaging system. In addition, only dedicated software will provide all the necessary functionality. This includes features such as averaging of ratio profiles over multiple metaphases, automatic corrections for unequal chromosome length, statistical analysis and validation, automatic chromosome axis determination, simultaneous display of ideograms, and proper background subtraction.

MULTICOLOR FISH

Multicolor FISH, also referred to as multiplex-FISH, M-FISH, multifluor-FISH, multispectral-FISH, color karyotyping, or spectral karyotyping®, is a technique by which a larger number of fluorochromes are used in what is otherwise a standard FISH procedure. Instead of the usual two or three fluorochromes, typically five to seven fluorochromes are used with multicolor FISH, and this may grow to nine or ten in the future *(26)*. In addition, probes can be labeled, not only with a single fluorochrome, but also with combinations of two or more fluorochromes. In this way, many colors can be produced (typically from 20 to about 40). See Chapter 15.

Multicolor FISH is another tool that absolutely requires the use of a computerized electronic imaging system, equipped with some fairly sophisticated hardware and software. Their task is to uniquely recognize which fluorochrome, or combination of fluorochromes, gave rise to all the observed probe fluorescence. Given the limited spectral bandwidth available for fluorescence imaging (roughly 300 nm to 800 nm), and the overlap between spectra of available fluorochromes, this is a significant challenge.

SUMMARY

All laboratory procedures were essentially manual at one time. Primitive centrifuges were hand operated, and even the earliest microscopes (examples of "new technology" in their own right) utilized mirrors to gather sunlight or candlelight before the discovery of electricity and the invention of the electric light bulb. Today, however, the typical clinical laboratory is dominated by technology, computers and automated instrumentation. These have improved laboratory practice in three basic ways:

- automation of tasks, which can free up technologist time, thereby improving efficiency and reducing costs,
- an increase in the speed (and sometimes accuracy) at which tasks can be performed, and
- performance of tasks that cannot be accomplished manually.

Nevertheless, the world of cytogenetics is still essentially one of manual manipulation and diagnosis. We have, however, seen that there are notable exceptions, assisting with both sample preparation and chromosome analysis. These fall into several basic categories: robotic harvesters, environmentally controlled drying chambers, and computerized imaging systems. We have further seen how the latter represents the single most significant example of automation in the cytogenetics laboratory. The major benefit of a computerized imaging system is a reduction in the amount of time required to complete each analysis. Whether automatically locating suitable metaphases, automating the karyotyping process, enabling the use of low-light fluorescence techniques, or eliminating the need for photomicrography and darkroom processing, computerized imaging systems can save valuable operator time.

As technology continues to advance, there seems little doubt that even more of the manual tasks required for chromosome analysis today will be automated in the laboratory of the future.

ACKNOWLEDGMENTS

The authors wish to thank Chris McAleer, Application Manager at Applied Imaging, for his substantial input to this manuscript.

REFERENCES

1. Spurbeck, J.L., Zinmeister, A.R., Meyer, K.J., and Jalal, S.M. (1996) Dynamics of chromosome spreading. *Am. J. Med. Genet.* **61,** 387–393.
2. Ledley, R.S. (1964) High-speed automatic analysis of biomedical images. *Science* **146,** 216–223.
3. Castleman, K.R., Meinyk, J., and Frieden, H.J. (1976) Computer assisted karyotyping. *J. Reprod. Med.* **17,** 53–57.
4. Rutovitz, D. (1983) Automatic chromosome analysis. *Pathologica (Suppl.)* **75,** 210–242.
5. Philip, J. and Lundsteen, C. (1985) Semi-automated chromosome analysis: a clinical test. *Clin. Genet.* **27,** 140–146.
6. Piper, J. and Lundsteen, C. (1987) Human chromosome analysis by machine. *Trends Genet.* **3,** 309–313.
7. Berry, R. and McGravan, L. (1987) Impact of computerized imaging system on a clinical cytogenetic laboratory: six month experience. In 25th Annual Somatic Cell Genetics Conference, New Mexico.
8. Graham, J. (1987) Automation of routine clinical chromosome analysis I. Karyotyping by machine. *Anal. Quant. Cytol. Histol.* **9,** 383–390.
9. Graham, J. and Pycock, D. (1987) Automation of routine clinical chromosome analysis II. Metaphase finding. *Anal. Quant. Cytol. Histol.* **9,** 391–397.
10. Lundsteen, C. and Martin, A.O. (1989) On the selection of systems for automated cytogenetic analysis. *Am. J. Med. Genet.* **32,** 72–80.
11. Lundsteen, C. and Philip, J. (1989) Automated cytogenetic analysis: accomplishments, present status, and practical future possibilities. *Clin. Genet.* **36,** 386–391.
12. Lundsteen, C. and Piper, J., eds. (1989) *Automation in Cytogenetics.* Springer-Verlag, Berlin.
13. Korthof, G. and Carothers, A.D. (1991) Tests of performance of four semi-automatic metaphase-finding and karyotyping systems. *Clin. Genet.* **40,** 441–451.
14. van Vliet, L.J., Young, I.T., and Mayall, B.H. (1990) The Athena semi-automated karyotyping system. *Cytometry* **11,** 51–58.
15. Bieber, F. (1994) Microscopy and image analysis. In *Current Protocols in Human Genetics.* John Wiley & Sons, New York, pp. 4.4.1–4.4.8.
16. Castleman, K.R. (1997) Cameras. In *Current Protocols in Cytometry.* John Wiley & Sons, New York, pp. 2.3.1–2.3.8.
17. Higgins, T.V. (1994) The technology of image capture. *Laser Focus World,* Dec., 53–60.
18. Aikens, R.S., Agard, D.A., and Sedat, J.W. (1989) Solid-state imagers for microscopy. *Methods Cell Biol.* **29,** 291–313.
19. Aikens, R.S. (1990) CCD cameras for video microscopy. In *Optical Microscopy for Biology.* John Wiley & Sons, New York, pp. 85–110.
20. Vrolijk, J., Sloos, W.C.R., Verwoerd, N.P., and Tanke, H.J. (1994) Applicability of a

noncooled video-rated CCD camera for detection of fluorescence in situ hybridization signals. *Cytometry* **15,** 2–11.

21. Stone, M.D. (1996) Endless storage. *PC Magazine,* **15,** 149–159.
22. Ravkin, I. and Temov, V. (1998) Automated microscopy system for detection and genetic characterization of fetal nucleated red blood cells on slides. *Proc. SPIE* **3260,** in press.
23. Pinkel, D., Straume, T., and Gray, J.W. (1986) Cytogenetic analysis using quantitative, high-sensitivity, fluorescence hybridization. *Proc. Natl. Acad. Sci. USA* **83,** 2934–2938.
24. Castleman, K.R., Riopka, T.P., and Wu, Q. (1996) FISH image analysis. *IEEE Eng. Med. Biol.* Jan., 67–75.
25. De Vries, S., Gray, J.W., Pinkel, D., Waldman, F.M., and Sudar, D. (1996) Comparative genomic hybridization. In *Current Protocols in Human Genetics.* John Wiley & Sons, New York, pp. 4.6.1–4.6.18.
26. Le Beau, M. (1996) One FISH, Two FISH, Red FISH, Blue FISH. *Nat. Genet.* **12,** 368–375.

III
Clinical Cytogenetics

Editor's Foreword to Section III

Cytogenetics is a laboratory discipline that has a wide variety of clinical applications. As such, it is utilized by a diverse assortment of specialists, many of whom are conversant only with those uses of chromosome analysis that are meaningful in a given setting.

This is of course appropriate, and in most cases works quite well. However, all health care providers are occasionally faced with cytogenetic scenarios that fall outside of their range of expertise. It is our goal in this section, therefore, to provide a comprehensive overview of the fundamental clinical issues involved in chromosome analysis, so that anyone who relies upon this science for one purpose will have a better understanding of its many other uses.

8
Autosomal Aneuploidy

Jin-Chen C. Wang, M.D.

INTRODUCTION

The term *aneuploidy* refers to cytogenetic abnormalities in which all or part of one or more chromosomes is added or deleted. Autosomal aneuploidy refers to all such abnormalities that do not involve the sex chromosomes. These can be either numerical (the topic of this chapter) or structural, the vast majority being trisomies, and may be present only in some cells (mosaic aneuploidy) or in all cells (nonmosaic). The incidence of autosomal aneuploidy in newborns is estimated to be 0.2% *(1)*. Many autosomal aneuploidies are incompatible with fetal survival and therefore have much higher incidences (approximately 27–30%) in spontaneous abortuses *(2–4)*. These are discussed below and covered in detail in Chapter 12.

Cytogenetic studies of human oöcytes and sperm reveal that the overall frequency of abnormalities is approximately 20% and 10%, respectively (reviewed in *ref. 5*). More than 90% of the abnormalities observed in oöcytes and less than 50% of those seen in sperm are numerical. Studies using fluorescence *in situ* hybridization (FISH) *(6)* or primed *in situ* labeling (PRINS) *(7,8)*, which do not require the presence of dividing cells, have shown that the frequency of autosomal aneuploidy in human sperm is relatively uniform for all chromosomes studied (chromosomes 3, 7, 8, 9, 10, 11, 13, 16, 17, 21), with a range of 0.26–0.34%. On the other hand, one study using FISH *(9)* reported a higher frequency of aneuploidy for chromosome 21 (0.29%) than for other chromosomes studied (0.08–0.19% for chromosomes 1, 2, 4, 9, 12, 15, 16, 18, 20). It is therefore possible that meiotic nondisjunction is random for all autosomes, with the possible exception of chromosome 21. However, this is not the case for aneuploidy actually observed in spontaneous abortuses or liveborns.

Trisomies for all autosomes have been reported in spontaneous abortuses *(3,10,11)*, including trisomy 1, which was reported in an eight cell human preembryo *(12)* and most recently in a clinically recognized pregnancy at 8–9 weeks post-LMP (last menstrual period) *(11)*. The observed frequency of each trisomy, however, varies greatly. For example, trisomy 16 accounts for approximately 30% of all autosomal trisomies in abortuses *(3)*. In liveborns, the only trisomies that have not been reported in either mosaic or nonmosaic form are those involving chromosomes 1, 6, and 11, although trisomies other than 13, 18, and 21 are rare. Autosomal monosomies, on the other hand, are extremely rare in both liveborns and recognized abortuses.

From: *The Principles of Clinical Cytogenetics*
Edited by: S. Gersen and M. Keagle © Humana Press Inc., Totowa, NJ

The supposition that, with the probable exception of trisomy 21, the frequencies of trisomy for each chromosome might be similar at the time of conception but differ greatly among abortuses and liveborns can be explained by the devastating effect of chromosomal imbalance. Many autosomal aneuploidies are so deleterious that they are lethal in the pre-embryonic stage and thus result in unrecognized and, therefore, unstudied spontaneous abortions. The lethality of a particular autosomal aneuploidy is shown to be correlated with the gene content of the chromosome involved *(10)*. Aneuploidies for "gene-rich" chromosomes are less likely to survive. Trisomies 13, 18, and 21, which involve chromosomes that are less "gene-rich," are therefore relatively "mild" and fetuses can survive to term.

This chapter addresses only those autosomal aneuploidies, both trisomies and monosomies, that have been observed in liveborns. Polyploidy, or changes in the number of *complete sets* of chromosomes, are also included, as are aneuploidies that are the result of supernumerary "marker" chromosomes.

MECHANISM AND ETIOLOGY

Errors in meiosis (nondisjunction) result in gametes that contain abnormal numbers of chromosomes and, following fertilization, produce aneuploid conceptuses. Using DNA markers, the parental origin of the additional chromosome in autosomal aneuploidies has been studied for trisomies 13, 14, 15, 16, 18, 21, and 22 *(2,13–25)*. These studies suggest that the proportion of trisomies of maternal origin varies among different chromosomes and that, with the exception of chromosome 18, nondisjunction in maternal meiosis stage I accounts for the majority of cases (**Table 1**).

The association between autosomal aneuploidy and maternal age has long been recognized. In 1933 Penrose demonstrated that maternal age was the key factor for the birth of Down syndrome children *(26)*. Why aneuploidy is maternal age dependent, and what constitutes the mechanism and etiology of chromosomal nondisjunction have been topics of much research, as summarized below.

Nondisjunction can occur during either meiosis I (MI) or meiosis II (MII). In MI, homologous chromosomes pair and form bivalents (see Chapter 2). Malsegregation of homologous chromosomes can occur in one of two ways. The first involves nondisjunction of the bivalent chromosomes with both homologues going to the same pole (**Fig. 1d,e**). This mechanism, as shown by Angell, may be a very rare occurrence *(27)*. The second type of error involves premature separation of the sister chromatids of one homologue, with subsequent improper distribution of the separated chromatids with the other homologue of the chromosome pair *(27a)*. In MII, sister chromatids separate. Malsegregation occurs when both chromatids go to the same pole (**Fig. 2g,h**). Recent cytogenetic studies of oöcytes, performed mostly on unfertilized or uncleaved specimens obtained from *in vitro* fertilization programs, have provided conflicting results regarding whether the frequency of aneuploidy actually increases with maternal age *(28–31)*, while a recent FISH study of human oöcytes using corresponding polar bodies as internal controls demonstrated that nondisjunction of bivalent chromosomes (MI error) does increase with maternal age *(32)*. More data are needed before a firm conclusion can be drawn. If confirmed, this latter study will provide direct evidence that a maternal age-dependent increase in the frequency of MI errors is the basis for the observation that the risk of having a trisomic offspring is greater in older women. In

Table 1
Parental and meiotic/mitotic origin of autosomal trisomies determined by molecular studies (number of cases)

Trisomy	Maternal					Paternal					Reference
	MI	MII	MI or MII	Mitotic	Total[a]	MI	MII	MI or MII	Mitotic	Total[a]	
13	4		17		21	1	1	1		3	20
14	3	4	2		9		2			2	20
15	21	3		3	27				5	5	*16 (UPD study)*
	10				10				2	2	*18 (UPD study)*
	5	2	8		15		1		1	2	*20*
16	56		6		62					0	21
18	11	17			56		1			6	*24*
					17					5	*17*
	16	≥35		3	61				2	2	*22*
21	9	1			22					3	*13*
					91					6	*14*
	128	38			188	2	7			9	*15*
						7	15		8	36	*19 (paternal study only)*
	174	58	79		311	9	15	8		32	*20*
	62	22			81		10			13	*23*
	67				97	4	4			10	*25*
22	6		11		17			2		2	20

[a]Total numbers may not add up because not all origins of error can be determined.

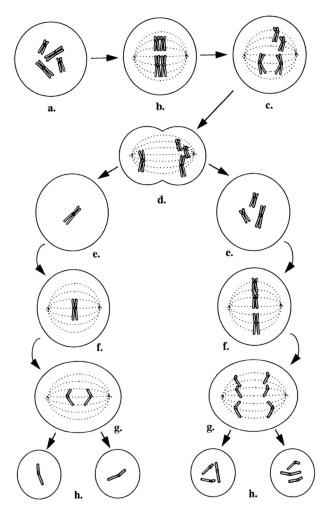

Fig. 1. Schematic representation of meiosis I nondisjunction. **a:** prophase I, **b:** metaphase I, **c:** anaphase I, **d:** telophase I (with both homologues of one chromosome pair segregating together) **e:** products of meiosis I, **f:** metaphase II, **g:** anaphase II, **h:** meiotic products—two gametes lack a chromosome and two gametes contain an additional chromosome.

contrast, the "relaxed selection" hypothesis assumes that older women are less likely to spontaneously abort trisomic conceptions *(33,34)*. However, data obtained from fetal death *(35)* and from comparison of frequencies of trisomy 21 between the time of chorionic villus sampling and the time of amniocentesis *(36)* suggest that selective miscarriage is actually enhanced with increasing maternal age. Other hypotheses have been proposed to explain the observed maternal age effect, but none have been substantiated.

If maternal MI nondisjunction does increase in older women, what then causes this? Different mechanisms have been proposed. One example is the "production line" hypothesis *(37)*. This theory proposes that oöcytes mature in adult life in the same order as the corresponding oögonia entered meiosis in fetal life. Oögonia that enter meiosis later in life may be more defective in the formation of chiasmata, and thus

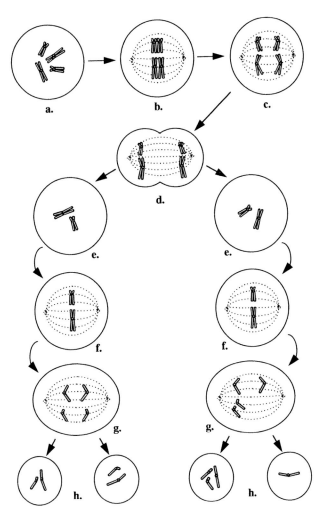

Fig. 2. Schematic representation of meiosis II nondisjunction. **a:** prophase I, **b:** metaphase I, **c:** anaphase I, **d:** telophase I, **e:** products of meiosis I, **f:** metaphase II, **g:** anaphase II, with both sister chromatids segregating together in the cell to the right, **h:** meiotic products—two gametes with a normal chromosome complement, one gamete lacking a chromosome and one gamete containing an additional chromosome.

more likely to undergo nondisjunction. The first direct cytologic support for this was provided by a study that examined the frequency of unpaired homologues in MI pachytene and diplotene in oöcytes obtained from abortuses at 13–24 weeks and 32–41 weeks of gestation *(38)*. Of the six chromosomes studied, the rate of pairing failures in early specimens (0–1.2%) was significantly lower than that in later specimens (1.3–5.5%). Further studies are needed to substantiate this observation.

One probable factor that predisposes gametes to nondisjunction is aberrant recombination (see Chapter 2). Data on recombination patterns are available for trisomies 15, 16, 18, and 21. Studies of chromosome 15 nondisjunction in uniparental disomy (see Chapter 16) revealed that there was a mild reduction in recombination in association with maternal nondisjunction, with an excess of cases that have zero or one crossover and a deficiency of cases that have multiple crossovers *(16)*. In contrast, in a study of trisomy 18, approximately one third [5/16] of maternal MI nondisjunctions were associated with a complete absence of recombination, whereas the remaining MI and all MII nondisjunctions appeared to have normal recombination *(22)*. Studies on trisomy 16 and trisomy 21 reported similar findings in these two trisomies. In trisomy 16, it was shown that recombination was reduced, but not absent, and that distribution of recombination was altered, with rare crossover in the proximal regions of the chromosome *(21)*. In trisomy 21, there was an overall reduction in recombination with an increase in both zero- and one-crossover in maternal MI nondisjunction *(39)*. Most recently, Lamb et al. showed that in maternal MI nondisjunction for chromosome 21, the average number of recombination events was decreased, especially in the proximal region of the long arm, whereas in MII nondisjunction there was an increase in the number of recombination events, particularly in the most proximal region of the long arm of the chromosome *(40)*. This altered recombination pattern was not maternal age dependent. Because homologous recombination occurs during MI, these data suggest that all nondisjunction events initiate in MI. These proximal recombinations may cause an "entanglement" effect resulting in premature separation of the sister chromatids of one or both homologues at MI. A random segregation of a single chromatid with its sister chromatid will result in an apparent MII nondisjunction. This study, together with one on trisomy 16, suggests that, at least for trisomies 16 and 21, distal chiasmata are less efficient in preventing nondisjunction in MI *(21)*. Based on these observations, Lamb et al. proposed a "two-hit" model *(41)*. They hypothesized that certain recombination configurations are less likely to be processed properly in older women. This could result from, for example, an age-dependent loss of spindle forming ability, thus explaining their observation for trisomy 21 that although an altered recombination pattern is not maternal age dependent, meiotic disturbance is age dependent *(42)*. The same argument was used by Hassold et al. to explain their findings with trisomy 16 *(21)*.

The possibility of the presence of a genetic predisposition to nondisjunction has also been proposed. To date, the only clear evidence for this has been that of a well-controlled study involving consanguineous families in Kuwait *(43)*. In that study, the relative risk for the occurrence of Down syndrome was approximately four times greater for closely related parents (first cousins, first cousins once removed, second cousins) than for unrelated parents. As consanguinity is usually perpetuated in certain families or sections of the population, these results were taken as evidence for the existence of an autosomal recessive gene that facilitates meiotic nondisjunction in homozygous parents. Thus, in a subgroup of trisomy 21 patients, nondisjunction may be genetically determined.

Our understanding of the mechanism and etiology of nondisjunction is still far from complete. It is possible that more than one mechanism contributes to the observed maternal age effect. Thus, "two-hit" and "production line" models, along with other hypothetical explanations, could explain a portion of the cases involving some chromosomes. Further studies are needed.

Nondisjunction occurring at mitosis, on the other hand, results in mosaicism, usually with both normal and abnormal cell lines.

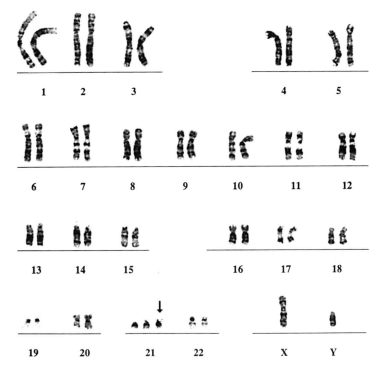

Fig. 3. Trisomy 21 Down syndrome male karyotype (47,XY,+21).

AUTOSOMAL TRISOMIES

Trisomy 21

Incidence

Trisomy 21 [47,XX or XY,+21] (**Fig. 3**) was the first chromosomal abnormality described in humans *(44)*. The phenotype was delineated by John Langdon Down (1828–1896) in 1866 and is referred to today as Down syndrome *(45)*. It is the most common single known cause of mental retardation. The frequency in the general population is approximately 1 in 700. Down syndrome is more frequent in males, with a male-to-female ratio of 1.2:1. A recent study using multicolor FISH showed that among sperm disomic for chromosome 21, significantly more were Y-bearing than X-bearing *(46)*. This finding was consistent with earlier reports showing an excess of males among trisomy 21 conceptuses that resulted from paternal meiotic errors *(19)*. This preferential segregation of the extra chromosome 21 with the Y chromosome contributes to a small extent to the observed sex ratio in trisomy 21 patients. Other mechanisms, such as *in utero* selection against female trisomy 21 fetuses, must also exist.

Trisomy 21 accounts for approximately 95% of all cases of Down syndrome. Mosaicism and Robertsonian translocations (see Chapter 9) comprise the remaining 5%. As described previously, the incidence of trisomy 21 in newborns is closely associated with maternal age (**Table 2**).

Table 2
Maternal age-specific risks for trisomy 21 at birth

Maternal Age (Years)	Incidence at Birth (1 in)	Maternal Age (Years)	Incidence at Birth (1 in)
15	1580	33	570
16	1570	34	470
17	1565	35	385
18	1555	36	307
19	1540	37	242
20	1530	38	189
21	1510	39	146
22	1480	40	112
23	1450	41	85
24	1400	42	65
25	1350	43	49
26	1290	44	37
27	1210	45	28
28	1120	46	21
29	1020	47	15
30	910	48	11
31	800	49	8
32	680	50	6

Modified from Cuckle et al. *(47)*. Data were based on eight pooled studies. Restriction of analysis to two studies with the most complete ascertainment yielded higher rates *(48)*.

Phenotype

The clinical phenotype of Down syndrome has been well described *(49,50)*. In brief, there is a characteristic craniofacial appearance with upward-slanting palpebral fissures, epicanthal folds, flat nasal bridge, small mouth, thick lips, protruding tongue, flat occiput, and small overfolded ears. Hands and feet are small and may demonstrate clinodactyly, hypoplasia of the midphalanx of the fifth finger, single palmar crease, and a wide space between the first and second toes. Hypotonia and small stature are common, and mental retardation is almost invariable. Cardiac anomalies are present in 40–50% of patients, most commonly endocardial cushion defects, ventricular septal defects (VSD), patent ductus arteriosus (PDA), and auricular septal defects (ASD). Other observed major malformations include duodenal atresia, annular pancreas, megacolon, cataracts, and choanal atresia. In addition, a 10- to 20-fold increase in the risk for leukemia has been observed in Down syndrome patients of all ages, with a bimodal age of onset in the newborn period and again at 3–6 years *(51)*. Moreover, a transient myeloproliferative disorder (TMPD) in the newborn period, characterized by a high spontaneous remission rate with occasional relapse, occurs more frequently in children with Down syndrome. Of interest is the observation of the presence of a trisomy 21 clone in association with TMPD in 15 phenotypically normal children, at least four of whom were determined to be constitutional mosaics for Down syndrome *(52)*.

Overall, the clinical phenotype is typically milder in mosaic Down syndrome patients, but there is no clear correlation between the percentage of trisomy 21 cells and the severity of clinical presentation. Mosaic patients can be as severe as nonmosaic trisomy 21 individuals.

Delineation of regions of chromosome 21 responsible for the Down syndrome phenotype has been attempted using molecular methods to study patients with partial trisomy 21 who present clinically with various features of the syndrome *(53–57)*. These studies suggest that the genes for CuZn superoxide dismutase *(SOD1)* and amyloid precursor protein *(APP)*, located proximal to band 21q22.1, may be excluded from a significant contribution to the Down syndrome phenotype, while parts of bands 21q22.2 and 21q22.3, including locus D21S55, may be the minimal region necessary for the generation of many Down syndrome features (see Chapter 3 for a discussion of band nomenclature). Studies by Korenberg et al. suggest that, instead of a single critical region, many chromosome 21 regions are responsible for various Down syndrome features *(58)*. They used a panel of cell lines derived from 16 partial trisomy 21 individuals to construct a "phenotypic map" correlating 25 Down syndrome features with regions of chromosome 21.

Recurrence

The empirical recurrence risk is about 1% in women under 30 years of age, and includes trisomies other than 21. For women over 30, the recurrence risk is not significantly different from the age specific risk *(59)*.

One study of families with two trisomy 21 children showed that three of thirteen had a parent who was mosaic for trisomy 21 (by cytogenetic studies), and two of thirteen had a parent who was potentially mosaic (by DNA polymorphism analysis) *(60)*. In a family with three trisomy 21 children, Harris et al. reported that the mother was mosaic for trisomy 21 in lymphocytes and skin fibroblasts *(61)*. In another single case report involving a family with four trisomy 21 children, the mother was found to have a trisomy 21 cell line in an ovarian biopsy specimen *(62)*. Thus, gonadal mosaicism in one parent is an important cause of recurrent trisomy 21 and should be looked for in families with more than one affected child. When present, the recurrence risk will be high and will depend on the proportion of trisomy 21 cells in the gonad.

The recurrence risk for mosaic trisomy 21 that results from mitotic nondisjunction should, in general, not be increased. However, several studies investigating the mechanism and origin of mosaic trisomy 21 have shown that in a relatively high proportion of cases (probably more than 50%), the mosaicism results from the loss of one chromosome 21 during an early mitotic division in a zygote with trisomy 21 *(63,64)*. In such cases, the recurrence risk for nondisjunction will be the same as for nonmosaic trisomy 21.

Trisomy 18

Incidence

Trisomy 18 [47,XX or XY,+18] was first described by Edwards et al. *(65)*. The incidence is 1 in 6000–8000 births. It is more frequent in females, with a male-to-female ratio of 1:3–4. The risk for trisomy 18 also increases with maternal age.

Phenotype

The most common features of trisomy 18 syndrome include mental and growth deficiencies, neonatal hypotonicity followed by hypertonicity, craniofacial dysmorphism (prominent occiput, narrow bifrontal diameter, short palpebral fissures, small mouth, narrow palate, low-set malformed ears, micrognathia), clenched hands with a tendency for the second finger to overlap the third and the fifth finger to overlap the fourth, short dorsiflexed hallux, hypoplastic nails, rocker bottom feet, short sternum, hernias, single umbilical artery, small pelvis, cryptorchidism, hirsutism, and cardiac anomalies (mainly ventricular septal defect, auricular septal defect, and patent ductus arteriosus). Recent studies show that median survival averages approximately 51 days, with 1-week survival at 35–45% *(66–69)*. Fewer than 10% of patients survive beyond the first year of life. A few patients over 10 years of age, all females with one exception *(70)*, have been described *(71,72)*; however, the presence of a normal cell line in these patients was not always looked for.

Mosaic trisomy 18 patients have, in general, milder phenotypes. At least six mosaic trisomy 18 patients, again all females, with normal intelligence and long-term survival have been reported *(73–78)*.

Two recent molecular studies, performed on a total of 10 patients with partial trisomy 18, suggest that the region proximal to band 18q12 does not contribute to the syndrome, while two critical regions, one proximal (18q12.1→q21.2) and one distal (18q22.3→qter), may work in cooperation to produce the typical trisomy 18 phenotype *(79,80)*. In addition, severe mental retardation in these patients may be associated with trisomy of the region 18q12.3→q21.1.

Recurrence

There are not enough data regarding the recurrence risk for trisomy 18. Single case reports of trisomy 18 in sibs (e.g., *81)*, and of trisomy 18 and a different trisomy in sibs or in prior or subsequent abortuses (e.g., *82,83*) are recorded. For genetic counseling purposes, a risk figure of less than 1% for another pregnancy with any trisomy is generally used and might be appropriate.

Trisomy 13

Incidence

Trisomy 13 [47,XX or XY,+13] was first described by Patau et al. in 1960 *(84)*. The incidence is estimated to be 1 in 12,000 births. It is seen slightly more in females than in males. Again, the risk for trisomy 13 increases with maternal age.

Phenotype

The most prominent features of trisomy 13 syndrome include the holoprosencephaly spectrum, scalp defects, microcephaly with sloping forehead, large fontanels, capillary hemangioma (usually on the forehead), microphthalmia, cleft lip, cleft palate, abnormal helices, flexion of the fingers, polydactyly, hernias, single umbilical artery, cryptorchidism, bicornuate uterus, cardiac abnormalities in 80% of patients (mostly ventricular septal defect, patent ductus arteriosus, auricular septal defect), polycystic kidneys, increased polymorphonuclear projections of neutrophils, and persistence of fetal hemoglobin. Prognosis is extremely poor, with a median survival of 2.5 days and a 6-month survival of 5% *(68)*. Severe mental deficiencies, failure to thrive, and seizures are seen in those who survive. Mosaic trisomy 13 patients are, again, in general less severely affected; however, the degree is very variable and can be as severe as in nonmosaic trisomy 13 individuals.

Development of a karyotype–phenotype correlation by studying partial trisomies for different segments of chromosome 13 has also been attempted *(85,86)*. These studies were based on cytogenetic methods and suggested that the proximal segment (13pter→q14) contributes little to the trisomy 13 phenotype, whereas the distal segment (13q14→qter or part of) is responsible for the complete trisomy 13 features. Molecular studies for a more accurate delineation of the breakpoints have not been done.

Recurrence

No empirical recurrence risk data are available. The risk is likely to be very low. A less than 1% risk is generally quoted for genetic counseling purposes, as in trisomy 18.

Trisomy 8

Trisomy 8 [47,XX or XY,+8] was first reported by Grouchy et al. in 1971 *(87)*. It is rare, with an unknown incidence. More than 100 cases have been reported in the literature *(88,89)*, most of them mosaics [47,+8/46]. The male-to-female ratio is 2–3:1.

Growth and the degree of mental deficiency are variable. Mild to severe retardation is seen, while a proportion of patients have normal IQs. Craniofacial dysmorphism (**Fig. 4**) includes prominent forehead, deep-set eyes, strabismus, broad nasal bridge, upturned nares, long upper lip, thick and everted lower lip, high arched or cleft palate, micrognathia, and large dysplastic ears with prominent antihelices. Skeletal abnormalities include a long, thin trunk, hemivertebrae, spina bifida, kyphoscoliosis, hip dysplasia, multiple joint contractures, camptodactyly, dysplastic nails, and absent or dysplastic patella. The presence of deep palmar and plantar furrows is characteristic. Renal and ureteral anomalies and congenital heart defects are common. A few cases of hematologic malignancy have been reported in mosaic trisomy 8 patients *(90,91)*. This is of particular interest because trisomy 8 is a frequently acquired cytogenetic abnormality in myeloid disorders. When studied, the abnormal cells in these patients appeared to have developed from the trisomic cell population. The significance of this is not clear, but the possibility remains that constitutional trisomy 8 may predispose individuals to myeloid neoplasia.

There is no direct correlation between the proportion of the trisomy 8 cells and the severity of the phenotype. The percentage of trisomic cells is usually greater in skin fibroblasts than in blood lymphocytes. In addition, the proportion in lymphocytes usually decreases with time.

The risk for recurrence is not known.

Trisomy 9

The first cases of trisomy 9 in either nonmosaic [47,XX or XY,+9] or mosaic [47,+9/46] form were reported in 1973 *(92,93)*. More than 40 cases of liveborns or term stillborns with trisomy 9 have been reported. Most were mosaics *(94–97)*. The male-to-female ratio is close to 1:1.

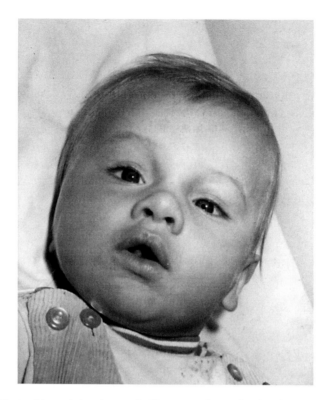

Fig. 4. An infant with mosaic trisomy 8. Note prominent forehead, strabismus, broad nasal bridge, upturned nares, long upper lip, and everted lower lip.

Clinical features include craniofacial anomalies (high narrow forehead, short upward slanting palpebral fissures, deep-set eyes, microphthalmia, low-set malformed auricles, bulbous nose, prominent upper lip, micrognathia), skeletal malformations (abnormal position/function of various joints, bone dysplasia, narrow chest, 13 ribs), overlapping fingers, hypoplastic external genitalia, and cryptorchidism. Cardiac anomalies are seen in more than 60% of cases, most frequently ventricular septal defects. Renal malformations are present in 40% of patients. The majority of patients die in the early postnatal period. With rare exceptions, all survivors have severe mental deficiency. Mosaic patients tend to survive longer, but the proportion of trisomy 9 cells does not predict the severity of the condition or the length of survival. It is possible that a normal cell line could be present in some tissues in apparently nonmosaic patients.

A recent study showed that the mean maternal age of women bearing trisomy 9 off-spring is significantly increased over that of the general population (*95*). This suggests that the occurrence of trisomy 9 may also be associated with advanced maternal age. The risk for recurrence is not known.

Trisomy 16

Trisomy 16 is the most frequently observed autosomal aneuploidy in spontaneous abortuses (see Chapter 12). Full trisomy 16 is almost always lethal during early embryonic or fetal development, although a single case of a stillborn at 35 weeks gestation has been recorded *(98)*.

Mosaic trisomy 16 individuals, however, may occasionally survive to term. At least 10 such cases have been reported *(99–104,* Hsu et al., *Am J Med Genet,* in press). Intrauterine growth retardation is invariable. A high maternal serum α-fetoprotein level during pregnancy was noted in more than 50% of cases, and elevated serum human chorionic gonadotropin (hCG) has also been observed. Congenital cardiac defects (mainly ventricular/atrial septal defects) were present in 60% of patients. Other clinical findings included postnatal growth retardation, mild developmental/speech delay, craniofacial asymmetry, ptosis, flat broad nasal bridge, low-set dysplastic ears, hypoplastic nipples, umbilical hernia, deep sacral dimple, scoliosis, nail hypoplasia, and single transverse palmar crease. Approximately 50% of the patients died within the first year of life. Long-term follow-up is not available; however, survival to more than 5 years to date has been observed (Hajianpour and Wang, personal observation).

The risk for recurrence is probably negligible.

Trisomy 20

Although mosaic trisomy 20 is one of the most frequent autosomal aneuploidies detected prenatally, its occurrence in liveborns is very rare *(105)*. The majority of prenatally diagnosed cases are not cytogenetically confirmed in postatal life. It appears that in conceptuses capable of surviving to the second trimester, trisomy 20 cells are largely confined to extraembryonic tissues. Only four liveborns with documented mosaic trisomy 20 have been reported and all were phenotypically normal at birth *(106–109)*. Long-term follow-up of these patients is not available. No case of liveborn nonmosaic trisomy 20 has been recorded.

Phenotypic abnormalities in abortuses with cytogenetically confirmed mosaic trisomy 20 include microcephaly, facial dysmorphism, cardiac defects, and urinary tract anomalies (megapelvis, kinky ureters, double fused kidney) *(110)*.

Trisomy 20 cells have been found in various fetal tissues including kidney, lung, esophagus, small bowel, rectum, thigh, rib, fascia, and skin *(105,110,111)*. Postnatally, they have been detected in cultured foreskin fibroblasts and urine sediments *(106–109)*. The detection of trisomy 20 cells in newborn cord blood has been reported in one case, but subsequent study of peripheral blood at 4 months of age produced only cytogenetically normal cells *(109)*. There are no other reports of trisomy 20 cells in postnatal blood cultures.

The risk for recurrence is probably negligible.

Trisomy 22

Trisomy 22 was first reported in 1971 *(112)*. Since then, more than 20 liveborns have been reported in the literature (reviewed in *ref. 113, 114–116)*. Although most cases were apparently nonmosaic full trisomies, the presence of an undetected, normal cell line confined to certain tissues cannot be excluded, as pointed out by Robinson and Kalousek *(117)*.

The most consistent phenotypic abnormalities include intrauterine growth retardation, low-set ears (frequently associated with microtia of varying degrees plus tags/pits), and midfacial hypoplasia. Other frequently seen abnormalities are microcephaly, hypertelorism with epicanthal folds, cleft palate, micrognathia, webbed neck, hypoplastic nails, anal atresia/stenosis and hypoplastic genitalia. Cardiac defects, complex in some cases, are seen in 80% of patients. Renal hypoplasia/dysplasia are also common. Most patients die in the first months of life. The longest survival reported is 3 years *(118)*. That patient had severe growth and developmental delay and died a few days before his third birthday.

Trisomy 22 cells can be detected in both blood lymphocytes and skin fibroblasts. In the mosaic case reported, 28% of lymphocytes and 100% of fibroblasts were trisomic for chromosome 22 *(115)*.

The risk for recurrence is unknown.

Other Rare Autosomal Trisomies

As noted in the introduction, mosaic or nomosaic autosomal trisomies for chromosomes other than 1, 6, and 11 have been reported in liveborns. Trisomies are detected much more frequently in spontaneous abortuses or in prenatal diagnostic specimens, following which elective terminations are often performed. Thus the occurrence of such trisomies in liveborns is extremely rare and only isolated case reports are available. The risks for recurrence for these rare trisomies are probably negligible.

A single case of possible mosaic trisomy 2, detected at amniocentesis and observed in a single cell of a foreskin fibroblast culture following the birth of a dysmorphic child, was reported *(119)*. Three cases of mosaic trisomy 3 have been reported *(10,120,121);* one of these, a mentally retarded woman, was alive at age 32. One case each of mosaic trisomy 4 *(122)* and mosaic trisomy 5 *(123)* have been reported. In both cases, the trisomic cells were detected in prenatal amniocytes and confirmed postnatally in skin fibroblasts, but not in blood lymphocytes. Both patients had multiple congenital anomalies. A few cases of trisomy 7, two of them reportedly nonmosaic, have been recorded *(124–127)*. All patients were phenotypically abnormal. A few cases of mosaic trisomy 10 have been reported *(reviewed in ref. 128)*. The clinical phenotype includes growth failure, craniofacial dysmorphism, deep palmar and plantar fissures, cardiac defects, and short survival.

At least five cases of trisomy 12 have been reported in liveborns; all were mosaics *(129–133)*. The earliest reported case was that of an infertile man, whereas the most recent case involved a neonatal death with multiple malformations. Trisomy 12 cells have been found in lymphocytes, skin fibroblasts, and urine sediments. In one case, interphase FISH performed on paraffin-embedded spleen tissue revealed three chromosome 12-specific hybridization signals in 5% of the cells examined *(133)*. However, it was not clear whether the control probe used in that study was actually included in the same reaction as an internal control.

At least ten cases of mosaic trisomy 14 have been reported *(reviewed in ref. 134)*. The most consistent phenotypic abnormalities were growth and mental retardation, broad nose, low-set dysplastic ears, micrognathia, congenital heart defects, and micropenis/cryptorchidism in males. Survival varied from days to more than 29 years. Trisomy 14 cells were detected in both lymphocytes and fibroblasts, with a generally higher percentage in lymphocytes. There was no clear correlation between the proportion of trisomic cells and the severity of the phenotype. In patients with body asymmetry, trisomic cells were usually limited to the atrophic side.

A total of seven cases of trisomy 15 have been recorded *(reviewed in refs. 135 and 136)*; two of them were reportedly nonmosaics *(137,138)*. Phenotypic abnormalities include hypotonia, various craniofacial dysmorphisms, minor skeletal anomalies, and congenital heart defects.

A single case of mosaic trisomy 17 was reported recently *(139)*. The trisomic cells were not seen in lymphocytes, but were found in high percentage in skin fibroblasts. Two cases of mosaic trisomy 19 are in the literature, one of them a stillborn male *(140,141)*.

AUTOSOMAL MONOSOMIES

As noted in the introduction, autosomal monosomies are extremely rare in either liveborns or abortuses, reflecting the severity of the genetic imbalance resulting from the loss of an entire chromosome. The only monosomies that have been reported are monosomy 21, mosaic monosomy 22, and a single case of a possible mosaic monosomy 20 *(142)*.

Monosomy 21

Mosaic monosomy 21 has been reported in four liveborns in the early literature *(143–146)*. The most prominent features include intrauterine growth retardation, postnatal growth and mental retardation, hypertonia, facial dysmorphism with downward slanting palpebral fissures, large low-set ears, and micrognathia. A more recent report described pathologic findings of an electively terminated 20-week female fetus after mosaic monosomy 21 was diagnosed by repeated amniocenteses *(147)*. The facial abnormalities described previously were present in this abortus. In addition, a complex cardiac malformation, malrotation of the bowel, uterus didelphys, small dysmature ovaries, and focal cystic dysplasia of the lung were noted.

Approximately 10 cases of nonmosaic monosomy 21 have been reported in liveborns *(reviewed in ref. 148, 149, 150)*. Some of these cases may actually represent partial monosomy 21 resulting from an undetected subtle translocation, explaining the observation that mosaic monosomy 21 is less commonly observed than apparently nonmosaic monosomy 21 and indicating that complete monosomy 21 is almost always incompatible with life. Most patients died before 2 years of age, although one male child survived to 11 years. The phenotypic features were quite like those observed in the mosaics, and included intrauterine growth retardation, postnatal growth and mental deficiencies, microcephaly, hypertelorism with downward slanting palpebral fissures, large low-set ears, prominent nose, cleft lip/palate, and micrognathia. Abnormal muscle tone, mostly hypertonia, was common. Cardiac anomalies were present in a few cases.

Monosomy 22

Three cases of mosaic monosomy 22 in liveborns have been reported *(151–153)*. All three were male. One was a 34-week premature birth with gastroschisis who died from intracranial hemorrhage shortly after birth. No dysmorphic features were noted, and autopsy was not performed *(153)*. The other two patients had growth and developmental deficiencies, microcephaly, and mild facial dysmorphism.

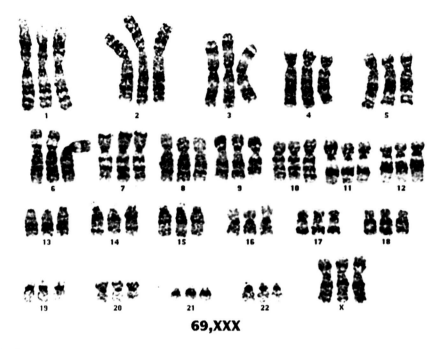

Fig. 5. Karyotype of a triploid fetus (69,XXX).

POLYPLOIDY

Polyploidies are numerical chromosome abnormalities with changes in the number of complete sets of chromosomes. They are usually incompatible with fetal survival and are extremely rare in liveborns.

Triploidy

The chromosome number in triploidy is $3n = 69$ (**Fig. 5**). It is estimated to occur in approximately 1% of all human conceptions and is found in 17–18% of all chromo-somally abnormal abortuses *(154,155)*. Only very rarely do triploid conceptuses survive to term. Two distinct phenotypes have been recognized *(156)*. One type presents as a relatively well-grown fetus with or without microcephaly, and an abnormally large and cystic placenta usually classified as partial hydatidiform moles. The parental origin of the extral haploid set of chromosomes in such cases is determined to be paternal (diandry) by analysis of cytogenetic heteromorphisms *(156,157)* or DNA polymor-phisms *(158)*. Diandry results from the fertilization of a normal ovum with either two sperm (dispermy) or a sperm that has a diploid chromosome complement resulting from a failure of meiotic division. The other type is characterized by severe intrauter-ine growth retardation with relative macrocephaly, and a small and noncystic placenta. The extra haploid set of chromosomes in such cases is maternal (digyny) *(156–159)*. Digyny can result from a failure of the first maternal meiotic division, generating a diploid egg, or from retention of the second polar body. Although the occurrence of triploidy does not appear to be associated with maternal age, digyny may play a major role in the generation of triploidy in the advanced maternal age group *(155)*. Earlier studies indicated that the majority of triploid conceptuses were diandric partial moles *(157,160)*. More recent studies based on DNA polymorphisms have suggested that a maternal contribution to triploidy may occur more frequently than was previously realized *(158)*. The sex chromosome complement in triploidy is either XXX or XXY, with XYY occurring only rarely. For example, the reported numbers of XXX:XXY:XYY cases in two studies performed on spontaneous abortuses were 82:92:2 *(3)* and 26:36:1 *(155)*, and in one study performed on amniotic fluid cells this ratio was 6:8:0 *(161)*. It has been suggested that 69,XYY triploid conceptuses are incompatible with significant embryonic development *(3)*.

The observation that the phenotype of triploidy depends on the parental origin of the extra set of chromosomes is an example of genomic imprinting, or the differential expression of paternally and maternally derived genetic material *(162,163)*. It correlates well with observations obtained from mouse embryo studies using nuclear transplantation techniques, which demonstrated that maternal and paternal genomes function differently and are both required for normal development *(164–166)* (see Chapter 16).

More than 50 cases of apparently nonmosaic triploidy, either 69,XXX or XXY, have been reported in liveborns. Most patients died shortly after birth. Eight patients with survival longer than 2 months have been reported *(reviewed in ref. 167)*, with the longest being $10^{1}/_{2}$ months *(168)*. In one of these, the origin of the extra set of chromosomes was determined to be paternal based on cytogenetic heteromorphisms only *(167)*. The most frequent phenotypic abnormalities include intrauterine growth retardation, hypotonia, craniofacial anomalies (macro/hydrocephalus, low-set dysplastic ears, broad nasal bridge), syndactyly, malformation of the extremities, adrenal hypoplasia, cardiac defects, and brain anomalies.

Mosaic triploidy (diploid/triploid mixoploidy) has been reported in approximately 20 patients. Triploid cells were found in both lymphocytes and fibroblasts, although in a number of cases the triploid cell line was limited to fibroblasts *(169)*. Patients with such mixoploidy are less severely affected than nonmosaics, and survival beyond 10 years has been observed. Usual clinical features include intrauterine retardation, psychomotor retardation, asymmetric growth, broad nasal bridge, syndactyly, genital anomalies, and irregular skin pigmentation *(170)*. Truncal obesity was seen in some patients *(171)*.

Mitotic nondisjunction cannot readily explain the occurrence of diploid and triploid cell lines in the same individual. One possible mechanism is double fertilization of an ovum by two sperm. Cytogenetic evidence for such a mechanism has been reported in at least one case *(172)*.

Tetraploidy

The chromosome number in tetraploidy is $4n = 92$. It is rarer than triploidy in spontaneous abortuses, seen in approximately 6–7% of such specimens with chromosome abnormalities *(154,155)*. Tetraploid conceptuses usually abort spontaneously early in gestation and only rarely do they survive to term. A probable origin of tetraploidy is chromosome duplication in the zygote resulting from a failure of cytoplasmic cleavage during first division. Other theoretically possible mechanisms require the occurrence of two independent, rare events and are thus highly unlikely.

At least eight apparently nonmosaic tetraploid liveborns have been reported *(reviewed in ref. 173)*. The sex chromosome complement was either XXXX or XXYY. No 92,XYYY or XXXY conceptuses have been reported. The most frequent abnormalities were growth and developmental delay, hypotonia, craniofacial anomalies (short palpebral fissures, low-set malformed ears, high arched/cleft palate, micrognathia), and contracture/structural abnormalities of the limbs, hands, and feet. Cardiac defects were present in three cases. Urinary tract abnormalities, such as hypoplastic kidneys, have also been recorded. Most patients died before 1 year of age. One girl had survived to 22 months at the time of report *(174)*.

Mosaic tetraploidy (diploid/tetraploid mixoploidy) has been reported in 12 liveborns *(reviewed in ref. 175,176)*. Tetraploid cells were seen in peripheral blood lymphocytes, skin fibroblasts, and bone marrow cells. In one severely malformed patient who died at 2 days of age, tetraploid cells were found in 95% of bone marrow cells *(177)*. In two females, aged 11 and 21 years, with severe intellectual handicaps and skin pigmentary dysplasia, tetraploid cells were found only in skin fibroblasts *(176)*; a decreasing proportion (with age) of tetraploid cells found in lymphocytes has been observed *(178)*. Overall, clinical features are similar to but less severe than those in nonmosaic tetraploidy patients. In addition to the longer term survivals mentioned previously *(176)*, survivals to 6 years at the time of reporting have also been recorded *(179)*.

PARTIAL AUTOSOMAL ANEUPLOIDIES

Partial duplication/deletion as a result of structural rearrangement is discussed in Chapter 9. Only those partial autosomal aneuploidies that result from the presence of a supernumerary chromosome are presented in this chapter.

Tetrasomy 8p

Tetrasomy 8p [47,XX or XY,+i(8)(p10)] usually results from the presence of a supernumerary isochromosome for the entire short arm of chromosome 8. All cases reported are mosaics, with a normal diploid cell line as well. The abnormal cell line was found in lymphocytes and skin fibroblasts. In some case, the origin of the abnormal isochromosome was confirmed by molecular cytogenetic (FISH) studies *(180–182)*. At least 10 cases have been reported (reviewed in ref. 182). A few patients died before the first year of life, but survival beyond 5 years was not uncommon. Weight and head circumference were normal at birth. The most frequently observed phenotypic features include mental retardation, speech and motor delay, dilatation of cerebral ventricles, mild facial dysmorphism (depressed nasal bridge, short nose, upturned nares, low-set and posteriorly rotated ears), and vertebral abnormalities. Agenesis of the corpus callosum was noted in five patients and cardiac defects in four. Deep palmar and plantar creases have also been reported. The phenotype resembles, to some degree, that of mosaic trisomy 8.

Tetrasomy 9p

Tetrasomy 9p [47,XX or XY,+i(9)(p10)], resulting from the presence of a supernumerary isochromosome, has been reported in at least 20 liveborns (reviewed in *ref. 183, 184, 185)*. The isochromosome consists of either the entire short arm of chromosome 9 as described previously, the entire short arm and part of the heterochromatic

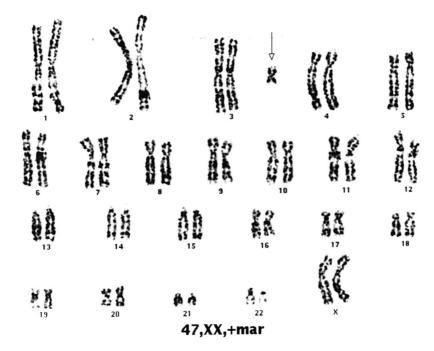

47,XX,+mar

Fig. 6. Tetrasomy 12p female karyotype. The marker chromosome was confirmed to be an i(12p) by FISH using a chromosome 12-specific α-satellite DNA probe. See Chapter 15, **Fig. 10**.

region of the long arm, or the entire short arm and part of the long arm extending to the euchromatic region. No consistent phenotypic differences have been observed among the three types. Both mosaic and apparently nonmosaic patients have been reported. The tetrasomy 9p cells were seen in both lymphocytes and skin fibroblasts. In contrast to tetrasomy 12p (described later), the 9p isochromosomes were present only in lymphocytes in five patients *(183,184,186–188)* and in fibroblasts at a much lower percentage than in lymphocytes in two patients *(189,190)*. The mechanism for this observed tissue-limited mosaicism for different chromosomes is not clear.

Survival is variable ranging from a few hours to beyond 10 years. The most frequent phenotypic abnormalities include low birthweight, growth and developmental delay, craniofacial anomalies (micophthalmia, low-set malformed ears, bulbous tip of the nose, cleft lip/palate, micrognathia), short neck, skeletal anomalies, joint contracture, nail hypoplasia, and urogenital anomalies. Cardiac defects are present in more than 50% of patients. Overall, nonmosaic patients are more severely affected. One patient who had the i(9p) present in 75% of lymphocytes but not in skin fibroblasts had only mild developmental delay and minor anomalies *(183)*.

Tetrasomy 12p

Tetrasomy 12p (Pallister–Killian syndrome) results from the presence of a supernumerary isochromosome for the entire short arm of chromosome 12 [i(12)(p10) or i(12p)] (**Fig. 6**). The syndrome was first described in 1977 by Pallister et al. in two adults, a 37-year-old man and a 19-year-old woman *(191)*. In 1981, Killian and Teschler-Nicola reported a 3-year-old girl with similar clinical manifestations *(192)*. Subsequently, more than 60 cases have been reported (reviewed in *ref. 193,194)*, and many more have been observed but not reported in the literature. All cases were mosaics, with a normal cell line in addition to cells containing i(12p). Maternal age for reported cases has been shown to be significantly higher than that for the general population *(195)*. This observation has been taken to suggest that the isochromosome arises from a meiotic error, and that the normal cell line results from subsequent loss of the i(12p) from some cells. Tissue specificity and both the *in vivo* and *in vitro* age-dependencies of the i(12p) have been well demonstrated *(reviewed in ref. 196)*. The i(12p) is found in a high percentage of skin fibroblasts and amniocytes, but is rarely seen in blood lymphocytes. The percentage of cells containing the isochromosome also decreases with age. The presence of tetrasomy 12p in 100% of bone marrow cells has been reported in at least two newborn infants *(197,198)*, and in only 6% of marrow cells in a $3\frac{1}{2}$ year-old child *(199)*. In lymphocytes, it has been found in fetal blood *(196,200)*, but has never been seen beyond childhood. In a case reported by Ward et al., the i(12p) was present in 10% of lymphocytes initially, but was not seen in these cells when the patient was 2 months old *(197)*. The isochromosome is more stable in skin fibroblasts and can be found in adults, usually at lower percentage than in younger patients. When fibroblast cultures were examined, the percentage of cells containing the isochromosome decreased with increasing numbers of cell passages *(196,197,199,201)*. One study using FISH showed that in lymphocytes, the i(12p) was present in a significantly higher proportion of interphase nuclei than in metaphase cells *(202)*. These authors proposed that lymphocytes containing i(12p) may fail to divide upon PHA stimulation. These observations suggest that tissue-limited mosaicism in Pallister–Killian syndrome may result from differential selection against cells containing i(12p) in different tissues, and that the selection can occur both *in vivo* and *in vitro*.

Many patients die shortly after birth, but survival to adulthood is possible. Clinically, a distinct pattern of anomalies is observed in these patients. Growth parameters at birth are usually normal. Profound hypotonia is present in the newborn period, while contractures develop later in life. Sparse scalp hair, especially bitemporally, is observed in infancy, with coarsening of facial features over time. Craniofacial dysmorphism includes prominent forehead, large malformed ears, hypertelorism, epicanthal folds, broad flat nasal bridge, short nose, upturned nares, long philtrum, thin upper lip, and high arched palate. Most patients have a generalized pigmentary dysplasia with areas of hyper- and hypopigmentation. Other abnormalities include short neck, macroglossia, micrognathia progressing to prognathia, accessory nipples, umbilical and inguinal hernias, and urogenital abnormalities. Severe mental retardation and seizure are seen in those who survive.

All cases are sporadic. The recurrence risk is probably negligible.

Tetrasomy 18p

Tetrasomy 18p [47,XX or XY,+i(18)(p10)] results from the presence of a supernumerary isochromosome for the entire short arm of chromosome 18. The syndrome was first described by Froland et al. in 1963 *(203)*, although identification of the marker as an i(18p) was not made until after the introduction of banding techniques in 1970.

Confirmation of the origin of the marker has been possible in recent years by FISH studies. Of interest is the finding of a loss of approximately 80% of chromosome 18 α-satellite DNA in the i(18p) in one case (204).

At least 50 cases have been reported *(205–208)*. Most are nonmosaics. The i(18p) is usually readily detectable in lymphocytes. Its presence in amniocytes *(209)* and cultured chorionic villus cells *(204)* has also been reported.

The most frequent clinical features include low birthweight, microcephaly, feeding problems, various degrees of psychomotor retardation, spasticity, seizures, craniofacial characteristics (oval shaped face, arched eyebrows, strabismus, low-set dysplastic ears, small pinched nose, small triangular mouth, high arched palate, micrognathia), narrow shoulders and thorax, small iliac wings, scoliosis, camptodactyly, and simian creases. Cardiac defects including ASD, VSD, and PDA have been observed in some cases. Urogenital anomalies including horseshoe kidneys, double ureter, and cryptorchidism have occasionally been seen.

It is not clear whether patients with tetrasomy 18p are born to mothers of increased age. Most of the reported cases are sporadic. The presence of i(18p) in maternal lymphocytes has been reported in three families. In two families, the mothers had an abnormal chromosome 18 with deletion of the short arm and a supernumerary i(18p), and thus were trisomic for 18p. The offspring inherited the normal chromosome 18 and the i(18p), and were therefore tetrasomic for 18p *(210,211)*. In the third family, the mother had low-level mosaicism for a supernumerary i(18p) and was mildly affected clinically. The child apparently had nonmosaic tetrasomy 18p and had the full clinical presentation of the syndrome *(212)*. Recurrence risk in such families will be high.

Other Partial Autosomal Aneuploidies

Supernumerary Marker Chromosomes

In addition to the tetrasomies described previously, partial autosomal aneuploidies can result from the presence of small supernumerary marker chromosomes of cytogenetically indeterminate origin. The frequency of such markers is approximately 0.7 per 1000 in newborns, based on a recent study *(213)*, and 0.8–1.5 per 1000 in prenatal specimens *(214–216)*. Because their cytogenetic origins are not initially known, these markers may or may not represent autosomal aneuploidy. Identification of such markers is now typically achieved using FISH, and is covered in Chapter 15.

These supernumerary markers are often classified as satellited or nonsatellited, and are frequently present in mosaic form. They are a heterogeneous group and the clinical significance of a marker depends on its origin and characteristics. Markers that contain only heterochromatin and/or the short arms of acrocentric chromosomes are generally of no phenotypic consequence. On the other hand, markers that contain euchromatin are generally not benign and can result in phenotypic abnormalities. Among these are the dicentric bisatellited markers which contain variable amounts of long arm euchromatin of an acrocentric chromosome.

Markers derived from all autosomes except chromosomes 5, 10, and 11 have been reported *(217–220)*. The most common marker is the so-called inverted duplication of chromosome 15, inv dup(15). This is an archaic misnomer that dates from an incorrect assessment of the mechanism of formation of such chromosomes, and represents a heterogeneous group of small markers consisting of two copies of the short arm of chromosome 15, with or without variable amounts of long arm material. They account for approximately 40% of all marker chromosomes *(220,221)*. The amount of the long arm euchromatin present in the marker dictates its phenotypic significance. A direct correlation has been observed between the presence of the Prader–Willi/Angelman syndrome regions (located at 15q11.2) on the marker and mental retardation or developmental delay *(222–224)*. Of particular interest is the observation of a few patients with this type of marker who present clinically with Prader–Willi syndrome *(223, 225–228)* or Angelman syndrome *(228,229)*. Molecular studies performed on some of these patients indicate that the abnormal phenotype results not from the presence of the marker, but from either uniparental disomy of the two normal chromosomes 15 *(223,228)* or a deletion of 15q11–q13 on one of the apparently cytogenetically normal 15s *(229)*.

Another type of marker chromosome that results in a clinically recognizable multiple congenital anomaly syndrome is the supernumerary bisatellited dicentric marker derived from chromosome 22. This marker contains two copies of a small segment of proximal long arm euchromatin (22q11.2), thus resulting in tetrasomy for 22q11.2. Clinically, these patients usually present with cat-eye syndrome *(230–232)*. Characteristic features include craniofacial anomalies (vertical coloboma of the iris, which gives the syndrome its name; coloboma of the choroid or optic nerve; preauricular skin tags/pits; down-slanting palpebral fissures) and anal atresia with rectovestibular fistula. Cardiac defects are present in more than one third of cases. Renal malformations include unilateral agenesis, unilateral or bilateral hypoplasia, or dysplasia. Other less frequent findings include microphthalmia, microtia, atresia of the external auditory canal, biliary atresia, and malrotation of the gut. Intelligence is usually low-normal to mildly deficient.

Other types of supernumerary markers, such as ring chromosomes derived from chromosome 22 resulting in either trisomy or tetrasomy for 22q11.2, can also cause various features of the cat-eye syndrome. The critical region of this syndrome has been shown to lie within a 2.1 Mb DNA segment defined distally by locus D22S57, and containing the ATP6E (the E subunit of vacuolar H-ATPase) gene *(233)*.

Clinically definable entities have not been observed for other markers, as each is typically unique. However, this may change as data concerning the composition of marker chromosomes accumulate through the use of FISH and other molecular technologies.

ACKNOWLEDGMENTS

I am grateful to Dr. Neil Lamb for sharing prepublication data on susceptible chiasmata configurations and maternal nondisjunction of chromosome 21, to Martha Keagle for preparation of the diagrams, to Dr. John Labavitch for reading this manuscript, and to Dr. Linda Randolph for providing the patient photograph. I also thank Jo Ann Rieger and Karen Rodriguez for their aid in preparation of the manuscript.

REFERENCES

1. Jacobs, P.A., Browne, C., Gregson, N., Joyce, C., and White, H. (1992) Estimates of the frequency of chromosome abnormalities detectable in unselected newborns using moderate levels of banding. *J. Med. Genet.* **29,** 103–108.

2. Hassold, T.J. and Takaesu, N. (1989) Analysis of non-disjunction in human trisomic spontaneous abortions. In *Molecular and Cytogenetic Studies of Non-Disjunction* (Hassold, T.J. and Epstein, C.J., eds.), Alan R. Liss, New York, pp. 115–134.

3. Warburton, D., Byrne, J., and Canki, N. (1991) *Chromosome Anomalies and Prenatal Development: An Atlas (Oxford Monographs on Medical Genetics No. 21).* Oxford University Press, New York.

4. Jacobs, P.A., Hassold, T.J., Henry, A., Pettay, D., and Takaesu, N. (1987) Trisomy 13 ascertained in a survey of spontaneous abortions. *J. Med. Genet.* **24,** 721–724.

5. Martin, R.H., Ko, E., and Rademaker, A. (1991) Distribution of aneuploidy in human gametes: comparison between human sperm and oöcytes. *Am. J. Med. Genet.* **39,** 321–331.

6. Guttenbach, M., Schakowski, R., and Schmid, M. (1994) Incidence of chromosome 3,7,10,11,17 and X disomy in mature human sperm nuclei as determined by nonradioactive *in situ* hybridization. *Hum. Genet.* **93,** 7–12.

7. Girardet, A., Coignet, L., Andreo, B., Lefort, G., Charlieu, J.P., and Pellestor, F. (1996) Aneuploidy detection in human sperm nuclei using PRINS technique. *Am. J. Med. Genet.* **64,** 488–492.

8. Pellestor, F., Girardet, A., Coignet, L., Andreo, B., and Charlieu, J.P. (1996) Assessment of aneuploidy for chromosomes 8, 9, 13, 16, and 21 in human sperm by using primed *in situ* labeling technique. *Am. J. Hum. Genet.* **58,** 797–802.

9. Spriggs, E.L., Rademaker, A.W., and Martin, R.H. (1996) Aneuploidy in human sperm: the use of multicolor FISH to test various theories of nondisjunction. *Am. J. Hum. Genet.* **58,** 356–362.

10. Kuhn, E.M., Sarto, G.E., Bates, B.G., and Therman, E. (1987) Gene-rich chromosome regions and autosomal trisomy. *Hum. Genet.* **77,** 214–220.

11. Hanna, J.S., Shires, P., and Matile, G. (1997) Trisomy 1 in a clinically recognized pregnancy. *Am. J. Med. Genet.* **68,** 98.

12. Watt, J.L., Templeton, A.A., Messinis, I., Bell, L., Cunningham, P., and Duncan, R.O. (1987) Trisomy 1 in an eight cell human pre-embryo. *J. Med. Genet.* **24,** 60–64.

13. Warren, A.C., Chakravarti, A., Wong, C., Slaugenhaupt, S.A., Halloran, S.L., Watkins, P.C., Metaxotou, C., and Antonarakis, S.E. (1987) Evidence for reduced recombination on the nondisjoined chromosome 21 in Down syndrome. *Science* **237,** 652–654.

14. Sherman, S.L., Takaesu, N., Freeman, S.B., Grantham, M., Phillips, C., Blackston, R.D., Jacobs, P.A., Cockwell, A.E., Freeman, V., Uchida, I., Mikkelsen, M., Kurnit, D.M., Buraczynska, M., Keats, B.J.B., and Hassold, T.J. (1991) Trisomy 21: association between reduced recombination and nondisjunction. *Am. J. Hum. Genet.* **49,** 608–620.

15. Antonarakis, S.E., Petersen, M.B., McInnis, M.G., Adelsberger, P.A., Schinzel, A.A., Binkert, F., Pangalos, C., Raoul, O., Slaugenhaupt, S.A., Hafez, M., Cohen, M.M., Roulson, D., Schwartz, S., Mikkelsen, M., Tranebjareg, L., Greenberg, F., Hoar, D.I., Rudd, N.L., Warren, A.C., Metaxotou, C., Bartsocas, C., and Chakravarti, A. (1992) The meiotic stage of nondisjunction in trisomy 21: determination by using DNA polymorphisms. *Am. J. Hum. Genet.* **50,** 544–550.

16. Robinson, W.P., Bernasconi, F., Mutirangura, A., Ledbetter, D.H., Langlois, S., Malcolm, S., Morris, M.A., and Schinzel, A.A. (1993) Nondisjunction of chromosome 15: origin and recombination. *Am. J. Hum. Genet.* **53,** 740–751.

17. Ya-gang, X., Robinson, W.P., Spiegel, R., Binkert, F., Ruefenacht, U., and Schinzel, A.A. (1993) Parental origin of the supernumerary chromosome in trisomy 18. *Clin. Genet.* **44,** 57–61.

18. Mutirangura, A., Greenberg, F., Butler, M.G., Malcolm, S., Nicholls, R.D., Chakravarti, A., and Ledbetter, D.H. (1993) Multiplex PCR of three dinucleotide repeats in the Prader–Willi/Angelman critical region (15q11-q13): molecular diagnosis and mechanism of uniparental disomy. *Hum. Mol. Genet.* **2,** 143–151.

19. Petersen, M.B., Antonarakis, S.E., Hassold, T.J., Freeman, S.B., Sherman, S.L., Avramopoulos, D., and Mikkelsen, M. (1993) Paternal nondisjunction in trisomy 21: excess of male patients. *Hum. Mol. Genet.* **2**, 1691–1695.

20. Zaragoza, M.V., Jacobs, P.A., James, R.S., Rogan, P., Sherman, S., and Hassold, T. (1994) Nondisjunction of human acrocentric chromosomes: studies of 432 trisomic fetuses and liveborns. *Hum. Genet.* **94**, 411–417.

21. Hassold, T., Merrill, M., Adkins, K., Freeman, S., and Sherman, S. (1995) Recombination and maternal age-dependent nondisjunction: molecular studies of trisomy 16. *Am. J. Hum. Genet.* **57**, 867–874.

22. Fisher, J.M., Harvey, J.F., Morton, N.E., and Jacobs, P.A. (1995) Trisomy 18: studies of the parent and cell division of origin and the effect of aberrant recombination on nondisjunction. *Am. J. Hum. Genet.* **56**, 669–675.

23. Zittergruen, M.M., Murray, J.C., Lauer, R.M., Burns, T.L., and Sheffield, V.C. (1995) Molecular analysis of nondisjunction in Down syndrome patients with and without atrioventricular septal defects. *Circulation.* **92**, 2803–2810.

24. Eggermann, T., Nothen, M.M., Eiben, B., Hofmann, D., Hinkel, K., Fimmers, R., and Schwanitz, G. (1996) Trisomy of human chromosome 18: molecular studies on parental origin and cell stage of nondisjunction. *Hum. Genet.* **97**, 218–223.

25. Yoon, P.W., Freeman, S.B., Sherman, S.L., Taft, L.F., Gu, Y., Pettay, D., Flanders, W.D., Khoury, M.J., and Hassold, T.J. (1996) Advanced maternal age and the risk of Down syndrome characterized by the meiotic stage of the chromosomal error: a population based study. *Am. J. Hum. Genet.* **58**, 628–633.

26. Penrose, L. (1933) The relative effects of paternal and maternal age in mongolism. *J. Genet.* **27**, 219–224.

27. Angell, R.R. (1997) First-meiotic-division nondisjunction in human oocytes. *Am. J. Hum. Genet.* **61**, 23–32.

27a. Angell, R.R., Xian, J., Keith, J., Ledger, W., and Baird, D.T. (1994) First meiotic division abnormalities in human oöcytes: mechanism of trisomy formation. *Cytogenet. Cell Genet.* **65**, 194–202.

28. Angell, R.R., Xian, J., and Keith, J. (1993) Chromosome anomalies in human oöcytes in relation to age. *Hum. Reprod.* **8**, 1047–1054.

29. Kumar, R.M. and Khuranna, A. (1995) The chromosome complement of human uncleaved oöcytes. *J. Obstet. Gynaecol.* **21**, 601–607.

30. Lim, A.S., Ho, A.T., and Tsakok, M.F. (1995) Chromosomes of oöcytes failing in-vitro fertilization. *Hum. Reprod.* **10**, 2570–2575.

31. Roberts, C.G. and O'Neill, C. (1995) Increase in the rate of diploidy with maternal age in unfertilized *in vitro* fertilization oöcytes. *Hum. Reprod.* **10**, 2139–2141.

32. Dailey, T., Dale, B., Cohen, J., and Munne, S. (1996) Association between nondisjunction and maternal age in meiosis-II human oöcytes. *Am. J. Hum. Genet.* **59**, 176–184.

33. Hook, E.B. (1983) Down syndrome rates and relaxed selection at older maternal ages. *Am. J. Hum. Genet.* **35**, 1307–1313.

34. Stein, Z., Stein, W., and Susser, M. (1986) Attrition of trisomies as a screening device: an explanation for the association of trisomy 21 with maternal age. *Lancet* **ii**, 944–946.

35. Warburton, D. (1989) The effect of maternal age on the frequency of trisomy: change in meiosis or in utero selection? In *Molecular and Cytogenetic Studies of Non-Disjunction* (Hassold T.J. and Epstein, C.J., eds.), Alan R. Liss, New York, pp. 165–181.

36. Kratzer, P.G., Golbus, M.S., Schonberg, S.A., Heilbron, D.C., and Taylor, R.N. (1992) Cytogenetic existence for enhanced selective miscarriage of trisomy 21 pregnancies with advancing maternal age. *Am. J. Med. Genet.* **44**, 657–663.

37. Henderson, S.A. and Edwards, R.G. (1968) Chiasma frequency and maternal age in mammals. *Nature* **218**, 22–28.

38. Cheng, E.Y., Chen, Y.J., and Gartler, S.M. (1995) A cytological evaluation of the pro-

duction line hypothesis in human oögenesis using chromosome painting. *Am. J. Hum. Genet.* **57**, A51.

39. Sherman, S.L., Petersen, M.B., Freeman, S.B., Hersey, J., Pettay, D., Taft, L., Frantzen, M., Mikkelsen, M., and Hassold, T.J. (1994) Non-disjunction of chromosome 21 in maternal meiosis I: evidence for a maternal age-dependent mechanism involving reduced recombination. *Hum. Mol. Genet.* **3**, 1529–1535.

40. Lamb, N.E., Feingold, E., Hassold, T.J., and Sherman, S.L. (1996) Examination of the underlying pattern of chromosomal exchange in meiosis leading to trisomy 21: evidence for initiation of all maternal meiotic errors at meiosis I. *Am. J. Hum. Genet.* **59**, A12.

41. Lamb, N.E., Freeman, S.B., Savage-Austin, A., Pettay, D., Taft, L., Hersey, J., Gu, Y., Shen, J., Saker, D., May, K.M., Avramopoulos, D., Petersen, M.B., Hallberg, A., Mikkelsen, M., Hassold, T.J., and Sherman, S.L. (1996) Susceptible chiasmata configurations of chromosome 21 predispose to nondisjunction in both maternal meiosis I and meiosis II. *Nature Genet.*, **14**, 400–405.

42. Hawley, R.S., Frazier, J.A., and Rasooly, R. (1994) Separation anxiety: the etiology of nondisjunction in flies and people. *Hum. Mol. Genet.* **3**, 1521–1528.

43. Alfi, O.S., Chang, R., and Azen, S.P. (1980) Evidence for genetic control of nondisjunction in man. *Am. J. Hum. Genet.* **32**, 477–483.

44. Lejeune, J., Gautier, M., and Turpin, R. (1959) Les chromosomes humains en culture de tissus. *C. R. Acad. Sci.* **248**, 602–603.

45. Down, J.L.H. (1866) Observations on an ethnic classification of idiots. *Clin. Lect. Rep., Lond. Hosp.* **3**, 259.

46. Griffin, D.K., Abruzzo, M.A., Millie, E.A., Feingold, E., and Hassold, T.J. (1996) Sex ratio in normal and disomic sperm: evidence that the extra chromosome 21 preferentially segregates with the Y chromosome. *Am. J. Hum. Genet.* **59**, 1108–1113.

47. Cuckle, H.S., Wald, N.J., and Thompson, S.G. (1987) Estimating a woman's risk of having a pregnancy associated with Down's syndrome using her age and serum alphafetoprotein level. *Brit. J. Obstet. Gynaecol.* **94**, 387–402.

48. Hecht, C.A. and Hook, E.B. (1994) The imprecision in rates of Down syndrome by 1-year maternal age intervals: a critical analysis of rates used in biochemical screening. *Prenat Diag.* **14**, 729–738.

49. Grouchy, J. de and Turleau, C., eds. (1984) *Clinical Atlas of Human Chromosomes.* John Wiley & Sons, New York.

50. Jones, K.L., ed. (1997) *Smith's Recognizable Patterns of Human Malformation.* W.B. Saunders, Philadelphia.

51. Fong, C.T. and Brodeur, G.M. (1987) Down's syndrome and leukemia: epidemiology, genetics, cytogenetics and mechanisms of leukemogenesis. *Cancer Genet. Cytogenet.* **28**, 55–76.

52. Bhatt, S., Schreck, R., Graham, J.M., Korenberg, J.R., Hurvitz, C.G., and Fischel-Ghodsian, N. (1995) Transient leukemia with trisomy 21: description of a case and review of the literature. *Am. J. Med. Genet.* **58**, 310–314.

53. Pellissier, M.C., Laffage, M., Philip, N., Passage, E., Mattei, M.G., and Mattei, J.F. (1988) Trisomy 21q223 and Down's phenotype correlation evidenced by *in situ* hybridization. *Hum. Genet.* **80**, 277–281.

54. McCormick, M.K., Schinzel, A., Petersen, M.B., Stetten, G., Driscoll, D.J., Cantu, E.S., Tranebjaerg, L., Mikkelsen, M., Watkins, P.C., and Antonarakis, S.E. (1989) Molecular genetic approach to the characterization of the "Down syndrome region" of chromosome 21. *Genomics* **5**, 325–331.

55. Rahmani, Z., Blouin, J.L., Creau-Goldberg, N., Watkins, P.C., Mattei, J.F., Poissonnier, M., Prieur, M., Chettouh, Z., Nicole, A., Aurias, A., Sinet, P., and Delabar, J. (1989) Critical role of the D21S55 region on chromosome 21 in the pathogenesis of Down syndrome. *Proc. Natl. Acad. Sci. USA* **86**, 5958–5962.

56. Delabar, J.M., Theophile, D., Rahmani, Z., Chettouh, Z., Blouin, J.L., Prieur, M., Noel, B., and Sinet, P.M. (1993) Molecular mapping of twenty-four features of Down syndrome on chromosome 21. *Eur. J. Hum. Genet.* **1,** 114–124.

57. Korenberg, J.R., Kawashima, H., Pulst, S.M., Ikeuchi, T., Ogasawara, N., Yamamoto, K., Schonberg, S.A., West, R., Allen, L., Magenis, E., Ikawa, K., Taniguchi, N., and Epstein, C.J. (1990) Molecular definition of a region of chromosome 21 that causes features of the Down syndrome phenotype. *Am. J. Hum. Genet.* **47,** 236–246.

58. Korenberg, J.R., Chen, X.N., Schipper, R., Sun, Z., Gonsky, R., Gerwehr, S., Carpenter, N., Daumer, C., Dignan, P., Disteche, C., Graham, J.M., Hugdins, L., McGillivray, B., Miyazaki, K., Ogasawara, N., Park, J.P., Pagon, R., Pueschel, S., Sack, G., Say, B., Schuffenhauer, S., Soukup, S., and Yamanaka, T. (1994) Down syndrome phenotypes: the consequences of chromosomal imbalance. *Proc. Natl. Acad. Sci. USA* **91,** 4997–5001.

59. Gardner, R.J.M. and Sutherland, G.R., eds. (1996) *Chromosome Abnormalities and Genetic Counseling. (Oxford Monographs on Medical Genetics No. 29).* Oxford University Press, New York.

60. Pangalos, C.G., Talbot, Jr., C.C., Lewis, J.G., Adlesberger, P.A., Petersen, M.B., Serre, J., Rethore, M., Blois, M. de, Parent, P., Schinzel, A.A., Binkert, F., Boue, J., Corbin, E., Croquette, M.F., Gilgenkrantz, S., Grouchy, J. de, Bertheas, M.F., Prieur, M., Raoul, O., Serville, F., Siffroi, J.P., Thepot, F., Lejeune, J., and Antonarakis, S.E. (1992) DNA polymorphism analysis in families with recurrence of free trisomy 21. *Am. J. Hum. Genet.* **51,** 1015–1027.

61. Harris, D.J., Begleiter, M.L., Chamberlin, J., Hankins, L., and Magenis, R.E. (1982) Parental trisomy 21 mosaicism. *Am. J. Hum. Genet.* **34,** 125–133.

62. Nielsen, K.G., Poulsen, H., Mikkelsen, M., and Steuber, E. (1988) Multiple recurrence of trisomy 21 Down syndrome. *Hum. Genet.* **78,** 103–105.

63. Niikawa, N. and Kajii, T. (1984) The origin of mosaic Down syndrome: four cases with chromosome markers. *Am. J. Hum. Genet.* **36,** 123–130.

64. Pangalos, C., Avramopoulos, D., Blouin, J., Raoul, O., deBlois, M., Prieur, M., Schinzel, A.A., Gika, M., Abazis, D., and Antonarakis, S.E. (1994) Understanding the mechanism(s) of mosaic trisomy 21 by using DNA polymorphism analysis. *Am. J. Hum. Genet.* **54,** 473–481.

65. Edwards, J.H., Harnden, D.G., Cameron, A.H., Crosse, V.M., and Wolff, O.H. (1960) A new trisomic syndrome. *Lancet* **i,** 787–790.

66. Carter, P., Pearn, J., Bell, J., Martin, N., and Anderson, N. (1985) Survival in trisomy 18. *Clin. Genet.* **27,** 59–61.

67. Young, I., Cook, J., and Mehta, L. (1986) Changing demography of trisomy 18. *Arch. Dis. Child.* **61,** 1035–1036.

68. Goldstein, H. and Nielsen, K. (1988) Rates and survival of individuals with trisomy 13 and 18. *Clin. Genet.* **34,** 366–372.

69. Root, S. and Carey, J.C. (1994) Survival in trisomy 18. *Am. J. Med. Genet.* **49,** 170–174.

70. Geiser, C.F. and Chindlera, N. (1969) Long survival in a male with trisomy 18 and Wilms tumor. *Pediatr.* **44,** 111.

71. Van Dyke, D.C. and Allen, M. (1990) Clinical management considerations in long-term survivors with trisomy 18. *Pediatrics* **85,** 753–759.

72. Mehta, L., Shannon, R.S., Duckett, D.P., and Young, I.D. (1986) Trisomy 18 in a 13 year old girl. *J. Med. Genet.* **23,** 256–257.

73. Beratis, N.G., Kardon, N.B., Hsu, L.Y.F., Grossmann, D., and Hirschhorn, K. (1972) Prenatal mosaicism in trisomy 18. *Pediatrics* **50,** 908–911.

74. Kohn, G. and Shohat, M. (1987) Trisomy 18 mosaicism in an adult with normal intelligence. *Am. J. Med. Genet.* **26,** 929–931.

75. Gersdorf, E., Utermann, B., and Utermann, G. (1990) Trisomy 18 mosaicism in an adult woman with normal intelligence and history of miscarriage. *Hum. Genet.* **84,** 298–299.

76. Sarigol, S.S. and Rogers, D.G. (1994) Trisomy 18 mosaicism in a thirteen-year-old girl with normal intelligence, delayed pubertal development, and growth failure. *Am. J. Med. Genet.* **50,** 94–95.

77. Butler, M.G. (1994) Trisomy 18 mosaicism in a 24-year-old white woman with normal intelligence and skeletal abnormalities. *Am. J. Med. Genet.* **53,** 92–93.

78. Collins, A.L., Fisher, J., Crolla, J.A., and Cockwell, A.E. (1995) Further case of trisomy 18 mosaicism with a mild phenotype. *Am. J. Med. Genet.* **56,** 121–122.

79. Mewar, R., Kline, A.D., Harrison, W., Rojas, K., Greenberg, F., and Overhauser, J. (1993) Clinical and molecular evaluation of four patients with partial duplications of the long arm of chromosome 18. *Am. J. Hum. Genet.* **53,** 1269–1278.

80. Boghosian-Sell, L., Mewar, R., Harrison, W., Shapiro, R.M., Zackai, E.H., Carey, J., Davis-Keppen, L., Hudgins, L., and Overhauser, J. (1994) Molecular mapping of the Edwards syndrome phenotype to two noncontiguous regions on chromosome 18. *Am. J. Hum. Genet.* **55,** 476–483.

81. FitzPatrick, D.R. and Boyd, E. (1989) Recurrences of trisomy 18 and trisomy 13 after trisomy 21. *Hum. Genet.* **82,** 301.

82. Stene, J. and Stene, E. (1984) Risk for chromosome abnormality at amniocentesis following a child with a non-inherited chromosome aberration. *Prenat. Diag.* **4,** 81–95.

83. Baty, B.J., Blackburn, B.L., and Carey, J.C. (1994) Natural history of trisomy 18 and trisomy 13:I. growth, physical assessment, medical histories, survival, and recurrence risk. *Am. J. Med. Genet.* **49,** 175–188.

84. Patau, K., Smith, D.W., Therman, E., Inhorn, S.L., and Wagner, H.P. (1960) Multiple congenital anomaly caused by an extra autosome. *Lancet* **i,** 790–793.

85. Tharapel, S.A., Lewandowski, R.C., Tharapel, A.T., and Wilroy, Jr., R.S. (1986) Phenotype–karyotype correlation in patients trisomic for various segments of chromosome 13. *J. Med. Genet.* **23,** 310–315.

86. Helali, N., Iafolla, A.K., Kahler, S.G., and Qumsiyeh, M.B. (1996) A case of duplication of 13q32-qter and deletion of 18p11.32-pter with mild phenotype: Patau syndrome and duplications of 13q revisited. *J. Med. Genet.* **33,** 600–602.

87. Grouchy, J. de, Turleau, C., and Leonard, C. (1971) Étude en fluorescence d'une trisomie C mosaique probablement 8: 46,XY/47,XY,?8+. *Ann. Genet.* **14,** 69–72.

88. Riccardi, V.M. (1977) Trisomy 8: an international study of 70 patients. *Birth Defects* **XIII,** 171.

89. Schinzel, A. (1983) Trisomy 8 mosaicism. In *Catalogue of Unbalanced Chromosome Aberrations in Man.* Walter de Gruyter, Berlin, pp. 325–328.

90. Grafter, U., Shabtai, F., Kahn, Y., Halbrecht, I., and Djaldetti, M. (1976) Aplastic anemia followed by leukemia in congenital trisomy 8 mosaicism. Ultrastructural studies of polymorphonuclear cells in peripheral blood. *Clin. Genet.* **9,** 134–142.

91. Hasle, H., Clausen, N., Pedersen, B., and Bendix-Hansen, K. (1995) Myelodysplastic syndrome in a child with constitutional trisomy 8 mosaicism and normal phenotype. *Cancer Genet. Cytogenet.* **79,** 79–81.

92. Feingold, M. and Atkins, L. (1973) A case of trisomy 9. *J. Med. Genet.* **10,** 184–187.

93. Haslam, R.H.A., Broske, S.P., Moore, C.M., Thomas, G.H., and Neill, C.A. (1973) Trisomy 9 mosaicism with multiple congenital anomalies. *J. Med. Genet.* **10,** 180–184.

94. Arnold, G.L., Kirby, R.S., Stern, T.P., and Sawyer, J.R. (1995) Trisomy 9: review and report of two new cases. *Am. J. Med. Genet.* **56,** 252–257.

95. Wooldridge, J. and Zunich, J. (1995) Trisomy 9 syndrome: report of a case with Crohn disease and review of the literature. *Am. J. Med. Genet.* **56,** 258–264.

96. Stoll, C., Chognot, D., Halb, A., and Luckel, J.C. (1993) Trisomy 9 mosaicism in two girls with multiple congenital malformations and mental retardation. *J. Med. Genet.* **30,** 433–435.

97. Cantu, E., Eicher, D.J., Pai, G.S., Donahue, C.J., and Harley, R.A. (1996) Mosaic vs.

nonmosaic trisomy 9: report of a liveborn infant evaluated by fluorescence in situ hybridization and review of the literature. *Am. J. Med. Genet.* **62,** 330–335.

98. Yancey, M.K., Hardin, E.L., Pacheco, C., Kuslich, C.D., and Donlon, T.A. (1996) Nonmosaic trisomy 16 in a third trimester fetus. *Obstet. Gynecol.* **87,** 856–860.

99. Gilbertson, N.J., Taylor, J.W., and Kovar, I.Z. (1990) Mosaic trisomy 16 in a live newborn infant. *Arch. Dis. Child.* **65,** 388–389.

100. Devi, A.S., Velinov, M., Kamath, M.V., Eisenfeld, L., Neu, R., Ciarleglio, L., Greenstein, R., and Benn, P. (1993) Variable clinical expression of mosaic trisomy 16 in the newborn infant. *Am. J. Med. Genet.* **47,** 294–298.

101. Lindor, N.M., Jalal, S.M., Thibodeau, S.N., Bonde, D., Sauser, K.L., and Karnes, P.S. (1993) Mosaic trisomy 16 in a thriving infant: maternal heterodisomy for chromosome 16. *Clin. Genet.* **44,** 185–189.

102. Garber, A., Carlson, D., Schreck, R., Fishcel-Ghodsian, N., Hsu, W., Oeztas, S., Pepkowitz, S., and Graham, J.M. (1994) Prenatal diagnosis and dysmorphic findings in mosaic trisomy 16. *Prenat. Diag.* **14,** 257–266.

103. Pletcher, B.A., Sanz, M.M., Schlessel, J.S., Kunaporn, S., McKenna, C., Bialer, M.G., Alonso, M.L., Zaslav, A.L., Brown, W.T., and Ray, J.H. (1994) Postnatal confirmation of prenatally trisomy 16 mosaicism in two phenotypically abnormal liveborns. *Prenat. Diag.* **14,** 933–940.

104. Hajianpour, M.J. (1995) Postnatally confirmed trisomy 16 mosaicism: follow-up on a previously reported patient. *Prenat. Diag.* **15,** 877–879.

105. Hsu, L.Y.F., Kaffe, S., and Perlis, T.E. (1991) A revisit of trisomy 20 mosaicism in prenatal diagnosis—an overview of 103 cases. *Prenat. Diag.* **11,** 7–15.

106. Miny, P., Karabacak, Z., Hammer, P., Schulte-Vallentin, M., and Holzgreve, W. (1989) Chromosome analyses from urinary sediment: postnatal confirmation of a prenatally diagnosed trisomy 20 mosaicism. *N. Engl. J. Med.* **320,** 809.

107. Park, J.P., Moeschler, J.B., Rawnsley, E., Berg, S.Z., and Wurster-Hill, D.H. (1989) Trisomy 20 mosaicism confirmed in a phenotypically normal liveborn. *Prenat. Diag.* **9,** 501–504.

108. Van Dyke, D.L., Roberson, J.R., Babu, V.R., Weiss, L., and Tyrkus, M. (1989) Trisomy 20 mosaicism identified prenatally and confirmed in foreskin fibroblasts. *Prenat. Diag.* **9,** 601–602.

109. Brothman, A.R., Rehberg, K., Storto, P.D., Phillips, S.E., and Mosby, R.T. (1992) Confirmation of true mosaic trisomy 20 in a phenotypically normal liveborn male. *Clin. Genet.* **42,** 47–49.

110. Hsu, L.Y.F., Kaffe, S., and Perlis, T.E. (1987) Trisomy 20 mosaicism in prenatal diagnosis—a review and update. *Prenat. Diag.* **7,** 581–596.

111. Punnett, H.H. and Kistenmacher, M.L. (1990) Confirmation of prenatally ascertained trisomy 20 mosaicism. *Prenat. Diag.* **10,** 136–137.

112. Hsu, L.Y.F., Shapiro, L.R., Gertner, M., Lieber, E., and Hirschhorn, K. (1971) Trisomy 22: a clinical entity. *J. Pediatr.* **79,** 12–19.

113. Bacino, C.A., Schreck, R., Fischel-Ghodsian, N., Pepkowitz, S., Prezant, T.R., and Graham, Jr., J.M. (1995) Clinical and molecular studies in full trisomy 22: further delineation of the phenotype and review of the literature. *Am. J. Med. Genet.* **56,** 359–365.

114. Hirschhorn, K., Lucas, M., and Wallace, I. (1973) Precise identification of various chromosomal abnormalities. *Ann. Hum. Genet.* **36,** 375–379.

115. Pridjian, G., Gill, W.L., and Shapira, E. (1995) Goldenhar sequence and mosaic trisomy 22. *Am. J. Med. Genet.* **59,** 411–413.

116. Ladonne, J., Gaillard, D., Carre-Pigeon, F., and Gabriel, R. (1996) Fryns syndrome phenotype and trisomy 22. *Am. J. Med. Genet.* **61,** 68–70.

117. Robinson, W.P. and Kalousek, D.K. (1996) Mosaicism most likely accounts for extended survival of trisomy 22. *Am. J. Med. Genet.* **62,** 100.

118. Kukolich, M.K., Kulharya, A., Jalal, S.M., and Drummond-Borg, M. (1989) Trisomy 22: no longer an enigma. *Am. J. Med. Genet.* **34,** 541–544.

119. Casey, J., Ketterer, D.M., Heisler, K.L., Daugherty, E.A., Prince, P.M., and Giles, H.R. (1990) Prenatal diagnosis of trisomy 2 mosaicism confirmed in foreskin fibroblasts. *Am. J. Hum. Genet.* **47,** A270.

120. Metaxotou, C., Tsenghi, C., Bitzos, I., Strataki-Benetou, M., Kalpini-Mavrou, A., and Matsaniotis, N. (1981) Trisomy 3 mosaicism in a live-born infant. *Clin. Genet.* **19,** 37–40.

121. Smith, S.C., Varela, M., Toebe, C., and Shapira, E. (1988) Trisomy 3 mosaicism: a case report. *Am. J. Hum. Genet.* **43,** A71.

122. Marion, J.P., Fernhoff, P.M., Korotkin, J., and Priest, J.H. (1990) Pre- and postnatal diagnosis of trisomy 4 mosaicism. *Am. J. Med. Genet.* **37,** 362–365.

123. Sciorra, L.J., Hux, C., Day-Salvadore, D., Lee, M.L., Mandelbaum, D.E., Brady-Yasbin, S., Frybury, J., Mahoney, M.J., and Dimaio, M.S. (1992) Trisomy 5 mosaicism detected prenatally with an affected liveborn. *Prenat. Diag.* 12, 477–482.

124. Yunis, E., Ramirez, E., and Uribe, J.G. (1980) Full trisomy 7 and Potter syndrome. *Hum. Genet.* **54,** 13–18.

125. Hodes, M.E., Gleiser, S., DeRosa, G.P., Yune, H.Y., Girod, D.A., Weaver, D.D., and Palmer, C.G. (1981) Trisomy 7 mosaicism and manifestations of Goldenhar syndrome with unilateral radial hypoplasia. *J. Craniofac. Genet. Dev. Biol.* **1,** 49–55.

126. Turleau, C., de Grouchy, J., Cabanis, M.O., Nihoul-Fekete, C., and Dufier, J.L. (1984) Trisomy 7. Internal intersexuality (masculine uterus) and severe abnormality of the anterior chamber of the eye. *Ann. Genet.* **27,** 115–117.

127. Pflueger, S.M., Scott, Jr., C.I., and Moore, C.M. (1984) Trisomy 7 and Potter syndrome. *Clin. Genet.* **25,** 543–548.

128. deFrance, H.F., Beemer, F.A., Senders, R.C., and Schaminee-Main, S.C. (1985) Trisomy 10 mosaicism in a newborn boy; delineation of the syndrome. *Clin. Genet.* **27,** 92–96.

129. Richer, C.L., Bleau, G., and Chapdelaine, A. (1977) Trisomy 12 mosaicism in an infertile man. *Can. J. Genet. Cytol.* **19,** 565–567.

130. Patil, S.R., Bosch, E.P., and Hanson, J.W. (1983) First report of mosaic trisomy 12 in a liveborn individual. *Am. J. Med. Genet.* **14,** 453–460.

131. Leschot, N.J., Wilmsen-Linders, E.J., van Geijn, H.P., Samsom, J.F., and Smit, L.M. (1988) Karyotyping urine sediment cells confirms trisomy 12 mosaicism detected at amniocentesis. *Clin. Genet.* **34,** 135–139.

132. Von Koskull, H., Ritvanen, A., Ammala, P., Gahmberg, N., and Salonen, R. (1989) Trisomy 12 mosaicism in amniocytes and dysmorphic child despite normal chromosomes in fetal blood sample. *Prenat. Diag.* **9,** 433–437.

133. Bischoff, F.Z., Zenger-Hain, J., Moses, D., Van Dyke, D.L., and Shaffer, L.G. (1995) Mosaicism for trisomy 12: four cases with varying outcomes. *Prenat. Diag.* **15,** 1017–1026.

134. Fujimoto, A., Allanson, J., Crowe, C.A., Lipson, M.H., and Johnson, V.P. (1992) Natural history of mosaic trisomy 14 syndrome. *Am. J. Med. Genet.* **44,** 189–196.

135. Bühler, E.M., Bienz, G., Straumann, E., and Bösch, N. (1996) Delineation of a clinical syndrome caused by mosaic trisomy 15. *Am. J. Med. Genet.* **62,** 109–112.

136. Milunsky, J.M., Wyandt, H.E., Huang, X.L., Kang, X.Z., Elias, E.R., and Milunsky, A. (1996) Trisomy 15 mosaicism and uniparental disomy (UPD) in a liveborn infant. *Am. J. Med. Genet.* **61,** 269–273.

137. Coldwell, S., Fitzgerald, B., Semmens, J.M., Ede, R., and Bateman, C. (1981) A case of trisomy of chromosome 15. *J. Med. Genet.* **18,** 146–148.

138. Kuller, J.A. and Laifer, S.A. (1991) Trisomy 15 associated with nonimmune hydrops. *Am. J. Perinatol.* **8,** 39–40.

139. Shaffer, L.G., McCaskill, C., Hersh, J.H., Greenberg, F., and Lupski, J.R. (1996) A clinical and molecular study of mosaicism for trisomy 17. *Hum. Genet.* **97,** 69–72.

140. Chen, H., Yu, C.W., Wood, M.J., and Landry, K. (1981) Mosaic trisomy 19 syndrome. *Ann. Genet.* **24,** 32–33.

141. Rethore, M.O., Debray, P., Guesne, M.C., Amedee-Manesme, O., Iris, L., and Lejeune, J. (1981) A case of trisomy 19 mosaicism. *Ann. Genet.* **24,** 34–36.

142. Wurster-Hill, D.H., Moeschler, J.M., Park, J.P., and McDermet, M.K. (1990) Mosaicism for monosomy 20 in a 30 month old boy. *Am. J. Hum. Genet.* **47,** A44.

143. Richmond, H.G., MacArthur, P., and Hunter, D. (1973) A "G" deletion syndrome anti-mongolism. *Acta Paediatr. Scand.* **62,** 216–220.

144. Mikkelsen, M. and Vestermark, S. (1974) Karyotype 45,XX,-21/46,XX,21q- in an infant with symptoms of G-deletion syndrome I. *J. Med. Genet.* **11,** 389–392.

145. Olinici, C.D., Butnariu, J., Popescu, A., and Giurgiuman, M. (1977) Mosaic 45,XY,-21/46,XY in a child with G deletion syndrome I. *Ann. Genet.* **20,** 115–117.

146. Abeliovich, D., Carmi, R., Karplus, M., Bar-Ziv, J., and Cohen, M.M. (1979) Monosomy 21: a possible stepwise evolution of the karyotype. *Am. J. Med. Genet.* **4,** 279–286.

147. Ghidini, A., Fallet, S., Robinowitz, J., Lockwood, C., Dische, R., and Willner, J. (1993) Prenatal detection of monosomy 21 mosaicism. *Prenat. Diag.* **13,** 163–169.

148. Wisniewski, K., Dambska, M., Jenkins, E.C., Sklower, S., and Brown, W.T. (1983) Monosomy 21 syndrome: further delineation including clinical, neuropathological, cytogenetic and biochemical studies. *Clin. Genet.* **23,** 102–110.

149. Pellissier, M.C., Philip, N., Voelckel-Baeteman, M.A., Mattei, M.G., and Mattei, J.F. (1987) Monosomy 21: a new case confirmed by in situ hybridization. *Hum. Genet.* **75,** 95–96.

150. Garzicic, B., Guc-Scekic, M., Pilic-Radivojevic, G., Ignjatovic, M., and Vilhar, N. (1988) A case of monosomy 21. *Ann. Genet.* **31,** 247–249.

151. Moghe, M.S., Patel, Z.M., Peter, J.J., and Ambani, L.M. (1981) Monosomy 22 with mosaicism. *J. Med. Genet.* **18,** 71–73.

152. Verloes, A., Herens, C., Lambotte, C., and Frederic, J. (1987) Chromosome 22 mosaic monosomy (46,XY/45,XY,-22). *Ann. Genet.* **30,** 178–179.

153. Lewinsky, R.M., Johnson, J.M., Lao, T.T., Winsor, E.J., and Cohen, H. (1990) Fetal gastroschisis associated with monosomy 22 mosaicism and absent cerebral diastolic flow. *Prenat. Diag.* **10,** 605–608.

154. Carr, D.H. and Gedeon, M. (1977) Population cytogenetics of human abortuses. In *Population Cytogenetics* (Hook E.B. and Porter I.H., eds.), Academic Press, New York, pp. 1–9.

155. Neuber, M., Rehder, H., Zuther, C., Lettau, R., and Schwinger, E. (1993) Polyploidies in abortion material decrease with maternal age. *Hum. Genet.* **91,** 563–566.

156. McFadden, D.E. and Kalousek, D.K. (1991) Two different phenotypes of fetuses with chromosomal triploidy: correlation with parental origin of the extra haploid set. *Am. J. Med. Genet.* **38,** 535–538.

157. Jacobs, P.A., Szulman, A.E., Funkhouser, J., Matsuura, J.S., and Wilson, C.C. (1982) Human triploidy: relationship between parental origin of the additional haploid complement and development of partial hydatidiform mole. *Ann. Hum. Genet.* **46,** 223–231.

158. McFadden, D.E., Kwong, L.C., Yam, I.Y., and Langlois, S. (1993) Parental origin of triploidy in human fetuses: evidence for genomic imprinting. *Hum. Genet.* **92,** 465–469.

159. Dietzsch, E., Ramsay, M., Christianson, A.L., Henderson, B.D., and de Ravel, T.J.L. (1995) Maternal origin of extra haploid set of chromosomes in third trimester triploid fetuses. *Am. J. Med. Genet.* **58,** 360–364.

160. Uchida, I.A. and Freeman, V.C.P. (1985) Triploidy and chromosomes. *Am. J. Obstet. Gynecol.* **151,** 65–69.

161. Gersen, S.L., Carelli, M.P., Klinger, K.W., and Ward, B.E. (1995) Rapid prenatal diagnosis of 14 cases of triploidy using FISH with multiple probes. *Prenat. Diag.* **15,** 1–5.

162. Hall, J.G. (1990) Genomic imprinting: review and relevance to human diseases. *Am. J. Hum. Genet.* **46,** 857–873.

163. Engel, E. and DeLozier-Blanchet, C.D. (1991) Uniparental disomy, isodisomy, and imprinting: probable effects in man and strategies for their detection. *Am. J. Med. Genet.* **40,** 432–439.

164. McGrath, J. and Solter, D. (1984) Completion of mouse embryogenesis requires both the maternal and paternal genomes. *Cell* **37,** 179–183.

165. Surani, M.A.H., Barton, S.C., and Norris, M.L. (1984) Development of reconstituted mouse eggs suggests imprinting of the genome during gametogenesis. *Nature* **308,** 548–550.

166. Surani, M.A.H., Barton, S.C., and Norris, M.L. (1986) Nuclear transplantation in the mouse: heritable differences between parental genomes after activation of the embryonic genome. *Cell* **45,** 127–136.

167. Niemann-Seyde, S.C., Rehder, H., and Zoll, B. (1993) A case of full triploidy (69,XXX) of paternal origin with unusually long survival time. *Clin. Genet.* **43,** 79–82.

168. Sherald, J., Bean, C., Bove, B., DelDuca Jr., V., Esterly, K.I., Karcsh, H.J., Munshi, G., Reamer, J.F., Suazo, G., Wilmoth, D., Dahlke, M.B., Weiss, C., and Borgaonkar, D.S. (1986) Long survival in a 69,XXY triploid male. *Am. J. Med. Genet.* **25,** 307–312.

169. Järvelä, I.E., Salo, M.K., Santavuori, P., and Salonen, R.K. (1993) 46,XX/69,XXX diploid-triploid mixoploidy with hypothyroidism and precocious puberty. *J. Med. Genet.* **30,** 966–967.

170. Tharapel, A.T., Wilroy, R.S., Martens, P.R., Holbert, J.M., and Summitt, R.L. (1983) Diploid-triploid mosaicism: delineation of the syndrome. *Ann. Genet.* **26,** 229–233.

171. Carakushansky, G., Teich, E., Ribeiro, M.G., Horowitz, D.D.G., and Pellegrini, S. (1994) Diploid/triploid mosaicism: further delineation of the phenotype. *Am. J. Med. Genet.* **52,** 399–401.

172. Dewald, G., Alvarez, M.N., Cloutier, M.D., Kelalis, P.P., and Gordon, H. (1975) A diploid–triploid human mosaic with cytogenetic evidence of double fertilization. *Clin. Genet.* **8,** 149–160.

173. Coe, S.J., Kapur, R., Luthardt, F., Rabinovitch, P., and Kramer, D. (1993) Prenatal diagnosis of tetraploidy: a case report. *Am. J. Med. Genet.* **45,** 378–382.

174. Lafer, C.Z. and Neu, R.L. (1988) A liveborn infant with tetraploidy. *Am. J. Med. Genet.* **31,** 375–378.

175. Wullich, B., Henn, W., Groterath, E., Ermis, A., Fuchs, S., and Zankl, M. (1991) Mosaic tetraploidy in a liveborn infant with features of the DiGeorge anomaly. *Clin. Genet.* **40,** 353–357.

176. Edwards, M.J., Park, J.P., Wurster-Hill, D.H., and Graham Jr., J.M. (1994) Mixoploidy in Humans: two surviving cases of diploid–tetraploid mixoploidy and comparison with diploid–triploid mixoploidy. *Am. J. Med. Genet.* **52,** 324–330.

177. Aughton, D.J., Saal, H.M., Delach, J.A., Rahman, Z.U., and Fisher, D. (1988) Diploid/tetraploid mosaicism in a liveborn infant demonstrable only in the bone marrow: case report and literature review. *Clin. Genet.* **33,** 299–307.

178. Quiroz, E., Orozco, A., and Salamanca, F. (1985) Diploid–tetraploid mosaicism in a malformed boy. *Clin. Genet.* **27,** 183–186.

179. Wittwer, B.B. and Wittwer, H.B. (1985) Information about diploid–tetraploid mosaicism in a six-year-old male. *Clin. Genet.* **28,** 567–568.

180. Fisher, A.M., Barber, J.C.K., Crolla, J.A., James, R.S., Lestas, A.N., Jennings, I., and Dennis, N.R. (1993) Mosaic tetrasomy 8p: molecular cytogenetic confirmation and measurement of glutathione reductase and tissue plasminogen activator levels. *Am. J. Med. Genet.* **47,** 100–105.

181. Schrander-Stumpel, C.T.R.M., Govaerts, L.C.P., Engelen, J.J.M., van der Blij-Philipsen, M., Borghgraef, M., Loots, W.J.G., Peters, J.J.M., Rijnvos, W.P.M., Smeets, D.F.C.M.,

and Fryns, J.P. (1994) Mosaic tetrasomy 8p in two patients: clinical data and review of the literature. *Am. J. Med. Genet.* **50,** 377–380.

182. Winters, J., Markello, T., Nance, W., and Jackson-Cook, C. (1995) Mosaic "tetrasomy" 8p: case report and review of the literature. *Clin. Genet.* **48,** 195–198.

183. Grass, F.S., Parke Jr., J.C., Kirkman, H.N., Christensen, V., Roddey, O.F., Wade, R.V., Knutson, C., and Spence, J.E. (1993) Tetrasomy 9p: tissue-limited idic(9p) in a child with mild manifestations and a normal CVS result. Report and review. *Am. J. Med. Genet.* **47,** 812–816.

184. Melaragno, M.I., Brunoni, D., Patricio, F.R., Corbani, M., Mustacchi, Z., dos Santos R. de C., and Lederman, H.M. (1992). A patient with tetrasomy 9p, Dandy–Walker cyst and Hirschsprung disease. *Ann. Genet.* **35,** 79–84.

185. Leichtman, L.G., Zackowski, J.L., Storto, P.D., and Newlin, A. (1996) Non-mosaic tetrasomy 9p in a liveborn infant with multiple congenital anomalies: case report and comparison with trisomy 9p. *Am. J. Med. Genet.* **63,** 434–437.

186. Ghymers, D., Hermann, B., Distèche, C., and Frederic, J. (1973) Tetrasomie partielle du chromosome 9, à l'état de mosaique, chez un enfant porteur de malformations multiples. *Hum. Genet.* **20,** 273–282.

187. Cuoco, C., Gimelli, G., Pasquali, F., Poloni, L., Zuffardi, O., Alicanta, P., Battaglino, G., Bernardi, F., Cerone, R., Cotellessa, M., Ghidoni, A., and Motta, S. (1982) Duplication of the short arm of chromosome 9. Analysis of five cases. *Hum. Genet.* **61,** 3–7.

188. Peters, J., Pehl, C., Miller, K., and Sandlin, C. (1982) Case report of mosaic partial tetrasomy 9p mimicking Klinefelter syndrome. *Birth Defects* **18,** 287–293.

189. Calvieri, F., Tozzi, C., Benincori, C., DeMerulis, M., Bellussi, A., Genuardi, M., and Neri, G. (1988) Partial tetrasomy 9 in an infant with clinical and radiological evidence of multiple joint dislocations. *Eur. J. Pediatr.* **147,** 645–648.

190. Papenhausen, P., Riscile, G., Miller, K., Kousseff, B., and Tedescto, T. (1990) Tissue limited mosaicism in a patient with tetrasomy 9p. *Am. J. Med. Genet.* **37,** 388–391.

191. Pallister, P.D., Meisner, L.F., Elejalde, B.R., Francke, U., Herrmann, J., Spranger, J., Tiddy, W., Inhorn, S.L., and Opitz, J.M. (1977) The Pallister mosaic syndrome. *Birth Defects Orig. Article Ser.* **3B,** 103–110.

192. Killian, W. and Teschler-Nicola, M. (1981) Case report 72: mental retardation, unusual facial appearance, abnormal hair. *Synd. Ident.* **7,** 6–17.

193. Bielanska, M.M., Khalifa, M.M., and Duncan, A.M.V. (1996) Pallister-Killian syndrome: a mild case diagnosed by fluorescence *in situ* hybridization. Review of the literature and expansion of the phenotype. *Am. J. Med. Genet.* **65,** 104–108.

194. Reynolds, J.F., Daniel, A., Kelly, T.E., Gollin, S.M., Stephan, M.J., Carey, J., Adkins, W.N., Webb, M.J., Char, F., Jimenez, J.F., and Opitz, J.M. (1987) Isochromosome 12p mosaicism (Pallister mosaic aneuploidy or Pallister–Killian syndrome): report of 11 cases. *Am. J. Med. Genet.* **27,** 257–274.

195. Wenger, S.L., Steele, M.W., and Yu, W.D. (1988) Risk effect of maternal age in Pallister i (12p) syndrome. *Clin. Genet.* **34,** 181–184.

196. Priest, J.H., Rust, J.M., and Fernhoff, P.M. (1992) Tissue specificity and stability of mosaicism in Pallister–Killian +i(12p) syndrome: relevance for prenatal diagnosis. *Am. J. Med. Genet.* **42,** 820–824.

197. Ward, B.E., Hayden, M.W., and Robinson, A. (1988) Isochromosome 12p mosaicism (Pallister–Killian syndrome): newborn diagnosis by direct bone marrow analysis. *Am. J. Med. Genet.* **31,** 835–839.

198. Wenger, S.L., Boone, L.Y., and Steele, M.W. (1990) Mosaicism in Pallister i(12p) syndrome. *Am. J. Med. Genet.* **35,** 523–525.

199. Peltomaki, P., Knuutila, S., Ritvanen, A., Kaitila, I., and de la Chapelle, A. (1987) Pallister–Killian syndrome: cytogenetic and molecular studies. *Clin. Genet.* **31,** 399–405.

200. Zakowski, M.F., Wright, Y., and Ricci, Jr., A. (1992) Pericardial agenesis and focal aplasia cutis in tetrasomy 12p (Pallister–Killian syndrome). *Am. J. Med. Genet.* **42,** 323–325.
201. Speleman, F., Leroy, J.G., Van Roy, N., DePaepe, A., Suijkerbuijk, R., Brunner, H., Looijenga, L., Verschraegen-Spae, M.R., and Orye, E. (1991) Pallister-Killian syndrome: characterization of the isochromosome 12p by fluorescent *in situ* hybridization. *Am. J. Med. Genet.* **41,** 381–387.
202. Reeser, S.L.T. and Wenger, S.L. (1992) Failure of PHA-stimulated i(12p) lymphocytes to divide in Pallister–Killian syndrome. *Am. J. Med. Genet.* **42,** 815–819.
203. Froland, A., Holst, G., and Terslev, E. (1963) Multiple anomalies associated with an extra small autosome. *Cytogenetics* **2,** 99–106.
204. Yu, L.C., Williams III, J., Wang, B.B.T., Vooijs, M., Weier, H.U.G., Sakamoto, M., and Ying, K.L. (1993) Characterization of i(18p) in prenatal diagnosis by fluorescence *in situ* hybridization. *Prenat. Diag.* **13,** 355–361.
205. Rivera, H., Moller, M., Hernandez, A., Enriquez-Guerra, M.A., Arreola, R., and Cantu, J.M. (1984) Tetrasomy 18p: a distinctive syndrome. *Ann. Genet.* **27,** 187–189.
206. Callen, C.F., Freemantle, C.J., Ringenbergs, M.L., Baker, E., Eyre, H.J., Romain, D., and Haan, E.A. (1990) The isochromosome 18p syndrome: confirmation of cytogenetic diagnosis in nine cases by *in situ* hybridization. *Am. J. Hum. Genet.* **47,** 493–498.
207. Back, E., Toder, R., Voiculescu, I., Wildberg, A., and Schempp, W. (1994) *De novo* isochromosome 18p in two patients: cytogenetic diagnosis and confirmation by chromosome painting. *Clin. Genet.* **45,** 301–304.
208. Eggermann, T., Nothen, M.M., Eiben, B., Hofmann, D., Hinkel, K., Fimmers, R., and Schwanitz, G. (1996) Trisomy of human chromosome 18: molecular studies on parental origin and cell stage of nondisjunction. *Hum. Genet.* **97,** 218–223.
209. Göcke, H., Muradow, I., Zerres, K., and Hansmann, M. (1986) Mosaicism of isochromosome 18p. cytogenetic and morphological findings in a male fetus at 21 weeks. *Prenat. Diag.* **6,** 151–157.
210. Taylor, K.M., Wolfinger, H.L. Brown, M.G., and Chadwick, D.L. (1975) Origin of a small metacentric chromosome: familial and cytogenetic evidence. *Clin. Genet.* **8,** 364–369.
211. Takeda, K., Okamura, T., and Hasegawa, T. (1989) Sibs with tetrasomy 18p born to a mother with trisomy 18p. *J. Med. Genet.* **26,** 195–197.
212. Abeliovich, D., Dagan, J., Levy, A., Steinberg, A., and Zlotogora, J. (1993) Isochromosome 18p in a mother and her child. *Am. J. Med. Genet.* **46,** 392–393.
213. Nielsen, J. and Wohlert, M. (1991) Chromosome abnormalities found among 34,910 newborn children: results from a 13-year incidence study in Arhus, Denmark. *Hum. Genet.* **87,** 81–83.
214. Sachs, E.S., Van Hemel, J.O., Den Hollander, J.C., and Jahoda, M.G.J. (1987) Marker chromosomes in a series of 10,000 prenatal diagnoses. Cytogenetic and follow-up studies. *Prenat. Diag.* **7,** 81–89.
215. Blennow, E., Bui, T-H., Kristoffersson, U., Vujic, M., Anneren, G., Holmberg, E., and Nordenskjold, M. (1994) Swedish survey on extra structurally abnormal chromosomes in 39,105 consecutive prenatal diagnoses: Prevalence and characterization by fluorescence in situ hybridization. *Prenat. Diag.* **14,** 1019–1028.
216. Brondum-Nielsen, K. and Mikkelsen, M. (1995) A 10-year survey, 1980–1990, of prenatally diagnosed small supernumerary marker chromosomes, identified by FISH analysis. Outcome and follow-up of 14 cases diagnosed in a series of 12,699 prenatal samples. *Prenat. Diag.* **15,** 615–619.
217. Callen, D.F., Eyre, H.J., Ringenbergs, M.L., Freemantle, C.J., Woodroffe, P., and Haan, E.A. (1991) Chromosomal origin of small ring marker chromosomes in man: characterization by molecular genetics. *Am. J. Hum. Genet.* **48,** 769–782.
218. Crolla, J.A., Dennis, N.R., and Jacobs, P.A. (1992) A non-isotopic in situ hybridization

study of the chromosomal origin of 15 supernumerary marker chromosomes in man. *J. Med Genet.* **29,** 699–703.

219. Plattner, R., Heerema, N.A., Howard-Peebles, P.N., Miles, J.H., Soukup, S., and Palmer, C.G. (1993) Clinical findings in patients with marker chromosomes identified by fluorescence in situ hybridization. *Hum. Genet.* **91,** 589–598.

220. Blennow, E., Nielsen, K.B., Telenius, H., Carter, N.P., Kristoffersson, U., Holmberg, E., Gillberg, C., and Nordenskjold, M. (1995) Fifty probands with extra structurally abnormal chromosomes characterized by fluorescence in situ hybridization. *Am. J. Med. Genet.* **55,** 85–94.

221. Buckton, K.E., Spowart, G., Newton, M.S., and Evans, H.J. (1985) Forty four probands with an additional "marker" chromosome. *Hum. Genet.* **69,** 353–370.

222. Leana-Cox, J., Jenkins, L., Palmer, C.G., Plattner, R., Sheppard, L., Flejter, W.L., Zackowski, J., Tsien, F., and Schwartz, S. (1994) Molecular cytogenetic analysis of inv dup(15) chromosomes, using probes specific for the Prader-Willi/Angelman syndrome region: clinical implications. *Am. J. Hum. Genet.* **54,** 748–756.

223. Chent, S-D., Spinner, N.B., Zackai, E.H., and Knoll, J.H.M. (1994) Cytogenetic and molecular characterization of inverted duplicated chromosomes 15 from 11 patients. *Am. J. Hum. Genet.* **55,** 753–759.

224. Crolla, J.A., Harvey, J.F., Sitch, F.L., and Dennis, N.R. (1995) Supernumerary marker 15 chromosomes: a clinical, molecular and FISH approach to diagnosis and prognosis. *Hum. Genet.* **95,** 161–170.

225. Wisniewski, L.P., Witt, M.E., Ginsberg-Fellner, F., Wilner, J., and Desnick, R.J. (1980) Prader–Willi syndrome and a bisatellited derivative of chromosome 15. *Clin. Genet.* **18,** 42–47.

226. Ledbetter, D.H., Mascarello, J.T., Riccardi, V.M., Harper, V.D., Airhart, S.D., and Strobel, R.J. (1982) Chromosome 15 abnormalities and the Prader-Willi syndrome: a follow-up report of 40 cases. *Am. J. Hum. Genet.* **34,** 278–285.

227. Mattei, J.F., Mattei, M.G., and Giraud, F. (1983) Prader–Willi syndrome and chromosome 15. A clinical discussion of 20 cases. *Hum. Genet.* **64,** 356–362.

228. Robinson, W.P., Wagstaff, J., Bernasconi, F., Baccichette, C., Artifoni, L., Franzoni, E., Suslak, L., Shih, L-Y., Aviv, H., and Schinzell, A.A. (1993) Uniparental disomy explains the occurrence of the Angelman or Prader–Willi syndrome in patients with an additional small inv dup(15) chromosome. *J. Med. Genet.* **30,** 756–760.

229. Spinner, N.B., Zackai, E., Cheng, S.D., and Knoll, J.H. (1995) Supernumerary inv dup(15) in a patient with Angelman syndrome and a deletion of 15q11-q13. *Am. J. Med. Genet.* **57,** 61–65.

230. Schinzel, A., Schmid, W., Fraccaro, M., Tiepolo, L., Zuffardi, O., Opitz, J.M., Lindsten, J., Zetterqvist, P., Enell, H., Baccichetti, C., Tenconi, R., and Pagon, R.A. (1981) The "cat eye syndrome": dicentric small marker chromosome probably derived from a no. 22 (tetrasomy 22pter to q11) associated with a characteristic phenotype. Report of 11 patients and delineation of the clinical picture. *Hum. Genet.* **57,** 148–158.

231. McDermid, H.E., Duncan, A.M., Brasch, K.R., Holden, J.J., Magenis, E., Sheehy, R., Burn, J., Kardon, N., Noel, B., and Schinzel, A. (1986) Characterization of the supernumerary chromosome in cat eye syndrome. *Science* **232,** 646–648.

232. Liehr, T., Pfeiffer, R.A., and Trautmann, U. (1992) Typical and partial cat eye syndrome: identification of the marker chromosome by FISH. *Clin. Genet.* **42,** 91–96.

233. Mears, A.J., El-Shanti, H., Murray, J.C., McDermid, H.E., and Patil, S.R. (1995) Minute supernumerary ring chromosome 22 associated with cat eye syndrome: further delineation of the critical region. *Am. J. Hum. Genet.* **57,** 667–673.

9

Structural Chromosome Rearrangements

Kathleen Kaiser-Rogers, Ph.D. and Kathleen Rao, Ph.D.

INTRODUCTION

The subject of structural chromosome rearrangements is an immense one, to which entire catalogs have been devoted. Indeed, there are theoretically an almost infinite number of ways in which chromosomes can reconfigure themselves from the normal, 23-pair arrangement we are familiar with. While we tend to think of the resulting structural rearrangements in terms of chromosome pathology, some rearrangements are fairly innocuous. In fact, a few such benign rearrangements (such as certain pericentric inversions of chromosome 9) are seen frequently enough to be considered polymorphic variants of no clinical significance.

In this chapter, we discuss and provide examples of the ways in which chromosome rearrangements can occur. We begin with an overview of general concepts that relate to all structural rearrangements and their association with human pathology. Each category of structural rearrangement is then dealt with as a unique entity in the second half of the chapter.

Mechanism of Formation

The exchange of genetic material between sister chromatids and/or homologous chromosomes is a normal occurrence in somatic and germ cells. These types of exchanges ensure mixing of the gene pool and may even be obligatory for normal cell division. It is only when exchanges occur between nonhomologous chromosomal regions that structural rearrangements result. Because chromosome breakage can theoretically occur anywhere within the human genome and the involved chromosome(s) can recombine in innumerable ways, the number of potential rearrangements that can result is immense. In practice, however, there appear to be particular areas of the genome that are more susceptible to breakage and rearrangement than others. The presence of a DNA sequence that is repeated elsewhere in the genome, a fragile site, and/or a particular secondary DNA structure may influence the likelihood that a particular chromosome region is involved in a structural rearrangement (1–4).

In theory, chromosome breakage, rearrangement, and reunion can occur during meiosis or mitosis. Meiotic errors, because they occur prior to conception, would be expected to be present in every cell in the resulting pregnancy. Postconception mitotic errors, in contrast, would be predicted to produce a mosaic pregnancy containing both

From: *The Principles of Clinical Cytogenetics*
Edited by: S. Gersen and M. Keagle © Humana Press Inc., Totowa, NJ

normal and abnormal cells. Interestingly, with the exception of mitotically unstable chromosomes such as rings or dicentrics, structural chromosome rearrangements are rarely seen in mosaic form. Although this observation suggests that many structural rearrangements may be formed during meiosis, ascertainment bias has likely played a role as well. Because mosaic individuals typically have milder phenotypes than comparable nonmosaics, they are less likely to be ascertained and karyotyped. This would be especially true of individuals carrying mosaic balanced rearrangements. Furthermore, mosaicism is difficult to detect, particularly when it is limited to a specific tissue or group of tissues, is present at a low level, and/or involves a subtle structural change.

In contrast to the maternal bias noted for numerical chromosome abnormalities, most structural chromosome rearrangements appear to be paternally derived *(5,6)*. Exactly why the male bias for *de novo* structural rearrangements exists is currently unknown. It has been suggested, however, that the lifelong mitotic proliferation of spermatogonial cells, compared to the finite number of mitotic divisions responsible for oögonial cell production in the female embryo, may promote the accumulation of mutations. In addition, studies on mouse and *Drosophila* suggest that male gametogenesis may be more sensitive to mutagens than oögenesis *(7)*.

Differentiating Between Balanced and Unbalanced Structural Rearrangements

Structural rearrangements are often divided into two general categories, balanced and unbalanced. Balanced rearrangements contain no net loss or gain of genetic information and the individuals who carry them are generally phenotypically normal. In contrast, additional and/or missing genetic material is observed in individuals who carry unbalanced rearrangements. Just as modifications in the amount of the various ingredients added to any recipe cause change in the final product, deviation from the normal disomic genetic complement results in a clinically affected individual.

Although it is easy to define balanced and unbalanced rearrangements, distinguishing between a truly balanced and an unbalanced rearrangement using traditional cytogenetic techniques is often impossible. The maximum level of resolution obtained using standard microscopy of G-banded prometaphase chromosomes is reported to be 2–5 megabases or $2–5 \times 10^6$ base pairs. This number will vary, however, depending on the quality of the chromosome preparations and the skill of the cytogeneticist examining the karyotypes. The ability to resolve or identify a rearrangement will also be influenced by the degree to which the banding pattern, overall size, and centromere location of an involved chromosome is altered. Obviously, the more apparent the change, the more likely it is to be detected. A number of molecular cytogenetic techniques such as fluorescence *in situ* hybridization (FISH), 24-color karyotyping, and comparative genomic hybridization (CGH) are currently being used to detect submicroscopic or otherwise cryptic rearrangements that cannot be detected using traditional cytogenetics (see Chapter 15).

Associated Risks

Once a structural chromosome rearrangement is detected, regardless of whether it is balanced or unbalanced, the subsequent steps to take depend on the type of specimen that was analyzed.

For prenatal samples or children, parental karyotypes should be obtained to assess whether the rearrangement has been inherited or represents a *de novo* mutation. If neither parent is found to be a carrier of the rearrangement, the most likely scenario is that it represents a *de novo* abnormality rather than an inherited one. However, nonpaternity and gonadal mosaicism must also be considered. Because the possibility of gonadal mosaicism can never be excluded, this family would be given a very low risk of having another child with the same structural abnormality. Prenatal testing would also be offered for all future pregnancies.

In contrast to the very low recurrence risk quoted to a couple with a child or pregnancy carrying a *de novo* rearrangement, the risk of chromosomally abnormal conceptions for an adult who carries a balanced structural rearrangement is much higher. In fact, for some familial rearrangements the risk can approach 50%. It is therefore imperative that these families are identified so that they can be given accurate genetic counseling regarding their reproductive risks and options. In situations where a familial rearrangement is identified, it must be remembered that it is not just the immediate family, but distant relatives as well who may be at risk for having children with unbalanced karyotypes and associated mental and/or physical abnormalities. By systematically karyotyping the appropriate individuals in each generation, all those with elevated reproductive risks can be identified and appropriately counseled regarding their risks and options. Although there has been some debate regarding the appropriateness of karyotyping the phenotypically normal minors of balanced carriers, 50% of whom would be expected to be balanced carriers themselves, there is a consensus that these children should be referred for appropriate genetic counseling when they reach reproductive age.

The situation becomes a bit more complex when chromosome analysis of a bone marrow or tumor specimen reveals an apparently balanced rearrangement, not known to be associated with any particular neoplasm, in all cells examined. In these cases, it is imperative to ascertain whether such a rearrangement represents a patient-specific acquired change (which can then be monitored during treatment, remission, relapse, or any change in disease aggression) or a constitutional abnormality present from birth.

The reasons for this are twofold. First, from the point of view of the physician treating the patient, the presence of any acquired cytogenetic change is significant. Alternatively, demonstrating that the rearrangement is constitutional can be considered "good news," because this means that there are, in fact, no acquired chromosomal changes. Second and equally important, however, is to consider the potential reproductive consequences for the extended family of what may be a familial rearrangement. Because it is necessary to focus on the treatment of the patient's cancer, and because many of these patients are elderly and well beyond childbearing age, reproductive issues associated with a familial chromosome rearrangement are frequently overlooked. It should be clear from this chapter, however, that these issues must be addressed.

Genetic counseling is covered in detail in Chapter 17.

De Novo *Rearrangements*

Every chromosome rearrangement was at one time a new or *de novo* rearrangement that carried the risks associated with an undefined entity. Children who carry unbalanced rearrangements, regardless of whether they represent new mutations or an unbalanced form of a familial rearrangement, almost inevitably demonstrate an abnormal phenotype. An imbalance is an imbalance regardless of how it arose.

In contrast, accurate predictions regarding the phenotype of a child or fetus that carries an apparently balanced *de novo* chromosome rearrangement are more difficult to make. In this situation we have no idea what has occurred at the molecular level within the rearrangement and we have no family members with the rearrangement from whom inferences can be made. The risk for an abnormal phenotype is therefore always higher for an individual with an apparently balanced *de novo*, rearrangement than for an individual who has inherited a similar rearrangement from a normal parent. Obviously these individuals also carry a significantly higher risk for phenotypic abnormalities than their chromosomally normal counterparts. Several population studies have shown, for example, that the incidence of *de novo*, apparently balanced rearrangements among mentally retarded individuals is approximately seven times that reported in newborns *(8)*. Apparently balanced *de novo*, rearrangements detected at amniocentesis are associated with a risk for congenital abnormalities that is two- to threefold that observed within the general population *(1)*.

A number of different mechanisms are thought to be responsible for the abnormal phenotypes observed in children with *de novo*, apparently balanced rearrangements. One possibility is that the translocation is not truly balanced. Structural rearrangements that appear balanced at the microscopic level may actually contain large duplications and/or deletions at the molecular level. Another possibility is that the rearrangement is "balanced," but a break has occurred within a critical gene or its surrounding regulatory sequences such that the gene product or its expression is altered. This scenario has been demonstrated in several patients with Duchenne muscular dystrophy, for example *(9)*. A position effect, in which the expression of a specific gene or group of genes is altered when the chromosome segment containing them is moved to a different location, could also result in an abnormal phenotype. Such an effect has been demonstrated in several X-autosome translocation chromosomes in which inactivation seems to spread from the inactive X chromosome into neighboring autosomal segments. This phenomenon has been documented in *Drosophila* and plants as well. Finally, the possibility that an individual's abnormal phenotype may be completely unrelated to his or her rearrangement must always be examined. Other nonchromosomal genetic disorders, prenatal exposures, birth trauma, etc. must all be considered.

Familial Rearrangements

Balanced structural rearrangements may pass through multiple generations of a family without being detected. When these families are ascertained it is usually due to the presence of infertility, multiple spontaneous pregnancy losses, or a clinically abnormal family member (**Fig. 1**). Meiotic events that result in cytogenetically unbalanced conceptions can explain the presence of all three occurrences within these families.

During normal meiosis, homologous chromosomes pair utilizing a mechanism of formation thought to depend, at least in part, upon interactions between their shared sequences. Under normal circumstances, all 23 pairs of homologous chromosomes align themselves to form 23 paired linear structures or bivalents that later separate and migrate to independent daughter cells. In cells carrying structurally rearranged chro-

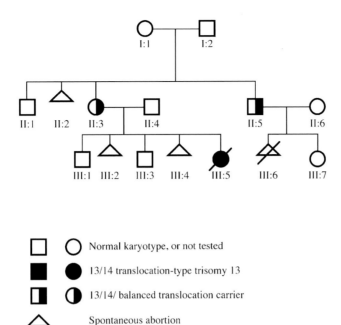

Fig. 1. A pedigree of a family in which a balanced Robertsonian (13;14) translocation is segregating. Multiple spontaneous abortions (see individuals II-2, III-2, and III-4), abnormal children (III-5), and infertility are frequently observed in families segregating a balanced rearrangement.

mosomes, pairing cannot occur in a simple linear fashion. Instead, complex pairing configurations are formed in an attempt to maximize pairing between homologous regions that now differ with regard to their chromosomal location and/or orientation (see "Translocations," "Inversions," "Insertions," and "Duplications," later). Chromosome malsegregation and/or particular recombination events within these complex configurations can then lead to unbalanced conceptions, many of which never implant or are spontaneously lost during gestation.

Cytogeneticists are frequently asked to make predictions regarding a balanced carrier's risk of producing an abnormal liveborn child. Although this is a legitimate question, it is in practice very difficult to answer accurately. One source of difficulty is the fact that, with very few exceptions, each family's rearrangement is unique. Therefore, unless a family is large and accurate information regarding the reproductive history and phenotype of each family member is available, typically no empiric data are available from which to obtain risk values. A second source of difficulty one encounters in assessing the reproductive risks associated with a particular balanced rearrangement is the breadth and complexity of the variables involved.

One important factor that is considered when assessing the reproductive risks of a carrier parent is the extent of imbalance demonstrated by the potential segregants. In general, the smaller the imbalance, the less severe the phenotype and the more likely the survival. An additional rule of thumb is that the presence of extra genetic material is less deleterious than the absence of genetic material. Another variable to be considered is the quality of the genetic information involved. Some chromosomes, such as 16 and 19, are infrequently involved in unbalanced structural rearrangements. Presumably this occurs because of the importance of maintaining a critical dosage for a gene or group of genes on these chromosomes. Conversely, imbalances involving other chromosomes such as 13, 18, 21, X, and Y appear to be more easily tolerated. In fact, a complete trisomy involving any of these chromosomes is survivable.

Each family's reproductive history can also provide important clues regarding the most likely outcome for an unbalanced pregnancy. As one might expect, those who have had a liveborn child with congenital abnormalities, especially those where an unbalanced form of the familial rearrangement has been documented, are at highest risk for having unbalanced offspring. In families or individuals in which multiple spontaneous abortions and/or infertility are noted, the risk for liveborn unbalanced offspring would be expected to be lower. In these families it is assumed that the unbalanced conceptions are being lost very early as unrecognized pregnancies (infertility) or later during gestation. Interestingly, the sex of the carrier parent also, in some cases, influences the risk of having unbalanced offspring. In situations where a sex bias does exist, the female carrier invariably possesses the higher risk. Why male carriers appear to produce fewer unbalanced offspring than their female counterparts is not known. Perhaps fewer unbalanced segregants form during spermatogenesis relative to oögenesis, and/or the selective pressure against unbalanced gametes is greater in the male, and/or imprinting effects may cause the unbalanced embryos of male carriers to be less viable than those of their female counterparts. Male infertility may also play a role *(7,10)*.

On rare occasions an abnormal phenotype is observed in an apparently balanced carrier of a familial rearrangement. Although some of these cases may simply represent coincidental events, other possible explanations exist as well. Very rarely, abnormal offspring resulting from uniparental disomy, or the inheritance of both homologous chromosomes from a single parent, has been documented in the offspring of balanced translocation carriers *(11;* see Chapter 16). Incomplete transmission of a partially cryptic rearrangement has also been observed in the abnormal offspring of a phenotypically normal carrier parent. Wagstaff and Herman, for example, describe a family in which an apparently balanced (3;9) translocation was thought to be segregating *(12)*. After the birth of two phenotypically abnormal offspring with apparently balanced karyotypes, molecular analysis demonstrated that the father's apparently balanced (3;9) translocation was actually a more complex rearrangement involving a cryptic insertion of chromosome 9 material into chromosome 8. Abnormal segregation of this complex rearrangement led to a cryptic deletion of chromosome 9 material in one sibling and a duplication of the same material in the other.

Phenotypic discrepancies between child and parent may also be explained by the presence of a recessive allele that is inherited from the chromosomally normal parent. While the parent is phenotypically normal due to the presence of a complementary normal allele on the homologous chromosome, the abnormal allele can be expressed in the offspring, who has no normal allele. The affected child inherits two mutant alleles; one mutant allele is inherited secondary to the balanced chromosome rearrangement, whereas the other is inherited from the cytogenetically normal parent.

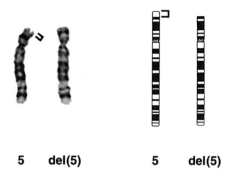

5   del(5)   5   del(5)

Fig. 2. A terminal deletion involving the distal short arm of chromosome 5 [del(5)(p15.3)]. Patients with similar deletions are said to have cri du chat or cat cry syndrome because of the characteristic cat-like cry present in many during infancy.

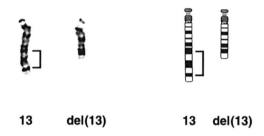

13   del(13)   13   del(13)

Fig. 3. An interstitial deletion involving the long arm of chromosome 13 [del(13)(q21.3q33)].

DELETIONS

Autosomal deletions that can be detected by traditional, high-resolution, or molecular cytogenetic methods produce monosomies that are generally associated with significant pathology. Some exceptions, however, do exist. Loss of the short arm material from acrocentric chromosomes during the formation of Robertsonian translocations, for example, has no impact on phenotype. Similarly, the striking size variation of heterochromatic regions in normal individuals suggests that loss of some, if not all, of this material is insignificant. There have even been reports of "benign" deletions in regions that have traditionally been considered euchromatic. Gardner and Sutherland catalog deletions of this type in bands 5p14, 11p12,11q14, 13q21, and 16q21 *(13)*.

Among deletions of pathologic significance, classic cytogenetic deletions that can be detected by routine methodology tend to be larger and associated with major malformations. Generally, large deletions have a more significant impact on phenotype and survival than smaller ones. The nature of the deleted material, however, also plays an important role in determining whether or not a specific deletion is viable. Thus, deletions of large segments of the short arms of chromosomes 4 and 5, and of the entire short arm of chromosome 18, are recurrent abnormalities among infants with major malformations, whereas deletions of similar size involving the short arms of chromosomes 17 and 19 are rarely, if ever, seen in liveborns *(14)*.

Classic deletions have traditionally been described as either terminal (**Fig. 2**) or interstitial (**Fig. 3**) based on chromosome banding patterns. A deletion is considered "terminal" if there is no discernible material beyond the site of initial breakage. Conversely, interstitial deletions have a proximal breakpoint, missing material, and a more distal breakpoint beyond which the chromosome continues with a normal banding pattern to its terminus. All stable chromosomes have telomeres comprised of the human consensus telomere sequence, TTAGGG. Chromosomes with apparent terminal deletions are no exception, and are assumed to have acquired "new" telomeres following the deletion event.

Although many deletions can be easily classified as terminal or interstitial, this is not always the case. This is illustrated in a study by Schwartz et al., who used molecular techniques to characterize 25 chromosomes with deletions that had been previously classified as terminal *(15)*. They found that five of these chromosomes maintained their original subtelomeric sequences, and concluded that at least some deletions that appear to be terminal by cytogenetic analysis are actually interstitial.

The use of high-resolution banding and molecular cytogenetic techniques has led to the identification of a new class of cytogenetic abnormality variously referred to as chromosomal microdeletions, contiguous gene syndromes, and, more recently, segmental aneusomy syndromes (SAS). These abnormalities are mostly very small interstitial deletions, often at or below the resolution of microscopic analysis, that recur with appreciable frequency and are associated with distinct clinical phenotypes. The term "microdeletion" is descriptive, but fails to include the minority category of "microduplications" (e.g., CMTIA; see also "Duplications" later) and the variable etiologies for some of the disorders. The term "contiguous gene syndromes" was introduced in 1986 to describe the involvement of multiple contiguous genes in the production of a clinical phenotype *(16)*. Although this terminology remains appropriate for some of the disorders in this new category, recent molecular investigations have shown that others are actually single-gene disorders, or the result of imprinting defects or uniparental disomy (see Chapter 16). In an effort to more accurately characterize the pathogenesis of these disorders, the term segmental aneusomy syndromes was proposed to imply that the phenotype is the result of "inappropriate dosage for critical genes within a genomic segment" *(17)*.

The phenotype of Williams syndrome, for example, which includes cardiovascular abnormalities, growth and developmental delays, infantile hypercalcemia, and dysmorphic facial features, results from a very small deletion (approximately 2 Mb) on the long arm of chromosome 7 *(18,19)*. At least two genes have been identified within this region that are believed to be responsible for specific aspects of the phenotype. Deletion of the elastin gene (*ELN*) accounts for the cardiovascular abnormalities, whereas loss of LIM-kinase 1, a novel kinase expressed in the brain, may explain the cognitive abnormalities in these patients *(18,20)*. Presumably, at least one other gene will be identified within the region to account for other aspects of the phenotype.

Molecular studies of the Williams syndrome deletions revealed the presence of flanking repeat sequences at the common deletion site. These repeats have provided recognition sites for unequal meiotic and mitotic exchange events that have resulted in the characteristic deletions *(19,21,22)*. Flanking repeats have also been found at the deletion sites of other SAS and probably account for the size consistency and the frequency of these disorders.

A partial listing of classic cytogenetic deletion syndrome (segmental aneusomy syndromes) can be found in **Table 1**.

DUPLICATIONS

The term "duplication" as applied to chromosome abnormalities implies the presence of an extra copy of a genomic segment resulting in a partial trisomy. A duplication can take many forms. It can be present in an individual as a "pure duplication," uncomplicated by other imbalances (**Fig. 4**), or in combination with a deletion or some other rearrangement. Examples of some types of rearrangements that involve duplications include isochromosomes, dicentrics, derivatives, recombinants, and markers. The origins and behavior of these abnormal chromosomes are discussed elsewhere in this chapter.

Tandem duplications represent a contiguous doubling of a chromosomal segment. The extra material can be oriented in the same direction as the original (a direct duplication), or in opposition (an inverted duplication). Most cytogenetically detectable tandem duplications in humans appear to be direct *(23)*.

Autosomal duplications produce partial trisomies and associated phenotypic abnormalities. Very few duplications, however, have occurred with sufficient frequency or been associated with such a strikingly characteristic phenotype that they have been recognized as defined clinical syndromes (**Table 2**). A few cases of distal 3q duplication have been reported in patients with features similar to Cornelia de Lange syndrome. However, these patients also have additional abnormalities not usually associated with the syndrome *(14)*. Paternally derived duplications of distal 11p have also been associated, in some cases, with Beckwith–Wiedemann syndrome *(24)*.

More intriguing, and perhaps more significant, is the emerging recognition of recurring duplications in regions associated with microdeletion syndromes. Many patients, for example, have been observed with apparent duplications in the proximal long arm of chromosome 15. These duplications are thought to encompass the same approximate region that is deleted in Prader–Willi and Angelman syndrome (PWS/AS). The clinical significance of these duplications has been difficult to assess, however, because of the wide variability in clinical findings among these patients. The observation that many of these patients have phenotypically normal relatives with the same duplication has further complicated the issue. Using molecular techniques, investigators have been able to demonstrate duplications in some of these families, but not others. Preliminary evidence suggests that a subset of these patients carry a maternally inherited recurring duplication that involves the imprinted region on 15q associated with PWS and AS. These duplications can be detected by FISH, and are associated with a phenotype that includes mental retardation and autism without major malformations *(25,26)*. Similarly, several patients have been reported who have a duplication of proximal 17 short arm, involving the same loci that are deleted in Smith–Magenis syndrome patients. Consistent clinical features in these duplication patients include growth and developmental delay *(27)*.

These observations, and others, suggest a common mechanism of origin for some duplications and deletions. Chromosomal duplications occur frequently in bacteria and *Drosophila,* and contribute significantly to human disease as well. A common feature

Table 1
Some Recurring Deletion Syndromes

Deletion Syndrome	Deleted Region	Key Clinical Feature
Wolf-Hirschhorn	4p[a]	Mental and growth retardation, microcephaly, hypertelorism, broad nasal bridge, down-turned mouth, cleft lip and/or palate, micrognathia, cryptorchidism, hypospadias.
Cri du Chat	5p[a]	Mental and growth retardation, cat-like cry in infancy, microcephaly, round face, hypertelorism, downslanting palpebral fissures.
Williams	7q11.23[b]	Mental retardation, short stature, supravalvular aortic stenosis, hypercalcemia, friendly disposition, hoarse voice, periorbital fullness, stellate pattern in the iris, anteverted nares, long philtrum, full lips.
Jacobsen	11q24.1- 11qter[a]	Mental and growth retardation, trigonocephaly, strabismus, cardiac defects, digit anomalies, thrombocytopenia.
Langer-Giedion (Tricho-Rhino-Phalangeal syndrome Type II)	8q24.11 – 8q24.13[a]	Mental and growth retardation, multiple exostoses, cone shaped epiphyses, fine scalp hair, bulbous nose, prominent ears, simple but prominent philtrum, loose redundant skin in infancy.
Angelman	maternal deletion of 15q11 – 15q13[a]	Mental and growth retardation, inappropriate laughter, ataxia and jerky arm movements, seizures, maxillary hypoplasia, deep-set eyes, large mouth with protruding tongue, widely spaced teeth, prognathia.
Prader-Willi	paternal deletion of 15q11 – 15q13[a]	Mental and growth retardation, hypotonia and feeding problems in infancy, later obesity associated with hyperphagia, narrow bifrontal diameter, almond shaped eyes, small hands and feet, hypogonadism, skin picking.
Miller-Dieker	17p13.3[a]	Mental and growth retardation, lissencephaly, microcephaly, bitemporal depression, long philtrum, thin upper lip, mild micrognathia, ear dysplasia, anteverted nostrils.
Smith-Magenis	17p11.2[a]	Mental retardation, behavioral problems, hyperactivity, sleep disturbance, decreased pain sensitivity, short stature, brachycephaly, midface hypoplasia, prognathism, fingertip pads, hoarse voice.
DiGeorge/Velo-Cardio-Facial (Shprintzen)	22q11.2[a]	Learning disabilities, short stature, overt or submucous cleft palate, velopharyngeal incompetence, prominent nose with squared nasal root and narrow alar base, conotruncal cardiac defects, and psychiatric disorders in some.
Kallmann[c]	Xp22.3[b]	Hypogonadotropic hypogonadism, eunuchoid habitus, anosmia or hyposmia, bimanual synkinesia.
Ichthyosis (X-linked)[c]	Xp22.3[b]	Hypertrophic ichthyosis, corneal opacities without impairment of vision.

Deletion is frequently visible (a) or typically not visible (b) using traditional cytogenetics. This has been seen in association with several other X-linked disorders when it occurs as part of a contiguous gene syndrome (c).

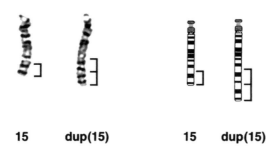

15   dup(15)   15   dup(15)

Fig. 4. A duplication involving the distal long arm of chromosome 15 [dup(15)(q24q26.3)]. This duplication was initially observed in the bone marrow of a patient with mental retardation and leukemia. By obtaining a peripheral blood karyotype we were able to demonstrate that the duplication was constitutional and apparently unrelated to his leukemia.

noted among the duplications of these diverse organisms is the presence of repetitive flanking DNA sequences. These repeats appear to serve as substrates for unequal exchange events following misalignment of sister chromatids or homologues. The products of these unequal exchanges include chromosomes with complementary duplications and deletions *(28)*. Brown et al. performed detailed molecular analyses on two patients with duplications of proximal 17p *(27)*. They concluded that the duplications seen in one of them probably resulted from an unequal sister chromatid exchange in the patient's mother. The duplicated region in both of these patients is thought to coincide with the region that is deleted in many Smith–Magenis patients.

INVERSIONS

Inversions are intrachromosomal rearrangements formed when a chromosome breaks in two places and the material between the two breakpoints reverses orientation. Inversions can be of two types: pericentric or paracentric. In pericentric inversions, the breakpoints lie on either side of the centromere and formation of the inversion often changes the chromosome arm ratio (centromere position) and alters the banding pattern of the chromosome (**Fig. 5, 6, and 8**). Paracentric inversions, on the other hand, have both breakpoints on the same side of the centromere, or within a single chromosome arm (Chapter 10, **Fig. 5**). In paracentric inversions, the centromere position does not change and the only clue to their presence is an alteration in the chromosome banding pattern. Prior to the development of banding techniques, the existence of paracentric inversions was theorized but could not be proven.

In those studies in which parents of a proband with an inversion have been karyotyped, the inversion is found in a parent as often as 85–90% of the time *(31,32)*. Thus most inversions, whether pericentric or paracentric, appear to be inherited.

Pericentric Inversions

Both recurring and unique pericentric inversions have been reported in humans. Some recurring inversions are considered normal variants. In these polymorphic inversions, a block of heterochromatin normally situated in the proximal long arm of the chromosome is inverted into the short arm of the chromosome. Such inversions are found in chromosomes 1, 9 (**Fig. 5**), and 16. A second group of apparently benign

Table 2
Some Recurring Duplication/Triplication Syndromes

Duplication/Triplication Syndrome	Duplicated/ Triplicated Region	Key Clinical Features
Duplication 3q	?3q26.3	A Cornelia de Lange-like phenotype that includes mental retardation, postnatal growth retardation, long philtrum, palate anomalies, anteverted nares, clinodactyly, talipes, renal and cardiac abnormalities.
Beckwith-Wiedemann	11p15.5 (paternal)	Macrosomia, macroglossia, organomegaly, omphalocele, ear creases, hypoglycemia, tumor susceptibility. Beckwith-Wiedemann patients with cytogenetic duplications are more likely to have learning difficulties.
Pallister-Killian	Mosaic tetrasomy 12p Usually secondary to an extra metacentric isochromosome	Mental retardation, streaks of hyper- and hypopigmentation, sparse anterior scalp hair, sparse eyebrows and eyelashes, prominent forehead, protruding lower lip, coarsening of face with age
Pseudodicentric 15 (Inverted duplicated 15)	Tetrasomy 15pter-15q13 Due to the presence of an extra pseudo-dicentric chromosome	Mental and growth retardation, autism, behavioral disturbance, seizures, low posterior hairline, epicanthal folds, low-set ears, strabismus. The smaller pseudodicentric 15 chromosomes may not cause phenotypic abnormalities.
Cat-eye	Tetrasomy 22q11.2 (occasionally trisomy) Usually secondary to an extra pseudodicentric or ring chromosome	Usually mild mental retardation, coloboma of the iris, downslanting palpebral fissures, preauricular tags and/or fistulas, anal atresia

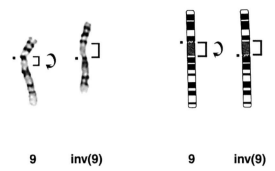

Fig. 5. This benign inversion of chromosome 9 [inv(9)(p11q13)] represents a pericentric inversion with breakpoints in both chromosome arms. The material between the two breakpoints has been inverted, the block of heterochromatin that normally sits in the long arm has been shifted to the short arm, and the banding pattern has been subtly changed. Because the breakpoints have not occurred symmetrically with respect to the centromere, the short arm to long arm ratio of the inverted chromosome has been altered as well.

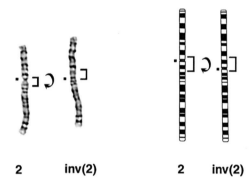

Fig. 6. Although this recurring pericentric inversion [inv(2)(p11q13)] is considered to be benign, individuals who carry this inversion may have a slightly increased risk for miscarriages.

recurring inversions, which have breakpoints very near the centromere in both the long and short arms, are found in chromosomes 2, 3, and 10, and in the Y chromosome. These variant forms have been observed in a large number of families and appear to segregate without deleterious effect. One group of investigators, however *(29),* have reported an increased risk for miscarriage among carriers of a pericentric inversion of chromosome 2 [inv (2) (p11q13), **Fig. 6**]. Other inversions have been observed in many families but are not without consequence. Of particular note is the inversion of chromosome 8 with breakpoints at p23 and q22, which has been seen in families of Mexican-American descent *(30).*

Unique inversions are those observed in a single individual or family. The clinical significance of these inversions must therefore be determined on a case-by-case basis; as described later, some inversions can impart substantial reproductive risk, depending on the chromosome segment involved.

Excluding the variant inversions discussed previously, the frequency of pericentric inversions in the human population has been estimated at 0.12–0.7% *(13).*

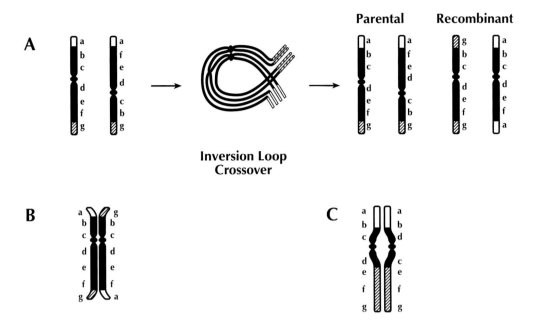

Fig. 7. Several models for meiotic pairing in a pericentric inversion heterozygote. **A:** An inversion loop containing a single crossover and the resulting parental and recombinant chromosomes. Note that only the material that is distal to the inversion breakpoints has been duplicated/deleted in each recombinant chromosome. **B, C:** Alternate models for pairing during which only partial pairing or synapsis occurs.

Meiotic Behavior and Risks for Carriers of Pericentric Inversions

To understand the reproductive risks of an inversion carrier (heterozygote), the meiotic behavior of inverted chromosomes must first be considered. During meiosis, homologous chromosomes pair in close association. During this pairing phase, genetic information is exchanged between homologues through a process known as crossing over or recombination (see Chapter 2). Crossing over appears to be a necessary step for orderly chromosome segregation, and is the mechanism that ensures human genetic individuality. A chromosome pair that consists of one normal chromosome and one chromosome with an inversion cannot achieve the intimate pairing of homologous regions necessary for normal meiosis through simple linear alignment. The classic model for pairing in an inversion heterozygote is the inversion or reverse loop demonstrated in **Fig. 7A**. In this model, the inverted segment forms a loop that can then pair with homologous regions on the normal chromosome. The noninverted portions of the chromosome (the chromosome segments distal to the inversion breakpoints) pair linearly with homologous regions on the normal chromosome. An odd number of crossovers between the same two chromatids within the inversion loop will result in the production of recombinant chromosomes, while an even number of crossovers between the same two chromatids within the inversion loop should result in the production of normal or balanced chromosomes.

Two types of recombinant chromosomes are formed when crossing over occurs between the inversion breakpoints. One recombinant will contain a duplication of the

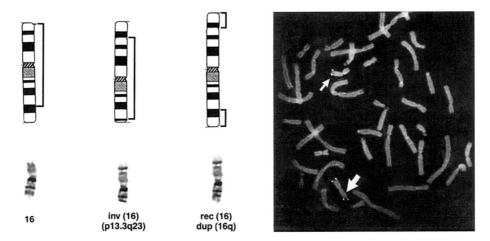

Fig. 8. Left: A normal 16, an inverted chromosome 16, and a recombinant chromosome 16 [rec(16)dup(16q)inv(16)(p13.3q23)] resulting from recombination within the inversion loop of the parental inversion carrier. The recombinant chromosome 16 is missing the material distal to the short arm breakpoint and contains a duplication of the material distal to the breakpoint within the long arm. **Right**: A metaphase that has been hybridized with a FISH probe specific for the subtelomeric region in the long arm of chromosome 16. A signal is seen on the distal long arm of the normal chromosome 16 (small arrow) and on both arms of the recombinant chromosome 16 (large arrow) confirming the duplication of long arm material.

material distal to the break point on the short arm, and a deletion of the material distal to the breakpoint in the long arm. The second recombinant is complementary to the first and contains a short arm deletion and a long arm duplication (**Fig. 8**). Both recombinants are known as duplication-deficiency chromosomes.

Alternate models for pairing in an inversion heterozygote are seen in **Figs. 7B, C**. In inversions with very small inverted segments (breakpoints are close to the centromere and the distal segments are large) the noninverted segments of both chromosomes may pair in linear fashion, with asynapsis or failure to pair in the small inverted segment. In this model, crossing over can only take place in the noninverted segments of the chromosomes, and thus abnormal recombinant chromosomes are not formed. In the opposite situation, where the inverted segment is very large relative to the size of the entire chromosome and the distal segments are small, pairing may occur only between the inversion breakpoints, and the distal material will remain unpaired. In this situation, a crossover between the inversion breakpoints will produce recombinant chromosomes in a manner similar to the reverse loop model discussed previously. Crossing over cannot take place in the segments distal to the inversion breakpoints because those regions do not pair.

Careful examination of the recombinant chromosomes produced when crossing over takes place between the breakpoints in a pericentric inversion reveals that the genetic imbalance always involves the material distal to the inversion breakpoints. Thus, large inversions have small distal segments and produce recombinant chromosomes with small duplications and deficiencies, while small inversions have large distal segments and produce recombinant chromosomes with large duplications and deficiencies. In general, then, large inversions are associated with a greater risk of producing abnormal liveborn offspring, as the recombinant chromosomes associated with them carry small duplications and deficiencies that have a greater probability of being compatible with survival. Furthermore, the larger the inversion, the greater the likelihood that a recombination event within the inversion loop will occur and form recombinant chromosomes. The opposite is true of small inversions with large distal segments, which are usually associated with a very low risk of liveborn abnormal offspring.

In addition to the size of the inverted segment, other factors must be considered when determining the reproductive risk associated with any given pericentric inversion. Because monosomies are generally more lethal than trisomies, with the exception of the sex chromosomes, only inversions that produce recombinants with a very small monosomy are associated with a high risk of abnormal offspring.

The nature of the genetic material in the inverted chromosomes can also be important. For instance, both trisomy and partial monosomy of chromosomes 13, 18, and 21 are seen in liveborn infants with birth defects and mental retardation. Once the duplications and deficiencies associated with the recombinants from a particular inversion are identified, review of the medical literature for evidence that these duplications and/or deficiencies are compatible with survival can aid in predicting the magnitude of the risk associated with that particular inversion.

Another clue to the level of risk associated with a given inversion is the manner in which the inversion was ascertained. If a balanced inversion is ascertained fortuitously, for instance during a prenatal chromosome study because of advanced maternal age, the risk associated with such an inversion is probably very low. On the other hand, an inversion that is ascertained through the birth of an infant with anomalies secondary to the presence of a recombinant chromosome is associated with a much higher risk, because the important question of whether the recombinant offspring is viable has already been answered. Careful examination of the family history in both types of ascertainment can provide additional important information in assessing risk.

Gardner and Sutherland reviewed several studies that contain data about the risks associated with pericentric inversions and estimated the risk for an inversion heterozygote to have an abnormal child secondary to a recombinant chromosome *(13)*. This risk was estimated to be 5–10% in families ascertained through an abnormal child and approximately 1% for families ascertained for any other reason. For families segregating very small inversions, the risk of having a liveborn recombinant child may be close to zero. In cases of recurring inversions, additional information about the risks can be gained from studying the literature. In the case of the inversion (8)(p23q22) mentioned earlier, for example, enough recombinant offspring have been observed to derive an empiric risk of 6% for a heterozygote to have a liveborn recombinant child *(33)*. Large inversions with distal segments that have been seen in liveborn children as monosomies or trisomies may be associated with high risk regardless of their mode of ascertainment in a particular family.

Paracentric Inversions

The presence of paracentric inversions in the human population was appreciated only after the advent of chromosome banding, and they are still reported less frequently than pericentric inversions. Their incidence has been estimated at 0.09–0.49 per thou-

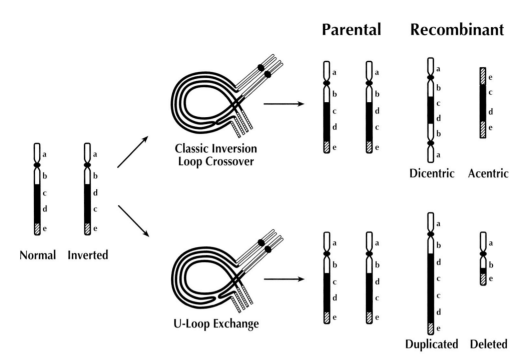

Fig. 9. The type of recombinant chromosome produced depends on which mechanism of chromosome exchange occurs within the paracentric inversion loop. A classic crossover within the inversion loop results in the formation of an acentric and a dicentric recombinant chromosome (**top**) while a U-type exchange produces only monocentric chromosomes (**bottom**).

sand *(31)*. Recurring paracentric inversions have been reported in the short arms of chromosomes 3 and 6 and in the long arms of chromosomes 7, 11, and 14. A recurring 11(q21q23) inversion has been observed in a large number of families in the Netherlands *(34)* and in Canadian Hutterites *(35)*.

Meiotic Behavior and Risk for Carriers of Paracentric Inversions

As with pericentric inversions, the classic solution to the problem of homologous pairing in paracentric inversions is the reverse loop. In this case, however, the centromeres are found in the segment distal to the inversion loop. On a theoretical basis, an odd number of crossovers within the inversion loop of a paracentric inversion should produce one dicentric and one acentric recombinant chromosome (**Fig. 9**). The dicentric recombinant is genetically unstable because each of the two centromeres could potentially orient toward opposite poles of the dividing cell. The material between the two centromeres would remain stretched between the poles of the two reorganizing daughter nuclei or break. Thus, with each cell division, the dicentric recombinant chromosome has a new opportunity to contribute a different and possibly lethal genetic imbalance to a new generation of cells. The acentric fragment, on the other hand, has no ability to attach to a spindle, because it lacks a centromere. Consequently, at cell division, it can be passively included in the daughter nuclei or be lost. Dicentric and acentric recombinant chromosomes are almost always lethal and are rarely found in liveborn children (see acentric chromosomes and dicentric chromosomes below).

Although dicentric and acentric recombinants are very rarely seen, there have been several reports of monocentric recombinants among the children of paracentric inversion carriers. Pettenati et al. identified 15 patients with monocentric recombinant chromosomes among 446 paracentric inversions they reviewed *(31)*. A variety of mechanisms have been proposed for the formation of these abnormal chromosomes with duplications and/or deletions, including breakage of dicentric recombinants, unequal crossing over, and abnormal U-loop exchanges similar to the one diagrammed in **Fig. 9**. All of these mechanisms involve abnormal processes of one type or another.

There is currently a fair amount of controversy surrounding the question of risks for having liveborn children with abnormalities secondary to the presence of a balanced paracentric inversion in a family. Pettenati et al. found a total of 17 recombinants (including the 15 monocentric recombinants discussed previously) among 446 inversions studied, but because of incomplete information and ascertainment bias they could not generate an accurate empiric risk estimate *(31)*. Gardner and Sutherland estimated that the risk for a paracentric inversion heterozygote to produce a child with one of the monocentric recombinants discussed earlier "lies in the range of 0.1–0.5%" *(13)*.

Many questions remain to be answered concerning the clinical significance of apparently balanced inherited paracentric inversions. Although the majority of these inversions have been ascertained incidentally, many have been found in children with birth defects and mental retardation. Some believe that familial paracentric inversions are relatively innocuous and carry a very small risk for abnormal offspring *(36,37)*, whereas others *(31)* express concern that not all of the associations of abnormal phenotypes with apparently balanced inherited paracentric inversions can be explained by coincidence and ascertainment bias.

DICENTRIC CHROMOSOMES

Any chromosome exchange in which the involved donor and recipient chromosome segments each contain a centromere will result in the formation of a chromosome with two centromeres. These chromosomes are referred to as dicentrics. The most common dicentric chromosomes are those that are derived from Robertsonian translocation events. Recombination within a paracentric inversion loop is also a well-documented method by which a dicentric chromosome can form (see "Inversions" earlier).

As one might suspect, the presence of two active centromeres in a single chromosome has the potential to wreak havoc during cell division. Normal segregation can occur only when the spindle apparatus from a single pole binds both centromeres of the dicentric chromosome. If instead the spindle apparatus from both poles independently bind only one of each of the two centromeres, the chromosome will be simultaneously pulled in opposing directions. As a result of this bipolar pulling, the chromosome may continue to straddle both daughter cells in a state of limbo until it is ultimately excluded from both. Alternatively, the chromosome may break, allowing some portion to go to each daughter cell. Regardless of which of these takes place, changes in the genetic content of the resulting sister cells will occur and mosaicism can result. Interestingly, not all dicentric chromosomes demonstrate mitotic instability. Some of these stable dicentric chromosomes appear to have closely spaced centromeres that function as a single large centromere *(38–40)*. The presence of one active and one inactive centromere is also frequently observed among stable dicentric chromosomes. These

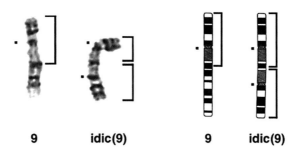

9 **idic(9)** **9** **idic(9)**

Fig. 10. A pseudoisodicentric chromosome involving the entire short arm and a portion of the long arm of chromosome 9. It appears to have one constricted active centromere (*upper dot*) and one unconstricted inactive centromere (*lower dot*). This chromosome was found in each of the cells of a phenotypically abnormal infant with the following karyotype; 47,XY,+psu idic(9)(q21.1).

"pseudodicentric" chromosomes contain two copies of the centromeric heterochromatin, but only the centromere with the primary constriction appears to bind the appropriate centromere proteins required for activity *(39,41)*. An example of a pseudoisodicentric chromosome 9 is shown in **Fig. 10**.

ACENTRIC CHROMOSOMES

Because the centromere is essential for chromosomal attachment to the spindle and proper segregation, chromosomes lacking this critical component are rapidly lost. Therefore, although single cells with acentric chromosomes or fragments are occasionally observed, individuals with constitutional karyotypes that include a true acentric chromosome are never seen. A handful of chromosomes with atypical centromeres have, however, been reported in the literature *(42–47)*. Like traditional centromeres, these too are denoted by the presence of a primary constriction. Interestingly, however, these "centromeres" are located in noncentromeric regions, and they interact with only a subset of the centromeric proteins that typically bind active centromeres. Furthermore, they do not react to stains specific for centromeric heterochromatin nor do they hybridize to centromere specific fluorescence *in situ* hybridization probes. These data suggest that the composition of these structures (neo-centromeres) may differ from that of a traditional centromere. Although these atypical centromeres appear to function reasonably well during cell division, the high incidence of mosaicism associated with chromosomes containing them suggests that they frequently do not function as well as their normal counterparts and are therefore lost during mitosis. It has been speculated that these atypical centromeres may represent ancient centromere sequences that have been reactivated as a consequence of chromosome rearrangement *(43,48)*.

ISOCHROMOSOMES

An isochromosome consists of two copies of the same chromosome arm joined in such a way that the arms form mirror images of one another. Isochromosomes may have a single centromere or may have two centromeres, in which case they are called isodicentric chromosomes. Individuals with 46 chromosomes, one of which is an isochromosome, are monosomic for the genes within the lost arm and trisomic for all

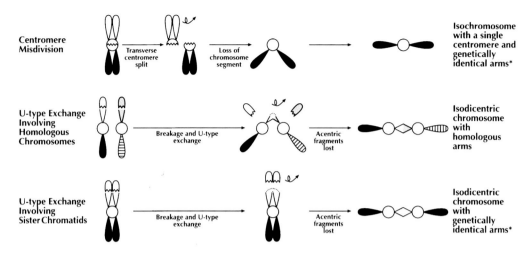

Fig. 11. Some of the mechanisms proposed for isochromosome formation. *Because recombination occurs during normal meiotic cell division, the arms of an isochromosome formed during meiosis would be identical only for markers close to the centromere.

genes present on the isochromosome. Tetrasomy for the involved chromosome segment is present when an isochromosome is present as an extra chromosome. In general, the smaller the isochromosome, the smaller the imbalance and the more likely the survival of the fetus or child that carries the isochromosome. It is therefore not surprising that, with few exceptions, the most frequently reported autosomal isochromosomes tend to involve chromosomes with small arms. Some of the more common chromosome arms involved in isochromosome formation include 5p, 8p, 9p, 12p, 18p, and 18q. The relatively large isochromosome involving the long arm of the X chromosome, shown in Chapter 10, **Fig. 3**, is the most common structural abnormality found in Turner syndrome patients.

Over the years a number of theories have been proposed to explain the mechanism of isochromosome formation *(38,49–52)*. One of the more popular proposals has been that isochromosome formation is the result of centromere misdivision (**Fig. 11**). Instead of splitting longitudinally to separate the two sister chromatids, the centromere was hypothesized to undergo a transverse split that separated the two arms from one another. Recent molecular studies, however, suggest that the breakage and reunion events required to form some isochromosomes may occur predominantly within the area adjacent to the centromere, rather than within the centromere itself *(38,53,54)*. The resulting chromosome, which appears monocentric at the cytogenetic level, would actually have two closely spaced centromeres and would more appropriately be called an isodicentric chromosome. Recently, other theories that invoke exchanges between homologous chromosomes have also been challenged as common mechanisms of isochromosome formation. Molecular evidence indicating that at least some isochromosomes are formed from genetically identical arms, rather than homologous arms, suggests that one predominant mechanism of isochromosome formation may rely on sister chromatid exchange *(52,53,55–58)*. Breakage and reunion involving the pericentromeric regions of sister chromatids, an event sometimes referred to as a sister chromatid U-type exchange, may therefore represent an important mechanism isochromosome formation. It is clear, how-

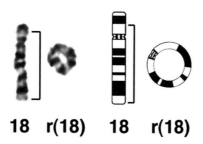

18   r(18)   18   r(18)

Fig. 12. A ring chromosome 18 [r(18)(p11.2q23)]. This ring chromosome is the result of fusion between two broken arms. The chromosome material distal to the breakpoints in each arm has been lost because it lacks a centromere.

ever, that other mechanisms of isochromosome formation also exist. Precisely which mechanism is found to predominate may largely depend on the chromosomal origin of the isochromosome, whether the chromosome is present in a disomic karyotype or represents an extra or supernumerary chromosome, and whether formation occurs during meiosis or mitosis. More extensive studies examining these issues are needed to establish a more complete understanding of isochromosome formation.

RING CHROMOSOMES

Autosomal ring chromosomes are rare, and usually arise *de novo* (**Fig. 12**). Reported frequencies range from 1 in 27,225 to 1 in 62,279 in consecutive newborn and prenatal diagnosis studies *(59)*. Rings have been reported for all chromosome pairs, although those involving chromosomes 13 and 18 are among the most common *(59)*. When ring chromosomes replace a normal homologue in a karyotype, they often represent a partial monosomy for both long and short arm material. When rings are present as supernumerary chromosomes, partial trisomies result.

Rings are traditionally thought to form as a result of breakage in both arms of a chromosome, with subsequent fusion of the ends and loss of the distal segments. Recent molecular studies, however, have suggested additional mechanisms. In a 1991 study, Callen et al. characterized 10 small supernumerary rings using FISH *(60)*. They found that some of the rings were missing specific satellite DNA sequences from one side of the centromere, suggesting that these rings originated from a "transverse misdivision of the centromere" combined with a U-type exchange on one of the chromosome arms *(60)*. In other studies, investigators have demonstrated that some rings form by telomere fusion, with no detectable loss of genetic material *(61)*. Yet other rings seem to be "breakdown" products of larger rings, and are composed of discontinuous sequences *(62)*.

One of the more striking characteristics of ring chromosomes is their instability. This instability is thought to result from sister chromatid exchanges that occur in the ring chromosome before cell division. Such exchanges are normal events that, because of the unique structure of the ring chromosome, lead to the formation of double sized dicentric rings and interlocking rings. Rings with even larger numbers of centromeres are also occasionally seen. The centromeres of these multicentric and interlocking rings can orient toward opposite poles during cell division. This can lead to breakage of the ring at anaphase, with subsequent generation of new ring structures. Alternatively, the entire ring chromosome can be lost. This active process of creating new cells with altered genetic material is termed "dynamic mosaicism" *(13,59)*. Not all ring chromosomes exhibit instability, however. Although the relationship between ring size and stability is not entirely clear, in most cases, smaller rings appear to be more stable than large rings *(13)*.

In addition to mosaicism, the genetic content and breakpoints of the rings will also have a significant impact on the patient's phenotype. A heterozygote with a partially deleted ring chromosome will have clinical findings associated with a partial monosomy. The specific phenotype of the individual will depend on both the amount and the nature of the deleted material. Similarly, for a patient with a supernumerary ring chromosome, the size of the ring, its genetic content, and the proportion of cells that contain the ring will all influence phenotype.

Another phenomenon that has the potential to impact on the phenotype of individuals with ring chromosomes is uniparental disomy (the inheritance of both copies of a chromosome pair from one parent, see Chapter 16). Petersen et al. described a patient with mosaicism for a normal cell line and a cell line in which one normal copy of chromosome 21 was replaced by a ring 21. Uniparental isodisomy for chromosome 21 was present in the cytogenetically normal cell line. The authors suggested that the isodisomy developed when the normal 21 was duplicated in a cell that had lost the ring ("monosomy rescue") *(63)*. Similarly, Crolla reported a patient with a supernumerary ring 6 in which the normal copies of chromosome 6 showed paternal isodisomy *(64)*.

One recurring phenotype seen in ring chromosome heterozygotes is the "ring syndrome," originally proposed by Cote et al. in 1981 *(65)*. These patients have 46 chromosomes, one of which is a ring chromosome with no detectable deletion. The ring is derived from one of the larger chromosomes in the karyotype, and the larger the chromosome, the more severe the phenotype. Typically, these patients have severe growth retardation without major malformations. Minor anomalies and mild to moderate mental retardation are often part of the picture. The ring syndrome is believed to result from instability of the ring chromosome. The larger chromosomes are thought to be more unstable than the smaller ones because they present more opportunities for sister chromatid exchange. The breakage that occurs during cell division generates new ring structures, most of which represent a more serious genetic imbalance than the previous forms and are thus less viable. This results in increased cell death, and contributes to growth failure and the disturbance of developmental pathways *(66)*. Kosztolanyi has proposed that this phenomenon may also contribute to the severity of the phenotype in patients who have ring chromosomes with obvious deletions *(66)*.

A 1991 literature review discovered 32 reported cases in which a ring chromosome was inherited from a carrier parent. The authors concluded that no more than 1% of ring chromosomes are inherited. Among the 32 patients with inherited rings, half had a phenotype similar to the carrier parent, while approximately one third were more severely affected *(67)*. In more than 90% of inherited ring chromosome cases, the carrier parent is the mother *(13)*.

In addition to the risks associated with ring instability, carriers of ring chromosomes may also be at risk for having children with other abnormalities involving the chromosome from which their ring is derived. There are at least three reports of carriers of a

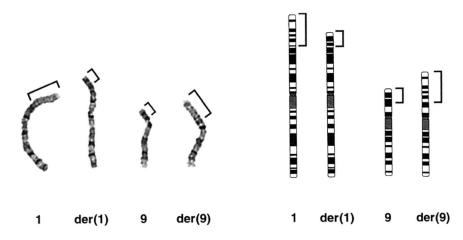

| 1 | der(1) | 9 | der(9) | | 1 | der(1) | 9 | der(9) |

Fig. 13. A balanced reciprocal translocation involving the short arm of chromosomes 1 and 9 [t(1;9)(p32.3;p21)]. The translocated segments of each chromosome have been bracketed.

ring chromosome 21 who had offspring with trisomy 21 secondary to a translocation or tandem duplication of chromosome 21 *(67)*.

RECIPROCAL AUTOSOMAL TRANSLOCATIONS

Reciprocal translocations represent one of the most common structural rearrangements observed in man. Estimates of the population frequency range from 1/673 to 1/1000 *(1,68)*. A reciprocal translocation forms when two different chromosomes exchange segments. In the example shown in **Fig. 13**, a balanced translocation involving chromosomes 1 and 9 has occurred. The distal short arm of chromosome 1 has replaced the distal short arm material on chromosome 9, and vise-versa. The individual who carries this balanced translocation is clinically normal. His rearrangement was identified when his wife had prenatal karyotyping because of advanced maternal age and a fetus with the same (1;9) translocation was found.

Although individuals who carry truly balanced reciprocal translocations are themselves clinically normal, they do have an increased risk for having children with unbalanced karyotypes secondary to meiotic malsegregation of their translocation. As discussed in the introduction to this chapter, during normal meiotic prophase all 23 sets of homologous chromosomes couple to produce 23 paired linear structures or bivalents that later separate and migrate to independent daughter cells. In a cell with a reciprocal translocation, 21 rather than 23 bivalents are formed. The remaining two derivative chromosomes involved in the reciprocal translocation and their normal homologues form a single pairing structure called a quadrivalent. The expected quadrivalent for the reciprocal (1;9) translocation described above is diagrammed in **Fig. 14**. Notice that the four chromosomes within the quadrivalent have arranged themselves such that pairing between homologous regions is maximized.

Segregation of the chromosomes within a quadrivalent can occur in multiple ways, most of which will result in chromosomally unbalanced gametes (also demonstrated in **Fig. 14**). Only a 2:2 segregation, during which the two alternating chromosomes within the quadrivalent travel together to the same daughter cell, yields chromosomally bal-

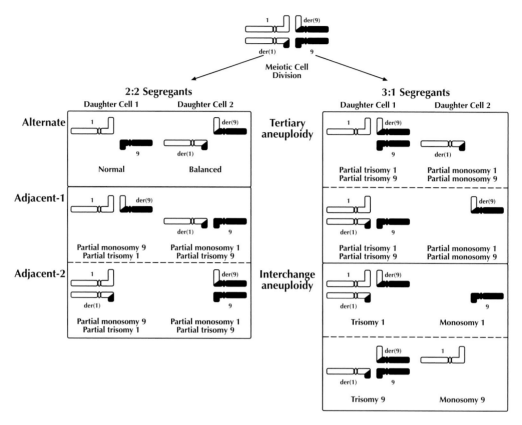

Fig. 14. The expected meiotic pairing configuration for the (1;9) translocation described in Fig. 13. Each of the 2:2 and 3:1 segregants typically produced during meiotic cell division are shown.

anced gametes (alternate segregation). In theory 50% of the resulting gametes will carry a normal chromosome complement while the other 50% will be balanced translocation carriers. Each of the remaining segregation patterns for a reciprocal translocation produces unbalanced gametes. A 2:2 segregation, during which two chromosomes with adjacent rather than alternate centromeres migrate to the same daughter cell, produces gametes with partial trisomies and monosomies (adjacent I and adjacent II segregation). Also, 3:1 and 4:0 segregations can occur, resulting in trisomies and monosomies or tetrasomies and nullisomies, respectively. Studies examining the sperm obtained from balanced reciprocal translocation carriers suggest that approximately equal numbers of alternate and adjacent segregants are generally formed and that these two groups represent the most common types of segregants. The remaining 3:1 and 4:0 segregants appear to be much more rare. Corresponding data are not available for female carriers, as large numbers of oöcytes are much more difficult to obtain and study than spermatocytes.

In addition to being inherited, reciprocal translocations can also occur as new or *de novo* mutations. As discussed in the introduction to this chapter, the risk for an abnormal outcome associated with a *de novo* apparently balanced rearrangement is always greater than that associated with an equivalent rearrangement that has been inherited from a normal parent. The actual risk associated with a *de novo* apparently "balanced"

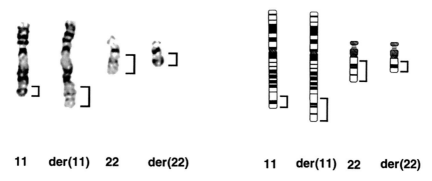

Fig. 15. A balanced reciprocal translocation involving the long arm of chromosomes 11 and 22 [t(11;22)(q23.3;q11.2)]. This is the only constitutional translocation that has been reported in multiple, apparently unrelated families.

translocation has been reported to be approximately 6–9% *(1)*. This is two to three times the overall rate of congenital abnormalities observed in the population.

The (11;22) Translocation

The (11;22) translocation, with breakpoints within 11q23.3 and 22q11.2, is unique because it represents the only recurring constitutional reciprocal translocation in humans (**Fig. 15**). More than 100 apparently unrelated families have been reported to date. It has yet to be determined, however, whether the ostensible reoccurrence of this translocation is best explained by the efficient transmission of a single ancient unique translocation through multiple generations or by multiple independent translocation events between two susceptible regions.

The presence of multiple families with the same apparent (11;22) translocation has made it possible to obtain good empiric data concerning viable segregants, expected phenotypes, and the various risks associated with this rearrangement. We know, for example, that a carrier's empiric risk for having a liveborn child with an unbalanced karyotype is 2–10% *(69,70)*. We also know that the unbalanced, liveborn offspring of (11;22) translocation carriers typically have 47 chromosomes—46 normal chromosomes plus an extra or supernumerary chromosome representing the derivative chromosome 22. These individuals are therefore trisomic for the distal long arm of chromosome 11 and the proximal long arm of chromosome 22. Mental retardation, congenital heart disease, malformed ears with preauricular skin tags and/or pits, a high arched or cleft palate, micrognathia, anal stenosis or atresia, renal aplasia, or hypoplasia, and genital abnormalities in males are common features shared by these unbalanced (11;22) segregants.

Balanced carriers of the (11;22) translocation are phenotypically normal with one possible exception. There is a single, unconfirmed report in the literature indicating that female carriers may have a predisposition to breast cancer *(71)*. Although cytogenetically the breakpoints involved in this translocation appear to be similar to those identified in the acquired chromosome rearrangements seen in Ewing's sarcoma, peripheral neuroepithelioma, and Askin tumor, molecular studies have shown that they differ *(72–74)*. The gene(s) and mechanisms responsible for the development of these

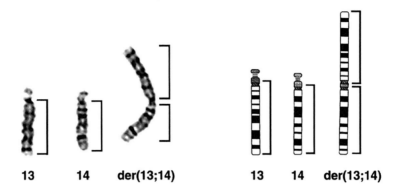

13   14   der(13;14)   13   14   der(13;14)

Fig. 16. This (13;14) translocation is the most common Robertsonian translocation observed in man [der (13;14)(p11.2;p11.2), sometimes described as der(13;14)(q10;q10); see Chapter 3].

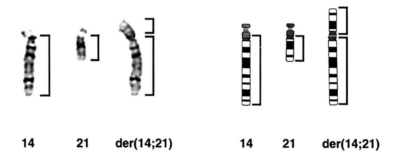

14   21   der(14;21)   14   21   der(14;21)

Fig. 17. Although less common than the (13;14) translocation, the Robertsonian (14;21) translocation is more clinically significant because the affected offspring of such a carrier are more likely to survive to birth. Their unbalanced offspring will inevitably have three copies of chromosome 21 long arm material or Down syndrome, a chromosome abnormality that is more compatible with survival than trisomy 13. [der(14;21)(p11.2;p11.2)].

neoplasms therefore have provided no clues regarding the etiology of breast cancer development in these patients.

ROBERTSONIAN TRANSLOCATIONS

A Robertsonian translocation occurs when the long arms of any two acrocentric chromosomes (13, 14, 15, 21, and 22) join to produce a single metacentric or submetacentric chromosome (**Figs. 16 and 17**). Although these translocations may in fact be reciprocal, the small complementary chromosome composed of short arm material is only occasionally seen, presumably because it is typically acentric and therefore lacks the stability conferred by a centromere *(75)*. Balanced carriers of Robertsonian translocations typically have 45 chromosomes rather than the usual 46. The only notable genetic material within the short arm region of each of these chromosomes is a nucleolar organizer region composed of multiple copies of the ribosomal RNA genes. Because this is redundant information, loss of this material from the two chromosomes involved in the translocation is not clinically significant. It has been suggested that the close association of these nucleolar organizer regions within the cell nucleus may promote the formation of Robertsonian translocations.

Since Robertsonian translocations were first described by W.R.B. Robertson in 1916, it has been recognized that these translocations are among the most common balanced structural rearrangements in the human population *(76)*. Numerous studies examining both spontaneous abortions and liveborn individuals indicate a frequency of approximately 1/1000 *(77–79)*. Although pairwise association of the five human acrocentric chromosomes can form 15 different Robertsonian translocations, these rearrangements do not occur with equal frequency and their mechanisms of formation appear to differ.

Nonhomologous Robertsonian Translocations

Approximately 95% of all Robertsonian translocations are formed between two nonhomologous or different chromosomes. Among this group, the (13;14) and (14;21) translocations are the most common and constitute approximately 75% and 10% of all nonhomologous Robertsonian translocations respectively (*13*; **Figs. 16 and 17**). Molecular studies performed to explore the origins of these rearrangements suggest that they occur predominantly during oögenesis.

Despite the monocentric appearance of many of these chromosomes, most are in fact dicentric *(80–82)*. The majority of these chromosomes therefore appear to form as a result of short arm fusion rather than centromere fusion or a combination of both. A single pair of short arm breakpoint regions has been observed in most (13;14) and (14;21) translocations, while multiple short arm breakpoint regions are utilized during formation of each of the remaining types of Robertsonian translocations *(80,83–85)*. Precisely where the breakpoint occurs within the short arm therefore seems to be dependent upon the type of Robertsonian translocations being formed. Nonrandom suppression of one centromere appears to provide mitotic stability to these dicentric chromosomes *(41,86)*.

Homologous Robertsonian Translocations

In contrast to nonhomologous Robertsonian translocations, *de novo* whole arm exchanges involving homologous or like chromosome pairs are very rare. They appear to be predominantly monocentric *(53,82)* and several of them have been shown to form postmeiotically *(58,87,88)*. Although historically all such rearrangements were collectively called homologous Robertsonian translocations, recent molecular studies have shown that approximately 90% of the chromosomes within this category may actually be isochromosomes composed of identical rather than unique homologous arms *(53,54,57,89)*. Molecular studies exploring the parental origin of *de novo* homologous Robertsonian translocations suggest that no parental bias exists. Equal numbers of maternally and paternally derived isochromosomes have been reported. True homologous Robertsonian translocations in balanced carriers appear to be composed of both a maternal and a paternal homologue suggesting a mitotic (postmeiotic) origin.

Reproductive Risks for Carriers of Robertsonian Translocations

Carriers of Robertsonian translocations are at risk for miscarriages and for offspring with mental retardation and birth defects associated with aneuploidy and rarely, uniparental disomy (UPD, the inheritance of both copies of a chromosome pair from one parent; see Chapter 16). The relative risk for each of these outcomes is a function of the sex of the heterozygous parent and/or the particular acrocentric chromosomes involved. In theory, all chromosome segregations within the carrier parent of a homologous Robertsonian translocation and all malsegregations within nonhomologous Robertsonian carriers produce monosomic or trisomic conceptions. Because all potential monosomies and most of the potential trisomies are lethal during the first trimester, miscarriage is not uncommon. Only those Robertsonian translocation chromosomes containing chromosomes 21 or 13 are associated with a significant risk for having liveborn trisomic offspring. Trisomy 22 occurring secondary to a Robertsonian translocation may also represent a rare possibility.

Occasionally, abnormal offspring with UPD have also been observed among the children of balanced Robertsonian translocation carriers *(11)*. UPD involving inherited as well as de novo homologous and nonhomologous translocations have been reported. Postzygotic correction of a trisomy through chromosome loss (trisomy rescue) is thought to represent the most likely mechanism for UPD in these offspring, although monosomy correction and gamete complementation may occur as well *(90)*. Current data indicate that UPD is most concerning when Robertsonian translocations containing chromosomes 14 or 15 are involved, as both chromosomes appear to have imprinted regions. Maternal and paternal UPD for chromosome 15 result in Prader–Willi syndrome and Angelman syndrome respectively *(91,92)*. Clinically abnormal offspring have also been documented in association with paternal *(93,94)* and maternal *(95,96)* UPD for chromosome 14. A single reported case of maternal UPD in a normal individual has created uncertainty regarding the association between maternal UPD 14 and phenotype *(94)*. While an abnormal phenotype is not likely to be directly associated with UPD for chromosomes 13, 21, and 22, residual disomy/trisomy mosaicism and recessive disease resulting from reduction to homozygosity through isodisomy may influence the phenotype of all UPD offspring *(90)* (see Chapter 16).

As discussed in the introduction, for some types of rearrangements the risk for unbalanced offspring appears to be significantly higher for a female carrier than a male carrier. This appears to be the case for nonhomologous Robertsonian translocations involving chromosome 21. In female carriers of these translocations, an unbalanced karyotype is detected in 13–17% of second trimester pregnancies *(13,97)*. For male carriers the same risk appears to be less than 2%. Precisely why male carriers appear to produce fewer unbalanced offspring than their female counterparts is not known.

JUMPING TRANSLOCATIONS

The term "jumping translocation" refers to dynamic or changing translocations that are rarely observed in constitutional karyotypes. It is used most often to describe a type of mosaicism in which a specific donor chromosome segment is translocated to two or more different recipient sites over the course of multiple mitotic cell divisions *(98–104)*. Jewett et al. have described an individual with four different cell lines in which long arm material of chromosome 15 was translocated to five different sites *(104)*. Within the child's main cell line, the chromosome 15 long arm segment was transferred to the distal long arm of chromosome 8. In additional cell lines, this same segment was transferred to the long arm of chromosome 12, the short arm of chromosome 6, or the short arm of chromosome 8.

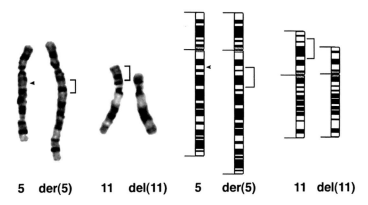

5   der(5)   11   del(11)   5   der(5)   11   del(11)

Fig. 18. Insertion. A portion of chromosome 11 short arm material has been inserted into the proximal long arm of chromosome 5 to produce an apparently balanced, inverted, interchromosomal insertion [ins(5;11)(q13.1;p15.3p13)]. The individual who carries this insertion was ascertained following the birth of a cytogenetically unbalanced child who inherited the derivative 5 but not the complementary derivative 11. (Figure courtesy of Dr. Frank S. Grass, Department of Pediatrics, Carolinas Medical Center.)

In other rare situations, families are described in which translocations involving a common donor chromosome segment but a different recipient chromosome are observed in parent and child *(105,106)*. Tomkins et al., for example, describe a mother and daughter with different, apparently balanced translocations involving the same short arm segment of chromosome 11 *(105)*. The mother carried an (11;22) translocation while the daughter carried a similar (11;15) translocation. In families like this, chromosome "jumping" appears to occur during gametogenesis rather than during mitosis as described previously.

The breakpoints observed in jumping translocations frequently involve regions known to contain repetitive DNA sequences such as telomeres, centromeres, and nucleolar organizers *(100,104,106,107)*. The location of breaks within these repetitive regions, and the suspicion that evolutionary chromosome rearrangements have distributed inactive forms of these sequences throughout the genome, suggest that recombination between homologous sequences may play a role. For now, however, the mechanism by which jumping translocations occur is unknown.

INSERTIONS

Insertions are complex three-break rearrangements that involve the excision of a portion of a chromosome from one site (two breaks) and its insertion into another site (one break). The orientation of the chromosomal material that has been moved can remain the same in relation to the centromere (a direct insertion) or be reversed (an inverted insertion). When the material is inserted into a different chromosome the insertion is considered interchromosomal, while with intrachromosomal insertions, material excised from one portion of a chromosome is reinserted into another portion of the same chromosome. An example of an interchromosomal insertion involving chromosomes 5 and 11 is shown in **Fig. 18**.

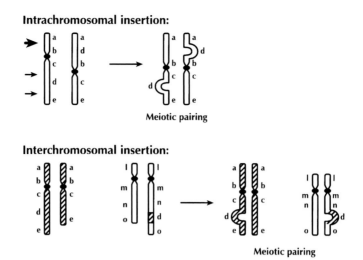

Fig. 19. Models for meiotic pairing during which partial pairing is observed between the insertion chromosome and its homologue.

Three-break rearrangements, of which insertions are an example, are extremely rare. Chudley et al. estimated that they occur ten times less frequently than two-break rearrangements, or in approximately 1 in 5000 live births *(108)*. Madan and Menko found only 27 reported cases of intrachromosomal insertions reported in the medical literature *(109)*. Although these complex rearrangements are rare, they can be associated with a very high risk for abnormal reproductive outcome.

Intrachromosomal Insertions

Intrachromosomal insertions can occur within a single chromosome arm or between chromosome arms. Direct within-arm insertions have occasionally been mistaken for paracentric inversions *(110,111)*.

During meiotic pairing, the inserted segment and its complementary region on the normal chromosome may loop out allowing synapsis, or pairing, of the rest of the chromosome (**Fig. 19**). A single crossover in the paired interstitial segments of such a bivalent results in the formation of recombinant chromosomes that are either duplicated or deleted for the inserted segment. The theoretical risk for the formation of such recombinant chromosomes could approach 50% for each meiosis, depending on the size of the interstitial segment. The risk for having a liveborn child with an unbalanced karyotype will depend, to some extent, on the viability of the duplications and deletions produced.

Alternatively, in the case of large inserted segments, complete pairing between the homologue with the insertion and its normal counterpart can be achieved through the formation of double-loop structures during meiosis. Crossing over or recombination in these fully synapsed chromosomes can result in the generation of chromosomes with duplications, deletions, or both. Madan and Menko, in their review of 27 cases, observed an overall 15% risk for each pregnancy that a carrier of an intrachromosomal insertion will have a liveborn child with an unbalanced karyotype *(109)*. This risk may differ greatly for individual insertions depending on the size of the inserted segment and the viability of the partial trisomies and monosomies produced by the abnormal recombinant chromosomes.

Interchromosomal Insertions

Interchromosomal insertions involve the movement of material from one chromosome to another. As discussed earlier, the inserted segment can be either direct or inverted relative to its original position. For relatively small inserted segments, it seems most likely that the homologues involved in the rearrangement will pair independently *(112)*. The inserted segment and its homologous region on the normal chromosome can loop out, allowing full pairing of the uninvolved segments of the bivalents (**Fig. 19**). Independent 2:2 segregation of the homologues in these two bivalents can result in the formation of four gamete types, two of which have a normal or balanced chromosome complement and two of which have an unbalanced complement, one duplicated and one deleted for the inserted segment. The theoretical risk, in this situation, would be 50% for producing a conceptus with an unbalanced karyotype. The risk for having a liveborn abnormal child would depend on the viability of the partial trisomy or partial monosomy of the inserted segment involved.

In the case of very long inserted segments, a quadrivalent containing an insertion loop may be formed, allowing complete pairing of the chromosomes involved in the rearrangement *(113)*. If no crossover occurs within the insertion loop, the consequences are the same as described earlier for nonpaired bivalents. If a crossover occurs within the insertion loop, however, recombinant chromosomes that would lead to the production of gametes with duplications and deletions may be formed. Once again, the risk for having a liveborn abnormal child will depend on the viability of the partial trisomies and monosomies produced.

Regardless of whether complete pairing is achieved between the chromosomes involved in an interchromosomal insertion or whether recombination takes place, compared to carriers of other chromosome rearrangements, an insertion carrier's risk of having an abnormal liveborn child is among the highest. Gardner and Sutherland reviewed the data from a number of individual case reports and found the average risk for having an abnormal child to be approximately 32% *(13)*. The theoretical risk approaches 50%.

COMPLEX CHROMOSOME REARRANGEMENTS

Although the definition of what constitutes a complex chromosome rearrangement (CCR) appears to vary somewhat, a rearrangement involving two or more chromosomes and at least three breakpoints is generally considered to be complex *(114)*. The more complex the rearrangement, the greater the number of chromosome breaks and the higher the probability that an essential gene has been interrupted or that genetic material has been lost or gained during its formation. It is therefore not surprising that CCRs are only rarely seen in constitutional karyotypes.

The majority of reported constitutional CCRs represent *de novo* events that appear to have occurred during spermatogenesis. The less frequently reported familial CCRs appear to be transmitted predominantly through females. As one might suspect, meiotic pairing and segregation can become quite complex in a CCR carrier. In theory, the more complex the rearrangement, the more elaborate the chromosome contortions required to optimize pairing between the rearranged chromosomes and their homologues. Similarly, the greater the number of involved chromosomes, the greater the potential number of unbalanced gametes. Thus it is somewhat surprising that a balanced CCR carrier's empiric risk for an unbalanced liveborn child does not appear to differ significantly from that of a simple balanced reciprocal translocation carrier. The risk for miscarriage among these carriers does, however, appear to be somewhat higher, suggesting that early loss of unbalanced pregnancies may partially explain this observation *(97,115–117)*. Selection against grossly unbalanced gametes at fertilization could also play a role. As discussed in the introduction to this chapter, the actual reproductive risks for any CCR carrier will vary depending upon the precise rearrangement involved as well as many other variables.

ACKNOWLEDGMENT

The ideograms in this chapter were reproduced with permission of S. Karger AG, Basel.

REFERENCES

1. Warburton, D. (1991) De novo balanced chromosome rearrangements and extra marker chromosomes identified at prenatal diagnosis: clinical significance and distribution of breakpoints. *Am. J. Med. Genet.* **49,** 995–1013.
2. Love, D.R., England, S.B., Speer, A., Marsden, R.F., Bloomfield, J.F., Roche, A.L., Cross, G.S., Mountford, C., Smith, T.J., and Davies, K.E. (1991) Sequences of junction fragments in the deletion-prone region of the Dystrophin gene. *Genomics* **10,** 57–67.
3. Giacalone, J.P. and Francke, U. (1992) Common sequence motifs at the rearrangement sites of a constitutional X/autosome translocation and associated deletion. *Am. J. Hum. Genet.* **50,** 725–741.
4. Cohen, O., Cans, C., Cuillel M., Gilardi, J.L., Roth, H., Mermet, M.-A. Jalbert, P., and Demongeot, J. (1996) Cartographic study: breakpoints in 1574 families carrying human reciprocal translocations. *Hum. Genet.* **97,** 659–667.
5. Olson, S.B., and Magenis, R.E. (1988) Preferential paternal origin of de novo structural chromosome rearrangements. In *The Cytogenetics of Mammalian Autosomal Rearrangements* (Daniel, A., ed.), Alan R. Liss, New York, pp. 583–589.
6. Chandley, A.C. (1991) On the parental origin of *de novo* mutations in man. *J. Med. Genet.* **28,** 217–223.
7. Chandley, A.C. (1988) Meiotic studies and fertility in human translocation carriers. In *The Cytogenetics of Mammalian Autosomal Rearrangements* (Daniel, A., ed.), Alan R. Liss, New York, pp. 361–382.
8. Warburton, D. (1982) De novo structural rearrangements: implications for prenatal diagnosis. In *Clinical Genetics: Problems in Diagnosis and Counseling* (Willey, A.M., Carter, T.P., Kelly, S., and Porter, ed.), Academic Press, New York, pp. 63–73.
9. Boyd, Y., Cockburn, D., Holt, S., Munro, E., Van Ommen, G.J., Gillard, B., Affara, N., Ferguson-Smith, M., and Craig, I. (1988) Mapping of 12 translocation breakpoints in the Xp21 region with respect to the locus for Duchenne muscular dystrophy. *Cytogenet. Cell Genet.* **48,** 28–34.
10. De Braekeleer, M. and Dao T.-N. (1991) Cytogenetic studies in male infertility. *Hum. Reprod.* **6,** 245–250.
11. James, R.S., Temple, I.K., Patch, C., Thompson, E.M., Hassold, T., and Jacobs, P.A. (1994) A systemic search for uniparental disomy carriers of chromosome translocations. *Eur. J. Hum. Genet.* **2,** 83–95.

12. Wagstaff, J. and Hemann, M. (1995) A familial "balanced" 3;9 translocation with cryptic 8q insertion leading to deletion and duplication of 9p23 loci in siblings. *Am. J. Hum. Genet.* **56,** 302–309.

13. Gardner, R.J.M. and Sutherland G.R. (1996) *Chromosome Abnormalities and Genetic Counseling* (Bobrow, M., Harper, P.S., Motulsky, A.G., and Scriver, C., eds.), Oxford University Press, New York.

14. Schinzel, A., ed. (1984) *Catalog of Unbalanced Chromosome Aberrations in Man.* Walter de Gruyter, New York.

15. Schwartz, S., Kumar, A., Becker, L.A., Crowe, C.A., Haren, J.M., Tsuchiya, K., Wandstrat, A.E., and Wolff, D.J. (1997) Molecular and cytogenetic analysis of de novo "terminal" deletions: implications for mechanism of formation. *Am. J. Hum. Genet. Suppl,* **61, A7.**

16. Schmickel, R.D. (1986) Contiguous gene syndromes: a component of recognizable syndromes. *J. Pediatr.* **109,** 231–241.

17. Budarf, M.L. and Emanuel, B.S. (1997) Progress in the autosomal segmental aneusomy syndromes (SASs): single or multi-locus disorders. *Hum. Mol. Genet.* **6,** 1657–1665.

18. Ewart, A.K., Morris, C.A., Atkinson, D., Jin, W., Sternes, K., Spallone, P., Stock, A.D., Leppert, M., and Keating M.T. (1993) Hemizygosity at the elastin locus in a developmental disorder, Williams syndrome. *Nat. Genet.* **5,** 11–16.

19. Osborne, L.R., Martindale, D., Scherer, S.W., Shi, X.-M., Huizenga, J., Heng, H.H.Q., Costa, T., Pober, B., Lew, L., Brinkman, J., Rommens, J., Koop, B., and Tsui, L.C. (1996) Identification of genes from a 500kb region at 7q11.23 that is commonly deleted in Williams-syndrome patients. *Genomics* **36,** 328–336.

20. Frangiskakis, J.M., Ewart, A.K., Morris, C.A., Mervis, C.A., Bertrand, J., Robinson, B.F., Klein, B.P., Ensing, G.J., Everett, L.A., Green, E.D., Proschel, C., Gutowski, N.J., Noble, M., Atkinson, D.L., Odelberg, S.J., and Keating, M.T. (1996) LIM-kinase 1 hemizygosity implicated in impaired visuospatial constructive cognition. *Cell* **86,** 59–69.

21. Dutly, F. and Schinzel, A. (1996) Unequal interchromosomal rearrangements may result in elastin gene deletions causing the Williams–Beuren syndrome. *Hum. Mol. Genet.* **5,** 1893–1898.

22. Urban, Z., Helms, C., Fekete, G., Csiszar, K., Bonnet, D., Munnich, A., Donis-Keller, H., and Boyd, C.D. (1996) 7q11.23 deletions in Williams syndrome arise as a consequence of unequal meiotic crossover. *Am. J. Hum. Genet.* **59,** 958–962.

23. Van Dyke, D.L. (1988) Isochromosomes and interstitial tandem direct and inverted duplications. In *The Cytogenetics of Mammalian Autosomal Rearrangements* (Daniel, A., ed.), Alan R. Liss, New York, pp. 635–666.

24. Brown, K.W., Gardner, A., Williams, J.C., Mott, M.G., McDermott, A., and Maitland, N.J. (1992) Paternal origin of 11p15 duplications in the Beckwith–Wiedemann syndrome. A new case and review of the literature. *Cancer Genet. Cytogenet.* **58,** 55–70.

25. Cook, E.H. Jr., Lindgren, V., Leventhal, B.L., Courchesne, R., Lincoln, A., Shulman, C., Lord, C., and Courchesne, E. (1997) Autism or atypical autism in maternally but not paternally derived proximal 15q duplication. *Am. J. Hum. Genet.* **60,** 928–934.

26. Browne, C.E., Dennis, N.R., Maher, E., Long, F.L., Nicholson, J.C., Sillibourne, J., and Barber, J.C.K. (1997) Inherited interstitial duplications of proximal 15q: genotype-phenotype correlations. *Am. J. Hum. Genet.* **61,** 1342–1352.

27. Brown, A., Phelan, M.C., Patil, S., Crawford, E., Rogers, C., and Schwartz, C. (1996) Two patients with duplication of 17p11.2: the reciprocal of the Smith–Magenis syndrome deletion? *Am. J. Med. Genet.* **63,** 373–377.

28. Lupski, J.R., Roth, J.R., and Weinstock, G.M. (1996) Chromosomal duplications in bacteria, fruit flies, and humans. *Am. J. Hum. Genet.* **58,** 21–27.

29. Dajalali, M., Steinbach, P., Bullerdiek, J., Holmes-Siedle, M., Verschraegen-Spae, M.R., and Smith, A. (1986) The significance of pericentric inversions of chromosome 2. *Hum. Genet.* **72,** 32–36.

30. Sujansky, E., Smith, A.C.M., Peakman, D.C., McConnell, T.S., Baca, P., and Robinson, A. (1981) Familial pericentric inversion of chromosome 8. *Am. J. Med. Genet.* **10,** 229–235.
31. Pettenati, M.J., Rao, P.N., Phelan, M.C., Grass, F., Rao, K.W., Cosper, P., Carroll, A.J., Elder, F., Smith, J.L., Higgins, M.D., Lanman, J.T., Higgins, R.R., Butler, M.G., Luthardt, F., Keitges, E., Jackson-Cook, C., Brown, J., Schwartz, S., Van Dyke, D.L., and Palmer, C.G. (1995) Paracentric inversions in humans: a review of 446 paracentric inversions with presentation of 120 new cases. *Am. J. Med. Genet.* **55,** 171–187.
32. Kaiser, P. (1984) Pericentric inversions: problems and significance for clinical genetics. *Hum. Genet.* **68,** 1–47.
33. Smith, A.C.M., Spuhler, K., Williams, T.M., McConnell, T., Sujansky, E., and Robinson, A. (1987) Genetic risk for recombinant 8 syndrome and the transmission rate of balanced inversion 8 in the hispanic population of the southwestern United States. *Am. J. Hum. Genet.* **41,** 1083–1103.
34. Madan, K., Pieters, M.H.E.C., Kuyt, L.P., van Asperen, C.J., de Pater, J.M., Hamers, A.J.H., Gerssen-Schoorl, K.B.J., Hustinx, T.W.J., Breed, A.S.P.M., Van Hemel, J.O., and Smeets, D.F.C.M. (1990) Paracentric inversion inv(11)(q21q23) in the Netherlands. *Hum. Genet.* **85,** 15–20.
35. Chodirker, B.N., Greenberg, C.R., Pabello, P.D., and Chudley, A.E. (1992) Paracentric inversion 11q in Canadian Hutterites. *Hum. Genet.* **89,** 450–452.
36. Sutherland, G.R., Callen, D.F., and Gardner, R.J.M. (1995) Paracentric inversions do not normally generate monocentric recombinant chromosomes. *Am. J. Med. Genet.* **59,** 390.
37. Warburton, D. and Twersky, S. (1997) Risk of phenotypic abnormalities in paracentric inversion carriers. *Am. J. Med. Genet.* **69,** 219.
38. Wolff, D.J., Miller, A.P., Van Dyke, D.L., Schwartz, S., and Willard, H.F. (1996) Molecular definition of breakpoints associated with human Xq isochromosomes: implications for mechanisms of formation. *Am. J. Med. Genet.* **58,** 154–160.
39. Schwartz, S. and Depinet, T.W. (1996) Studies of "acentric" and "dicentric" marker chromosomes: implications for definition of the functional centromere. *Am. J. Hum. Genet. Suppl.* **59(4),** A14.
40. Sullivan, B.A. and Willard, H.F. (1996) Functional status of centromeres in dicentric X chromosomes: evidence for the distance-dependence of centromere/kinetochore assembly and correlation with malsegregation in anaphase. *Am. J. Hum. Genet. Suppl.* **59(4),** A14.
41. Sullivan, B.A. and Schwartz, S. (1995) Identification of centromeric antigens in dicentric Robertsonian translocations: CENP-C and CENP-E are necessary components of functional centromeres. *Hum. Mol. Genet.* **4(12),** 2189–2197.
42. Magnani, I., Sacchi, N., Darfler, M., Nisson, P.E., Tornaghi, R., and Fuhrman-Conti, A.M. (1993) Identification of the chromosome 14 origin of a C-negative marker associated with a 14q32 deletion by chromosome painting. *Clin. Genet.* **43,** 180–185.
43. Voullaire, L.E., Slater, H.R., Petrovic, V., and Choo, K.H.A. (1993) A functional marker centromer with no detectable alpha-satellite, satellite III, or CENP-B protein: activation of a latent centromere. *Am. J. Hum. Genet.* **52,** 1153–1163.
44. Blennow, E., Telenius, H. de Vos, D., Larsson, C., Henriksson, P., Johansson, O., Carter, N.P., and Nordenskjold, M. (1994) Tetrasomy 15q: two marker chromosomes with no detectable alpha-satellite DNA. *Am. J. Hum. Genet.* **54,** 877–883.
45. Osashi, H., Wakui, K., Ogawa, K., Okano, T., Niikawa, N., and Fukushima, Y. (1994) A stable acentric marker chromosome: possible existence of an intercalary ancient centromere at distal 8p. *Am. J. Hum. Genet.* **55,** 1202–1208.
46. Maraschio, P., Tupler, R., Rossi, E., Barbierato, L., and Uccellatore, F. (1996) A novel mechanism for the origin of supernumerary marker chromosomes. *Hum. Genet.* **97,** 382–386.
47. Van den Enden, A., Verschraegen-Spae, M.R., Van Roy, N., Decaluwe, W., and De Praeter, C. (1996) Mosaic tetrasomy 15q25→qter in a newborn infant with multiple anomalies. *Am. J. Med. Genet.* **63,** 482–485.

48. du Sart, D., Cancilla, M.R., Earle, E., Mao, J.-I., Saffery, R., Tainton, K.M., Kalitsis, P., Martyn, J., Barry, A.E., and Choo, K.H.A. (1997) A functional neo-centromere formed through activation of a latent human centromere and consisting of non-alpha-satellite DNA. *Nat. Genet.* **16**, 144–153.

49. Darlington, C.D. (1939) Misdivision and the genetics of the centromere. *J. Genet.* **37**, 341–364.

50. Therman, E., Sarto, G.E., and Patau, K. (1974) Apparently isodicentric but functionally monocentric X chromosome in man. *Am. J. Hum. Genet.* **26**, 83–92.

51. Phelan, M.C., Prouty, L.A., Stevenson, R.E., Howard-Peebles, P.N., Page, D.C., and Schwartz, C.E. (1988) The parental origin and mechanisms of formation of three dicentric X chromosomes. *Hum. Genet.* **80**, 81–84.

52. Lorda-Sanchez, I., Binkert, F., Maechler, M., and Schinzel, A. (1991) A molecular study of X isochromosomes: parental origin, centromeric structure and mechanisms of formation. *Am. J. Hum. Genet.* **49**, 1034–1040.

53. Shaffer, L.G., McCaskill, C., Haller, V., Brown, J.A., and Jackson-Cook, C.K. (1993) Further characterization of 19 cases of rea(21q21q) and delineation as isochromosomes or Robertsonian translocations in Down syndrome. *Am. J. Med. Genet.* **47**, 1218–1222.

54. Shaffer, L.G., McCaskill, C., Han, J.-Y., Choo, K.H.A., Cutillow, D.M., Donnenfeld, A.E., Weiss, L., and Van Dyke, D.L. (1994) Molecular characterization of de novo secondary trisomy 13. *Am. J. Hum. Genet.* **55**, 968–974.

55. Callen, D.F., Mulley, J.C., Baker, E.G., and Sutherland, G.R. (1987) Determining the origin of human X isochromosomes by use of DNA sequence polymorphisms and detection of an apparent i(Xq) with Xp sequences. *Hum. Genet.* **77**, 236–240.

56. Harbison, M., Hassold, T., Kobryn, C., and Jacobs, P.A. (1988) Molecular studies of the parental origin and nature of human X isochromosomes. *Cytogenet. Cell Genet.* **47**, 217–222.

57. Antonarakis, S.E., Adelsberger, P.A., Petersen, M.B., Binkert, F., and Schinzel, A.A. (1990) Analysis of DNA polymorphisms suggests that most de novo dup(21q) chromosomes in patients with Down syndrome are isochromosomes and not translocations. *Am. J. Hum. Genet.* **47**, 968–972.

58. Robinson, W.P., Bernasconi, F., Basaran, S., Yuksel-Apak, M., Neri, G., Serville, F., Balicek, P., Haluza, R., Farah, L.M.S., Luleci, G., and Schinzel, A.A. (1994) A somatic origin of homologous Robertsonian translocations and isochromosomes. *Am. J. Hum. Genet.* **54**, 290–302.

59. Wyandt, H.E. (1988) Ring autosomes: identification, familial transmission, causes of phenotypic effects and in vitro mosaicism. In *The Cytogenetics of Mammalian Autosomal Rearrangements* (Daniel, A., ed.), Alan R. Liss, New York, pp. 667–696.

60. Callen, D.F., Eyre, H.J., Ringenbergs, M.L., Freemantle, C.J., Woodroffe, P., and Haan, E.A. (1991) Chromosomal origin of small ring marker chromosomes in man: characterization by molecular cytogenetics. *Am. J. Hum. Genet.* **48**, 769–782.

61. Pezzolo, A., Gimelli, G., Cohen, A., Lavaggetto, A., Romano, C., Fogu, G., and Zuffardi, O. (1993) Presence of telomeric and subtelomeric sequences at the fusion points of ring chromosomes indicates that the ring syndrome is caused by ring instability. *Hum. Genet.* **92**, 23–27.

62. Fang, Y.-Y., Eyre, H.J., Bohlander, S.K., Estop, A., McPherson, E., Trager, T., Riess, O., and Callen, D.F. (1995) Mechanisms of small ring formation suggested by the molecular characterization of two small accessory ring chromosomes derived from chromosome 4. *Am. J. Hum. Genet.* **57**, 1137–1142.

63. Petersen, M.B., Bartsch, O., Adelsberger, P.A., Mikkelsen, M., Schwinger, E., and Antonarakis, S.E. (1992) Uniparental isodisomy due to duplication of chromosome 21 occurring in somatic cells monosomic for chromosome 21. *Genomics* **13**, 269–274.

64. Crolla, J. (1997) FISH and molecular studies of autosomal supernumerary marker chro-

mosomes excluding those derived from chromosome 15: review of the literature. *Am. J. Med. Genet.* **75,** 367–381.

65. Cote, G.B., Katsantoni, A., and Deligeorgis, D. (1981) The cytogenetic and clinical implications of a ring chromosome 2. *Ann. Genet.* **24,** 231–235.

66. Kosztolanyi, G. (1987) Does "ring syndrome" exist? An analysis of 207 case reports on patients with a ring autosome. *Hum. Genet.* **75,** 174–179.

67. Kosztolanyi, G., Mehes, K., and Hook, E.B. (1991) Inherited ring chromosomes: an analysis of published cases. *Hum. Genet.* **87,** 320–324.

68. Van Dyke, L.L., Weiss, L., Robertson, J.R., and Babu, V.R. (1983) The frequency and mutation rate of balanced autosomal rearrangements in man estimated from prenatal genetic studies for advanced maternal age. *Am. J. Hum. Genet.* **35,** 301–308.

69. Iselius, L., Lindsten, J., Aurias, A., Fraccaro, M., Bastard, C., Botelli, A.M., Bui, T.-H., Caufin, D., Dalpra, L., Delendi, N., Dutrillaux, B., Fukushima, Y., Geraedts, J.P.M., De Grouchy, J., Gyftodimou, J., Hanley, A.L., Hansmann, I., Ishii, T., Jalbert, P., Jingeleski, S., Kajii, T., von Koskull, H., Niikawa, N., Noel, B., Pasquali, F., Probeck, H.D., Robinson, A., Roncarati, E., Sachs, E., Scappaticci, S., Schwinger, E., Simoni, G., Veenema, H., Vigi, V., Volpato, S., Wegner, R.-D, Welch, J.P., Winsor, E.J.T., Zhang, S., and Zuffardi, O. (1983) The 11q;22q translocation: A collaborative study of 20 new cases and analysis of 110 families. *Hum. Genet.* **64,** 343–355.

70. Zackai, E.H. and Emanuel, B.S. (1980) Site-specific reciprocal translocation, t(11;22)(q23;q11), in several unrelated families with 3:1 meiotic disjunction. *Am. J. Med. Genet.* **7,** 507–521.

71. Lindblom, A., Sandelin, K., Iselius, L., Dumanski, J., White, I., Nordenskjold, M., and Larsson, C. (1994) Predisposition for breast cancer in carriers of constitutional translocation 11q;22q. *Am. J. Hum. Genet.* **54,** 871–876.

72. Griffin, C.A., McKeon, C., Isreal, M.A., Gegonne, A., Ghysdael, J., Stehelin, D., Douglass, E.C., Green, A.A., and Emanuel, B.S. (1986) Comparison of constitutional and tumor-associated 11;22 translocations: nonidentical breakpoints on chromosomes 11 and 22. *Proc. Natl. Acad. Sci. USA* **83,** 6122–6126.

73. Delattre, O., Grunwald, M., Bernard, A., Grunwald, D., Thomas, G., Frelat, G., and Aurias, A. (1988) Recurrent t(11;22) breakpoint mapping by chromosome flow sorting and spot-blot hybridization. *Hum. Genet.* **78,** 140–143.

74. Budarf, M., Sellinger, B., Griffin, C., and Emanuel, B.S. (1989) Comparative mapping of the constitutional and tumor-associated 11;22 translocation. *Am. J. Hum. Genet.* **45,** 128–139.

75. Schmutz, S.M. and Pinno, E. (1986) Morphology alone does not make an isochromosome. *Hum. Genet.* **72,** 253–255.

76. Robertson, W. (1916) Taxonomic relationships shown in the chromosome of Tettigidae and Acrididae: V-shaped chromosomes and their significance in Acrididae, Locustidae, and Gryllidae: chromosomes and variation. *J. Morphol.* **27,** 179–331.

77. Hamerton, J.L., Canning, N., Ray, M., and Smith, S. (1975) A cytogenetic survey of 14,069 newborn infants: incidence of chromosomal abnormalities. *Clin. Genet.* **8,** 223–243.

78. Jacobs, P.A. (1981) Mutation rates of structural chromosomal rearrangements in man. *Am. J. Hum. Genet.* **33,** 44–54.

79. Nielsen, J. and Wohlert, M. (1991) Chromosome abnormalities found among 34,910 newborn children: results from a 13-year incidence study in Arhus, Denmark. *Hum. Genet.* **87,** 81–83.

80. Earle, E., Shaffer, L.G., Kalitsis, P., McQuillan, C., Dale, S., and Choo, K.H.A. (1992) Identification of DNA sequences flanking the breakpoint of human t(14q21q) Robertsonian translocations. *Am. J. Hum. Genet.* **50,** 717–724.

81. Gravholt, C.H., Friedrich, U., Caprani, M., and Jorgensen, A.L. (1992) Breakpoints in Robertsonian translocations are localized to satellite III DNA by fluorescence in situ hybridization. *Genomics* **14,** 924–930.

82. Wolff, D.J. and Schwartz, S. (1992) Characterization of Robertsonian translocations by using fluorescence in situ hybridizations. *Am. J. Med. Genet.* **50,** 174–181.

83. Han, J-Y., Choo, K.H.A., and Shaffer, L.G. (1994) Molecular cytogenetic characterization of 17 rob(13q14q) Robertsonian translocations by FISH, narrowing the region containing the breakpoints. *Am. J. Hum. Genet.* **55,** 960–967.

84. Page, S.L., Shin, J-C., Han, J-Y., Choo, K.H.A., and Shaffer, L.G. (1996) Breakpoint diversity illustrates distinct mechanisms for Robertsonian translocation formation. *Hum. Mol. Genet.* **5(9),** 1279–1288.

85. Sullivan, B.A., Jenkins, L.S., Karson, E.M., Leana-Cox, J., and Schwartz, S. (1996) Evidence for structural heterogeneity from molecular cytogenetic analysis of dicentric Robertsonian translocations. *Am. J. Hum. Genet.* **59,** 167–175.

86. Sullivan, B.A., Wolff, D.J., and Schwartz, S. (1994) Analysis of centromeric activity in Robertsonian translocations: implications for a functional acrocentric hierarchy. *Chromosoma* **103,** 459–467.

87. Blouin J-L., Binkert, F. and Antonarakis, S.E. (1994) Biparental inheritance of chromosome 21 polymorphic markers indicates that some Robertsonian translocations t(21;21) occur postzygotically. *Am. J. Med. Genet.* **49,** 363–368.

88. Robinson, W.P., Bernasconi, F., Dutly, F., Lefort, G., Romain, D.R., Binkert, F., and Schinzel, A.A. (1996) Molecular studies of translations and trisomy involving chromosome 13. *Am. J. Med. Genet.* **61,** 158–163.

89. Shaffer, L.G., Jackson-Cook, C.K., Meyer, J.M., Brown, J.A., and Spence, J.E. (1991) A molecular genetic approach to the identification of isochromosomes of chromosome 21. *Hum. Genet.* **86,** 375–382.

90. Ledbetter, D.H. and Engel, E. (1995) Uniparental disomy in humans: development of an imprinting map and its implications for prenatal diagnosis. *Hum. Mol. Genet.* **4,** 1757–1764.

91. Nicholls, R.D., Knoll, J.H.M., Butler, M.G., Karam, S., and Lalande, M. (1989) Genetic imprinting suggested by maternal heterodisomy in non-deletion Prader–Willi syndrome. *Nature* **342,** 281–285.

92. Malcolm, S., Clayton-Smith, J., Nichols, M., Robb, S., Webb, T., Armour, J.A.L., Jeffreys, A.J., and Pembrey, M.E. (1991) Uniparental paternal disomy in Angelman syndrome. *Lancet* **337,** 694–697.

93. Wang, J.-C.C., Passage, M.B., Yen, P.H., Shapiro, L.J., and Mohandas, T.K. (1991). Uniparental heterodisomy for chromosome 14 in a phenotypically abnormal familial balanced 13/14 Robertsonian translocation carrier. *Am. J. Hum. Genet.* **48,** 1069–1074.

94. Papenhausen, P.R., Mueller, O.T., Johnson, V.P., Sutcliff, M., Diamond, T.M., and Kousseff, B.G. (1995) Uniparental isodisomy of chromosome 14 in two cases: an abnormal child and a normal adult. *Am. J. Med. Genet.* **59,** 271–275.

95. Healey, S., Powell, F., Battersby, M., Chenevix-Trench, G., and McGill, J. (1994) Distinct phenotype in maternal uniparental disomy of chromosome 14. *Am. J. Med. Genet.* **51,** 147–149.

96. Temple, I.K., Cockwell, A., Hassold, T., Pettay, D., and Jacobs, P. (1991) Maternal uniparental disomy for chromosome 14. *J. Med. Genet.* **28,** 511–514.

97. Boué, A. and Galano, P. (1984) A collaborative study of the segregation of inherited chromosome structural rearrangements in 1356 prenatal diagnoses. *Prenat. Diagn.* **4,** 45–67.

98. Greenberg, F., Elder, F.B., and Ledbetter, D.H. (1987) Neonatal diagnosis of Prader–Willi syndrome and its implications. *Am. J. Med. Genet.* **28,** 845–856.

99. Rivera, H., Zuffardi, O., and Gargantini, L. (1990) Nonreciprocal and jumping translocations of 15ql-qter in Prader–Willi syndrome. *Am. J. Med. Genet.* **37,** 311–317.

100. Park, V.M., Gustashaw, K.M., and Wathen, T.M. (1992) The presence of interstitial telomeric sequences in constitutional chromosome abnormalities. *Am. J. Hum. Genet.* **50,** 914–923.

101. Duval, E., van den Enden, A., Vanhaesebrouck, P., and Speleman, F. (1994) Jumping translocation in a newborn boy with dup(4q) and severe hydrops fetalis. *Am. J. Med. Genet.* **52,** 214–217.
102. von Ballestrem, C.L., Boavida, M.G., Zuther, C., Carreiro, M.H., David, D., and Gal, A. (1996) Jumping translocation in a phenotypically normal female. *Clin. Genet.* **49,** 156–159.
103. Gross, S.J., Tharapel, A.T., Phillips, O.P., Shulman, L.P., Pivnich, E.K., and Park, V.M. (1996) A jumping Robertsonian translocation: a molecular and cytogenetic study. *Hum. Genet.* **98,** 291–296.
104. Jewett, T., Marnane, D., Rao, P.N., and Pettenati, M.J. (1996) Evidence of telomere involvement in a jumping translocation involving chromosome 15 in a mildly affected infant. *Am. J. Hum. Genet. Suppl.* **59(4),** A121.
105. Tomkins, D.J. (1981) Unstable familial translocations: a t(11;22)mat inherited as a t(11;15). *Am. J. Hum. Genet.* **33,** 745–751.
106. Farrell, S.A., Winsor, E.J.T., and Markovic, V.D. (1993) Moving satellites and unstable chromosome translocations: clinical and cytogenetic implications. *Am. J. Med. Genet.* **46,** 715–720.
107. Giussani, U., Facchinetti, B., Cassina, G., and Zuffardi, O. (1996) Mitotic recombination among acrocentric chromosomes' short arms. *Ann. Hum. Genet.* **60,** 91–97.
108. Chudley, A.E., Bauder, F., Ray, M., McAlpine, P.J., Pena, S.D.J., and Hamerton, J.L. (1974). Familial mental retardation in a family with an inherited chromosome rearrangement. *J. Med. Genet.* **11,** 353–363.
109. Madan, K. and Menko, F.H. (1992). Intrachromosomal insertions: a case report and a review. *Hum. Genet.* **89,** 1–9.
110. Hoegerman, S.F. (1979). Chromosome 13 long arm interstitial deletion may result from maternal inverted insertion. *Science* **205,** 1035–1036.
111. Callen, D.F., Woollatt, E., and Sutherland, G.R. (1985). Paracentric inversions in man. *Clin. Genet.* **27,** 87–92.
112. Walker, A.P. and Bocian, M. (1987). Partial duplication 8q12→q21.2 in two sibs with maternally derived insertional and reciprocal translocations: case reports and review of partial duplications of chromosome 8. *Am. J. Med. Genet.* **27,** 3–22.
113. Meer, B. Wolff, G., and Back, E. (1981) Segregation of a complex rearrangement of chromosomes 6, 7, 8, and 12 through three generations. *Hum. Genet.* **58(2),** 221–225.
114. Pai, G.S., Thomas, G.H., Mahoney, W., and Migeon, B.R. (1980) Complex chromosome rearrangements; Report of a new case and literature review. *Clin. Genet.* **18,** 436–444.
115. Gorski, J.L., Kistenmacher, M.L., Punnett, H.H., Zackai, E.H., and Emanuel, B.S. (1988) Reproductive risks for carriers of complex chromosome rearrangements: analysis of 25 families. *Am. J. Med. Genet.* **29,** 247–261.
116. Daniel, A., Hook, E., and Wulf, G. (1989) Risks of unbalanced progeny at amniocentesis to carriers of chromosomal rearrangements. *Am. J. Med. Genet.* **31,** 14–53.
117. Batista, D.A.S., Pai, S., and Stetten, G. (1994) Molecular analysis of a complex chromosomal rearrangement and a review of familial cases. *Am. J. Med. Genet.* **53,** 255–263.

10

Sex Chromosomes and Sex Chromosome Abnormalities

Cynthia M. Powell, M.D.

INTRODUCTION

It can be argued that the sex chromosomes are the most important pair of chromosomes given their role in determining gender and, therefore, allowing for reproduction and procreation. Considered together, sex chromosome aneuploidies are the most common chromosome abnormalities seen in liveborn infants, children, and adults. Physicians in many specialties including pediatrics, obstetrics and gynecology, endocrinology, internal medicine, and surgery commonly encounter individuals with sex chromosome abnormalities. There has been a great deal of misinformation in the past regarding outcomes and developmental profiles of these patients, leading to bias and discrimination. This chapter attempts to provide a summary of information regarding the sex chromosomes, sex chromosome abnormalities, and disorders of sexual development with normal chromosomes.

THE X AND Y CHROMOSOMES

Role in Sexual Differentiation

Genetic sex is established at the time of fertilization and is dependent on whether an X- or Y-bearing sperm fertilizes the X-bearing egg. The type of gonads that develop (gonadal sex) is determined by the sex chromosome complement (XX or XY). Before the seventh week of embryonic life, the gonads of both sexes appear identical *(1)*. Normally, under the influence of the Y chromosome, the immature gonad becomes a testis. In the absence of the Y chromosome, the gonad differentiates into an ovary. The term phenotypic sex refers to the appearance of the external genitalia and in some disorders may not correspond to the genetic or gonadal sex (see "Sex Reversal" later).

X Chromosome Inactivation

There are thousands of genes on the X chromosome, but few on the Y chromosome. The explanation for the fact that males survive quite nicely with only one X chromosome while females have two involves a concept called "dosage compensation," and is termed the Lyon hypothesis after its proponent, Dr. Mary Lyon *(2)*.

From: *The Principles of Clinical Cytogenetics*
Edited by: S. Gersen and M. Keagle © Humana Press Inc., Totowa, NJ

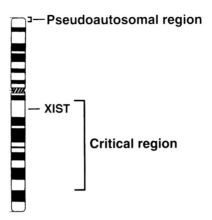

Fig. 1. Ideogram of the X chromosome showing the pseudoautosomal region on Xp, the location of the X-inactive-specific transcript (*XIST*) gene, and the critical region on Xq.

In somatic cells in females only one X chromosome is active. X inactivation occurs early in embryonic life, beginning about 3 days after fertilization, and is completed by the end of the first week of development. The inactivation is random between the two X chromosomes. Either the maternal or paternal X may be inactivated, and after one X has become inactive, all the daughter cells from that original cell have the same inactive X. A gene that seems to control X inactivation, *XIST* (X-inactive-specific transcript) has been identified and sequenced. It is located at the X-inactivation center (XIC) at band Xq13 (**Fig. 1**). Only the inactive X expresses this gene. Several genes have been identified that escape inactivation and remain active on both X chromosomes in females.

Early evidence for the existence of the inactive X was the observation of the Barr body, named for the Canadian cytologist Murray Barr *(3)*. This is a dark-staining chromatin body, present in one copy in normal females, which is the condensed, inactive X chromosome. Normal males have no Barr body. Initially, a buccal smear was obtained from patients to look for Barr bodies. Because of improved methods for looking at sex chromosomes and the inaccuracy of the buccal smear technique, it is now considered an obsolete test. The sex chromatin body in polymorphonuclear leukocytes takes the form of the "drumstick," seen attached to the nucleus in approximately 2% of these cells in women, but not in men *(4)*.

Techniques for detecting the inactive X are based on the fact that it is late replicating. The most commonly used method involves the use of bromodeoxyuridine (BrdU) *(5)*. Newer methods for detecting the inactive X involve molecular techniques.

Pseudoautosomsal Regions

The distal region of the short arms of the X and Y chromosomes contain highly similar DNA sequences. During normal meiosis in the male, crossing over occurs between these regions. Because this resembles the crossing over that occurs between autosomes, these regions have been termed pseudoautosomal (**Fig. 1**). There is also a region of homology at distal ends of Xq and Yq, which have been observed to associate during male meiosis, with proven recombination events *(6)*.

The Y Chromosome

The testis-determining factor (TDF) that leads to differentiation of the indifferent gonads into testes is located on the short arm of the Y chromosome. TDF was mapped by molecular analysis of sex-reversed patients (chromosomally female but phenotypically male and vice versa), and the gene *SRY* (sex-determining region Y) was identified in 1990 *(7)*. It is located on the short arm of the Y at band p11.3. Deletions and mutations in this gene have been found in some 46,XY females.

A gene controlling spermatogenesis, termed the azoospermia factor (*AZF*), was first proposed by Tiepolo and Zuffardi in 1976 *(8)*. In studies of men with azoospermia or severe oligospermia, deletions in different intervals of Yq11 have been found and up to three spermatogenesis loci have been hypothesized *(9)*, but as yet no gene has been identified.

A locus for susceptibility to gonadoblastoma (*GBY*) has been proposed on the Y chromosome based on the high incidence of gonadoblastoma in females with 45,X/46,XY mosaicism or XY gonadal dysgenesis *(10)*. Deletion mapping has localized this putative gene to a region near the centromere, but has raised the possibility of multiple *GBY* loci dispersed on the Y chromosome *(11,12)*.

NUMERICAL ABNORMALITIES OF THE SEX CHROMOSOMES

Introduction

Numerical abnormalities of the sex chromosomes are one of the most common types of chromosomal aneuploidies, with a frequency of 1 in 500 live births. This may be due to the fact that abnormalities of sex chromosomes have less severe clinical abnormalities and are more compatible with life as compared to autosomal disorders. Reasons for this include inactivation of all additional X chromosomes and the small number of genes on the Y.

Sex chromosomes abnormalities are more commonly diagnosed prenatally than autosomal aneuploidies. It is important for women undergoing amniocentesis and chorionic villus sampling to be informed about the possibility of detecting a sex chromosome abnormality, and, if this occurs, to be given accurate information about the disorder.

Turner Syndrome

45,X (and its variants) occurs in 1 in 5000 live female births, but is one of the most common chromosome abnormality in spontaneous abortions. Ninety-nine percent of 45,X conceptuses result in spontaneous loss, usually by 28 weeks. Although 45,X is quite lethal in the fetus, those that survive to term have relatively minor problems. The reasons for this are not known, although it has been speculated that all conceptions that survive have some degree of undetected mosaicism for a normal cell line. The older medical literature sometimes referred to the Turner syndrome karyotype as 45,XO. *This terminology is incorrect and should not be used; there is no O chromosome.*

The exact cause remains unclear. There are probably one or more genes on the X chromosome that, when present in a haploid state early in embryogenesis, lead to the phenotype of Turner syndrome. *ZFX* is a current candidate gene.

Fig. 2. Eight-year-old girl with Turner syndrome. (Courtesy of Marsha Davenport, University of North Carolina, Chapel Hill, NC.)

Clinical features of Turner syndrome in newborns may include decreased mean birthweight (average weight 2800 g), posteriorly rotated ears, neck webbing, and edema of hands and feet, although many are phenotypically normal. Older children and adults with Turner syndrome have proportionate short stature and variable dysmorphic features which may include downslanting eyes, posteriorly rotated ears, low posterior hairline, webbed neck, short fourth metacarpals, cubitus valgus, and a broad chest (**Fig. 2**). Congenital heart defects, especially coarctation of the aorta, and structural renal anomalies are common and should be checked for. Without hormonal supplementation, there is primary amenorrhea and lack of secondary sex characteristics. The gonads are generally streaks of fibrous tissue. Intelligence is average to above average, although there may be specific problems in spatial perception. Treatment should include estrogen hormone and consideration of growth hormone, which may increase ultimate height. It is recommended that these patients be followed by endocrinologists familiar with Turner syndrome.

Turner Syndrome Variants

Slightly more than half of all patients with Turner syndrome have a 45,X karyotype. The remainder exhibit mosaicism and structural abnormalities of the X chromosome.

MOSAICISM

Mosaicism for 45,X and another cell line is found in 15–20% of patients with Turner syndrome. A 46,XX cell line may modify the phenotypic features of the syndrome. As mentioned previously, to explain why 99% of 45,X conceptions terminate in miscar-

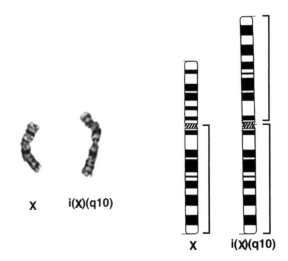

Fig. 3. Normal X chromosome and isochromosome Xq: 46,X,i(X)(q10) in a patient with Turner syndrome. Brackets indicate region of duplication on Xq.

riage, it has been proposed that all surviving 45,X fetuses have some degree of mosaicism, although this has not been proven.

Patients with 45,X/46,XY mosaicism may have external genitalia ranging from normal male to ambiguous to normal female. A structurally abnormal Y chromosome is often present. A study of 92 prenatally diagnosed cases found that 95% had normal male genitalia. Abnormal genitalia included hypospadias, micropenis, and abnormal scrotum. In those fetuses where pathologic studies were possible, 27% had abnormal gonadal histology, classified as dysgenetic gonads. The percentage of mosaicism found in amniotic fluid samples was a poor predictor of the phenotype *(13)*.

The risk of gonadoblastoma is 15–20% in phenotypic females with 45,X/46,XY mosaicism, and is also increased in patients with ambiguous external genitalia and intraabdominal gonads. Surgical removal of gonads is recommended. For those patients with a male phenotype and external testes, the risk of neoplasm is not as high, but frequent physical and ultrasound examinations are recommended *(14),* and a gonadal biopsy may also be helpful.

ISOCHROMOSOME X

An isochromosome X (**Fig. 3**), consisting of two copies of the long arm (missing all or most of the short arm), is seen in 18% of patients, either as a single cell line, or, more commonly, in mosaicism with a 45,X cell line. These patients are phenotypically indistinguishable from those with pure 45,X, although there have been some reports of an increased risk of autoimmune problems in patients with an isochromosome X. The X isochromosomes may be mono- or dicentric.

RING X

A subset of patients have one normal X chromosome and a ring X, most often as a mosaic cell line. Although some have typical features of Turner syndrome, others have mental retardation and congenital malformations. Studies have shown that the more severely affected patients have smaller rings that are deleted for *XIST*. It has been hypothesized that the lack of *XIST* causes the ring to fail to inactivate, thus causing functional disomy for genes present on the ring, resulting in phenotypic abnormalities *(15,16)*. Larger rings have *XIST* present and are preferentially inactivated.

Marker Chromosomes in Patients with Turner Syndrome

It is important to identify the origin of a marker chromosome in a patient with Turner syndrome, due to the risk of gonadoblastoma if it is made up of Y material (see 45,X/46,XY mosaicism earlier). This can be done either with FISH or molecular techniques. One study, using the polymerase chain reaction (PCR) and Southern blot analysis found that 40% of 45,X patients had SRY sequences *(17)*. Most patient samples produced only a faint signal, indicating a low percentage of cells with Y-chromosome material (or contamination). A more recent study found evidence of Y mosaicism in only 3.4% of patients with Turner syndrome who were cytogenetically nonmosaic *(18)*. The clinical significance of these findings regarding the risk of gonadoblastoma is not known.

Origin of the X Chromosome in Turner Syndrome

In approximately 75% of patients with 45,X, the X chromosome is maternal in origin *(19,20)*. Although phenotypic differences have not been found between Turner patients with a maternal or paternal X chromosome, recent reports suggest that there may be some cognitive differences. This has been theorized to occur on the basis of an imprinted X-linked locus *(21)*.

47,XXX

This is the most frequent sex chromosome abnormality present at birth in females, occurring in 1 in 1000 live female births. It was first described in 1959 by Jacobs et al. *(22)*. Unfortunately, the term originally used for this cytogenetic abnormality was "super female" which gives a misconception of the syndrome and is no longer in use.

Origin

Most 47,XXX conceptions result from maternal nondisjunction at meiosis I, and so there is an association with increased maternal age. Two of the X chromosomes are inactivated, and abnormalities may result from three active X chromosomes early in embryonic development, prior to X inactivation and/or from genes on the X chromosome that escape inactivation.

Phenotype

In contrast to the result of a 45,X karyotype, there is not a recognizable syndrome in females with an extra X chromosome. The majority are physically normal, although there is a slight increase in the frequency of minor anomalies. Mean birthweight is at the 40th percentile, mean birthlength is at the 70th percentile, and mean birth head circumference is at the 30th percentile *(23)*. In general, as adults these women have tall stature, with an average height of 172 cm. Pubertal development is normal and most have normal fertility, although a small number have ovarian dysfunction *(24)*. There is a small but slightly increased risk of chromosomally abnormal offspring of 47,XXX women *(25,26)*. Although they do not, remarkably, appear to be at significantly increased risk of having XXX or XXY children, prenatal diagnosis should be offered for all pregnancies. Most have good health, although one study found that 25% had recurrent nonorganic abdominal pain as teenagers *(27)*.

Development

Females with 47,XXX are represented in mentally retarded and psychotic institutionalized populations at four to five times greater a rate than would be expected based on the incidence in newborns *(24,28)*. However, there is a great deal of variability in this syndrome, and some women have normal intelligence and are well adjusted. Precise predictions regarding an individual child's prognosis are not possible *(23)*.

Many studies of 47,XXX females have ascertainment bias; however, in a group of 11 females with 47,XXX ascertained at birth by unbiased screening of all newborns who were then followed into adulthood, most had serious patterns of dysfunction *(29)*. Most showed early delays in motor, language, and cognitive development and were described as shy, withdrawn, and immature, with poor coordination *(23)*. Full scale IQ was 26 points lower than in normal sibling controls, average IQ was in the 85–90 range, and patients were at the 24th percentile in academic achievement scores, but mental retardation was rare. Nine of the 11 needed special education in high school either full or part time, and fewer than half completed high school, but two achieved As and Bs and one excelled in math. Most did not participate in extracurricular activities. They were described as socially immature. All had heterosexual orientation. Compared to individuals with other types of sex chromosome abnormalities, 47,XXX females seem to have the most psychological problems, including depression and occasionally psychoses. However, one woman attended college and many have been able to overcome psychological problems and become independent, hold jobs, and marry. Stability of the home environment was somewhat related to outcome but not to such an extent as is seen in other sex chromosome disorders *(29)*. In a study of five girls ranging in age from 7 to 14 years with 47,XXX diagnosed prenatally, only one had motor and language delays and learning problems; the others had normal IQs (range 90–128) and were doing well in school *(30)*. Reasons for the difference between the two groups may be the higher socioeconomic status and greater stability of the prenatally diagnosed group.

Variants with Additional X Chromosomes

48,XXXX

As compared to 47,XXX there is almost always mild to moderate mental retardation with average IQ of 60, ranging from less than 30 to 75 *(31)*; one patient was reported with a normal IQ *(32)*. Phenotypic features include mild hypertelorism, epicanthal folds, micrognathia, and midface hypoplasia *(33)*. Tall stature is common, with adult women achieving an average height of 169 cm *(31)*. Skeletal anomalies include radio-ulnar synostosis and fifth finger clinodactyly. Incomplete development of secondary sex characteristics may occur, with scant axillary and public hair and small breasts, and some patients have gonadal dysgenesis *(34)*. Speech and behavioral problems are common. Fertility is reduced, although some have had chromosomally normal offspring.

49,XXXXX

Phenotypic features seen in penta-X females include microcephaly, short stature, upslanting palpebral fissures, low hairline, and coarse, Down syndrome-like facial features. Congenital heart and renal anomalies have been reported *(35,36)*. Most patients have moderate mental retardation (IQ 20–75, average IQ 50) *(31,37)* and are described as shy and cooperative *(31)*. There have been no reports of pregnancy in women with this chromosomal aneuploidy *(31)*. Nondisjunction in successive meiotic divisions is the probable mechanism, and molecular studies have shown that, at least in some cases, the extra X chromosomes are all maternally derived *(38,39)*.

47,XXY (Klinefelter Syndrome)

Klinefelter syndrome was the first sex chromosome abnormality to be described and is the most common cause of hypogonadism and infertility in males *(37)*. It is found in approximately 1 in 1000 newborn males.

Origin

In 53% of patients, the extra chromosome arises at paternal meiosis I, 34% at maternal meiosis I, 9% at maternal meiosis II, and 3% from postzygotic errors. There is an association with increased maternal age in those with maternal meiosis I errors *(40)*.

Phenotype

Males with 47,XXY have taller than average stature, with mean height at the 75th percentile and may have a eunuchoid build with long limbs and pear-shaped hips, although there is a great deal of phenotypic variability *(27)*. Testicular and penile size is usually small during childhood, although prepubertal phenotype is often unremarkable. Gynecomastia occurs in up to 50% of 47,XXY males during adolescence. Most enter puberty normally, although there is usually inadequate testosterone production and most require testosterone supplementation. Testes are small in adulthood. Almost all have infertility with absent spermatogenesis, tubular hyalinization, and Leydig cell hyperplasia. Most are diagnosed in adulthood, with a chief complaint of infertility.

Development

IQs are lower than in controls, with the average between 85 and 90, although there is a wide range *(27)*. Verbal IQ is usually lower than performance IQ. The majority require special help in school, especially in the areas of reading and spelling. They are often described as awkward, with mild neuromotor deficits, shy, immature, restrained, reserved, and lacking confidence *(27)*. A group of 13 males with Klinefelter syndrome, diagnosed as newborns and followed into adulthood, were said to have struggled through adolescence with limited academic success but were able to function independently in adulthood *(29)*. Most needed at least some special education help in school; nine completed high school and four went to college. All were heterosexual. Some had psychological problems, including conduct disorder and depression and difficulties with psychosocial adjustment. A stable and supportive family environment was found to correlate with better outcome *(29)*. In a group of five boys ranging in age from 7 to 14 years who had been prenatally diagnosed with 47,XXY karyotypes, there were fewer language and motor deficits in childhood and all were doing well in school. IQs ranged from 90 to 131. The reason for the better outcomes may be due to environmental and other genetic factors *(30)*.

Variants with Additional X or Y Chromosomes

48,XXYY

This is the most common variant of Klinefelter syndrome *(31)*. The incidence is estimated at 1 in 50,000 male births *(41)*. Men are tall statured with adult height above 6 feet, a eunuchoid body habitus, and long thin legs. There is hypergonadotropic hypogonadism and individuals have small testes and sparse body hair. Gynecomastia occurs frequently *(31)*.

Most 48,XXYY patients have mild mental retardation, although IQs ranging from 60 to 111 have been reported. Psychosocial and behavior problems are generally more severe than in 47,XXY individuals, although patients without significant behavior problems have been reported *(31)*.

The 48,XXYY variant is not associated with advanced parental age. Nondisjunction in both the first and second male meiotic divisions leading to an XYY sperm has been hypothesized *(42)*.

48,XXXY

This is a rare condition, associated with more abnormal features and mental retardation than 47,XXY. It was first described by Barr in 1959 *(43)*. Stature is normal to tall and facial features include epicanthal folds, hypertelorism, protruding lips, prominent mandible, and radioulnar synostosis and fifth finger clinodactyly. Hypergonadotropic hypogonadism and hypoplastic penis are found in 25% of patients. There is also infertility, and testosterone therapy has shown to be beneficial. Gynecomastia is common, as is mild to moderate mental retardation, with most patients in the 40–60 IQ range, although an individual with an IQ of 79 has been reported *(31)*. Most have speech delay, slow motor development, and poor coordination. Behavior is immature, with personalty traits described as passive, pleasant, placid, and cooperative *(31)*.

49,XXXXY

More than 100 cases of 49,XXXXY have been described in the literature since this karyotype was first reported by Fraccaro et al. in 1960 *(44)*. Common features include low birthweight; short stature in some patients; craniofacial features consisting of round face in infancy; coarsening of features in older age, with hypertelorism, epicanthal folds, and prognathism *(42)*; and a short, broad neck *(31)*. Congenital heart defects are found in 15–20% *(31,45)*, with patent ductus arteriosus being the most common defect described. Skeletal anomalies include radioulnar synostosis, genu valgus, pes cavus, and fifth finger clinodactyly. Muscular hypotonia and hyperextensible joints are present. Genitalia are hypoplastic, and cryptorchidism with hypergonadotropic hypogonadism is common *(31)*.

Mild to moderate mental retardation is seen in most patients, with IQs ranging from 20 to 78. Language development is most severely impaired. Behavior has been described as timid, shy, pleasant, anxious, and irritable *(31)*. Testosterone replacement therapy has been found to be beneficial in some patients, with improvement in attention and behavior *(46)*.

49,XXXYY

Only five cases of liveborn males with this sex chromosome abnormality have been described. Physical features include tall stature, dysmorphic facial features, gynecomastia, and hypogonadism. All have had moderate to severe mental retardation, with passive but occasionally aggressive behavior and temper tantrums. One patient had autistic-like behavior *(31)*.

47,XYY

One in 1000 males has an extra Y chromosome. This arises through nondisjunction at paternal meiosis II. Males with 47,XYY tend to have normal birth length and weight, but when older, most are above the 75th percentile in height. Minor anomalies are found in 20% of patients, but the rate of major malformations is not increased. Most infants are normal in appearance *(47)*. Patients often have severe facial acne. Pubertal development is usually normal, although onset of puberty in one group of patients studied was approximately 6 months later than average for males with no sex chromosome abnormality. Sexual orientation is heterosexual. Individuals are described as somewhat awkward and have minor neuromotor deficits *(27)*. Most have normal fertility and are able to father children.

Intelligence is normal, although there is an increased incidence of learning disabilities. There have been two groups of patients with 47,XYY studied long term, one diagnosed in a newborn screening program and the second diagnosed prenatally. The latter group of patients comes from families with an above-average socioeconomic status. The first group had an IQ range of 93–109, and all required part-time special education. The second group had an IQ range of 109–147, and all were reported to be getting As and Bs in school. IQ is usually somewhat lower than in siblings *(27,30)*. Most older boys attend college or have jobs after high school.

Hyperactive behavior, distractibility, temper tantrums, and a low frustration tolerance are reported in some boys in late childhood and early adolescence. Aggressiveness is not common in older boys. Although early studies raised the possibility of an increase in criminal behavior in these individuals, recent studies have shown that although there are a higher percentage of males with 47,XYY in prisons than in the general population, there was not an increase in violent behavior in these individuals.

The condition is clearly variable. Most blend into the population as normal individuals. Better outcomes seem to be associated with a supportive, stable environment.

Variants with Additional Y Chromosomes

48,XYYY

There is no consistent phenotype for males with two extra Y chromosomes. Mild mental retardation to low normal intelligence and sterility have been described *(48,49)*.

49,XYYYY

Only three cases have been reported. Facial features include hypertelorism, low-set ears, and micrognathia. Skeletal abnormalities include radioulnar synostosis, scoliosis, and clinodactyly. Mental retardation and speech delay, along with impulsive and aggressive behavior, were reported *(31,50,51)*.

Origin of Extra Chromosomes

The extra chromosomes in polysomy X syndromes most likely arise from sequential nondisjunction events during either maternal or paternal gametogenesis. Studies using polymorphic microsatellite DNA markers have shown a maternal origin of extra X chromosomes in 30 cases of 49,XXXXY and 49,XXXXX *(38,39,52–55)*. There does not appear to be a maternal age effect. Two cases of 48,XXYY have been shown to arise from paternal nondisjunction *(39,56)*.

For cases of 48,XXXY studied with Xg blood groups or other polymorphic markers, two are maternal and five are paternal in origin *(19,39,53,57–59)*. Nondisjunction at first and second meiotic divisions is proposed versus fertilization of an ovum by an XY sperm followed by postzygotic nondisjunction, because mosaicism has not been detected in these patients *(39)*.

SEX CHROMOSOME ANEUPLOIDY AND AGE

Increased aneuploidy with advancing age was first reported by Jacobs et al. in 1961 *(60)*. This was subsequently found to be due to loss of the X chromosome in females and of the Y in males *(61)*. Premature centromere division in the X chromosome and loss through anaphase lag and formation of a micronucleus is the proposed mechanism in females *(62,63)*. This is supported by the finding that hyperdiploidy for the X is much less common than monosomy, which would not be expected if nondisjunction were the mechanism *(64)*. It is usually the inactive X chromosome that is missing in X monosomic cells *(65)*. X-chromosome aneuploidy is not observed in bone marrow preparations from older women, but is seen in phytohemagglutinin (PHA)-stimulated peripheral blood lymphocytes. Although some early studies suggested an increase in sex chromosome aneuploidy in women with a history of reproductive loss, recent studies have shown that this is probably not true and that it is purely a phenomenon of aging *(64,66)*.

In a prospective study of 11 women from 83 to 100 years of age, Jarvik et al. found a fourfold increase in X chromosome loss after 5 years, compared with the initial level *(67)*. Galloway and Buckton found a 10-fold increase in X chromosome aneuploidy in women from age 25 to 35 compared with those between 65 and 75 *(61)*. Between 30 and 55 the rate of hypodiploid cells was 3–5% in females, increasing to 8% at age 70. This holds true for loss of any chromosome, but the most common was loss of an X chromosome. It should be noted that there is variability of sex chromosome loss between individuals of the same age, so that what is "normal" aneuploidy at a specific age is impossible to predict. This makes it difficult to interpret the clinical implications of X chromosome loss seen in an older woman who may also have features of Turner syndrome. Because the age-related loss is limited to peripheral blood lymphocytes, analysis of other tissues such as skin fibroblasts may be helpful in clarifying these situations.

Age-associated loss of the Y chromosome in men is found more often in bone marrow than in peripheral blood, and approaches the rate of X chromosome loss generally seen in peripheral blood in women *(64)*. Studies of bone marrow preparations have shown that Y chromosome loss was restricted to males over age 40–50, with a frequency of 8–10% *(68,69)*.

Most studies comparing age-related sex chromosome aneuploidy were done on metaphase preparations and are, therefore, at risk for preparation aneuploidy. Guttenbach et al. performed an *in situ* study of lymphocyte interphase nuclei to look at sex chromosome loss and aging. In males the rate was 0.05% up to age 15, 0.24% in 16 to 20-year olds, and then steadily increased to 1.34% at age 76–80. The mean value of monosomy X cells in females was 1.58% in 0- to 5-year-olds, and increased to 4–5% in women over 65. Only women over 51 years old showed a distinct age correlation. This study also found no difference in aneuploidy rates between cultured and noncultured cells *(70)*.

These findings should be considered when analyzing peripheral blood chromosomes in older females and bone marrow from older men, to avoid misinterpretation of normal age-related aneuploidy as clinically significant mosaicism or acquired changes.

STRUCTURAL ABNORMALITIES OF THE X CHROMOSOME

In addition to the isochromosome Xq commonly found in patients with Turner syndrome, the X chromosome can be involved in translocations, both balanced and unbalanced, and can also have deletions and duplications. Structural abnormalities of the X in males are generally associated with more severe phenotypic manifestations than in females. This is partly explained by preferential inactivation of the structurally abnormal X in cases of duplications or deletions or in unbalanced X;autosome translocations in females. In cases of balanced X;autosome translocations, there is usually preferential inactivation of the normal X chromosome. Theories explaining this are discussed in the following paragraphs. High-resolution chromosome analysis should be performed for females manifesting X-linked disorders to look for a structural X abnormality.

X;Autosome Translocations

Balanced Translocations

In females, balanced translocations involving the X chromosome and an autosome may lead to primary or secondary ovarian failure and variable Turner syndrome-like features if the translocation occurs within the critical region of Xq13-q26 *(71,72)*. In general, the normal X becomes preferentially inactivated in approximately 75% of patients, probably due to selection against the functional X chromosome disomy and autosomal monosomy which would result if the translocated X was inactivated *(73,74)*. When the translocation disrupts a gene located on the X chromosome, a female with such a translocation may manifest a disease condition *(75,76)*. Several X-linked genes have been mapped in this way.

The majority of females with balanced X;autosome translocations with breakpoints above the X inactivation center at Xq13 are phenotypically normal *(77,78)*. However, owing to variability in X inactivation from one person to another with the same X;autosome translocation, it is possible for a phenotypically normal mother to have a daughter with phenotypic abnormalities and mental retardation even though both carry the same such rearrangement. This is because of skewed inactivation in the former and random inactivation in the latter, leading to functional X disomy and functional autosomal monosomy, and is estimated to occur in approximately 25% of females with X;autosome translocations *(75)*.

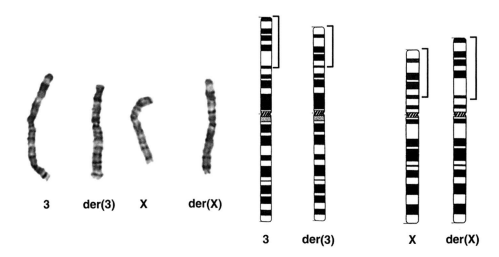

Fig. 4. Balanced reciprocal translocation between the short arms of chromosomes X and 3: 46,X,t(X;3)(p11.3;p21.2). Brackets indicate regions involved in translocation on the derivative chromosomes. The patient was a 30-year-old clinically normal, fertile female who had a daughter with an unbalanced translocation consisting of a normal X, the derivative X, and two normal chromosomes 3 (partial monosomy Xp and partial trisomy 3p).

A fertile woman with a balanced X;autosome translocation is at risk for having offspring with an unbalanced rearrangement (**Fig. 4**). There is also the risk that even balanced offspring could be abnormal due to random or skewed inactivation of the abnormal X in a female child or by disruption of a functional gene on the X in a male. The risk for a female with a balanced X;autosome translocation to have a liveborn child with a structural and/or functional aneuploidy has been estimated at 20–40% *(74)*. Phenotypic abnormalities may range from mild effects to severe mental retardation and birth defects.

Males with balanced X;autosome translocations may be normal except for infertility, but may also have severe genital abnormalities. As noted previously there is also a risk of an X-linked recessive disorder due to disruption of a gene.

Unbalanced Translocations

In females with unbalanced X;autosome translocations, the abnormal X is generally inactive if the X-inactivation center is present, probably secondary to selection against cells with an autosomal imbalance and functional X disomy. If the X-inactivation center is not present in the translocated segment, phenotypic abnormalities usually result from such imbalances and can include mental retardation and multiple congenital anomalies *(79)*. There have also been patients described who have unbalanced X;autosome translocations but no phenotypic abnormalities and only mild behavioral problems *(80)*.

Deletions of Xp

Males with deletions of the short arm of the X show contiguous gene syndromes characterized by different combinations of phenotypes, depending upon the location and length of the deletion *(81)*. X-linked ichthyosis, Kallmann syndrome (anosmia and hypogonadism), mental retardation, and chondrodysplasia punctata (skeletal dysplasia) are seen in males with deletions involving distal Xp. Deletions in Xp21 may cause a contiguous gene syndrome of Duchenne muscular dystrophy, retinitis pigmentosa, adrenal hypoplasia, mental retardation, and glycerol kinase deficiency *(82)*. Larger Xp deletions in males are lethal.

In females, there is usually preferential inactivation of the structurally abnormal X. Females with Xp deletions do not usually manifest any of the recessive disorders due to presence of a normal X chromosome although almost all have short stature and some have Turner syndrome phenotypic features. A putative gene, termed *SHOX*, within the pseudoautosomal region on Xp has recently been described *(83,84)* and may explain the short stature in these individuals. Females with terminal deletions at Xp11 usually have complete ovarian failure, whereas those with terminal deletions originating at Xp21 are more likely to show premature rather than complete ovarian failure *(85)*.

Deletions of Xq

Deletions of the long arm of the X lead to variable phenotypic outcomes. Forty-three percent of women with Xq deletions have short stature *(86)*. In a study by Geerkens et al. it was found that women with breakpoints in Xq13 to Xq25 had both normal and short stature, suggesting a variable inactivation of growth genes in Xp or proximal Xq *(87)*. Deletions in various regions of the long arm are sometimes associated with gonadal dysgenesis or premature ovarian failure. Females with terminal deletions originating at Xq13 are more likely to have complete ovarian failure, whereas those with deletions at Xq24 may have premature ovarian failure *(85)*. Clinical features of Turner syndrome are less common in Xq as compared to Xp deletions *(88)*. Large Xq deletions in males are not compatible with survival.

Xp duplications

Duplications of Xp involving bands p21.2 to 22.2, plus a Y chromosome have been reported in patients who were phenotypic females, suggesting a sex determining gene locus on Xp *(89,90)*. These patients also had mental retardation and multiple anomalies. This area of the X has been termed the dosage-sensitive sex reversal (*DSS*) region (see "46,XY Females" later). Both normal and abnormal phenotypes, and normal fertility as well as amenorrhea, have also been reported in females with Xp duplications and one normal X chromosome *(89–91)*. The abnormal phenotype, including Turner syndrome features, short stature, seizures, amenorrhea, but normal intelligence was seen in a female with complete inactivation of the duplicated X, suggesting that random inactivation was not the cause *(91)*.

A dicentric inverted duplication of most of the short arm of the X [dic inv dup(X)(qter→p22.3::p22.3→cen:)] has been reported in a mother and daughter with short stature, mental retardation, and dysmorphic features. The mother had the duplicated X as the inactive X in all cells, but the daughter had the duplicated X active in 11% of lymphocytes *(92)*.

Xq Duplications

Males with duplications of the long arm of the X usually have significant mental retardation and birth defects due to functional disomy of the duplicated regions. Most

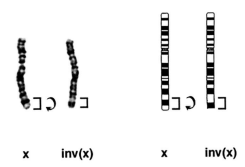

x   inv(x)   x   inv(x)

Fig. 5. Distal paracentric inversion of Xq: inv(X)(q26.?3q28) in a woman with normal phenotype and fertility. Brackets indicate region involved in inversion.

females with Xq duplications have normal phenotypes and are ascertained after the birth of an abnormal male child. However, there have been females with phenotypic abnormalities and Xq duplications. Reasons for this may be random or nonskewed X inactivation, duplication of dosage-sensitive genes, incomplete inactivation of a portion of the duplicated segment, or an imprinting effect *(93)*.

Inversions of the X Chromosome

Paracentric Inversions

Paracentric inversions of the X chromosome (**Fig. 5**) are relatively rare. There has been a wide range of phenotypes described. In general, when long arm paracentric inversions involve the critical region at Xq13–26, females have some degree of ovarian dysfunction *(94)*. When the inversion is outside the critical region, normal phenotype and fertility have been reported *(95)*, although there are exceptions to this *(96)*. There has also been variability in mental function in females, with some having mental retardation and others with normal intelligence, even in the same family *(97)*. Males may be phenotypically normal, or have mental retardation *(96)*. Fertility in males is also variable *(98)*.

Pericentric Inversions

Pericentric inversions of the X have been reported in females with gonadal dysgenesis, with mental retardation *(99)*, and with normal intelligence *(100)*. Other families have been reported with no gonadal dysfunction in females carrying such an inversion *(101,102)*. Keitges et al. reported dizygotic twins with the same pericentric X inversion (p11;q22) *(99)*. One twin was phenotypically normal with normal intelligence and menses, and had random inactivation of the X. The other was mildly mentally retarded, and had psychiatric problems, irregular menses, minor anomalies, and selective inactivation of the inverted X. Proposed explanations for these findings include different normal Xs, a nondetectable deletion or duplication in the abnormal twin, or chance. This also raises the likelihood that the replication pattern of the inverted X is a better predictor of fertility than the breakpoints. Interestingly, females with random X inactivation are more likely to have normal fertility than those with skewed inactivation of the inversion X *(99)*. Offspring of females with pericentric inversions are at risk for inheriting a recombinant chromosome.

There have been reports of males with normal development and phenotype and a maternally inherited pericentric inversion, although X-linked disorders have been found to segregate with pericentric inversions of the X, presumably by disruption or deletion of a gene by the inversion *(103,104)*.

Isodicentric X Chromosomes

Isodicentric X chromosomes are formed by the fusion of two X chromosomes *(105)*. The phenotypic effects are variable and dependent on whether the chromosomes are fused at long or short arms, and whether or not there is a deletion. Patients with isodicentric X chromosomes joined at their short arms exhibit short or normal stature, gonadal dysgenesis, and occasionally Turner syndrome features, whereas those with long arms joined are normal or above average in stature and have gonadal dysgenesis *(106)*. Explanations for the phenotype of short stature when the short arms are joined is most likely secondary to deletion of the distal short arm at the region of the putative short stature gene. Mechanisms to explain formation of terminal rearrangements between homologous chromosomes include:

1. breakage and deletion of a single chromosome followed by rejoining of sister chromatids,
2. breakage and deletion of two homologous chromosomes at the same breakpoints followed by interchromosomal reunion,
3. terminal fusion without chromatin loss between sister chromatids or homologous chromosomes *(107)*.

The isodicentric X is almost always late replicating. The second centromere is nonfunctional, making it a pseudodicentric chromosome *(108)*.

STRUCTURAL ABNORMALITIES OF THE Y CHROMOSOME

Structural abnormalities of the Y chromosome that lead to deletion of the proximal long arm may be associated with azoospermia, infertility, and short stature. Marker chromosomes derived from Y chromosomes are important to detect due to the risk of gonadoblastoma in females with Turner syndrome. Fluorescence *in situ* hybridization (FISH) probes have improved the ability to recognize marker Y chromosomes.

Translocations Involving the Y Chromosome

The Y chromosome can be involved in translocations with any other chromosome (another Y, an X, or an autosome).

X;Y Translocations

Hsu reviewed 51 reported cases of X;Y translocations, 47 with a derivative X and 4 with a derivative Y *(109)*. The X;Y translocations with a derivative X were divided into seven types, with the most common types involving translocations of a portion of Yq11 to Yqter onto Xp22.

Patients with type 1, in which there is a normal Y chromosome, were phenotypic males. For those with reported heights (14 of 15 reported), all were short, presumably due to nullisomy for the short stature gene on Xp22.3. Eleven cases with information available on skin condition showed evidence of ichthyosis, presumably due to nullisomy for the steroid sulfatase gene on Xp22. All 12 of the patients for whom

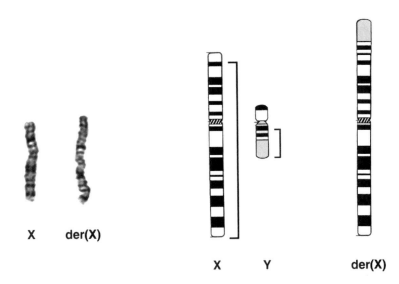

Fig. 6. Derivative X chromosome consisting of a small terminal Xp deletion and translocation of Yq: 46,X,der(X)t(X;Y)(p22.3?1;q11.2) mat. This was seen in a 5-year-old girl with short stature who had inherited the chromosome from her mother, who also had short stature but was otherwise normal. Brackets indicate regions on X and Y making up the derivative X.

information on intelligence was provided were mentally retarded. Minor facial anomalies, including flat nasal bridge and hypertelorism, were also reported. Four patients had short limbs, compatible with the diagnosis of chondrodysplasia punctata, presumed secondary to nullisomy for the X-linked chondrodysplasia punctata gene on Xp (*CDPX*). In two adult males azoospermia and small testes were reported.

With probes for the *STS* and Kallman syndrome regions on Xp, it is now possible to use FISH to delineate the extent of deletions of Xp22. This will be important in helping to predict phenotype, especially in prenatally diagnosed cases.

Type 2 patients had a translocation of Yq11 to Yqter onto Xp22, with one normal X chromosome: 46,X,der(X)t(X;Y)(Xqter→Xp22::Yq11→Yqter) (**Fig. 6**). Most of these women were ascertained through sons with a type 1 translocation. All 25 reported cases were phenotypic females, and 17 of 22 with height information were short. Most had proven fertility or reportedly had normal ovaries.

Type 3 patients had one normal X chromosome and a second sex chromosome that was dicentric, consisting of major portions of both X and Y: 46,X,dic(X;Y)(Xqter→Xp22::Yp11→Yqter). All three patients reported were phenotypic males, and had short stature and hypogonadism or azoospermia.

Type 4 patients had a portion of Yq translocated to band p11 of the second X chromosome. Of one type 4 case reported, the patient was a phenotypic female, with short stature, streak gonads, and secondary amenorrhea.

Types 5 and 6 patients had varying amounts of Yq material translocated to Xq22; of two patients described, both had streak gonads.

Type 7 had a dicentric chromosome: 46,X,dic(X;Y)(Xpter→Xq22::Yp11→Yqter), and the one case reported was a phenotypic female with streak gonads, normal stature, and secondary amenorrhea.

Four cases were reported of X;Y translocations with a derivative Y, which Hsu classified into four types. All involved a portion of Xp22 (three cases) or Xq28 (one case) translocated to Yq11, and all patients had normal stature, hypogonadism with hypoplastic male external genitalia or ambiguous genitalia, mental retardation, and various dysmorphic features *(109,110)*.

One case has been reported of a 45,X male with an X;Y translocation, in which distal Yp was translocated to Xp: 45,der(X)t(X;Y)(Xqter→Xp22.3::Yp11→Ypter). The patient had short stature, a short broad neck, broad chest, wide spaced nipples, short metacarpals and slight cubitus valgus, normal male external genitalia but small testes, and normal intelligence *(111)*.

It should be noted that the presence or absence of a 45,X cell line in addition to one with an X;Y translocation can be of significance concerning the development of external genitalia. When a 45,X cell line is present, there is an increased likelihood of a female phenotype with features of Turner syndrome *(109)*.

Xp;Yp translocations involving the testis determining factor may be found in XX males or, rarely, XY females with sex reversal. These translocations are usually not seen with cytogenetic analysis and require molecular probes for diagnosis *(81)* (see later).

Y;Autosome Translocations

In a review of more than 130 cases of Y;autosome translocations, Hsu reported that the most common involved translocation of the fluorescent heterochromatic region of Yq to the short arm of a D group (13 to 15) or G group (21 and 22) chromosome. Most of these are familial, and an otherwise normal 46,XX or 46,XY karyotype with this translocation is associated with a normal phenotype. Chromosomes 15 and 22 are most commonly involved: t(Yq12;15p) and t(Yq12;22p). When the translocation is familial, it is unlikely to have any phenotypic effects, and fertility is not affected. When the diagnosis is made prenatally in a 46,XX,der(D group chromosome) or der(G group chromosome) fetus and the translocation can only be found in male relatives, the possibility of the presence of Yp material in the derivative chromosome cannot be ruled out *(109)*. Molecular studies using Yp probes are indicated in such situations.

Translocations have been reported for all autosomes except 11 and 20. Twenty-nine of 50 cases that did not involve a D or G group chromosome involved a reciprocal translocation, of which 27 were associated with a male phenotype and 2 with a female phenotype. Of the 27 without another chromosome abnormality, 25 were male and 2 female. Eighty percent of the adult males had azoospermia/oligospermia or infertility, although there was bias of ascertainment, making the true risk of infertility in males with a balanced Y;autosome translocation unknown. Four of the patients were infants or boys with mental retardation and/or multiple congenital abnormalities. Two of the 27 patients had female phenotypes, with gonadal dysgenesis and streak gonads. A small Yp deletion or 45,X mosaicism could not be ruled out in these patients.

Hsu also reviewed 21 cases with unbalanced Y;autosome translocations, of which 13 had a male phenotype *(109)*. Two of five adult males had azoospermia or hypogonadism; the other three were phenotypically normal and fertile. Eight were infants or children with abnormalities secondary to autosomal aneusomy. Six patients were phenotypic females, five with gonadal dysgenesis and one with Turner syndrome features; three had developed gonadoblastoma.

Males with 45,X and Y;autosome translocations involving all of Yp or a portion of distal Yp may have azoospermia or infertility, although some have normal fertility *(109)*. The presence of Yp in a Y;autosome translocation explains the male sex determination.

Yp Deletions

Individuals with deletions of the short arm of the Y are usually phenotypic females. Most have streak gonads with Turner syndrome features, especially lymphedema, but normal stature *(109)*. This is in contrast to females with 46,XY "pure" gonadal dysgenesis who do not have features of Turner syndrome.

Yq Deletions

Deletions involving the heterochromatic portion of Yq are compatible with normal genital development and sexual differentiation (see "Y Chromosome Polymorphisms"). Larger deletions involving the euchromatic portion of Yq may cause azoospermia *(112)*. When detected prenatally or in a young patient, the father should be tested to see whether the deleted Y is an inherited or a *de novo* abnormality. Hsu reviewed 52 cases of 46,X,del(Y)(q11); 48 were phenotypic males and most were infertile with azoospermia or oligospermia *(109)*. Based on patients with Yq deletions, the azoospermia factor *(AZF)* was hypothesized (see earlier). Males with these deletions may have short or normal stature. No patients had gonadoblastoma. Of the three patients who were phenotypic females, two had streak or dysgenetic gonads, and two had normal stature. One patient had ambiguous external genitalia with left testis and right streak gonad, normal stature, and Turner syndrome features. Mosaicism for a 45,X cell line could not be ruled out *(109)*.

Y Isochromosomes

In most cases of isochromosome for Yp or Yq, the abnormal chromosome is dicentric and present in mosaic fashion.

i(Yp)

Phenotypic features reported in patients with isochromosome for the short arm of Y include ambiguous genitalia and Turner syndrome features with normal stature, although a partial Yq deletion could not be ruled out *(109,113)*. Other patients were phenotypic males. One who was an adult had short stature, mental retardation, and facial anomalies *(114)*.

i(Yq)

Hsu reviewed seven reported cases with nonmosaic, monocentric isochromosome Yq. All were phenotypic females (expected due to the absence of *SRY*), with sexual infantilism and streak gonads. Approximately half had Turner syndrome features and short stature *(109)*.

Isodicentric Y

The dicentric Y is among the most commonly detected structural abnormality of the Y chromosome (**Fig. 7**). Most (91%) are found in mosaic form, usually with the other cell line being 45,X. It is therefore difficult to know the phenotypic effects of a dicen-

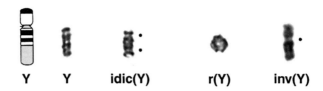

Fig. 7. (Left to Right) Normal Y; isodicentric Y consisting of two copies of the short arm, centromere, and proximal long arm (q11.2); ring Y, and pericentric inversion of the Y: inv(Y)(p11q11). Dots indicate location of centromeres.

tric Y alone. Some dicentric Ys have the breakpoint in the long arm, with duplication of the proximal long arm and entire short arm, whereas others have the break in the short arm with the proximal short arm and entire long arm duplicated.

Most reported patients have short stature, and external genitalia may be female, ambiguous, or male. Gonadoblastoma has been reported in females with a dicentric Y cell line. Males often have hypospadias. Azoospermia is common in phenotypic males with an isodicentric Y. Again, this has been proposed to be due to loss of the *AZF* gene *(109,115)*. Mental retardation has been reported in a few patients, although there is a bias of ascertainment in postnatally diagnosed cases and there are very few reports of prenatally diagnosed cases *(115)*.

The presence of a 45,X cell line in addition to any cell line with an isochromosome Y or isodicentric Y leads to variable phenotypic manifestations, ranging from phenotypic male with azoospermia to ambiguous genitalia to phenotypic females with typical or atypical Turner syndrome features *(109)*.

Ring Y Chromosome

The brightly fluorescent region of Yq is usually deleted during formation of a ring Y (**Fig. 7**), making this an unreliable tool for identification. The most accurate way to determine origin of a ring sex chromosome in a patient with a 46,X,–X or Y,+r karyotype is with FISH, using probes for X and Y. In a review of 34 cases with r(Y), 25 had a 45,X cell line. Nine cases were nonmosaic: eight were phenotypic males, one of whom had proven azoospermia. Other variable features described included small testes, small penis, hypospadias, and short stature. One patient was a phenotypic female with streak gonads and sexual infantilism. Of cases with mosaicism, phenotype varied from normal male to ambiguous genitalia to normal female. Phenotypes were similar to the nonmosaic cases *(109)*.

Y Chromosome Polymorphisms

Heterochromatic Length

The Y chromosome varies in size in the normal male population, due to variability in size of the heterochromatic portion of Yq (Yqh or Yq12). This is not associated with phenotypic abnormalities or infertility (see "Yq Deletions").

Satellited Y Chromosome

The presence of satellites on the end of the long arm of the Y chromosome is considered to be a normal variant not associated with phenotypic abnormalities. Transmission through several generations has been reported. These chromosomes are thought to arise from translocations involving the short arm of an acrocentric autosome *(116)*. All have an active nucleolar organizer region.

At least one case of satellited short arm of the Y chromosome has been reported, and was present in a phenotypically normal, fertile male *(117)*.

Inverted Y

Pericentric inversion of the Y chromosome [inv(Y)(p11q11), (**Fig. 7**)] is estimated to occur as a normal variant in 0.6 in 1000 males. A very high frequency of 30.5% was found in the Gujerati Muslim Indian population of South Africa. It is not associated with any phenotypic or reproductive abnormalities *(118)*.

DISORDERS OF SEXUAL DEVELOPMENT WITH "NORMAL" SEX CHROMOSOMES

Although visible structural abnormalities of the sex chromosomes are often associated with phenotypic abnormalities of the internal or external genitalia, there are other disorders of sexual development in which the sex chromosomes may appear structurally normal.

Complete Sex Reversal

46,XX Males

This is a genetically heterogeneous group of conditions involving individuals who have bilateral testes while lacking a Y chromosome. Most have normal external genitalia, although 10–15% have some degree of genital ambiguity, cryptorchidism, and/or hypospadias. These individuals are more likely to be diagnosed in childhood *(119)*. Others present in adulthood with infertility or gynecomastia. Most have small testes and some signs of androgen deficiency, similar to Klinefelter syndrome patients *(120)*. The majority have Y sequences including the *SRY* gene, most often found translocated to one of the X chromosomes, but as many as one third are truly Y-negative. Explanations for sex reversal in the latter include mosaicism for either a Y-bearing cell line or other autosomal or sex-linked genes important for sex determination.

Most *SRY*-positive XX males have normal male external genitalia, whereas those lacking Y-derived sequences are more likely to have ambiguous genitalia *(121,122)*. There have been familial cases of 46,XX males, suggesting autosomal recessive inheritance *(123)*. There have also been families reported with both XX males and XX true hermaphrodites, so that there may be a common origin for both *(121)*. Others have also found evidence that full virilization requires the expression of a second Y-linked gene, near *SRY*, which may be expressed outside the testis *(124,125)*.

The pseudoautosomal regions of Xp and Yp pair during male meiosis, and there sometimes may be unequal interchange of material extending beyond the pseudoautosomal boundaries. This theory has been used to explain the origin of XX males with SRY and other portions of Yp translocated to Xp *(126)*.

46,XY Females

Some 46,XY females have loss of *SRY*, whereas others may have complete androgen insensitivity ("testicular feminization"). Malformation syndromes such as Smith–

Lemli–Opitz and campomelic dysplasia also produce female or ambiguous genitalia with a 46,XY karyotype. These are due to mutations or deletions in the autosomal genes *DHCR7* (7-dehydrocholesterol reductase) and *SOX-9*, respectively. There is also a dosage-sensitive region on Xp (*DSS*) that, when duplicated, leads to female external genitalia in a 46,XY individual. A candidate gene is *DAX-1* (dosage-sensitive sex reversal/adrenal hypoplasia congenita/critical region on the X chromosome, gene 1) *(127,128)*. Mutations in this gene are associated with congenital adrenal hypoplasia and hypogonadotropic hypogonadism *(129,130)*.

True Hermaphroditism

This is a rare condition where both testicular and ovarian tissue is present either as separate structures or as an ovotestis. A few patients who were chimeras with 46,XX and 46,XY cell lines arising from the fusion of two zygotes have been described *(120)*, although not all 46,XX/46,XY individuals have true hermaphroditism *(131)*. At least 50% of true hermaphrodites are 46,XX with no Y DNA *(120)*.

Gonadal neoplasia and breast cancer have been reported in these patients *(131,132)*.

Pseudohermaphroditism

Patients with pseudohermaphroditism have gonadal tissue of one sex but ambiguous external genitalia. Female pseudohermaphroditism, in which there is a 46,XX chromosome complement, is usually due to congenital adrenal hyperplasia. It is critical to identify these patients early due to the risk of hypovolemic shock in untreated 21-hydroxylase deficiency, the most common type of congenital adrenal hyperplasia. This is an autosomal recessive condition and prenatal diagnosis and treatment are possible.

Male pseudohermaphroditism (46,XY) has many causes, including partial androgen insensitivity or Reifenstein syndrome, which is an X-linked condition *(133)*. Genetic males with 5α-reductase deficiency have ambiguous genitalia at birth but have normal virilization at puberty. This is an autosomal recessive condition *(14)*. Drash syndrome is a condition with Wilms tumor, aniridia, and male pseudohermaphroditism *(134)*. Many other malformation syndromes associated with male pseudohermaphroditism have been described *(135–137)*. Appropriate clinical evaluation and endocrine, cytogenetic, and imaging studies are critical in differentiating the various forms. Early studies suggested that sex-of-rearing different from genetic sex did not make a difference in terms of gender identification and adjustment. Recently, however, there have been reports of major difficulties for these individuals, forcing physicians to reexamine their treatment of such patients and emphasizing the need for more long-term studies *(138)*.

ACKNOWLEDGMENT

The ideograms in this chapter were reproduced with permission of S. Karger AG, Basel.

REFERENCES

1. Moore, K.L. and Persaud, T.V.N. (1993) In *The Developing Human,* 5th Edition. WB Saunders, Philadelphia, pp. 281–285.
2. Lyon, M.F. (1961) Gene action in the X-chromosome of the mouse (*Mus musculus* L.) *Nature* **190,** 372–373.

3. Moore, K.L. and Barr, M.L. (1954) Nuclear morphology, according to sex, in human tissues. *Acta Anat.* **21,** 197–208.
4. Davidson, W.M. and Smith, D.R. (1954) A morphological sex difference in the polymorphonuclear neutrophil leucocytes. *Br. Med. J.* **II,** 6–7.
5. Latt, S.A. (1974) Sister chromatid exchanges, indices of human chromosome damage and repair: detection of fluorescence and induction of mitomycin C. *Proc. Natl. Acad. Sci. USA* **71,** 3162–3166.
6. Freije, D., Helms, C., Watson, M.S., and Donis-Keller, H. (1992) Identification of a second pseudoautosomal region near the Xq and Yq telomeres. *Science* **258,** 1784–1787.
7. Sinclair, A.H., Berta, P., Palmer, M.S., Hawkins, R., Griffiths, B.L., Smith, M., Foster, J.W., Frichauf, A.M., Lovell-Badge, R., and Goodfellow, P.N. (1990) A gene from the human sex-determining region encodes a protein with homology to a conserved DNA-binding motif. *Nature* **346,** 240–245.
8. Tiepolo, L. and Zuffardi, O. (1976) Localization of factors controlling spermatogenesis in the nonfluorescent portion of the human Y chromosome long arm. *Hum. Genet.* **34,** 119–124.
9. Vogt, P.H., Edelmann, A., Kirsch, S., Henegariu, O., Hirschmann, P., Kiesewetter, F., Kohn, F.M., Schill, W.B., Farah, S., Ramos, C., Hartmann, M., Hartschuh, W., Meschede, D., Behre, H.M., Castel, A., Nieschlag, E., Weidner, W., Grone, H.J., Jung, A., Engel, W., and Haidl, G. (1996) Human Y chromosome azoospermia factors (AZF) mapped to different subregions in Yq11. *Hum. Mol. Genet.* **5,** 933–943.
10. Page, D.C. (1987) Hypothesis: a Y-chromosomal gene causes gonadoblastoma in dysgenetic gonads. *Development* **101** (Suppl), 151–155.
11. Tsuchiya, K., Reijo, R., Page, D.C., and Disteche, C.M. (1995) Gonadoblastoma: molecular definition of the susceptibility region on the Y chromosome. *Am. J. Hum. Genet.* **57,** 1400–1407.
12. Salo, P., Kaarianinen, H., Petrovic, V., Peltomaki, P., Page, D.C., and de la Chapelle, A. (1995) Molecular mapping of the putative gonadoblastoma locus on the Y chromosome. *Genes Chromosomes Cancer* **14,** 210–214.
13. Chang, H.J., Clark, R.D., and Bachman, H. (1990) The phenotype of 45,X/46,XY mosaicism: an analysis of 92 prenatally diagnosed cases. *Am. J. Hum. Genet.* **46,** 156–167.
14. Simpson, J.L. (1996) Disorders of the gonads, genital tract, and genitalia. In *Emery and Rimoin's Principles and Practice of Medical Genetics* (Rimoin, D.L., Connor, J.M., and Pyeritz, R.E., eds.), Churchill Livingstone, New York, pp. 1477–1500.
15. Migeon, B.R., Luo, S., Stasiowski, B.A., Jani, M., Axelman, J., Van Dyke, D.L., Weiss, L., Jacobs, P.A., Yang-Feng, T.L., and Wiley, J.E. (1993) Deficient transcription of *XIST* from tiny ring X chromosomes in females with severe phenotypes. *Proc. Natl. Acad. Sci. USA* **90,** 12025–12029.
16. Wolff, D.J., Brown, C.J., Schwartz, S., Duncan, A.M.V., Surti, U., and Willard, H.F. (1994) Small marker X chromosomes lack the X inactivation center: implications for karyotype/phenotype correlations. *Am. J. Hum. Genet.* **55,** 87–95.
17. Kocova, M., Siegel, S.F., Wenger, S.L., Lee, P.A., and Trucco, M. (1993) Detection of Y chromosome sequences in Turner's syndrome by Southern blot analysis of amplified DNA. *Lancet* **342,** 140–143.
18. Chu, C.E., Donaldson, M.D.C., Kelnar, C.J.H., Smail T.J., Greene, S.A., and Connor, J.M. (1997) Low level mosaicism in Turner's syndrome-genotype/phenotype correlation. *Am. J. Hum. Genet.* **61,** A35.
19. Sanger, R., Tippett, P., and Gavin, J. (1971) Xg groups and sex abnormalities in people of northern European ancestry. *J. Med. Genet.* **8,** 417–426.
20. Mathur, A., Stekol, L., Schatz, D., MacLaren, N.K., Scott, M.L., and Lippe, B. (1991) The parental origin of the single X chromosome in Turner syndrome: lack of correlation with parental age or clinical phenotype. *Am. J. Hum. Genet.* **48,** 682–686.

21. Skuse, D.H., James, R.S., Bishop, D.V.M., Coppin, B., Dalton, P., Aamodt-Leeper, G., Bacarese-Hamilton, M., Creswell, C., McGurk, R., and Jacobs, P.A. (1997) Evidence from Turner's syndrome of an imprinted X-linked locus affecting cognitive function. *Nature* **387,** 705–708.

22. Jacobs, P.A., Baikie, A.G., Court Brown, W.M., MacGregor, T.N., MacLean, N., and Harnden, D.G. (1959) Evidence for the existence of the human "super female." *Lancet* **ii,** 423–425.

23. Linden, M.G., Bender, B.G., Harmon, R.J., Mrazek, D.A., and Robinson, A. (1988) 47,XXX: What is the prognosis? *Pediatrics* **82,** 619–630.

24. Barr, M.L., Sergovich, F.R., Carr, D.H., and Shaver, E.L. (1969) The triplo-X female: an appraisal based on a study of 12 cases and a review of the literature. *Can. Med. Assoc. J.* **101,** 247–258.

25. Fryns, J.P., Kleczowska, A., Petit, P., and van den Berghe, H. (1983) X chromosome polysomy in the female: personal experience and review of the literature. *Clin. Genet.* **23,** 341–349.

26. Neri, G. (1984) A possible explanation for the low incidence of gonosomal aneuploidy among the offspring of triplo-X individuals. *Am. J. Med. Genet.* **18,** 357–364.

27. Robinson, A., Bender, B.G., and Linden, M.G. (1991) Summary of clinical findings in children and young adults with sex chromosome anomalies. *Birth Defects Orig. Article Ser.* **26,** 225–228.

28. Hook, E.B. (1979) Extra sex chromosomes and human behavior: the nature of evidence regarding XYY, XXY, XXYY, and XXX genotypes. In *Genetic Mechanisms of Sexual Development* (Vallet, H.L. and Porter, I.Y., eds.), Academic Press, New York, pp. 437–463.

29. Bender, B.G., Harmon, R.J., Linden, M.G., and Robinson, A. (1995) Psychosocial adaptation of 39 adolescents with sex chromosome abnormalities. *Pediatrics* **96,** 302–308.

30. Robinson, A., Bender, B.G., and Linden, M.G. (1992) Prognosis of prenatally diagnosed children with sex chromosome aneuploidy. *Am. J. Med. Genet.* **44,** 365–368.

31. Linden, M.G., Bender, B.G., and Robinson, A. (1995) Sex chromosome tetrasomy and pentasomy. *Pediatrics* **96,** 672–682.

32. Blackston, R.D., Grinzaid, K.S., and Saxe, D.F. (1994) Reproduction in 48,XXXX women. *Am. J. Med. Genet.* **52,** 379.

33. Telfer, M.A., Richardson, C.E., Helmken, J., and Smith, G.F. (1970) Divergent phenotypes among 48,XXXX and 47,XXX females. *Am. J. Hum. Genet.* **22,** 326–335.

34. Carr, D.H., Barr, M.L., and Plunkett, E.R. (1961) An XXXX sex chromosome complex in two mentally defective females. *Canad. Med. Assoc. J.* **84,** 131–137.

35. Archidiacono, N., Rocchi, M., Valente, M., and Filippi, G. (1979) X pentasomy: a case review. *Hum. Genet.* **52,** 69–77.

36. Sergovich, F., Vilenberg, C., and Pozaonyi, J. (1971) The 49,XXXXX condition. *J. Pediatr.* **78,** 285.

37. Jones, K.L., ed. (1997) *Smith's Recognizable Patterns of Human Malformation.* WB Saunders, Philadelphia.

38. Deng, H.-X., Abe, K., Kondo, I., Tsucahara, M., Inagaki, H., Hamada, I., Fukushima Y., and Niikawa, N. (1991) Parental origin and mechanisms of formation of polysomy X: an XXXXX case and four XXXXY cases determined with RFLPs. *Hum. Genet.* **86,** 541–544.

39. Leal, C.A., Belmont, J.W., Nachtman, R., Cantu, J.M., and Medina, C. (1994) Parental origin of the extra chromosomes in polysomy X. *Hum. Genet.* **94,** 423–426.

40. Jacobs, P.A., Hassold, T.J., Whittington, E., Butler, G., Collyer, S., Keston, M., and Lee, M. (1988) Klinefelter's syndrome: an analysis of the origin of the additional sex chromosome using molecular probes. *Ann. Hum. Genet.* **52,** 93–109.

41. Sorensen, K., Nielsen, J., Jacobsen, P., and Rolle, T. (1978) The 48,XXYY syndrome. *J. Ment. Defic. Res.* **22,** 197–205.

42. Gorlin, R.J. (1977) Classical chromosome disorders. In *New Chromosomal Syndromes* (Yunis, J.J., ed.), Academic Press, New York, pp. 59–117.

43. Barr, M.L., Shaver, E.L., Carr, D.H., and Plunkett, E.R. (1959) An unusual sex chromosome pattern in three mentally deficient subjects. *J. Ment. Defic. Res.* **3,** 78–87.

44. Fraccaro, M., Kaijser, K., and Lindsten, J. (1960) A child with 49 chromosomes. *Lancet* **ii,** 899–902.

45. Karsh, R.B., Knapp, R.F., Nora, J.J., Wolfe, R.R., and Robinson, A. (1975) Congenital heart disease in 49,XXXXY syndrome. *Pediatrics* **56,** 462–464.

46. Sheridan, M.K., Radlinski, S.S., and Kennedy, M.D. (1990) Developmental outcome in 49,XXXXY Klinefelter syndrome. *Dev. Med. Child Neurol.* **32,** 528–546.

47. Robinson, A., Lubs, H.A., Nielsen, J., and Sorensen, K. (1979) Summary of clinical findings: profiles of children with 47,XXY, 47,XXX and 47,XYY karyotypes. *Birth Defects Orig. Article Ser.* **15,** 261–266.

48. Townes, P.L., Ziegler, N.A., and Lenhard, L.W. (1965) A patient with 48 chromosomes (XYYY) *Lancet* **i,** 1041–1043.

49. Hunter, J. and Quaife, R. (1973) A 48,XYYY male: a somatic and psychiatric description. *J. Med. Genet.* **10,** 80–82.

50. Sirota, L., Zlotogora, Y., Shabtai, F., Halbrecht, I., and Elian, E. (1981) 49,XYYYY. A case report. *Clin. Genet.* **19,** 87–93.

51. Sirota, L., Shaghapour, S.E., Elitzur, A., and Sirota, P. (1986) Neurodevelopmental and psychological aspects in a child with 49,XYYYY karyotype. *Clin. Genet.* **30,** 471–474.

52. Villamar, M., Benitez, J., Fernandez, E., Ayuso, C., and Ramos, C. (1989) Parental origin of chromosomal nondisjunction in a 49,XXXXY male using recombinant-DNA techniques. *Clin. Genet.* **36,** 152–155.

53. Hassold, T., Pettay, D., May, K., and Robinson, A. (1990) Analysis of non-disjunction in sex chromosome tetrasomy and pentasomy. *Hum. Genet.* **85,** 648–650.

54. Huang, T.H.M., Greenberg, F., and Ledbetter, D.H. (1991) Determination of the origin of nondisjunction in a 49,XXXXY male using hypervariable dinucleotide repeat sequences. *Hum. Genet.* **86,** 619–620.

55. David, D., Marquez, R.A., Carreiro, M.H., Moreira, I., and Boavida, M.G. (1992) Parental origin of extra chromosomes in persons with X chromosome tetrasomy. *J. Med. Genet.* **29,** 595–596.

56. Rinaldi, A., Archidiacono, N., Rocchi, M., and Filippi, G. (1979) Additional pedigree supporting the frequent origin of XXYY from consecutive non-disjunction in paternal gametogenesis. *J. Med. Genet.* **16,** 225–226.

57. Greenstein, R.M., Harris, D.J., Luzzatti, L., and Cann, H.W. (1979) Cytogenetic analysis of a boy with the XXXY syndrome: origin of the X-chromosomes. *Pediatrics* **45,** 677–685.

58. Pfeiffer, R.A. and Sanger, R. (1973) Origin of 48,XXXY: the evidence of the Xg blood groups. *J. Med. Genet.* **10,** 142.

59. Lorda-Sanchez, I., Binkert, F., Hinkel, K.G., Moser, H., Rosenkranz, W., Maechler, M., and Schinzel, A. (1992) Uniparental origin of sex chromosome polysomies. *Hum. Hered.* **42,** 193–197.

60. Jacobs, P.A., Court Brown, W.M., and Doll, R. (1961) Distribution of human chromosome counts in relation to age. *Nature* **191,** 1178–1180.

61. Galloway, S.M. and Buckton, K.E. (1978) Aneuploidy and aging: chromosome studies on a random sample of the population using G-banding. *Cytogenet. Cell Genet.* **20,** 78–95.

62. Fitzgerald, P.H. and McEwan, C.M. (1977) Total aneuploidy and age-related sex chromosome aneuploidy in cultured lymphocytes of normal men and women. *Hum. Genet.* **39,** 329–337.

63. Ford, J.H., Schultz, C.J., and Correll, A.T. (1988) Chromosome elimination in micronuclei: a common cause of hypodiploidy. *Am. J. Hum. Genet.* **43,** 733–740.

64. Stone, J.F. and Sandberg, A.A. (1995) Sex chromosome aneuploidy and aging. *Mutat. Res.* **338,** 107–113.

65. Abruzzo, M.A., Mayer, M., and Jacobs, P.A. (1985) Aging and aneuploidy: evidence for the preferential involvement of the inactive X chromosome. *Cytogenet. Cell Genet.* **39,** 275–278.

66. Nowinski, G.P., Van Dyke, D.L., Tilley, B.C., Jacobsen, G., Babu, V.R., Worsham, M.J., Wilson, G.N., and Weiss, L. (1990) The frequency of aneuploidy in cultured lymphocytes is correlated with age and gender but not with reproductive history. *Am. J. Hum. Genet.* **46,** 1101–1111.

67. Jarvik, L.F., Yen, F.-S., Fu, T.-K, and Matsuyama, S.S. (1976) Chromosomes in old age: a six year longitudinal study. *Hum. Genet.* **33,** 17–22.

68. Sakurai, M. and Sandberg, A.A. (1976) The chromosomes and causation of human cancer and leukemia XVIII, the missing Y in acute myeloblastic leukemia (AML) and Ph^1-positive chronic myelocytic leukemia (CML). *Cancer* **38,** 762–769.

69. United Kingdom Cancer Cytogenetics Group (1992) Loss of Y chromosome from normal and neoplastic bone marrows. *Genes Chromosomes Cancer* **5,** 83–88.

70. Guttenbach, M., Koschorz, B., Bernthaler, U., Grimm, T., and Schmid, M. (1995) Sex chromosome loss and aging: in situ hybridization studies on human interphase nuclei. *Am. J. Hum. Genet.* **57,** 1143–1150.

71. Therman E., Laxova, R., and Susman, B. (1990) The critical region on the human Xq. *Hum. Genet.* **85,** 455–461.

72. Powell, C.M., Taggart, R.T., Drumheller, T.C., Wangsa, D., Qian, C., Nelson, L.M., and White, B.J. (1994) Molecular and cytogenetic studies of an X;autosome translocation in a patient with premature ovarian failure and review of the literature. *Am. J. Med. Genet.* **52,** 19–26.

73. Schmidt, M. and Du Sart D. (1992) Functional disomies of the X chromosome influence the cell selection and hence the X inactivation pattern in females with balanced X-autosome translocations: a review of 122 cases. *Am. J. Med. Genet.* **42,** 161–169.

74. Gardner, R.J.M. and Sutherland, G.R., eds. (1996) *Chromosome Abnormalities and Genetic Counseling.* Oxford University Press, New York.

75. Merry, D.E., Lesko J.G., Siu, V., Flintoff, W.F., Collins, F., Lewis, R.A., and Nussbaum, R.L. (1990) DXS165 detects a translocation breakpoint in a woman with choroideremia and a de novo X;13 translocation. *Genomics* **6,** 609–615.

76. Giacalone, J.P. and Francke, U. (1992) Common sequence motifs at the rearrangement sites of a constitutional X/autosome translocation and associated deletion. *Am. J. Hum. Genet.* **50,** 725–741.

77. Hatchwell, E. (1996) Hypomelanosis of Ito and X;autosome translocations: a unifying hypothesis. *J. Med. Genet.* **33,** 177–183.

78. Mattei, M.G., Mattei, J.F., Ayme, S., and Giraud, F. (1982) X-autosome translocation: cytogenetic characteristics and their consequences. *Hum. Genet.* **61,** 295–309.

79. Summitt, R.L., Tipton, R.E., Wilroy, R.S., Martens P.M., and Phelan, J.P. (1978) X-autosome translocations: a review. *Birth Defects Orig. Article Ser.* **14,** 219–247.

80. Petit, P., Hilliker, C., Van Leuven, F., and Fryns, J.-P. (1994) Mild phenotype and normal gonadal function in females with 4p trisomy due to unbalanced t(X;4)(p22.1;p14). *Clin. Genet.* **46,** 304–308.

81. Ballabio, A. and Andria, G. (1992) Deletions and translocations involving the distal short arm of the human X chromosome: review and hypotheses. *Hum. Mol. Genet.* **1,** 221–227.

82. Dunger, D.B., Davies, K.E., Pembrey, M., Lake, B., Pearson, P., and Williams, D. (1986) Deletion on the X chromosome detected by direct DNA analysis in one of two unrelated boys with glycerol kinase deficiency, adrenal hypoplasia, and Duchenne muscular dystrophy. *Lancet* **i,** 585–587.

83. Rao, E., Weiss, B., Fukami, M., Rump, A., Niesler, B., Mertz, A., Muroya, K, Binder, G., Kirsch, S., Winkelmann, M., Nordsiek, G., Heinreich, U., Breuning, M.H., Panke, M.B., Rosenthal, A., Ogata, T., and Rappold, G.A. (1997) Pseudoautosomal deletions encompassing a novel homeobox gene cause growth failure in idiopathic short stature and Turner-syndrome. *Nat. Genet.* **16,** 54–63.

84. Ellison, J.W., Wardak, Z., Young, M.E., Robey, P.G., Laigwebster, M., and Chiong, W. (1997) PHOG, a candidate gene for involvement in the short stature of Turner syndrome. *Hum. Mol. Genet.* **6,** 1341–1347.

85. Tharapel, A.T., Anderson, K.P., Simpson, J.L., Martens, P.R., Wilroy R.S., Llerena J.C., and Schwartz, C.E. (1993) Deletion (X)(q26.1→q28) in a proband and her mother: molecular characterization and phenotype-karyotypic deductions. *Am. J. Hum. Genet.* **52,** 463–471.

86. Therman, E. and Susman, B. (1990) The similarity of phenotypic effects caused by Xp and Xq deletions in the human female: a hypothesis. *Hum. Genet.* **85,** 175–183.

87. Geerkens, C., Just, W., and Vogel W. (1994) Deletions of Xq and growth deficit: a review. *Am. J. Med. Genet.* **50,** 105–113.

88. Skibsted, L., Westh, H., and Niebuhr, E. (1984) X long arm deletions. A review of nonmosaic cases studied with banding techniques. *Hum. Genet.* **67,** 1–5.

89. Bernstein, R., Jenkins, T., Dawson, B., Wagner, J., DeWald, G., Koo, G.C., and Wachtel, S.S. (1980) Female phenotype and multiple abnormalities in sibs with a Y chromosome and partial X chromosome duplication:H-Y antigen and Xg blood group findings. *J. Med. Genet.* **17,** 291–300.

90. Scherer, G., Schempp, W., Baccichetti, C., Lenzini, E., Bricarelli, F.D., Carbone, L.D.L., and Wolf, U. (1989) Duplication of an Xp segment that includes the ZFX locus causes sex inversion in man. *Hum. Genet.* **81,** 291–294.

91. Wyandt, H.E., Bugeau-Michaud, L., Skare, J.C., and Milunsky, A. (1991) Partial duplication of Xp: a case report and review of previously reported cases. *Am. J. Med. Genet.* **40,** 280–283.

92. Tuck-Muller, C.M., Martinez, J.E., Batista, D.A., Kearns, W.G., and Wertelecki, W. (1993) Duplication of the short arm of the X chromosome in mother and daughter. *Hum. Genet.* **91,** 395–400.

93. Aughton, D.J., AlSaadi, A.A., Johnson, J.A., Transue, D.J., and Trock, G.L. (1993) Dir dup(X)(q13→qter) in a girl with growth retardation, microcephaly, developmental delay, seizures, and minor anomalies. *Am. J. Med. Genet.* **46,** 159–164.

94. Madan, K. (1995) Paracentric inversions: a review. *Hum. Genet.* **96,** 503–515.

95. Neu, R.L., Brar, H.S., and Koos, B.J. (1988) Prenatal diagnosis of inv(X)(q12q28) in a male fetus. *J. Med. Genet.* **25,** 52–60.

96. Abeliovich, D., Dagan, J., Kimchi-Sarfaty, C., and Zlotogora, J. (1995) Paracentric inversion X(q21.2q24) associated with mental retardation in males and normal ovarian function in females. *Am. J. Med. Genet.* **55,** 359–362.

97. Herr, H.M., Horton, S.J., and Scott, C.I. (1985) De novo paracentric inversion in an X chromosome. *J. Med. Genet.* **22,** 140–153.

98. Phillips, K.K., Kaiser-Rogers, K.A., Eubanks, S., and Rao, K.W. (1997) Prenatal diagnosis of an inversion X(q26.?3q28) in a family with a fertile female and apparently fertile male. *Am. J. Hum. Genet.* **61,** A138.

99. Keitges, E.A., Palmer, C.G., and Weaver, D.D. (1982) Pericentric X inversion in dizygotic twins who differ in X chromosome inactivation and menstrual cycle function. *Hum. Genet.* **62,** 210–213.

100. Kristensen H., Friedrich, U., Larsen, G., and Therkelsen, A. J. (1975) Structural X-chromosome abnormality in a female with gonadal dysgenesis. *Hum. Genet.* **26,** 133–138.

101. Maeda, T., Ohno, M., Takada, M., Nishida, M., Tsukioka, K., and Tomita, H. (1979)

Turner's syndrome with a duplication-deficiency X chromosome derived from a maternal pericentric inversion X chromosome. *Clin. Genet.* **15,** 259–266.

102. Nikolis, J. and Stolevic, E. (1978) Recombinant chromosome as a result of a pericentric inversion of X chromosome. *Hum. Genet.* **45,** 115–122.

103. Buckton, K.E., Newton, M.S., Collyer, S., Lee, M., and Spowart, G. (1981) Phenotypically normal individuals with an inversion (X)(p22q13) and the recombinant (X), dup q. *Ann. Hum. Genet.* **45,** 159–168.

104. Brothman, A.R., Newlin, A., Phillips, S.E., Kinzie, G.Q., and Leichtman, L.G. (1993) Prenatal detection of an inverted X chromosome in a male. *Clin. Genet.* **44,** 139–141.

105. Zakharov, A.F., and Baranovskaya, L.I. (1983) X-X chromosome translocations and their karyotype-phenotype correlations. In *Cytogenetics of the Mammalian X Chromosome, Part B: X Chromosome Anomalies and Their Clinical Manifestations* (Sandberg, A.A., ed.), Alan R. Liss, New York, 261–279.

106. Barnes, I.C.S., Curtis, D.J., and Duncan, S.L.B. (1986) An isodicentric X chromosome with short arm fusion in a woman without somatic features of Turner's syndrome. *J. Med. Genet.* **24,** 428–431.

107. Rivera, H., Sole, M.T., Garcia-Cruz, D., Martinez-Wilson, M., and Cantu, J. M. (1984) On telomere replication and fusion in eukaryotes: apropos of a case of 45,X/46,X,ter rea(X;X)(p22.3;p22.3). *Cytogenet. Cell Genet.* **38,** 23–28.

108. Sarto, G.E. and Therman, E. (1980) Replication and inactivation of a dicentric X formed by telomeric fusion. *Am. J. Obstet. Gynecol.* **136,** 904–911.

109. Hsu, Y.F. (1994) Phenotype/karyotype correlations of Y chromosome aneuploidy with emphasis on structural aberrations in postnatally diagnosed cases. *Am. J. Med. Genet.* **53,** 108–140.

110. Bardoni, B., Floridia, G., Guioli, S., Peverali, G., Anichini, C., Cisternino, M., Casalone, R., Danesino, C., Fraccaro, M., Zuffardi, O., and Camerino, G. (1993) Functional disomy of Xp22-pter in three males carrying a portion of Xp translocated to Yq. *Hum. Genet.* **91,** 333–338.

111. Weil, D., Portnoi, M.-F., Levilliers, J., Wang, I., Mathieu, M., Taillemite, J.-L, Meier, M., Boudailliez, B., and Petit, C. (1993) A 45,X male with an X;Y translocation: implications for the mapping of the genes responsible for Turner syndrome and X-linked chondrodysplasia punctata. *Hum. Mol. Genet.* **2,** 1853–1856.

112. Fryns, J.P., Kleczkowska, A., and Van den Berghe, H. (1985) Clinical manifestations of Y chromosome deletions. In *The Y Chromosome. Part B: Clinical Aspects of Y Chromosome Abnormalities* (Sandberg, A.A., ed.), Alan R. Liss, New York, pp. 151–170.

113. Fechner, P.Y., Smith, K.D., Jabs, E.W., Migeon, C.J., and Berkovitz, G.D. (1992) Partial gonadal dysgenesis in a patient with a marker Y chromosome. *Am. J. Med. Genet.* **42,** 807–812.

114. Bardoni, B., Zuffardi, O., Guioli, S., Ballabio, A., Simi, P., Cavalli, P., Grimoldi, M.G., Fraccaro, M., and Camerino, G. (1991) A deletion map of the human Yq11 region: implications for the evolution of the Y chromosome and tentative mapping of a locus involved in spermatogenesis. *Genomics* **11,** 443–451.

115. Tuck-Muller, C.M., Chen, H., Martinez, J.E., Shen, C.C., Li, S., Kusyk, C., Batista, D.A.S., Bhatnagar, Y.M., Dowling, E., and Wertelecki, W. (1995) Isodicentric Y chromosome: cytogenetic, molecular and clinical studies and review of the literature. *Hum. Genet.* **96,** 119–129.

116. Schmid, M., Haaf, T., Solleder, E., Schempp, W., Leipoldt, M., and Heilbronner, H. (1984) Satellited Y chromosomes: structure, origin, and clinical significance. *Hum. Genet.* **67,** 72–85.

117. Lin, C.L., Gibson, L., Pober, B., and Yang-Feng, T.L. (1995) A de novo satellited short arm of the Y chromosome possibly resulting from an unstable translocation. *Hum. Genet.* **96,** 585–588.

118. Bernstein, R., Wadee, A., Rosendorff, J., Wessels, A., and Jenkins, T. (1986) Inverted Y chromosome polymorphism in the Gujerati Muslim Indian population of South Africa. *Hum. Genet.* **74,** 223–229.

119. de la Chapelle, A. (1972) Analytic review: nature and origin of males with XX sex chromosomes. *Am. J. Hum. Genet.* **24,** 71–105.

120. Robinson, A. and de la Chapelle, A. (1996) Sex chromosome abnormalities. In *Emery and Rimoin's Principles and Practice of Medical Genetics* (Rimoin, D.L., Connor, J.M., and Pyeritz, R.E., eds.), Churchill Livingstone, New York, pp. 973–997.

121. Abbas, N.E.,Toublanc, J.E., Boucekkine, C., Toublanc, M., Affara, N.A., Job, J.-C, and Fellous, M. (1990) A possible common origin of "Y negative" human XX males and XX true hermaphrodites. *Hum. Genet.* **84,** 356–360.

122. Ferguson-Smith, M.A., Cooke, A., Affara, N.A., Boyd, E., and Tolmie, J.L. (1990) Genotype-phenotype correlations in XX males and the bearing on current theories of sex determination. *Hum. Genet.* **84,** 198–202.

123. de la Chapelle, A., Koo, G.C., and Wachtel, S.S. (1975) Recessive sex-determining genes in human XX male syndrome. *Cell* **15,** 837–842.

124. Bogan, J.S. and Page, D.C. (1994) Ovary? Testis? A mammalian dilemma. *Cell* **76,** 603–607.

125. Lopez, M., Torres, L., Mendez, J.P., Cervantes, A., Perez-Palacios, G., Erickson, R.P., Alfaro, G., and Kofman-Alfaro, S. (1995) Clinical traits and molecular findings in 46,XX males. *Clin. Genet.* **48,** 29–34.

126. Therkelson, A.J. (1964) Sterile man with chromosomal constitution, 46,XX. *Cytogenetics* **3,** 207–218.

127. Zanaria, E., Bardoni B., Dabovic, B., Calvari, V., Fraccaro, M., Zuffardi, O., and Camerino, G. (1995) Xp duplications and sex reversal. *Philos. Trans. Roy. Soc. Lond. Ser. B. Biol. Sci.* **350,** 291–296.

128. McCabe, E. (1996) Sex and the single DAX1: too little is bad, but can we have too much? *J. Clin. Invest.* **98,** 881–882.

129. Guo, W., Mason, J.S., Stone, C.G., Morgan, S.A., Madu, S.I., Baldini, A., Lindsay, E.A., Biesecker, L.G., Copeland, K.C., Horlick, M.N.B., Pettigrew, A.L., Zanaria, E., and McCabe, E.R.B. (1995) Diagnosis of X-linked adrenal hypoplasia congenita by mutation analysis of the *DAX-1* gene. *JAMA* **274,** 324–330.

130. Muscatelli, F., Strom, T.M., Walker, A.P., Zanaria, E., Recan, D., Meindl, A., Bardoni, B., Guioli, S., Zehetner, G., Rabl, W., Schwartz, H.P., Kaplan, J.C., Camerino, G., Meitinger, T., and Monaco, A.P. (1994) Mutations in the *DAX1* gene give rise to both X-linked adrenal hypoplasia congenita and hypogonadrotrophic hypogonadism. *Nature* **372,** 672–676.

131. Simpson, J.L. (1978) True hermaphroditism. Etiology and phenotypic considerations. *Birth Defects Orig. Article Ser.* **14,** 9–35.

132. Verp, M.S. and Simpson, J.L. (1987) Abnormal sexual differentiation and neoplasia. *Cancer Genet. Cytogenet.* **25,** 191–218.

133. Pinsky, L., Kaufman, M., and Chudley, A.E. (1985) Reduced affinity of the androgen receptor for 5α-dihydrotestosterone but not methyltrienalone in a form of partial androgen resistance. Studies on cultured genital skin fibroblasts. *J. Clin. Invest.* **75,** 1291–1296.

134. Eddy, A.A. and Mauer, M. (1985) Pseudohermaphroditism, glomerulopathy, and Wilms tumor (Drash syndrome): frequency in end-stage renal failure. *J. Pediatr.* **87,** 584–587.

135. Opitz, J.M. and Howe, J.J. (1969) The Meckel syndrome. *Birth Defects Orig. Article Ser.* **5,** 167–172.

136. Greenberg, F., Gresik, M.W., Carpenter, R.J., Law, S.W., Hoffman, L.P., and Ledbetter, D.H. (1987) The Gardner–Silengo–Wachtel or genito–palato–cardiac syndrome: male pseudohermaphroditism with micrognathia, cleft palate, and conotruncal cardiac defects. *Am. J. Med. Genet.* **26,** 59–64.

137. Ieshima, A., Koeda, T., and Inagaka, M. (1986) Peculiar face, deafness, cleft palate, male pseudohermaphroditism, and growth and psychomotor retardation; a new autosomal recessive syndrome? *Clin. Genet.* **30,** 136–141.
138. Diamond, M. and Sigmundson, H.K. (1997) Management of intersexuality—guidelines for dealing with persons with ambiguous genitalia. *Arch. Pediatr. Adol. Med.* **151,** 1046–1050.

11
Prenatal Cytogenetics

Linda Marie Randolph, M.D.

INTRODUCTION AND HISTORY
Amniocentesis

Amniocentesis, the transabdominal or transcervical puncture of the uterus for the purpose of removing amniotic fluid, has been practiced since the 1930s (1). It was used in the early 1950s in the prenatal evaluation of Rh sensitization (2).

A key event that laid the foundation for prenatal cytogenetic analysis was the discovery of the ability to determine gender on the basis of the incidence of the sex chromatin body observed in the nuclei of oral mucosa smears (3,4). In 1956, James (5) described the use of amniotic fluid sediment to determine fetal sex by Papanicolaou and Giemsa stains, and Fuchs and Riis (6) showed in amniotic fluid of term pregnancies that they could accurately determine the fetal sex in 20 of 21 cases. It is of interest that they concluded, "Although transabdominal puncture of the uterus has been carried out often for therapeutic and experimental reasons without accidents, mere curiosity does not justify the procedure, and its practical value is probably limited in the human. If the results are confirmed in animals, however, it might become of great significance in veterinary practice."

Other investigators confirmed the accurate determination of fetal gender by similar procedures, staining amniotic fluid obtained at term by various techniques (7,8).

In 1966, Steele and Breg demonstrated, in a study of amniotic fluid obtained from women because their fetuses were at risk for erythroblastosis fetalis, that human amniotic cells could be cultured and the chromosomes analyzed (9). They foresaw that this " . . . would allow more practical genetic counseling of mothers with high risks of having children with chromosome abnormalities or inborn errors of metabolism."

Further refinement of the technique and timing of amniocentesis was demonstrated in a 1967 article by Jacobson and Barter (10), and they proposed that the optimal timing of amniocentesis is 16 weeks after performing the procedure from 5 weeks to term in 85 women. Of these 85, 57 were successfully cultured. In a thoughtful discussion after the article, Edward C. Hughes noted that, "Speculation might go so far as to suggest that, although chromosome constitution cannot be changed, a specific DNA that would carry the coding information lacking in certain diseases might replace the missing element," and in the same discussion, S.R.M. Reynolds pointed out that " . . . in the future there will be even more refined methods of evaluating gene abnormalities in which the karyotype appears normal."

From: *The Principles of Clinical Cytogenetics*
Edited by: S. Gersen and M. Keagle © Humana Press Inc., Totowa, NJ

In 1968, Nadler and Gerbie described the use of amniocentesis for the detection of cytogenetic and biochemical abnormalities in 155 women at increased risk for these disorders. They reported a highly successful culture rate of 97% and uniformity of timing of the procedure, from 13 to 18 weeks (11).

By 1986, more than a quarter of a million amniocenteses had been performed for cytogenetic analysis (12), and the number to date is probably in the millions.

Chorionic Villus Sampling

Although techniques for transcervical (13) and transabdominal (14) placental biopsy, or late chorionic villus sampling (CVS), were described in the 1960s for the diagnosis of hydatidiform mole, the first article describing a technique for fetal genetic diagnosis was published in 1968 (15). Mohr developed techniques for sampling fetal cells no later than the third month of pregnancy by a transabdominal approach. Due primarily to the absence of real-time ultrasound, low culture success rate and the risks of endo-scopic approaches, as described by Kullander and Sandahl in 1973 (16) and by Hahnemann in 1974 (17), the technique was not widely used in the United States. In Kullander and Sandahl's experience, 19 of 39 specimens (48.7%) were successfully cultured, which they described as a "high percentage." In Hahnemann's experience, there was a 38% success rate, with causes of failure being puncture or biopsy of the amniotic membrane and bleeding. The optimal time of performing the procedure was the 10th week of gestation, and although the procedure had a low success rate in terms of obtaining tissue, the culture success rate was 91%. All but one of the pregnancies was terminated by previous intention, and in the one continuing pregnancy the new-born was normal.

In China, transcervical CVS was widely used in the 1970s as a method of fetal sex prediction and selection. A report of the Chinese experience was published in 1975 (18). The accuracy of their fetal sex prediction, based on X chromatin, was 94%. Efforts to replicate this success were unsuccessful for several years (19,20).

In their 1981 article, Niazi et al. reported an improved technique using trypsin for culturing trophoblastic cells obtained by transcervical CVS, minimizing the risk of maternal cell admixture in fetal cells (21).

The first use of real-time ultrasound scanning in CVS was reported in 1982 by Kazy et al. (22). Of their 165 patients, 139 had biopsies performed prior to induced abortion, and in 26 patients, biopsy was performed for genetic reasons. Of the eventual 13 continuing pregnancies, none was spontaneously aborted, and all 11 babies who had been born to date were normal. Fetal sex prediction by X chromatin was accurate in all cases. This was the first study that brought CVS out of the experimental category and into the world of a promising prenatal diagnostic test. As of 1995, 158,774 CVS cases from 146 centers worldwide had been entered in the World Health Organization-spon-sored CVS registry in Philadelphia. This number would be much larger if all of the hundreds of centers around the world that perform CVS provided their data to the reg-istry (23).

Percutaneous Umbilical Cord Sampling (PUBS)

In the early 1970s, in an effort to develop a method for prenatal diagnosis of hemoglobinopathies, investigators sought to establish safe techniques for fetal and/or placental blood sampling. In his preliminary report of 1973, Valenti demonstrated in 11 women scheduled for abortion that, in the second trimester, a surgical "endoamnioscope" with a flexible needle introduced through it could be operated under direct vision *(24)*. This required regional or general anesthesia and an abdominal wall incision. Three of the women had umbilical cord puncture, and the blood obtained was shown to be of fetal origin. Hobbins and Mahoney performed fetoscopy in 34 women scheduled for abortion *(25)*. Local anesthesia was provided, and the cannula was smaller than the endoscope used by Valenti. In eight of these, successful blood sampling of a placental vessel was attempted and achieved. However, in only one of these cases was the composition of the blood 100% fetal. Placentocentesis was essentially replaced by cordocentesis thereafter.

Daffos et al. *(26)* demonstrated in 50 women referred for abortion that by using local anesthetic, real-time ultrasound, and puncture of the umbilical vein, pure fetal blood was obtained in 46 cases. Sixteen of the 50 women underwent abortion 2–10 days later, by which time none of these fetuses had died. Twelve other women delivered healthy babies, and 22 pregnancies were ongoing.

The technique was later applied, by the same group, in 606 samplings of 562 women with a variety of indications for prenatal diagnosis *(27)*. Complications were seen in 15%, including a 2% rate of fetal death or spontaneous abortion. By obtaining larger volumes of fetal blood, these investigators were able to perform physiologic and hematologic assays that helped provide the basis for normal values in fetal blood, and they showed that PUBS deserved a place in the prenatal diagnostic-testing world.

The Incidence of Chromosome Abnormalities

Combining surveys from 1969 to 1982 of 68,159 liveborn babies, 1 in 156 live births were found to have a major chromosome abnormality (see **Table 1**) *(28)*. The most common remains trisomy 21, or Down syndrome, with an incidence from these surveys of one in 833 live births. The next most common are sex chromosome aneuploidies, with one XYY or XXY per 1000 male liveborns and one XXX per 1000 female liveborns. Because nonbanded chromosome preparations were used in the early survey years (from 1969 to 1975), it was thought, when Giemsa banding (G-banding) was introduced, that the incidence of chromosome abnormalities would be found to be higher. However, in a 1980 study by Buckton et al. of 3,993 newborns, no significant difference in the frequency of rearrangements or of other chromosome aneuploidies was found *(29)*.

It is clear that the incidence of most fetal chromosome abnormalities increases with maternal age. Data for women ages 35 through 49 were compiled by Hook based on North American collaborative studies and the New York State registry *(30)*. His analysis of the data indicated a 30% differential between the rates observed at amniocentesis and those seen at birth, a figure which is still valid almost 20 years later.

In 1982 Schreinemachers et al. analyzed data on the results of 19,675 prenatal cytogenetic diagnoses on women aged 35 and over for whom there was there no known cytogenetic risk for a chromosome abnormality except parental age *(31)*. The expected rates at amniocentesis of clinically significant cytogenetic abnormalities by maternal age were obtained and compared with previously estimated rates by maternal age in live births. A differential between amniocentesis and live birth incidences was shown

Table 1
Chromosome Abnormalities in Surveys of 68,159 Liveborn Babies

Type of abnormality	Total abnormalities (%)
Sex chromosomes, males	
47,XYY	45 (0.103)
47,XXY	45 (0.103)
Other	32 (0.073)
Sex chromosomes, females	
45,X	6 (0.024)
47,XXX	27 (0.109)
Other	9 (0.036)
Autosomal trisomies	
47,+21	82 (0.120)
47,+18	9 (0.013)
47,+13	3 (0.004)
Other	2 (0.002)
Structural balanced arrangements	
Robertsonian translocation	
der(D;D)(q10;q10)[a]	48 (0.070)
der(D;G)(q10;q10)[b]	14 (0.020)
Reciprocal and insertional translocation	64 (0.093)
Inversion[c]	13 (0.019)
Structural unbalanced arrangements	
Robertsonian	
Reciprocal and insertional	5 (0.007)
Inversion	9 (0.013)
Deletion	1 (0.001)
Supernumerary	5 (0.007)
Other	14 (0.020)
	9 (0.013)
Total abnormalities	442 (0.648)
Total babies surveyed	
Males	43,612
Females	24,547

Data from ref. *(28)*.

[a]DqDq refers to Robertsonian translocations involving chromosomes 13, 14, and/or 15.

[b]DqGq refers to Robertsonian translocations involving chromosomes 13, 14 or 15 and 21 or 22.

[c]Excludes common pericentric inversion of chromosome 9.

for trisomies 21, 18, and 13, but not for 47,XXY and 47,XYY (see **Table 2**). In the following year, Hook confirmed and refined the differences in the incidences for trisomies 21, 13, and 18, and also found a difference between fetal and newborn rates of 47,XXY, 47,XYY, 45,X and 45,X/46,XX but not for 47,XXX (32) (see **Table 3**). Con-

Table 2
Maternal Age-Specific Rates (%) for Chromosome Abnormalities

Maternal age (years)	From liveborn studies[a]			From amniocenteses			From CVS	
	47,+21[b]	47,+21[c]	All chromosome abnormalities[b]	47,+21[b]	47,+21[c]	All chromosome abnormalities[b]	47,+21[c]	All chromosome abnormalities[d]
33	0.16	—	0.29	0.24	—	0.48	—	—
34	0.20	—	0.36	0.30	—	0.66	—	—
35	0.26	—	0.49	0.40	0.31	0.76	—	0.78
36	0.33	0.35	0.60	0.52	0.80	0.95	0.42	0.80
37	0.44	0.43	0.77	0.67	0.73	1.20	0.68	2.58
38	0.57	0.42	0.97	0.87	0.84	1.54	0.45	3.82
39	0.73	0.79	1.23	1.12	1.03	1.89	2.05	2.67
40	0.94	1.21	1.59	1.45	1.50	2.50	1.20	3.40
41	1.23	2.67	2.00	1.89	2.92	3.23	3.12	6.11
42	1.56	4.28	2.56	2.44	3.05	4.00	2.88	8.05
43	2.00	1.82	3.33	3.23	1.52	5.26	1.20	5.15
44	2.63	—	4.17	4.00	2.50	6.67	2.63	10.00
45	3.33	—	5.26	5.26		8.33	8.33	7.14

[a]Estimated liveborn statistics (31).
[b]Data compiled from 19,675 genetic amniocenteses (31).
[c]Data compiled from 3041 CVS, 7504 amniocenteses, and 13,139 with no test (32a). These are observed prevalences.
[d]Data compiled by L. Hsu (28).

Table 3
Fetal Deaths Subsequent to Amniocentesis

Abnormalities	Number	Fetal deaths Proportion (%)	95% confidence interval (%)
47,+21	73	30.1	19.0–42.0
47,+18	25	68.0	46.5–85.1
47,+13	7	42.9	9.9–81.6
47,XXX	39	0.0	0.0–9.0
47,XXY	37	8.1	0.8–11.0
47,XYY	33	3.0	0.08–15.8
45,X	12	75.0	42.8–94.5
45,X/46,XX	19	10.5	1.3–33.1
Balanced translocations and inversions	71	2.8	0.3–9.8
Markers, variants, fragments	27	0.0	0.0–12.8

Data from ref. *(32)*.

Note: Proportion refers to the number of fetal losses compared to the total number of fetuses diagnosed with the given abnormality.

Table 4
The Incidence of *De Novo* Balanced Structural Rearrangements in 337,357 Genetic Amniocenteses

De novo rearrangement	Number of cases	Percentage
Reciprocal translocation	176	0.047
Robertsonian translocation	42	0.011
Inversion	33	0.009
Supernumerary small marker chromosome	162	0.040
Satellited marker	77	0.020
Nonsatellited marker	85	0.023
Total	413	0.109

Data from ref. *(33)*.

trary to what was found in other studies, there was no significant maternal age effect in the incidence of fetal death of chromosomally abnormally fetuses.

The incidence of *de novo* balanced structural rearrangements in 337,357 amniocenteses was reported by Warburton *(33)*. The results are shown in **Table 4**.

Spontaneous Abortions

It is a well-established fact that the incidence of major chromosome abnormalities is much higher in first-trimester spontaneously aborted fetuses than later in pregnancy and at birth. The incidences in various studies range from 20% to 60%, with the average in pooled data of unselected spontaneous abortions being 41% *(28,33a)* (see **Table 5**). A cautionary note in consideration of this high incidence range is that the

Table 5
Frequencies of Chromosome Abnormalities in Unselected Spontaneous Abortions

Number of abortuses studied	Number of abortuses (%) with chromosome aberrations	Different types of chromosome abnormalities (% of all chromosome abnormalities)					
		Autosomal trisomy	45,X	Triploid	Tetraploid	Other	Reference
8841	3613 (40.87%)	1890 (52.29%)	689 (19.06%)	586 (16.21%)	119 (5.51%)	249 (6.89%)	(28)[a]
3300	1312 (39.8%)	645 (49.2%)	201 (15.3%)	198 (15.1%)	78 (5.9%)	190 (14.5%)	(33a)

[a]Data compiled from more than 10 studies.

Table 6
Frequency of Autosomal Trisomy
for Each Human Chromosome
Among Aborted Specimens

Trisomy Chromosome	Number of trisomies (%)
1	0
2	34 (5.2)
3	6 (0.93)
4	15 (2.3)
5	5 (0.78)
6	5 (0.78)
7	27 (4.2)
8	23 (3.6)
9	18 (2.8)
10	11 (1.7)
11	0
12	2 (0.31)
13	53 (8.2)
14	32 (5.0)
15	52 (8.1)
16	202 (31.3)
17	4 (0.62)
18	23 (3.6)
19	0
20	18 (2.8)
21	54 (8.4)
22	55 (8.5)
Total	645 (100)

Data from ref. *(33a)*.

tissue cultured and analyzed may not represent the fetus. It has been shown that 45,X cells and some lethal trisomies seen in chorionic villus samples may not be seen in the fetus, so this may lead to spurious elevation of estimates of chromosome abnormalities in spontaneous abortion tissue *(34)*. Notwithstanding this caveat, the following frequencies of chromosome abnormalities are reported in spontaneous abortions: autosomal trisomies comprise the largest group of 52% of chromosome abnormalities, followed by 45,X at 19%, triploidy at 16%, and tetraploidy at 6% *(28)*.

The association between advanced maternal age and the incidence of trisomies has been demonstrated in spontaneous abortions. Of interest is that 45,X appears to be associated with younger maternal age, with about one third of 45,X spontaneous abortions coming from women 20–24 years of age *(35)*. The distribution of trisomies is quite different from that seen at birth or even at amniocentesis, with 30% being trisomy 16, compared to almost negligible rates of trisomy 16 at amniocentesis *(33a;* see **Table 6**).

Table 7
Frequencies of Chromosome Abnormalities in Stillbirths
and Neonatal Deaths; Combined Data from Three Studies

Macerated stillbirths		Nonmacerated stillbirths		Neonatal deaths		Total	
Number karyotyped	*Abnormal*	*Number karyotyped*	*Abnormal*	*Number karyotyped*	*Abnormal*	*Number karyotyped*	*Abnormal*
59	7 (11.86%)	215	9 (4.18%)	549	33 (6.0%)	823	52 (6.31%)

Data from ref. *(28)*.

Stillbirths and Neonatal Deaths

Fetal loss from 28 weeks on in pregnancy is defined as stillbirth, and neonatal death refers to death occurring within the first 4 weeks after birth. Chromosome studies in such cases have shown that the incidence of chromosome abnormality is about 10 times that in the rest of the population. Combining three studies of stillbirths and neonatal deaths, of those in which chromosome analysis was performed, 52 of 823 (6.3%) studied had a chromosome abnormality. Of these 823, 59 macerated stillbirths were studied, of which seven (11.9%) had a chromosome abnormality. Of 215 nonmacerated stillborns, nine (4.2%) were chromosomally abnormal, and of 549 neonatal deaths, 33 (6.0%) had a chromosome abnormality *(28)*. Given the value it provides families in terms of understanding more about their losses and in providing recurrence risks, it is recommended that consideration of chromosome analysis be given in all such cases (see **Table 7**).

PRENATAL CYTOGENETIC DIAGNOSIS

Genetic Amniocentesis

With increased public awareness, number of practitioners, laboratory capacity, proportion of women greater than 35 having babies, and use of maternal serum screening, the utilization rate of amniocentesis has grown. It was estimated that in 1974, 3,000 women underwent genetic amniocentesis *(36)*, and the number now is in the millions. The increased utilization has extended to women of lower socioeconomic status who previously did not have access to or finances for the procedure *(37)*. With improvements in laboratory procedures, including sterile technique, plasticware, enriched cell culture media, and automated harvesting and imaging systems, the turnaround time for reporting results of an amniocentesis has dropped dramatically, from several weeks in the 1970s and 1980s to less than a week in some laboratories today. The cost of the laboratory test has dropped as well owing to increased efficiency and competition. Thus prenatal diagnosis by amniocentesis has become and probably will remain by far the most common mode of prenatal diagnosis until such time as a reliable, cost-effective noninvasive procedure is developed.

The accuracy of amniocentesis for the detection of recognized chromosome abnormalities is greater than 99%. Diagnostic accuracy has been enhanced by the recent use of fluorescence *in situ* hybridization (FISH) and biotinylated chromosome-specific probes. These are of particular value in marker chromosome, translocation, and deletion cases, when microscopic findings require further study for clarification *(38–45)*.

Conventional Amniocentesis—15–24 Weeks of Gestation

Mid-trimester, defined here as the 15th through the 24th week of gestation, is by far the most common time period for performing the amniocentesis procedure. Culture of amniotic fluid cells is optimal in this time period *(45a,46),* both from the perspective of rapidity of cell growth (and therefore sample turnaround time) and because the culture failure rate is less than 0.5% in experienced laboratories.

The risks associated with mid-trimester amniocentesis include leakage of fluid, cramping, bleeding, infection, and miscarriage. The risk of miscarriage following mid-trimester amniocentesis is related to practitioner experience, number of needle insertions, size of the needle, and other factors *(47).* The appropriate risk figure to provide patients is still debated. In spite of the millions of amniocentesis procedures performed and the importance of an accurate risk figure to provide patients, there has been only one large prospective controlled study performed regarding the risks of amniocentesis. In this article, known as "the Danish study," 4606 women comprised the final study population *(48).* Of these, half were randomized to have amniocentesis, and the other half were randomly assigned to the control, nonamniocentesis group. At the conclusion of the study, it was found that the total rate of spontaneous abortion was 1.7% in the study group and 0.7% in the control group ($p < 0.05$). When the women with a high maternal serum α-fetoprotein level were considered, it was found that they had a relative risk of spontaneous abortion after amniocentesis of 8.3 compared to women with a normal maternal serum α-fetoprotein level. This equated to an overall relative risk of 2.3. Other factors found to increase the risk of spontaneous abortion were transplacental passage of the needle (relative risk of 2.6) and discolored amniotic fluid (relative risk 9.9).

An important and often overlooked component of providing risk assessments to patients is the underlying incidence and timing of pregnancy losses. A prospective study of 220 ultrasonographically normal pregnancies in women recruited prior to conception (in order to avoid bias of selection) found a pregnancy loss rate after 8 weeks of 3.2% *(49).* Other studies have shown a maternal age factor in the loss rate *(34).* The prevalence of trisomies is about 50% higher at 16 weeks compared to term pregnancies (ibid.), so selection against chromosomally abnormal abortuses is still occurring at 16 weeks. The incidence of spontaneous pregnancy loss after 16 weeks is 1%.

Some genetic counselors and amniocentesis practitioners counsel patients regarding the risk of the amniocentesis relative to the risk of a fetal chromosome abnormality and in effect use this as a decision point. In this way, a woman with a risk of fetal chromosome abnormality of 1 in 200 might be inclined to decline amniocentesis if the risk of miscarriage as a result of the procedure was quoted as 1 in 100 and the risks compared during the counseling session. A maternal age of 35 has traditionally been used as a cutoff for the definition of advanced maternal age, because the risk of a fetal chromosome abnormality at this age is roughly equivalent to the originally reported risk of miscarriage as a result of the amniocentesis. This is not sound reasoning because the burdens of the risks are quite different—one burden being the potential lifetime task of caring for an individual with mental retardation and physical/health problems and the other being miscarriage of a potentially healthy fetus *(50).*

Early Amniocentesis

Interest in early amniocentesis (EA) has risen in recent years, due in large part to the continued desire to provide and receive prenatal diagnosis at an earlier gestation without some of the risks and limitations associated with chorionic villus sampling, which are outlined in the following paragraphs. An increase in sophistication in ultrasound technology has also made earlier imaging of fetuses more feasible and has added to the confidence level of the physicians performing the procedure. Adding to this is the opportunity to measure amniotic fluid α-fetoprotein and acetylcholinesterase, which is not possible with CVS. One center reported a rise in EA procedures from 3.2% of their 495 procedures in early 1985 to 6.5% of 980 procedures in late 1987 *(51)*.

Early amniocentesis is usually described as one that occurs before 15 weeks' gestation. It has been shown that the earlier a prenatal diagnosis procedure is performed, the higher the fetal loss rate is *(52)*. One should therefore further divide the periods at which amniocentesis is performed to provide better comparative data for a variety of procedures because " . . . true risks . . . appear to be a function of gestational age and less related to the procedure performed" *(52)*.

Although the procedure by which EA is performed is similar to that of mid-trimester amniocentesis, practitioners report several challenges unique to EA. The placenta is more widely spread, the amniotic fluid volume is lower, and the amniotic membrane is not yet totally adherent to the uterine wall, leading to the "tenting" reported by some physicians *(53)*.

BACKGROUND

In one study conducted from 1979 through 1986, 4,750 amniocenteses were performed, 541 of which were performed before the 15th week since the last menstrual period *(54)*. Outcome data were available for 298 women, of whom 108 were under 35 years of age. Fetal loss within 2 weeks of the procedure was seen in five pregnancies, all in the 14th week, when 228 of the 308 women had the procedure. When all spontaneous fetal losses were accounted for, there were eleven spontaneous abortions (3.6%), two stillbirths (0.7%), and one neonatal death (0.3%), resulting in a total post-procedure loss rate of 14/298 (4.7%). No culture failures were seen. The needle gauge was 20, and no difference in outcome was seen in transplacental versus placental passage.

In 1988, the combined experience of six groups, including the study mentioned previously, was reviewed *(55)*. The total loss rate in 1240 pregnancies of known outcome ranged from 1% to 4.7%. Cell culture and amniotic fluid α-fetoprotein measurements were satisfactory. The conclusion was that EA is feasible, but that other safety issues had not been adequately addressed, such as congenital orthopedic anomalies and neonatal pulmonary compromise, which had been seen in some babies born after mid-trimester amniocentesis *(56)*.

Several other studies were published in the early 1990s *(57–62)*. In one study, 505 amniocentesis procedures were performed between 11 and 15 weeks' gestation. In all but three pregnancies, follow-up information was available, including 16 fetal losses (3.1%)—10 in the two weeks after the procedure and 6 within the 28th week. The authors reported a significantly higher risk for fetal loss when the amniocentesis was performed at the 11th–12th week of gestation compared with the 13- to 15-week group.

The fetal loss rate between the 12- to 13-week and the 14- to 15-week groups showed no statistically significant difference. They concluded that early amniocentesis is "a valid alternative to traditional amniocentesis" (57).

In their 1990 article, Elejalde et al. performed a prospective controlled study involving 615 amniocenteses performed between weeks 9 and 16 of gestation, and they reviewed previous EA studies (58). Their results showed that amniocentesis after the 9th week of pregnancy does not appear to differ significantly in its complications and outcome from the results of the same procedure at 15 to 16 weeks or later. The issue of pseudomosaicism was also addressed and is covered more fully later in this chapter.

Penso et al. in 1990 (59) performed amniocentesis in 407 women between gestational ages of 11–14 weeks and compared the safety and accuracy with data obtained from collaborative studies of amniocentesis performed later in the second trimester. Theirs was the first report to provide information regarding neonatal outcome associated with EA. The spontaneous abortion rate within 4 weeks of the procedure was 2.3%, and the fetal loss rate was 6.4%. Orthopedic postural deformities, including club feet, scoliosis, and congenital dislocation of the knees and hips were seen in eight newborns, three of whose mothers had post-amniocentesis leakage of amniotic fluid. A total of 10 women in the study (2.6%) had post-procedure fluid leakage. It appeared that the orthopedic deformities might be related to a post-procedure history of amniotic fluid loss. They concluded that the accuracy, risks, and complications were similar to those associated with traditional amniocentesis.

In 1990, Hanson et al. reported their increased practitioner experience and use of continuous ultrasonographic guidance in EA of gestations from 10 to 14 weeks (60). The needle gauge was changed from the 20-gauge used in their 1987 study to 22, and the volume of fluid removed was generally less. Pregnancy outcome was reported for 523 patients, of whom 12 (2.3%) had a post-procedural loss. This compared favorably with their previously reported loss rate of 4.7%. Of eight women with post-procedure amniotic fluid leakage, one had a baby at term with a dislocated knee. Another experienced fetal death 3 weeks after the amniocentesis, and the rest had normal term deliveries.

In a smaller series, 105 EA procedures were performed (61). There were two pregnancy losses in the 64 patients for whom outcome information was available at the time of publication, and four congenital anomalies were seen in the 66 delivered babies: one imperforate anus, one hemangioma of the tongue, and two cases of positional talipes that required no treatment. These were apparently unrelated to amniotic fluid leakage.

Crandall et al. (62) retrospectively studied 693 consecutive EA (prior to 15 weeks) cases, which had a spontaneous abortion rate (to 28 weeks' gestation) of 1.5%, compared with a nonrandomized, later control group of 1386 women having traditional amniocentesis, whose spontaneous abortion rate was 0.6%, a statistically significant difference. In their review of background risk of pregnancy loss in the second trimester, they concluded that "at least some of the pregnancy loss subsequent to early amniocentesis is independent of the procedure but the risk may be minimally higher than that for standard amniocentesis." There were no significant differences in congenital anomalies in the EA group (1.8%) vs the traditional amniocentesis group (2.2%). Interestingly, in the EA group, 4 of the 12 abnormalities involved congenital hip dislocation/subluxation or club feet, and three of the 30 congenital anomalies seen in the traditional amniocentesis group were congenital hip dislocation or club feet. They concluded that EA is a "relatively safe prenatal diagnostic test and an alternative to CVS and later amniocentesis." See **Table 8** for a comparison of fetal loss rates.

In all these studies the investigators concluded that, apart from a higher rate of pseudomosaicism seen in some EA cases, the laboratory analysis of EA specimens compares favorably in validity and reliability compared to traditional amniocentesis specimens. This was confirmed in two laboratory studies of a combined 1805 EA specimens of 10–14 weeks' gestation *(63,64)*. The culture success rate was 99.8% for EA vs 100% for traditional amniocentesis in one study and 98.6% for EA vs 99.9% for traditional amniocentesis in the other study. The turnaround times for reporting results were 1–2 days longer in the EA group. In one study, the EA group showed a significant increase in the number of structural and numerical single-cell abnormalities and an increase in numerical multiple-cell abnormalities compared to amniocenteses performed at 16–18 weeks. These were dealt with by examining parallel cultures.

More recent studies are mixed in their conclusions. Diaz Vega's group performed 181 amniocenteses at 10–12 weeks' gestation and reported a fetal loss rate within 2 weeks of the procedure of 0.5%, with a total fetal loss rate during pregnancy of 1.6% *(65)*. However, the culture success rate was only 94.5% overall, with one culture failure out of three 10-week amniotic fluid specimens.

Brumfield's group performed a retrospective matched-cohort study using a study group of 314 patients who had amniocentesis at 11–14 weeks vs a control group of 628 women who had amniocentesis at 16–19 weeks *(66)*. With the same practitioners, ultrasound equipment, and technique, they found a significant difference in the fetal loss rate within 30 days of amniocentesis (2.2% versus 0.2%) in the EA group compared to the later-amniocentesis group. This was attributed at least in part to higher post-procedure amniotic fluid leakage (2.9% vs 0.2%) and vaginal bleeding (1.9% vs 0.2%) rates. The culture success rates were not reported.

Whether transplacental needle passage is a factor in fetal loss after EA was examined by Bravo et al. *(66a)*. They reviewed 380 consecutive EA procedures performed for advanced maternal age and found that transplacental needle passage had occurred in 147 cases (38.7%). Although the frequency of "bloody taps" was significantly increased in this group, there was no difference in fetal loss rates (3.4% in both groups, including stillbirths).

In Wilson's review *(67)*, he states that there have been no studies that have adequately addressed the critical question of the safety of EA relative to traditional amniocentesis, pointing out that to date only two randomized trials had been performed, and they differed in their methodologies and their conclusions. He also stated that procedures at less than 13 weeks' gestation should be considered experimental. Certainly the cumulative experience with 13–14 week EA procedures is much greater than that with under-13-week EA procedures. In addition, the two randomized EA studies he cites evaluate 11- to 12-week gestations, and thus are not comparable to the 13- to 14-week gestation studies.

Comparison of Early Amniocentesis with Chorionic Villus Sampling

To compare first-trimester prenatal diagnostic modalities, a number of investigators have published studies comparing CVS with early amniocentesis. Shulman et al. *(68)*

Table 8
Outcome in Early (11–14 Weeks) Amniocentesis Studies

Group	Study period	No. of patients with outcome data in EA group	Fetal loss rate within 2 weeks of procedure (%)	Fetal loss rate (%)[a], week(s) gestation at time of amniocentesis	Total fetal loss rate (%)	Needle gauge	Comments
Hanson (1987)[54]	1979–1986	298	1.7	5/80(6.3), 11–13 weeks; 5/228 (2.2), 14 weeks	4.7	20	Loss rate was 3.3% if patients with preamniocentesis history of bleeding were eliminated
Johnson and Godmilow (1988)[55]	Review of six studies including Hanson[54], 1979–1987	1240	N/A[b]	N/A[b]	1–4.7	22 in 5 centers; 20 in 1 center[54]	
Stripparo (1990)[57]	1987–1988	397	1.98[e]	9/208 (4.3), 11–13 weeks; 0/176 (0), 14 weeks[d]	3.9	22	
Penso (1990)[59]	1986–1989	389	0.8[c]	6/365 (1.6), 11–13 weeks; 3/42 (7.1), 14 weeks	3.96	22	Three of eight newborns with postural deformities born after post-amniocentesis fluid leak
Hanson (1990)[60]	1986–1987	517	0.8	6/272 (2.2), 11–13 weeks; 5/255 (1.96), 14 weeks	2.5	20	
Crandall (1994)[62]	1988–1993	681	0.9	13/681 (1.9), 11–14 weeks 13/1,342 (0.97), 15–22 weeks	1.9% for EA; 0.97% for conventional amniocentesis	22; sometimes 25	EA was compared to conventional amnio; spontaneous abortion rate was significantly higher in EA group. In the EA group 0.6% had hip dislocation or club feet compared to 0.22% in conventional amnio group.

[a]This figure includes spontaneous abortions, stillbirths, and neonatal deaths.
[b]NA = Not available.
[c]Fetal loss within 4 weeks.
[d]Data based on status at 28 weeks' gestation.
[e]One hundred and eight 15-week gestation amnios were included in this figure.

reported on 500 women, half of whom had transabdominal CVS (TA CVS) from 1986 to 1988, and half of whom had EA from 1987 to 1991. Of the EA specimens, all but 11 were obtained from weeks 12–14, and the rest were from weeks 9–11. Of the continuing pregnancies, loss rates of 3.8% and 2.1% for EA and TA CVS, respectively, were seen. This was not statistically significant. The culture failure rates for both procedures was 0.8%. This study has limited applicability inasmuch as the numbers were small and the patients not randomized, and the time intervals were different. Although all procedures were listed as initial cases, the relative degree of prior individual practitioner experience in the two procedures was not addressed.

Later in 1994, Nicolaides et al. *(69)* reported on a prospective, partially randomized study comparing EA and TA CVS in 1870 women. The spontaneous loss rate was significantly higher after EA at 5.3% than with the CVS group (1.2%). The rate of successful sampling was the same at 97.5%. Culture failure occurred in 2.3% of the EA group, compared to 0.5% in the CVS group. Confined or true mosaicism was seen in 1.2% of the CVS group, compared to 0.1% of the EA group. The authors concluded that although EA and CVS are equally likely to produce valid cytogenetic results, CVS would probably become the "established technique" owing to the 2–3% excess risk of fetal loss in the EA group.

In response to this study, Saura et al. *(69a)* stated that EA could be a "true alternative" to CVS after the 13th week, when the disadvantages of culture failure and fetal losses decrease. Bombard et al. *(69b)* reported one loss in 121 procedures (0.83%) performed by one practitioner at 10–13 weeks using a 22-gauge needle. They suggested that Nicolaides' higher EA fetal loss rate could be related to the needle gauge and the multiple practitioners in his study, compared to one practitioner in Bombard's center.

Similar results were reported by Vandenbussche et al., who, in a partially randomized study, reported eight fetal losses among 120 EA procedures, compared to none among the 64 CVS patients with a follow-up of 6 or more weeks *(70)*.

Another response to these reports proposed the idea that the main drawback to the studies was the very small numbers of EA procedures performed and the evident greater practitioner experience with CVS than with EA. The authors reported a spontaneous abortion rate after EA of 1% up to week 24 on the basis of 1800 pregnancies. The culture failure rate was 0.3% for gestations ranging to 10 weeks 4 days *(71)*.

An important consideration raised by some investigators *(67,71)* is that the banding quality of amniocentesis specimens of any gestation is generally superior to that of CVS specimens, which increases the informativeness of the cytogenetic analysis. The fact that amniotic fluid AFP levels and multiples of the median have been established in many laboratories down to 12 or 13 completed weeks of gestation adds another advantage to the diagnostic power of EA compared to CVS *(72)*.

Specimen Requirements

The volume of amniotic fluid obtained for prenatal diagnosis varies with the stage of gestation, with 15–20 mL conventionally removed by mid-trimester amniocentesis practitioners. In one report, data from several small studies was pooled and the volume of amniotic fluid for weeks 10–20 was calculated *(58)* (see **Table 9**). At gestations under 15 weeks, many practitioners have adopted the practice of removing 1 mL per

Table 9
Volume of Amniotic Fluid (mL) Calculated Using
All the Values for a Given Week from Published Data

Week	n	Mean	SD	Range
10	7	29.7	11.2	18–33
11	9	53.5	16.4	64–76
12	13	58.0	23.4	35–86
13	13	71.4	21.3	38–98
14	14	124.1	42.1	95–218
15	15	136.8	43.7	64–245
16	16	191.2	59.7	27–285
17	20	252.6	98.5	140–573
18	4	289	150	70–410
19	14	324.5	65.2	241–470
20	3	380	39	355–425

Data from ref. *(58)*.

week of gestation, and others have found excellent culture success rate and turnaround time with less fluid removed. For example, one group withdrew 4–12 mL in gestations of 9–14 weeks and obtained a 100% culture success rate in 222 specimens *(73)*, while others withdrew 5–8 mL in pregnancies of 10 weeks and 4 days to 13 weeks and 6 days for an overall culture success rate of 99.7% *(71)*. It has been observed that the total cell numbers rise exponentially from 8 to 18 weeks' gestation, but the number of viable cells increases only slightly during that time *(67)*. This probably explains the comparable culture success rate of EA compared to mid-trimester amniocentesis.

Chorionic Villus Sampling

Associated Risks, Limitations, Benefits, Turnaround Time

Risks associated with CVS have been extensively studied. Perhaps the issue receiving the most attention in the past few years was raised by Boyd et al. involving one case *(74)*, and then more extensively by Firth et al. *(75)*, who reported five babies with severe limb abnormalities out of 289 pregnancies in which TA CVS had been performed at 56–66 days' gestation. Four of these had oromandibular-limb hypogenesis syndrome. They hypothesized that CVS undertaken up to 66 days' gestation may be associated with an increase in the risk of oromandibular-limb hypogenesis syndrome and other limb reduction defects. This report generated many others, with mixed conclusions.

A flurry of letters to the editor of *Lancet* in 1991 followed Firth's report. Reporting evidence to support the association between CVS and limb reduction defects were Mastroiacovo *(76)* and Hsieh *(77)*. Monni et al. suggested that the incidence and severity of limb defects were related to the gauge of the needle, because they used a 20-gauge needle whereas Firth used an 18-gauge needle. In a series of 525 CVS procedures done before 66 days' gestation, no severe limb defects were seen, and only two mild defects were seen in 2227 procedures that were done later *(78)*. Mahoney *(79)* then reported on two multicenter studies that compared transcervical CVS with amniocentesis, and another comparing transabdominal CVS with transcervical CVS. Of 9588 pregnancies studied, 88% of the CVS procedures were performed after 66 days' gestation. Significant limb-reduction defects were present in seven babies. Two of these defects were longitudinal, and five were transverse. Another baby had minor reduction defects of the toes. They compared these abnormalities to those reported to the British Columbia registry and found no significant increase in these birth defects. The timing of the CVS procedures that resulted in babies with abnormalities ranged from 62 to 77 days' gestation.

Similar conclusions were reached in a study in which 12,863 consecutive CVS procedures were performed *(80)*. Five limb reduction defects were seen, which were found not to be significantly different from the incidence observed in the British Columbia registry of birth defects. Of the 12,863 procedures, 2367 were done at 56–66 days, and one of the limb defects was seen in this group. The authors observed no gestational time-sensitive interaction related to CVS and postulated that this was due to their larger experience base.

In 1993, Jahoda et al. reported on 4300 consecutive transabdominal and transcervical CVS cases for which newborn follow-up information was obtained *(81)*. Of the 3973 infants born in this group, three (0.075%) had a terminal transverse limb defect. Two of these occurred in the transcervical CVS group sampled before 11 weeks' gestation (1389 patients), and the other one was in the transabdominal CVS group, sampled after 11 weeks (2584 patients). The authors found the latter figure to be comparable to the prevalence figure given in population studies. They concluded that postponement of CVS to the late first or early second trimester of pregnancy would contribute to the safety of the procedure.

In the same year, a report of the National Institute of Child Health and Human Development Workshop on Chorionic Villus Sampling and Limb and Other Defects was issued *(82)*. The conclusions, based on a review of the literature, were mixed; some concluded that exposure to CVS appeared to cause limb defects, while others did not. All agreed that the frequency of oromandibular-limb hypogenesis appeared to be more common among CVS-exposed infants. This seemed to correlate with CVS performed earlier than 7 weeks post-fertilization (9 weeks post-last menstrual period). Whether or not a distinctive type of limb defect was associated with CVS could not be determined, and it also was unclear whether the CVS-exposed infant had an increased frequency of other malformations, including cavernous hemangiomas.

A five-center retrospective cohort study was performed by the Gruppo Italiano Diagnosi Embrio–Fetali to examine this issue, with results published in 1993 *(83)*. Of 3430 pregnancies in which CVS had been performed, outcome information was available for 2759. Of these, three had transverse limb reduction defects, two among 804 CVS procedures performed at 9 weeks and one among 1204 CVS procedures performed at 10 weeks. There were no limb reduction defects noted in 2192 amnioceteses with completed follow-up performed during the same study period. The authors concluded that performing CVS at less than 10 weeks' gestation "should be discouraged until further evidence against this association can be obtained," while noting that their follow-up rate was only 80%.

Hsieh et al. *(84)* surveyed 165 obstetric units in Taiwan regarding the incidence of limb defects with and without CVS. Of these, 67 hospitals responded, representing 78,742 deliveries. The incidence of limb defects was found to be 0.032% in the general population and 0.294% in the CVS population. The abnormalities seen in the CVS group included amelia, transverse reductions, adactylia, and digit hypoplasia, much like the abnormalities reported by Firth et al. *(75)*. The 25 limb abnormalities in the non-CVS group involved syndactyly or polydactyly. In addition, oromandibular-limb hypogenesis was seen in four of 29 CVS cases with limb abnormalities but in none of the non-CVS cases with limb abnormalities. The severity of the post-CVS limb abnormalities appeared to correlate with timing of the procedure, and the authors recommended performing CVS only after 10 full gestational weeks to minimize the risks.

In 1995, Olney et al. reported on a United States multistate case-controlled study comprising the years 1988–1992 *(85)*. The case population was 131 babies with nonsyndromic limb deficiency born to women 35 and older, and control subjects were 131 babies with other birth defects. These were drawn from a total of 421,489 births to women greater than 34 years of age. The odds ratio for all types of limb deficiency after CVS was 1.7, and for transverse digital deficiency an odds ratio of 6.4 after CVS was observed. They estimated the absolute risk for transverse digital deficiency in babies after CVS was one per 2900 births (0.03%).

Froster and Jackson reported on outcome data in a World Health Organization (WHO) study on limb defects and CVS in 1996 *(86)*. From 1992 to 1994, 77 babies or fetuses with limb defects from 138,996 pregnancies exposed to CVS were reported to the WHO CVS registry. This group represented the entire experience of 63 European and American centers reporting to the registry. They found that the overall incidence of limb defects in the CVS cohort did not differ from that in the general population, and they did not see a different pattern of distribution of limb defects between the groups. No correlation between limb reduction defects and gestational age was identified. They indicated that other studies finding an association between limb defects and CVS are confusing because of different methodologies and interpretations, and that the numbers reported are too small to draw firm conclusions.

Larger numbers were collected by Kuliev et al. *(87)*, who summarized the accumulated experience of 138,996 cases of CVS from the same 63 centers that report cases to the World Health Organization CVS registry. They reported an overall incidence of limb reduction defects after CVS of 5.2–5.7 per 10,000, compared with 4.8–5.97 per 10,000 in the general population. They also found no difference in the pattern distribution of limb defects after CVS, and similarly concluded that their data provide no evidence for any risk for congenital malformation caused by CVS.

Maternal Age—a Confounder?

Because CVS is usually performed on women 35 and older, the issue of whether the limb deficiencies seen after CVS were related to maternal age was raised by Halliday et al. *(88)* in a study from Victoria, Australia. A congenital malformations registry maintained there was reviewed by a medical geneticist, who classified all cases using the International Classification of Diseases, 9th edition (REF). All babies born with limb defects in 1990–1991 were identified, and the number of those whose mothers had had amniocentesis, CVS, or no invasive study was known. Excluding babies with chromosome abnormalities, recognized inherited syndromes, or amniotic bands, they found a twofold relative risk of having a baby with a limb deficiency of any type among women at age 35 or older, compared to women under 35. They also discuss the difficulty in interpreting studies of limb defects and CVS, as others had *(86)*, pointing out the importance of 100% follow-up, inclusion of all recognized cases of limb deficiencies (induced abortions as well as all other births), recognition of the heterogeneity of the condition, and the different risk estimates at different gestational ages.

A subsequent study found no maternal age confounding effect in interpretation of CVS/transverse limb deficiency data *(89)*. The authors analyzed the maternal age-specific rates of transverse limb deficiencies in the Italian Multicentric Birth Registry and used a case-control model for maternal age. No difference in the relative risk was seen between the 35-and-older group, whether or not CVS had been performed, and the under-35 group. The risk estimate for transverse limb defects associated with CVS was 12.63 and did not change after stratification for maternal age or for gestational age.

Since 1991, the utilization of CVS has dropped significantly *(90,91)*, due in large part to the concern regarding limb deficiencies.

Fetal Loss in CVS

In the first large controlled study of the safety of CVS, Rhoads et al. *(92)* reported on seven centers' experience with transcervical CVS in 2235 women compared to that of 651 women who had amniocentesis at 16 weeks' gestation. They found an overall excess loss rate of 0.8% in the CVS group after statistical adjustments for gestational age and maternal age. CVS procedures in which more than one attempt was made were associated with a substantially higher loss rate, supporting the observation by Silver et al. and others that increased operator experience is a key factor in assessing the risks of CVS *(93)*. Silver's group found that the number of placental passes and increased sample weight/aspiration attempt ratio may be more sensitive indicators of competence than the fetal loss rate.

Results of a randomized international multicenter comparison of transabdominal and transcervical CVS with second-trimester amniocentesis were reported in 1991 *(94)*. Outcome information was available for 1609 singleton pregnancies in the CVS group and 1592 in the amniocentesis group. Thirty-one centers participated, and the numbers of cases submitted ranged from 4 to 1709. Significantly fewer surviving newborns were seen in the CVS group than in the amniocentesis group (4.6% difference, $p < 0.01$). Most of the difference was in the significantly greater number of spontaneous fetal deaths before 28 weeks: 86/1528 in the successfully sampled CVS group and 25/1467 of the successfully sampled amniocentesis group (rate difference 2.9%, $p < 0.02$).

In a report from the Centers for Disease Control, an overall risk of spontaneous abortion attributed to CVS is reported from a literature survey as 0.5–1.0%, compared to 0.25–0.50% for amniocentesis procedures *(95)*.

In the WHO study, registry participants reported a spontaneous pregnancy loss rate after transabdominal or transcervical CVS of 2.5–3.0%, with several large-volume operators having loss figures of less than 2% *(87)*. This risk was deemed comparable to that of amniocentesis.

Transabdominal vs Transcervical CVS

Efficacy and risks associated with transabdominal CVS (TA CVS) and transcervical CVS (TC CVS) have been studied at several centers *(94,96–98).* The majority of CVS had been performed transcervically until the late 1980s, when more centers began using TA CVS to avoid cervical microorganisms and to reach placentae more easily. In their pilot study in 1988, Smidt-Jensen and Hahnemann *(98)* reported on 100 TA CVS cases at 8–12 weeks' gestation followed to term, compared to 200 amniocentesis cases. In all CVS cases, a sample was successfully obtained and cultured, and the fetomaternal complication rates were found not to be significantly different from those of previous TC CVS reports.

Transabdominal CVS (**Fig. 1**) has been increasingly used in recent years compared to TC CVS. Brambati et al. *(96)* reported on efficiency and risk factors in 2411 patients, 1501 of whom had TC CVS and 910 of whom had TA CVS. The two approaches had comparable success rates and complication rates, but TA CVS was considered easier to learn and less likely to be contraindicated by clinical and anatomical conditions. Subsequently, this group published results of a randomized clinical trial of TA- and TC CVS *(97).* All CVS procedures were performed by the same practitioner, who had prior similar experience in both techniques. The procedures were found to be equally effective, although TA CVS required significantly fewer insertions. The authors concluded " . . . transabdominal and transcervical CVS appear equally effective, and by and large the choice may be based on the operator's preferences."

Confined Placental Mosaicism

Chromosomal mosaicism is characterized by the presence of two or more karyotypically different cell lines within one individual. Confined placental mosaicism (CPM) is defined as a discrepancy between the chromosomal constitutions of placental and embryonic/fetal tissues. CPM results from viable mitotic mutations occurring in the progenitor cells of trophoblast or extraembryonic mesoderm during early embryonic development. In 1983, Kalousek and Dill *(99)* reported on numerical discrepancies between the karyotypes of fetal and placental cells, either full trisomies or mosaic aneuploidies, and similar reports followed *(100).* Based on six cases in which placental/CVS cells had a different chromosome constitution from that of amniotic fluid cells, the authors concluded that the results of cytogenetic analysis from placental tissue may not be representative of the fetus. Their figures, though small, were similar to the 2% incidence of this phenomenon as previously reported *(101).* Since then, others have found CPM to occur in 0.8–2% of viable pregnancies studied by CVS at 9–11 weeks' gestation *(102–108)* and in 0.1% or less in amniocentesis specimens *(69,103).*

The outcomes of pregnancies in which CPM is diagnosed vary from apparently normal outcomes to severe intrauterine growth restriction (IUGR), although few follow-up reports are yet available in the literature. Kalousek et al. *(109)* found six cases of IUGR among 17 gestations with CVS-detected CPM, five in liveborns and one associated with intrauterine death. They noted that others had found a 22% fetal loss rate among pregnancies with CPM. Wolstenhome et al. found 73 cases of CPM in 8004 CVS specimens from women referred for advanced maternal age, previous child with aneuploidy or family history thereof *(108).* Comparison at delivery with the control population did not show a marked increase in adverse pregnancy outcome. In 108 other

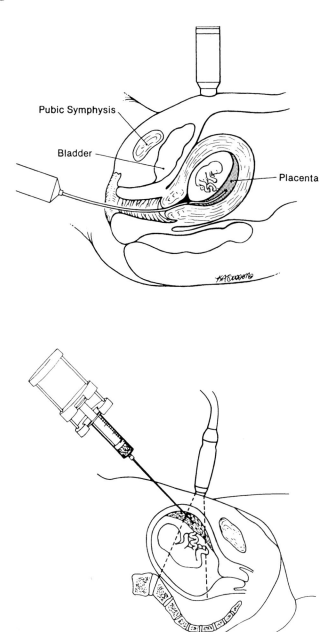

Fig. 1. Illustration of transcervical and transabdominal chorionic villus sampling. **Upper:** Transcervical CVS. A flexible catheter is introduced into the chorionic villi, or future placenta. **Lower:** Transabdominal CVS. A spinal needle is inserted through the abdominal wall for sampling. [From Elias and Simpson *(219)*. Reprinted with permission.]

cases referred for ultrasound detection of isolated IUGR, seven were shown to have CPM involving the following chromosomes: 2 and 15 [1], 9 [1], 16 [3], del(13) [1], and 22 [1].

Hahnemann and Vejerslev *(110)* evaluated cytogenetic outcomes of 92,246 successfully karyotyped CVS specimens from 79 laboratories from 1986 to 1994. CVS mosa-

Table 10
Distribution of Specific Single Autosomal Trisomies
in Each of the Groups of Mosaicism/Discrepancy in
Chorionic Villus Tissue

Trisomy	CPM (No. of cases)[a]	True fetal mosaicism (No. of cases)
2	11	
3	10	
5	3	
7	32	
8	11	1
9	9	1
10	6	
11	1	
12	2	1
13	15	2
14	3	
15	11	1
16	11	
17	1	
18	29	4
20	12	1
21	22	9
22	3	
ALL	192	20

[a]Includes all types of confined placental mosaicism, including direct-only, long-term culture-only and both. Data from ref. *110*.

icism or nonmosaic fetoplacental discrepancy was found in 1415 (1.5%) of the specimens. **Table 10** shows the mosaic and nonmosaic chromosome findings seen. Their work on several cell lineages indicated that mosaic or nonmosaic trisomies found in cytotrophoblasts, with a normal karyotype in the villus mesenchyme, were not seen in fetal cells. However, if such trisomies were seen on direct preparations, a risk of fetal mosaic or nonmosaic trisomy existed. They recommended amniocentesis in all pregnancies involving mosaic autosomal trisomy in villus mesenchyme.

Uniparental Disomy in Confined Placental Mosaicism

When a conceptus is trisomic, this aneuploidy will be "corrected" if by chance there is early loss of one of the trisomic chromosomes. Depending upon the parental origin of the trisomy and of the chromosome that is lost, this can lead to an apparently normal diploid cell line with uniparental disomy (both chromosomes in a pair from one parent) for that chromosome. Because most trisomies are maternally derived, the disomy seen is often maternal, as was the case in two previously reported cases of trisomy 15 mosaicism seen at CVS in which the neonates subsequently manifested Prader–Willi syndrome due to maternal disomy 15 *(108)*. The authors also note the reports of several cases of chromosome 16 CPM-associated IUGR in which maternal disomy 16 was seen in most of the cases. The evaluation of parental disomy in all CPM cases involving chromosome 15 should be offered, and this recommendation may extend to other chromosomes as more information becomes available.

For a thorough discussion of UPD, refer to Chapter 16.

Interphase FISH in Confined Placental Mosaicism

Interphase fluorescence *in situ* hybridization (FISH) can be useful for the diagnosis of CPM, given that interphase FISH is rapid and has the great advantage of not requiring growing, dividing cells to obtain results. Harrison et al. *(102)* examined the placentae of 12 pregnancies in which nonmosaic trisomy 18 had been diagnosed and found significant levels of mosaicism, confined to the cytotrophoblast, in 7 of the 12. Based on their observation that most of the mosaic results were seen in stillborn or newborn trisomy 18 babies, and on the fact that the great majority of trisomy 18 conceptuses spontaneously abort, they suggested that a normal diploid trophoblast component in placental tissue may be necessary to facilitate the prolonged survival of trisomy 18 conceptuses.

Schuring-Blom et al. *(111)* used FISH to document CPM in three pregnancies in which mosaic trisomy 8, mosaic trisomy 10, and nonmosaic monosomy X were observed following CVS, but which were found to be chromosomally normal at amniocentesis. In all three cases, FISH showed the presence of the mosaic cell line confined to one part of the placenta.

Henderson et al. *(112)* performed a cytogenetic analysis using a "mapping" technique of nine term placentae after CPM had been diagnosed, and found tissue-specific and site-specific patterns of mosaicism. In addition to metaphase chromosome analysis, they employed interphase FISH to examine several areas of the placentae. Noting that the outcomes of pregnancies are highly variable after CPM is diagnosed, they proposed a wider study involving extensive analysis of term placentae when CPM is diagnosed, to obtain more information regarding the outcome of such pregnancies.

Direct and Cultured Preparations

Direct CVS preparations involve the rapid metaphase analysis of villous cytotrophoblastic tissue. Cultured preparations involve the extraembryonic mesoderm, or chorionic stroma cells, in the villi. Some laboratories use only cultured cell preparations, and others utilize both methods. Investigations into the outcomes of pregnancy after CVS support the use of both techniques to maximize the accuracy of the test *(104,106,107)*. These studies documented false-negative and false-positive results using direct and cultured preparations, and the first two groups concluded that results from both direct and cultured techniques were necessary in a substantial number of cases to accurately predict the fetal karyotype. In one study *(104)*, long-term culture was advocated as having higher diagnostic accuracy, and the direct method was said to be a useful adjunct to the culture method. In part at least, the finding that both techniques add to the diagnostic accuracy appears to be related to the nonrandom findings of some trisomies in direct vs long-term cultured tissues. Trisomy 2 is seen more in cultured cells, and trisomy 3 is more often seen in direct preparations *(107,108)*. False-positive trisomy 7 or 18 can occur with either technique.

Maternal cell contamination (MCC) in CVS is generally due to the lack of complete separation of chorionic villi from maternal decidua, and it is reported in an estimated 1.0–1.8% of cases *(104,106,107)*. The MCC reported in these studies is about half of the figures above, reflecting the XX/XY admixtures, and is doubled to account for the likely equal incidence of MCC in female fetuses. MCC occurs more often in cultured cells than in direct preparations, thus underscoring the importance of using both methods in a full CVS cytogenetic analysis. In one report *(104)*, the rate of MCC was significantly higher in specimens obtained by the transcervical method (2.16%) than in samples obtained by the transabdominal method (0.79%).

A note of caution is prudent here. Generally, when there is a discrepancy between the direct and the cultured preparations, a subsequent amniocentesis is considered to provide the "true" result. However, a case of mosaic trisomy 8 reported by Klein et al. *(113)* illustrates the fact that a true low-level tissue-specific mosaicism can exist. In this case, the CVS showed a normal direct preparation, and mosaic trisomy 8 in culture. Subsequent amniocentesis showed normal chromosomes, but peripheral blood cultures of the newborn showed trisomy 8 mosaicism. Therefore, when considering amniocentesis or PUBS as follow-up studies because of possible CPM observed in CVS, one needs to weigh factors such as the specific aneuploidy involved, the likelihood of detecting it using a given sampling technique, and the risks of the additional invasive procedure.

Specimen Requirements

The minimum amount of chorionic villus material necessary to obtain diagnostic results and the transport medium should be established in advance with the laboratory. In general, a minimum of 10 mg of tissue is needed to obtain both a direct and a cultured cell result, and 20 mg is ideal. If possible, the specimen should be viewed through a dissecting microscope to ensure that villi are present. The specimen should be transported at ambient temperature to the cytogenetics laboratory as soon as possible.

Percutaneous Umbilical Blood Sampling (PUBS)

Risks, Limitations, and Benefits

Percutaneous umbilical blood sampling (PUBS) is also known as periumbilical blood sampling, fetal blood sampling, or cordocentesis. The largest series in the literature regarding risks of PUBS *(114)* included outcomes of 1260 diagnostic cordocenteses among three fetal diagnosis centers and 25 practitioners. A fixed needle guide was used in this study, and prospective data were compared to the published experience of large centers that use a freehand technique, where a 1–7% fetal loss rate has been reported. The procedure-related loss rate at a mean gestation of 29.1 ± 5 weeks at the time of sampling was 0.9%, leading to the conclusion that technique is a variable in the loss rate for cordocentesis.

PUBS experience at an earlier gestation was described by Orlandi et al. *(115)* in 1990, who pointed out that, while cordocentesis was a technique largely confined to the middle of the second trimester to term, in their experience it could be performed as early as the 12th week with acceptable results. They evaluated the outcomes of 500 procedures performed between 12 and 21 weeks for thalassemia study [386], chromosome analysis [97], fetomaternal alloimmunization [10], and infectious disease diagnosis [7]. One practitioner performed the procedures, and the volume of blood obtained ranged from 0.2 to 2.0 mL, depending on the gestational age. Of the 370 pregnancies not electively terminated and for which outcome information was available, the fetal loss rate was 5.2% for fetuses of 12–18 weeks' gestation and 2.5% between 19 and 21 weeks. Indicators of adverse outcome included cord bleeding, fetal bradycardia, prolonged procedure time, and anterior insertion of the placenta. Fetal bradycardia is a commonly reported complication after PUBS and is associated with a higher likelihood of fetal loss. In a review of 1400 pregnancy outcomes after PUBS, the overall incidence of recognizable fetal bradycardia was estimated at 5% *(116)*. It was significantly more likely to occur when the umbilical artery was punctured. Boulot et al. *(117)* performed 322 PUBS and noted fetal bradycardia, usually transitory, in 7.52% of their cases. Fetal bradycardia occurred in 2.5% of cases with normal outcome and in 12.5% of cases of fetal loss in one study *(115)*, while in another, 11 of 12 fetal losses were associated with prolonged fetal bradycardia *(116)*.

The underlying fetal pathology is a significant factor in fetal loss rate. Of these 12 losses, 10 were fetuses with a chromosome abnormality or severe fetal growth restriction. In gestations from 17 to 38 weeks, Maxwell et al. *(118)* compared the loss rates within 2 weeks of the procedure with the indications. Of 94 patients having prenatal diagnosis with normal ultrasound findings, one pregnancy of the 76 that were not electively terminated was lost. Of the group with structural fetal abnormalities, 5 in 76 were lost, and in the group of 35 with nonimmune hydrops, 9 were lost. It is important to take this factor into account when counseling patients before the procedure.

It has been said that no other fetal tissue " . . . can yield such a broad spectrum of diagnostic information (cytogenetic, biochemical, hematological) as fetal blood" *(115)*. As a means of fetal karyotyping, it has the advantage of generating results in 2 to 4 days, compared to 6 to 14 or more for amniotic fluid and CVS cells. When pseudomosaicism or mosaicism is seen in amniotic cell cultures, PUBS can provide valuable additional information regarding the likelihood of true mosaicism *(119–122)* and thereby assist the couple in their decision making.

Although pseudomosaicism in amniotic fluid cell cultures is usually associated with normal chromosome analysis after PUBS, the absence of trisomic cells in fetal blood does not guarantee that mosaicism has been definitely excluded *(123)*. For example, fetal blood karyotyping is not useful for the evaluation of mosaic or pseudomosaic trisomy 20. For further discussion of mosaicism, see "Special Issues" later in this chapter, and also in Chapter 8.

Because PUBS is associated with a significantly higher fetal loss rate than other prenatal diagnostic procedures, use of this technique should be recommended and provided with great care and only in certain high-risk situations such as those mentioned previously.

Specimen Requirements

Ideally, 1–2 mL of blood should be obtained and put into a small sterile tube containing sodium heparin. Results can usually be obtained from 0.5 mL, and in some cases 0.2 mL, so even small amounts obtained should not be discarded. A Kleihauer–Betke test may be useful in evaluating the possibility of maternal cell admixture, particularly when a 46,XX karyotype results.

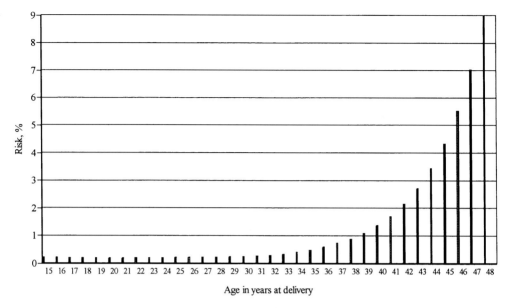

Fig. 2. Risk of chromosomally normal women to deliver chromosomally abnormal offspring *(123a).*

INDICATIONS FOR PRENATAL CYTOGENETIC DIAGNOSIS

Advanced Maternal Age

Advanced maternal age, generally defined in the United States as 35 or older at delivery, is the most common indication for prenatal cytogenetic diagnosis. For women in this age group, this indication alone provides the advantage of greater than 99% accuracy for detection of chromosome abnormalities. The chief disadvantage lies in the fact that, overall, it results in the detection of only 20% of chromosomally abnormal fetuses, given that 80% of chromosomally abnormal babies are born to women under age 35. Advanced maternal age is the most significant determinant of the risk of a chromosome abnormality for all trisomies, structural rearrangements, marker chromosomes, and 47,XXY (Klinefelter syndrome). Maternal age is not a factor in 45,X (Turner syndrome), triploid (69 chromosomes instead of 46), tetraploid (92 chromosomes instead of 46), or 47,XYY karyotypes.

Very young women are also at increased risk of fetal chromosome abnormality. A 15-year-old has a 1 in 454 risk of having a term infant with a chromosome abnormality, compared to a 1 in 525 risk for a 20-year-old and a 1 in 475 risk for a 25-year-old woman *(123a;* see **Fig. 2**).

Women 31 and Older with Twin Pregnancies

A 31-year-old with a twin gestation of unknown zygosity has a risk comparable to that of a 35-year-old woman; this is calculated as follows: Given that two thirds of such twins are dizygotic, the risk that one or the other has a chromosome abnormality is about 5/3 times that of a singleton pregnancy for that age. Thus, given that a 31-year-old woman's risk is 1 in 384 at term for any chromosome abnormality, if she is carrying twins of unknown zygosity, the risk that one or the other has a chromosome

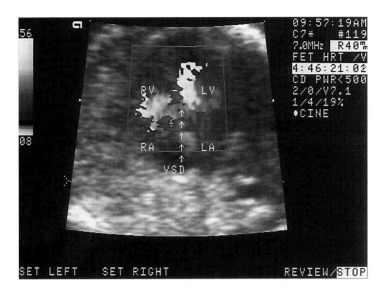

Fig. 3. Ultrasound image of a ventricular septal defect (indicated as VSD by arrows) in a 17-week-gestation fetus. RV = right ventricle. LV = left ventricle. RA = right atrium. LA = left atrium. (Courtesy of Greggory DeVore, M.D.)

abnormality is 5/3 × 1/384, or one in 231. This is between the risk of a 34-year-old (1 in 243) and that of a 35-year-old.

The risk of a chromosome abnormality is not significantly greater for monozygotic pregnancies compared to singletons. For pregnancies known to be dizygotic, the risk that one or the other twin has a chromosome abnormality is about twice that of a singleton.

Abnormal Fetal Ultrasound Findings

Many fetal ultrasound findings are associated with an increased risk for chromosome abnormalities. This list will continue to grow as the skill of practitioners and the resolution of ultrasound machines improve, and also as the search for indicators of increased risk other than advanced maternal age continues.

Heart Abnormalities

STRUCTURAL HEART ABNORMALITIES

Structural heart abnormalities are a well-established risk factor for chromosome abnormalities. Postnatal data indicate a frequency of chromosome abnormalities in infants with congenital heart diseases to be 5–10%, and 2 to 8 per 1000 live births have a structural cardiac abnormality *(124)*. Prenatal data indicate that up to 32–48% of fetuses with cardiac abnormalities are chromosomally abnormal *(124–126)*. The difference between prenatal and postnatal data probably reflects the high incidence of *in utero* demise in fetuses with chromosome abnormalities.

The most frequent prenatally and postnatally diagnosed heart abnormality is the ventricular septal defect (VSD, see **Fig. 3**), followed by tetralogy of Fallot (TOF), right or left hypoplastic heart, and transposition of the great arteries. Many investigators use the four-chamber view to evaluate the fetal heart, with an 80–92% sensitivity claimed by this method *(125)*. However, the four-chamber view alone will not detect TOF or transposition of the great arteries, and detects only approximately 59% of heart abnormalities.

Extracardiac abnormalities are seen, depending on the gestational ages at which the ultrasound evaluations are performed and what is considered an abnormality, in 36–71% of fetuses with heart abnormalities *(124–126)*. The presence of extracardiac abnormalities increases the risk of a chromosome abnormality from 32–48% to 50–71%.

Conotruncal heart abnormalities are those related to faulty conotruncal septation, or division, of the single primitive heart tube into two outflow tracts that in turn result from the fusion of two swellings that arise in the truncal region at 30 days' gestation. With increasing awareness of the strong association between conotruncal heart abnormalities and chromosome 22q11 deletions or microdeletions, it is now recommended that FISH analysis of this region be performed when a conotruncal heart abnormality is seen on fetal ultrasound and fetal chromosomes are normal. In five patients whose fetuses had fetal cardiac abnormalities and a prenatal diagnosis of 22q11 deletion [del(22)(q11)], the heart abnormalities included TOF with absent pulmonary valve, pulmonary atresia with VSD, truncus arteriosus, and left atrial isomerism with double outlet right ventricle. One of the fetuses had an absent kidney, and the others had isolated cardiac abnormalities *(127)*.

In a retrospective study of clinical and cytogenetic databases at St. Louis Children's Hospital, 70 patients who had absent pulmonary valve, TOF, TOF/pulmonary atresia, interrupted aortic arch, truncus arteriosus, or anomalous origin of the pulmonary artery had chromosome analysis and FISH for del(22)(q11) *(128)*. Of these, 33 had the deletion, and 9 of these had no other reported clinical abnormalities, although not all had been evaluated by a geneticist. Based on these results, the investigators recommended cytogenetic analysis with FISH for del(22)(q11) for patients with any of the cardiac abnormalities listed previously, except for anomalous origin of the pulmonary artery. It is likely that recommendations for del(22)(q11) testing will both expand and become more specific in the next few years as the phenotypes associated with del(22)(q11) become better understood.

INTRACARDIAC ECHOGENIC FOCI

Echogenic lesions within the fetal cardiac ventricles have been recognized since 1987, when they were described in the left ventricles of 3.5% of fetuses examined by ultrasound *(129,* see **Fig. 4**). The foci were attributed to thickening of the chordae tendinae. Others have reported a 20% incidence of left ventricular echogenic foci and right ventricular foci in 1.7% *(130)*.

The association between left ventricular echogenic foci and chromosome abnormalities was noted in a study of 2080 fetuses at 18–20 weeks' gestation; 33, or 1.6%, had an echogenic focus. Four of these had chromosome abnormalities (two trisomy 18, one 45,X and one trisomy 13). All had other abnormalities, including heart defects *(131)*.

The natural history of intracardiac echogenic foci was studied in a cohort of 1139 patients *(132)*. Echogenic foci were seen in 41 of 1139 fetuses, or 3.6%. In 38, the foci were in the left ventricle; in two, they were in the right ventricle; and in one, they were in both. None of these fetuses had other abnormalities. The echogenic foci were again seen in the 27 newborns having echocardiograms up to 3 months of age. The authors

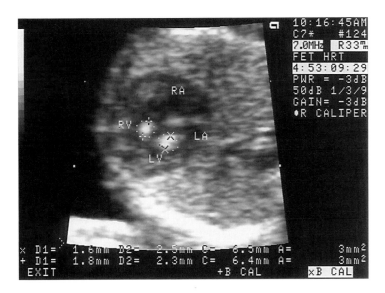

Fig. 4. Ultrasound image of intracardiac echogenic foci (indicated by +'s, x's, and circles of dots) in a 16-week-gestation fetus with trisomy 13. No other abnormalities are detected. RV = right ventricle. LV = left ventricle. RA = right atrium. LA = left atrium. (Courtesy of Greggory DeVore, M.D.)

pointed out the key clinical significance of echogenic lesions is that they should be differentiated from intracardiac tumors and ventricular thrombi.

The outcomes of 25,725 ultrasound examinations were reported in a retrospective study from 12 to 24 weeks' gestation *(133)*. Echogenic intracardiac foci were seen in 44 cases (0.17%). Of the 35 fetuses with left-sided isolated foci, all had uneventful neonatal courses. In nine others, multiple foci were seen, involving the right ventricle in five cases. Of these, two had uneventful courses, but the other seven had additional findings, including five with structural or functional cardiac disease (including one with trisomy 13), one with GM_1 gangliosidosis, and one with echogenic bowel and missed abortion. The article includes a useful discussion of the various possible causes of the echogenic foci, and the authors conclude by agreeing with the consensus that isolated left ventricular echogenic foci are a benign finding, but other intracardiac echogenic findings may not be.

Two subsequent publications, in contrast, found a significantly increased risk of trisomy 21 in fetuses with an intracardiac echogenic focus. In a study by Bromley et al. *(134)* of 1334 high-risk second-trimester patients, 66 (4.9%) had an echogenic intracardiac focus. Four of 22 (18%) trisomy 21 fetuses had an echogenic focus, compared with 62 (4.7%) of 1312 fetuses without trisomy 21. The presence of this finding increased the risk of trisomy 21 fourfold. In two of the trisomy 21 fetuses, no other ultrasound abnormalities were seen.

In a retrospective blinded study of pregnancies at 15–21 weeks' gestation, Norton et al. found an echogenic focus of unspecified location in the heart in five of 21 (24%) trisomy 21 fetuses compared to four of 75 (5%) controls, yielding an odds ratio for trisomy 21 of 5.5 (1.12< OR <28.4) when an echogenic focus is seen *(135)*.

The variations in reported incidences of intracardiac echogenic foci probably reflect the differences in definition of echogenic foci and in ultrasound machines. That some centers report an association between the foci and an increased incidence of trisomy 21 and other chromosome abnormalities, and others do not, may reflect differences in the populations studied—whether small or large, whether high risk or not. Overall it appears advisable to discuss the finding with the patient and counsel her of the probable increased risk of trisomy 21 and possibly other chromosome abnormalities even if no other abnormal ultrasound findings are present.

Renal Pyelectasis

Renal pyelectasis is mild dilation of the renal pelvis. A possible link between fetal renal pyelectasis and trisomy 21 was described in 1990 *(136)*. This led to other studies with conflicting results. In 1996, Wickstrom et al. *(137)* published a prospective study of 7481 patients referred for prenatal ultrasound evaluation. Of these, 121 (1.6%) had isolated fetal pyelectasis (defined as ≥ 4 mm before 33 weeks' gestation and ≥ 7 mm at 33 weeks' gestation). This compares with prevalences of 1.1–18% in other studies. Of the 121, 99 karyotypes were available. One of these was trisomy 21, and another was mosaic 47,XYY/46,XY. Based on maternal age and the baseline risk for trisomy 21 in the population, the authors calculated a relative risk of 3.9 for trisomy 21 when isolated renal pyelectasis is seen and a 3.3-fold increase in risk for all chromosomal abnormalities in the presence of isolated fetal pyelectasis.

Corteville et al. *(138)* studied 5944 fetuses for the presence of pyelectasis, defined as an anteroposterior renal pelvic diameter of 4 mm or greater before 33 weeks or 7 mm or greater after 33 weeks, the same definition as was used by Wickstrom et al. *(137)*. Pyelectasis was seen in 4 of 23 (17.4%) of trisomy 21 fetuses and in 120 of 5876 (2%) of normal controls. This was statistically significant at $p < 0.001$. When fetuses with other ultrasound abnormalities were excluded, the predictive value of pyelectasis fell from 1 in 90 to 1 in 340. They recommended that amniocentesis should be reserved for those cases presenting other risk factors such as advanced maternal age, abnormal maternal serum screening results, or other ultrasound abnormalities. They did not adjust the risk for trisomy using maternal age.

In a literature review study, Vintzileos and Egan *(139)* found that isolated pyelectasis was not associated with an increased risk for trisomy 21 unless other markers were present, such as those noted previously *(138)*.

Degani et al. *(140)* evaluated the recurrence rate of fetal pyelectasis in subsequent pregnancies. They studied 420 women with two consecutive normal uncomplicated pregnancies screened at 15–24 weeks by ultrasound. Pyelectasis was defined as a fetal pelvis of 4 mm or more in its anteroposterior dimension. Of 64 women with fetuses with pyelectasis, 43 (67%) had a recurrence in the next pregnancy. Compared with normal fetuses, those with pyelectasis had a relative risk of 6.1 to have a recurrence (95% confidence interval = 4.3–7.5, $p < 0.001$). This study has implications for determining the clinical significance of pyelectasis. In this regard, Johnson et al. *(141)* studied 56 pregnant women with fetal pyelectasis or cystic lesions identified from 7500 ultrasound examinations. They found that none of 50 kidneys 15 mm or smaller in anteroposterior diameter had obstruction, and 11 of 14 (79%) of kidneys larger than 15 mm were obstructed or showed vesicoureteral reflux on postnatal examination. Noting that other studies have found the need for intervention in the child after a prenatal ultrasound finding of 10-mm dilation, they recommended complete radiologic evaluation after birth for infants with pelvic diameters exceeding 10 mm. For children with mild to moderate unilateral hydronephrosis, evaluation may be delayed for 1–2 weeks, because oliguria in the first 2 days of life leads to an underestimation of the degree of hydronephrosis.

Choroid Plexus Cysts

The existence of choroid plexus cysts (CPCs) has become recognized, along with several other fetal ultrasound findings, due to improvements in ultrasound imaging. CPCs were first described in 1984 *(142)*. The choroid plexuses are round or oval anechoic structures within the choroid plexus of the lateral ventricle derived from neuroepithelial folds. CPCs are seen in 0.18–2.3% of pregnancies *(143)*. These cysts usually disappear in the second trimester in normal pregnancies but may also disappear in chromosomally abnormal pregnancies *(144)*.

The first association between CPCs and fetal trisomy 18 was published in 1986 by Nicolaides et al. *(145)*. In the intervening years, many publications on the association between CPCs and chromosome abnormalities have come out. Consensus has been reached as to the positive association between CPCs and chromosome abnormalities. However, investigators have differed in their conclusions as to whether an isolated CPC confers a risk of chromosome abnormality high enough to warrant amniocentesis *(146–150)* or whether the risk is not high enough to routinely recommend amniocentesis unless other risk factors are present *(143,151–153)*. Gross et al. prospectively studied patients at their institution and reviewed literature to include a meta-analysis of other studies prospectively done with more than 10 cases of CPCs. From these data, they estimated the risk of trisomy 18 in fetuses with isolated CPCs to be 1 in 374. From the incidence of trisomy 18 and of isolated CPCs, plus these data, they estimated the positive predictive value of CPCs with trisomy 18 in the general prenatal population to be 1 in 390.

Nyberg et al. reviewed 47 consecutive cases of trisomy 18 and found that 12 of 47 fetuses (25%) had CPCs, two of whom had no other ultrasound abnormality *(154)*. Although trisomy 18 is the chromosome abnormality most often associated with CPCs, seen in about three fourths of aneuploid fetuses with CPCs *(151)*, trisomy 21, mosaic trisomy 9 *(152)*, triploidy *(143,146)*, 47,XXY, and 45,X/46,XX [(146), trisomy 13 *(143)*, unbalanced (3;13) translocation *(149)*, and cri du chat syndrome [del(5p)] *(155)* have also been seen in fetuses with CPC.

Shields et al. *(146)* include mention of two issues in CPC, namely size and uni- vs bilaterality. They conclude, based on a review of the literature, that neither size nor laterality plays a part in the risk assessment. Size varies with gestational age, and laterality can be difficult to determine due to near-field artifact on ultrasound examination. These conclusions were also reached by Meyer et al. *(156)* in a retrospective review of 119 pregnancies with CPCs.

On balance, counseling regarding isolated CPCs cannot be undertaken in a vacuum. A young woman with a negative triple marker screen for trisomies 18 and 21 and no

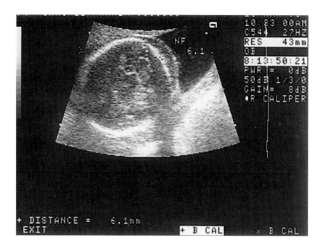

Fig. 5. Ultrasound image of increased nuchal fold (NF) measuring 6.1 mm (indicated by +'s) in a second-trimester fetus. (Courtesy of Greggory DeVore, M.D.)

other ultrasound abnormalities is much less likely to be carrying a fetus with trisomy 18 than is a 39-year-old woman with a triple marker screen result positive for trisomy 18 and no other fetal ultrasound abnormalities.

Even without other ultrasound abnormalities and with normal chromosomes, CPCs are frightening to prospective parents, who often are concerned about a "hole in my baby's head." It is important to explain their significance in a balanced way, to indicate that in the majority of fetuses they are an incidental finding and that they are likely to disappear before birth. Results of a follow-up study (mean 35.5 ± 16.2 months) on 76 children who as fetuses were found to have CPCs are also reassuring; no effect on development was found as measured by the Denver II Developmental Screening Test *(157)*.

Nuchal Thickening

Nuchal thickening/folds are caused by abnormal lymphatic development and obstruction. The fluid collects in the posterior neck fold, causing the appearance of a nuchal membrane separation on ultrasound examination (see **Fig. 5**). With resolution of the fluid collection, a nuchal fold or thickening develops. Nuchal membranes have been recorded as early as 9 weeks' gestation.

Nuchal folds and cystic hygromas have been known to be associated with chromosome abnormalities since 1966, with an incidence of chromosome abnormalities ranging from 22% to more than 70% in various series *(158)*. Based on 22 other studies, plus their own data, Landwehr et al. found 32% of 1649 karyotyped fetuses with nuchal folds or membranes and/or cystic hygromas had a chromosome abnormality. These included 207 cases of trisomy 21, 108 cases of trisomy 18, 30 cases of trisomy 13, 131 cases of 45,X, and 48 other chromosome abnormalities. This study included first- and second-trimester ultrasound scans, which employ different criteria for nuchal thickness.

In a 12-center study designed to determine the sensitivity and specificity of second trimester soft-tissue nuchal fold measurement for the detection of trisomy 21, 3308 fetuses of 14–24 weeks' gestation were evaluated *(159)*. Using 6 mm as a cutoff, a nuchal skin fold was seen in 8.5% of chromosomally normal fetuses and in 38% of those with trisomy 21. A false-positive rate below 5% was obtained by 81% of the investigators. The authors concluded that this sign is useful in skilled hands in the second trimester, but it does not appear suitable for population screening because of the high variability in the results among the investigators.

A nuchal thickness cutoff of 4 mm was chosen by Nadel et al. *(160)* in a study of 71 fetuses of 10–15 weeks' gestation, of whom 63 were karyotyped. Abnormal karyotypes were found in 31 of 37 hydropic fetuses and in 12 of 26 nonhydropic fetuses. The nonhydropic fetuses also had no septations in the hygromas. Of the fetuses with septated hygromas, 22 had chromosome analysis, and 19 had abnormal chromosomes. Of fetuses with hydrops and no septations, 11 of the 14 had abnormal chromosomes.

There have been several first-trimester ultrasound studies of nuchal thickening. Van Vugt et al. *(161)* karyotyped 102 first-trimester fetuses with a nuchal translucency of 3 mm or more and found 46% had an abnormal karyotype: 19 had trisomy 21, 9 had trisomy 18, 13 had 45,X, one had 47,XXX, and 5 had other chromosome abnormalities. Multiple logistic regression analysis was used to take into account data modifiers such as gestational age and maternal age. They examined the septated vs the nonseptated nuchal translucencies. Septa were seen in 45 (44%) of the fetuses, of whom 36 (80%) had chromosome abnormalities. Of 57 fetuses with no septation, 11 (19%) had abnormal chromosomes. This compared to a 56% incidence of chromosome abnormalities in first-trimester fetuses with septation and 23% incidence of chromosome abnormalities in first-trimester fetuses without septation in Landwehr's study *(158)*.

In 1015 fetuses of 10–14 weeks' gestation with nuchal fold thicknesses of 3 mm, 4 mm, 5 mm, and > 5 mm, Pandya et al. found incidences of trisomies 21, 18, and 13 to be approximately 3 times, 18 times, 28 times, and 36 times higher than the respective numbers expected on the basis of maternal age alone *(162)*. This corresponded to risks of one of these chromosome abnormalities to be 5%, 24%, 51%, and about 60%, respectively.

Using a 4-mm cutoff in fetuses of 9–13 weeks, Comas et al. detected 57.1% of aneuploidies with a false-positive rate of 0.7% and a positive predictive value of 72.7% *(163)*.

Szabó et al. evaluated 2100 women under 35 years of age by ultrasound at 9–12 weeks' gestation *(164)*. Women were offered CVS if the nuchal fold was 3 mm or greater. They found an incidence of first-trimester nuchal fold to be 1.28% in women under 35, with a corresponding percentage of chromosome abnormalities being 0.43%. This indicated a 1 in 3 risk for chromosome aneuploidy in this age group when a thickened nuchal fold was seen.

Given that nuchal thickening is clearly associated with chromosome abnormalities, most commonly trisomy 21, and that it is the most common abnormal ultrasound finding in the first trimester, ultrasound evaluation of nuchal thickness in the first trimester promises to be an increasingly important early screening tool to evaluate an increased risk of aneuploidy. Combining it with maternal serum markers will further increase the power of detection of aneuploidy in the first trimester *(165)*.

Cystic Hygroma and Cytogenetic Evaluation of Cystic Hygroma Fluid

Women whose second- or third-trimester fetuses have large cystic hygromas may not have an easily accessible fluid pocket in which to perform an amniocentesis. In such cases, paracentesis of the hygroma may yield a cytogenetic result, and at fetal demise or delivery, chorionic villous or placental cell cultures may prove beneficial in obtaining chromosomal diagnosis. The yield from amniocentesis is still the greatest, so if it can be accomplished, this is still the procedure of choice for cytogenetic diagnosis in such cases *(166)*.

Short Humerus or Femur

Measurement of the long bones of the fetus does not require the same level of expertise as evaluating more subtle structural malformations. Thus, because shortness of the long bones is associated with an increased risk of chromosome abnormalities and because the length is relatively easy to measure, several investigators have focused on this finding as a way of increasing or decreasing a woman's *a priori* risk of having a fetus with a chromosome abnormality.

Shortness of the humerus and the tibia may have greater sensitivity in detecting trisomy 21 than shortness of the femur and fibula, as was found in a prospective study of 515 patients between 14 and 23 weeks' gestation who were at increased risk for a chromosome abnormality because of age or triple marker screening results or both *(167)*. Tables of risk for trisomy 21 for maternal age and maternal serum screening positivity were developed that take into account all four long bones' being normal vs one, two, three, or four bone lengths' being normal. Use of this approach led to the conclusion that if all long bone lengths are normal, amniocentesis may not be recommended to women under age 40. Others have not found femur length to be reliable in ultrasound screening of trisomy 21 *(168,169)*, although humerus length does appear to be associated *(169)*. The positive predictive value for trisomy 21 in women with risks of one in 500 and one in 1000 was found to be 2.3 and 1.2%, respectively.

A significant confounder, however, is that long bone length varies with race, and this factor has not been taken into account in most studies. In a fetal biometry study of Asians, the long bone lengths were measured in more than 6000 fetuses, and the conclusion was that the reference charts derived should be used in all Asian fetuses *(170)*. Thus use of fetal biometric measures should be cautiously interpreted with racial factors in mind.

Hyperechoic Bowel

Hyperechoic bowel (HEB), also known as echogenic bowel and hyperechogenic fetal bowel, is a qualitative ultrasound finding of unclear significance. It has been described as a normal variant with an incidence of 0.2–0.56%, as reviewed by several authors *(171–174)*. It is also associated with several adverse outcomes, including fetal chromosome abnormalities, fetal cytomegalovirus infection, other infections, cystic fibrosis (CF), intrauterine growth restriction, fetal demise, and intestinal obstruction possibly related to CF *(171,172,174–182)*. The presence of coexisting elevated maternal serum α-fetoprotein increases the risk of adverse outcome, particularly fetal IUGR and demise *(172,182)*. See **Table 11**. The studies referenced above describe the finding of HEB on second-trimester ultrasound examination. Third-trimester HEB associated with trisomy 21 has also been reported in a fetus in which the second-trimester scan did not show HEB *(183)*.

The incidence of HEB in second-trimester fetuses with trisomy 21 is 7% *(184)*. The relative risk of adverse outcome in isolated HEB is 6.5 *(172)*.

Table 11
Clinical Outcome of Second-Trimester Finding of Isolated Bright Hyperechoic Bowel[a]

	Scioscia (171)[b]	*Nyberg* (172)[c]	*Bromley* (174)[b]	*Slotnick* (181)[c]	*Muller* (176)[b]	*MacGregor* (177)[c]
1. No. of cases with isolated bright HEB	18	64	42	102	182	45
2. No. of cases with normal outcome (%)	13 (72)	41 (75)	26 (62)	—	111 (67)	34 (76)
3. No. of cases with chromosome abn. (%)	2 (11)[d]	7 (11)[g]	0	5 (4.9)[h]	8 (4.5)	0/16 (0)
4. No. of cases with cystic fibrosis mutations (%)	0[e]/17 (0)	NT[f]	NT[f]	7/65 (11)	10/116[i] (8.6)	2/15 (13)
5. No. of cases with infections (%)	NT[f]	1 (1.6)	—	—	7/? (?)	2/45 (4)
6. No. of cases with IUGR (%)	1 (5.6)	6 (9.3)	8 (19)	—	10/121 (8)	NR[j]
7. No. of cases with nonelective demise (%)	2 (11)	3 (4.7)	15 (36)	—	24/104 (23)	3/45 (6.7)

[a]This table excludes fetuses with ultrasound abnormalities other than isolated HEB.
[b]Retrospective study.
[c]Prospective study.
[d]Both trisomy 21.
[e]Seven CF mutations tested.
[f]Not tested.
[g]Five trisomy 21, one 47,XXX, one trisomy 13.
[h]All trisomy 21.
[i]One ΔF508 homozygote, 9 heterozygotes; 7 of the 9 were unaffected, and the other 2 had no follow-up information. One to eight mutations tested.
[j]Not reported.

Part of the reported variation in outcome of HEB is due to different degrees of brightness of the finding and also to intermachine and interobserver variability (see **Fig. 6**). Grades of echogenicity, from 0 (isoechoic) to 3 (bonelike density) have been used *(172,181)*, but even those compare the finding to different fetal parts—liver vs iliac crest, for example. The more hyperechoic, the higher the risks. Another reason for variability in reported outcomes relates to the *a priori* risks. For example, Caucasian non-Hispanic patients have a much higher *a priori* risk of CF than individuals of other races.

What causes the finding of HEB? One group *(182)* commented on the decreased microvillar enzymes in amniotic fluid in pregnancies affected by trisomy 21, trisomy 18, and CF. It was thought that the low levels in CF may be due to delayed passage of meconium, and in trisomy 18 and 21, the delayed passage may be due to decreased bowel motility or abnormal meconium. Fetuses with intraamniotic bleeding have a four- to sevenfold increase in HEB *(174,184)*. These investigators hypothesized that swallowing of amniotic fluid containing heme pigments after intraamniotic bleeding seemed to be the cause of the echogenicity.

Other Ultrasound Markers of Aneuploidy

A summary of several series of ultrasound studies indicating risks of chromosome abnormalities in association with specific ultrasound findings is shown in **Table 12**.

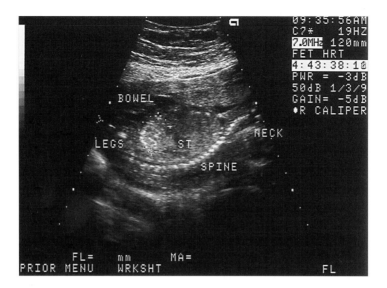

Fig. 6. Fetal ultrasound finding of hyperechoic bowel, as shown by the patchy white area to the left of "ST." Note the brightness of the finding is comparable to that of the spine, indicated at the bottom of the image. (Image courtesy of Greggory Devore, M.D.)

Clearly, some ultrasound markers in isolation indicate a significant risk of chromosome abnormality, and others may not achieve significance unless other ultrasound abnormalities or other maternal risk factors are present.

In the past 10 years, medicine in the United States has evolved from recommending amniocentesis to women 35 and older to refining risks based upon a variety of ultrasound and maternal serum screening markers. This has led to increased detection of chromosome abnormalities while not significantly increasing the use of amniocentesis, as some women 35 and older now have their *a priori* risks altered downward and choose not to have amniocentesis as a result. Several scoring indices have been developed to provide individualized risk assessments *(169,185–187)*. Unfortunately, anyone with an ultrasound machine in the office may do an ultrasound examination, and the ranges of expertise and resolution vary significantly among practitioners. Optimally, each practitioner should develop his or her own index based on the prospective evaluation of a large series of patients. These indices will be much more valid in that practice than those derived from the literature.

Positive Maternal Serum Marker Screen

High Maternal Serum α-Fetoprotein

The association between an elevated level (2.0 or 2.5 multiples of the median) of maternal serum α-fetoprotein (AFP) and fetal neural tube defects has been known for many years. More recently, the presence of an unexplained elevated level of maternal serum AFP has been found to be associated with an increased risk for fetal chromosome abnormalities, with an incidence of 10.92 per 1000 amniocenteses *(188)*. Of these, fetal sex chromosome abnormalities were seen in 47%. Thus, although some practitioners discourage patients from having an amniocentesis with an elevated AFP and a normal ultrasound study, the facts that sex chromosome abnormalities other than 45,X

Table 12
Ultrasound Markers of Fetal Aneuploidy

Finding	Risk(s) of aneuploidy if isolated finding	Risk(s) of aneuploidy if other ultrasound abnormalities are present	Comment
Structural heart defect	32–48%[124-126]	50–71%[124-126]	
Intracardiac echogenic foci	Not increased[131]; four-to fivefold baseline risk[134,135]	1.6-fold baseline risk[131]	Isolated left ventricular foci appear more likely to be benign than multiple or right-sided foci (133).
Renal pyelectasis (≥4 mm before 33 weeks and ≥7 mm at 33 weeks)	Not increased[139]; 3.3-fold increase over baseline[137]; one in 340[138]	One in 90[138]	In ref. 138, no adjustment was made for maternal age.
Choroid plexus cyst(s)	Not increased[143,151-153]; one in 374 for trisomy 18[151]; 1–2%[146]; 0.6%[147]; 2.4%[156]; 3.1%[148]; 4%[149]; one in 82[150], one in 150[151]	8.2%[146]; 3.5%[147]; 5%[156]; 9.5%[148]; one in three[151]	No adjustment for maternal age in ref. 150.
Septated nuchal membrane, 9–20 weeks; >3 mm, <15 weeks and >5 mm, 15–20 weeks	56–60%[158]	Not studied	Ref. 158 is a retrospective database analysis. All pregnancies included in study had isolated nuchal finding and karyotype.
Simple nuchal membrane, 9–20 weeks; >3 mm, <15 weeks and >5 mm, 15–20 weeks	10–25%[158]	Not studied	
Nuchal fold, >5 mm, 15–20 weeks	19–33%[158]	Not studied	

(continued)

295

Table 12 (Continued)

Finding	Risk(s) of aneuploidy if isolated finding	Risk(s) of aneuploidy if other ultrasound abnormalities are present	Comment
Nuchal fold, >5 mm, 15–20 weeks	19–33%[158]	Not studied	
Nuchal thickening, 10–15 weeks, ≥4 mm	46%[160]	84% if hydrops and/or septations are present[160]	Best outcome was in nonseptated, nonhydropic; worst was in septated/hydropic fetuses
Nuchal thickening, 9–15.5 weeks, ≥3 mm Septated Simple	80%[161] 19%[161]; 27-fold risk for 34-year-old women and ninefold risk for women 35 and older	Other abnormalities only reported for chromosomally normal fetuses	
Short humerus, femur	Positive predictive value for trisomy 21 in women with risks of one in 500 and one in 1000 = 2.3 and 1.2%, respectively, for short humerus. For short femur, some studies found very little increased risk.	Increased to variable degrees	If all long bones are of normal length and other ultrasound findings normal, some feel amniocentesis is not indicated in women under 40. Racial factors should be considered in any long bone measurement.

and its related karyotypes have no significant associated ultrasound abnormalities, and that they are quite common (with incidences of 47,XXX, 47,XXY and 47,XYY each one in 500 liveborns), support consideration of amniocentesis in this group.

Low Maternal Serum AFP and Triple Marker Screening

The association between low maternal serum AFP and fetal Down syndrome was established in 1984 *(189),* and in 1987, the association between high maternal serum human chorionic gonadotropin (hCG) *(190)* and low unconjugated estriol *(191)* and fetal Down syndrome was established. These three substances, or markers, are combined now in what is commonly known as triple marker screening (TMS). Hundreds of thousands of women in the United States have TMS in the second trimester of pregnancy, with a resultant increase in detection of trisomy 21 before age 35 and what appears to be a decrease in the incidence of Down syndrome births due to abortion of affected fetuses. The overall detection of trisomy 21 with TMS is about 65% with a mid-trimester risk cutoff of one in 190, with a much lower detection in young women (about 44% in 18-year-olds) and a much higher detection in older women (about 78% in 36-year-olds) *(192).*

TMS detects 60% of trisomy 18 fetuses as well, when a mid-trimester risk cutoff of one in 100 is used *(193).*

Less recognized is the fact that TMS detects many chromosome abnormalities nonspecifically, for unknown reasons. Thus for every trisomy 21 fetus found by TMS, a fetus with a different chromosome abnormality is also detected *(194).* This is important to keep in mind when counseling patients.

Other maternal serum markers have been studied, but none is used as commonly as TMS *(195).* Holding promise for first-trimester detection of trisomy 21 are free β-hCG and pregnancy-associated plasma protein A (PAPP-A) *(196,197),* which in combination with nuchal translucency measurements in the first trimester may prove to be the most sensitive first-trimester screening methods. Now the challenge is to have more patients present for prenatal care in the first trimester!

Previous Pregnancy or Child with a Chromosome Abnormality

Having a previous pregnancy or child with certain chromosome abnormalities produces an increased risk of a future fetal chromosome abnormality *(198).* The reasons for this are not known. There is some evidence that suggests that one or more recessive genes may predispose couples to nondisjunction.

Chromosome abnormalities known to increase the future risk of aneuploidy include all nonmosaic trisomies, 47,XXY, structural rearrangements, and marker chromosomes. Genetic counseling is suggested for couples who have had a pregnancy or child with any such karyotype, and ultrasound plus amniocentesis or CVS are recommended for consideration in future pregnancies.

Not known to be associated with an increased recurrence risk are 47,XYY, triploidy, tetraploidy, and 45,X. However, couples who have undergone the experience of having a pregnancy with one of these findings may experience anxiety and may wish to have genetic counseling, ultrasonography, and prenatal chromosome analysis.

Mosaicism presents complicated counseling issues. It is prudent to apprise the couple of this and offer them the opportunity for prenatal diagnosis, as the risk of recurrence may be increased. Mosaicism is discussed later.

Other Indications for Prenatal Diagnosis

Pregnancy at Increased Risk for an X-Linked Disorder

In the past, fetal sex determination by amniocentesis was the only reliable method of avoiding an X-linked disorder not associated with fetal structural abnormalities detectable by ultrasound. For a growing list of X-linked conditions, however, prenatal diagnosis is now available through linkage analysis, direct DNA studies, or enzymatic analysis of amniocytes. It is strongly advised to consult a genetics professional to inquire about availability of testing for a given disorder, given the rapidity of advances in the field.

Previous Pregnancy or Child with Open Neural Tube Defect

Rates of open neural tube defects (NTDs) vary geographically. In California, NTDs occur in 1.05 per 1000 Hispanic women and 0.66 per 1000 Asian women, with non-Hispanic Caucasians falling between *(91)*. The risk of recurrence of an isolated NTD is 3–5%. Folic acid supplementation of 0.4 mg/day periconceptionally decreases the risk by 50–70% *(199)*, so the increased fortification of grains with 1.4 mg of folate per pound of enriched cereal-grain products by the Food and Drug Administration was announced in 1997. Having a previous affected pregnancy or child merits offering genetic counseling, ultrasonography, and amniocentesis. Such women are advised to take 4 mg of folate periconceptionally. All women of childbearing age, particularly those at increased risk for NTDs, should receive information about folate supplementation.

Chromosome abnormalities are associated with spina bifida and encephalocele *(200)* but do not appear to any significant degree to be associated with isolated anencephaly.

Chromosome Rearrangement in Either Member of a Couple

Some balanced structural rearrangements predispose a couple to an increased risk of fetal chromosome abnormality. The risk depends on the rearrangement and how it was ascertained.

For balanced reciprocal translocations, if the rearrangement was ascertained through multiple spontaneous abortions, the risk of having a child with abnormal chromosomes is 1.4–4.8%, with the lower risk associated with a paternal carrier. If it was ascertained by a previous child or stillborn with unbalanced chromosomes, the risk increases to 19.8–22.2% *(201)*.

For balanced Robertsonian translocations, the risk of unbalanced chromosomes in the fetus is much less and appears to be negligible where chromosome 21 is involved if the translocation is paternal *(201)* (see **Table 13**).

Most pericentric inversions (see Chapter 9), except the population variant inv(9), are associated with an increased risk of unbalanced offspring due to deletions/duplications, and individuals with such inversions should be offered amniocentesis. The risk of unbalanced offspring depends on the length of the inversion segments *(202)* (see **Table 14**). Whether this recommendation applies to individuals with the common pericentric inv(2) is debatable. This inversion is so common that some cytogenetics laboratories do not report it.

Paracentric inversions in a carrier parent may give rise to acentric fragments or dicentric chromosomes, either of which would be expected to be lethal *in utero*. However, amniocentesis is generally to be recommended, given the possibility of viability

Table 13
Prenatal Results for Robertsonian Translocations Involving Chromosomes 13, 14, or 15 and 21

Robertsonian translocation type	Maternal carrier				Paternal carrier				Grand total
	Balanced	Normal	Unbalanced	Total	Balanced	Normal	Unbalanced	Total	
der(13;21)(q10;q10)	6	4	2 (16.7%)	12	5	4	0	9	21
der(14;21)(q10;q10)	36	25	10 (14.1%)	71	9	13	1 (4.3%)	23	94
der(15;21)(q10;q10)	5	4	0	9	4	2	0	6	15
Total	47	33	12 (13.0%)	92	18	19	1 (2.6%)	38	130

Data from ref. (201).

Table 14
Prenatal Results for Pericentric Inversions (n = 173)

Method of ascertainment	Maternal carrier				Paternal carrier				Grand total
	Balanced	Normal	Unbalanced	Total	Balanced	Normal	Unbalanced	Total	
Through term unbalanced progeny	6	1	1 (12.5%)	8	2	3	0	5	13
Through recurrent miscarriages	10	4	0	14	4	2	0	6	20
Other	63	4	2 (2.9%)	69	68	3	0	71	140
Total	79	9	3 (3.3%)	91	74	8	0	82	173

Data from ref. 201.

of a fetus with structurally unbalanced chromosomes, and the occasional difficulty in distinguishing between a paracentric inversion and an insertion *(203)*.

Because of the observation that marker chromosomes can interfere with meiosis, leading to aneuploidy, prenatal diagnosis is also recommended to individuals with marker chromosomes, even when these apparently confer no adverse phenotypic effect.

Men with 47,XYY karyotypes usually have normal fertility and may be at increased risk for chromosomally unbalanced offspring. Some of the reported chromosome abnormalities occurring in pregnancies of 47,XYY males include markers, trisomy 21, 47,XYY, and others *(204)*.

A widely debated indication for prenatal diagnosis is a low level of mosaicism (3% or less) for 45,X/46,XX in the mother. This chromosome finding is quite common in unselected populations, and the bulk of evidence suggests that the 45,X cell line may be an artifact of culture or an age-related phenomenon. Data do not support an increased risk of chromosomally abnormal offspring in this population.

Pregnancy Exposed to Valproic Acid or Carbamazepine

These two anticonvulsant agents are known to be associated with an increased risk of neural tube defects *(205–207)*, so amniocentesis for measurement of amniotic fluid AFP and acetylcholinesterase is generally offered to women who took these medications between 3 months before and 1 month after conception. There is no known increased risk for chromosome abnormalities, but if amniocentesis is being performed, it is prudent to perform cytogenetic analysis on the specimen.

Advanced Paternal Age

A body of old literature in genetics suggests an increased risk of fetal chromosome abnormality with advanced paternal age, but the most carefully constructed analyses do not support this association *(208–211)*. Advanced paternal age is not associated with fetal chromosome abnormalities. It is, however, associated with an exponentially increased risk of some autosomal dominant new mutations in the offspring, including achondroplasia and thanatophoric dysplasia *(212)*. Although there is no consensus, it is prudent to offer detailed fetal ultrasound examination in pregnancies involving men 45 years and older.

Special Issues

True Mosaicism and Pseudomosaicism

Mosaicism, or the presence of two or more cell lines in culture, is one of the most complex and challenging issues in prenatal diagnosis. There are three levels of mosaicism in amniotic fluid and CVS culture—levels I, II, and III. Level I is defined as a single-cell abnormality. Level II is defined as a multiple-cell abnormality or (with an *in situ* culture method) a whole-colony abnormality in one culture not seen in any other cell cultures. Level III mosaicism is "true" mosaicism—the presence of a second cell line in two or more independent cultures. The incidences of these in amniotic cell cultures range from 2.5 to 7.1% for level I, 0.6–1.1% for level II, and 0.1–0.3% for level III mosaicism *(212a–212c)*.

The origin of the mosaic cell line cannot be determined without molecular studies. In general, however, it appears that the majority of 45,X/45,XX cases occur after a normal disomic fertilization, most mosaic trisomies are due to postzygotic loss of the

Table 15
Outcome of Cases with Rare Autosomal
Trisomy Mosaicism Diagnosed in Amniocytes

Type	Abnormal outcomes / Total no. of cases	Abnormal phenotype (no. with IUGR)[a]	Fetal demise or stillborn
46/47,+2	10/11 (90.9%)	7 (2)	3
46/47,+3	1/2 —	1	0
46/47,+4	1/2 —	1	0
46/47,+5	2/5 (40.0%)	2 (2)	0
46/47,+6	0/3 —	0	0
46/47,+7	1/8 (12.5%)	1	0
46/47,+8	1/14 (7.1%)	1	0
46/47,+9	14/25 (56.0%)	14(2)	0
46/47,+11	0/2 —	0	0
46/47,+12	6/23 (26.1%)	4	2
46/47,+14	2/5 (40.0%)	2	0
46/47,+15	6/11 (54.5%)	6 (3)	0
46/47,+16	15/21 (71.4%)	15 (8)	0
46/47,+17	0/7 —	0	0
46/47,+19	0/1 —	0	0
46/47,+22	7/11 (63.6%)	6 (2)	1

Data from ref. *(216a)*.
[a]IUGR = intrauterine growth restriction.

trisomic chromosome and, for trisomy 8, most cases are due to somatic gain of the third chromosome 8 postzygotically *(213)*.

Besides the level of mosaicism, the chromosome involved is an important consideration. True mosaicism has been reported in liveborns with almost all trisomies *(33a)*. However, true mosaicism for trisomies 8, 9, 21, 18, 13, X, and Y and for monosomies X and Y has potentially great significance. For chromosomes 8 and 9, mosaicism is the most common form in which trisomies occur in liveborns, perhaps because the full trisomy is not compatible with fetal survival in the majority of cases *(214,215)*. Even one cell with trisomy 8 may be significant. For trisomies of chromosomes 13, 18, 21, X, and Y and monosomy X and Y, mosaicism has been fairly commonly reported, and the clinical manifestations may vary from no apparent abnormality, at least in the newborn period, to more characteristic features of the full trisomy. The degree of mosaicism is not related to the outcome *(28)*. See **Table 15** for incidences of mosaicism for specific chromosomes.

Mosaicism for trisomies 12 and 20 poses unique problems. For both of these trisomies, mosaicism has been reported that appeared to have no discernible effect on the fetus or liveborn, and yet in other cases the mosaicism was associated with an abnormal outcome. A case report and survey of a decade of literature *(216)* showed a total of 13 reported cases in which trisomy 12 mosaicism was observed in amniocytes. In nine cases, the pregnancy was terminated, and in seven of the nine, no phenotypic abnormalities were reported. One fetus was not described, and one had only two lobes in each lung and appeared otherwise normal. In seven cases, confirmatory cytogenetic studies on skin, blood, rib, placenta, kidney, liver, lung, and/or villi was performed, and in the six cases in which fetal tissue was known to be cultured, five showed confirmation of mosaic trisomy 12.

In five cases in which the pregnancy was continued after diagnosis of trisomy 12 mosaicism in amniocytes, the diagnosis was confirmed in urinary cells or skin in two children. One of them had mild dysmorphic features with near normal development at 3 years, and the other was dysmorphic and died in the first weeks of life with cardiac abnormalities. In the other three, the diagnosis was not confirmed in fetal skin and/or blood; one had normal development at 5 months and the other two died in the newborn period with heart, kidney, vertebral, tracheo–esophageal, and other abnormalities.

It is interesting to note that the terminated fetuses were described as normal, and the liveborns were almost all abnormal. This was not related to degree of mosaicism. It may be due to unrecognized abnormalities in second trimester fetuses, or there may have been a bias toward reporting live births with congenital abnormalities.

Outcomes of 144 cases of trisomy 20 mosaicism *(28)* indicate that 112 of 123 cases (91%) were associated with a normal phenotype; 18 of these were abortuses. In most cases, the cells with trisomy 20 are extraembryonic or largely confined to the placenta. Of the 11 abnormal outcomes, three were in liveborns and eight in abortuses. Three abortuses with urinary tract abnormalities and two with heart abnormalities represent the only consistent, serious abnormalities associated with such mosaicism. Of 21 children followed for 1–2 years, all were normal except for two with borderline psychomotor delay. It was also apparent that attempted cytogenetic confirmation of the finding should not be limited to analysis of fetal blood, because trisomy 20 has not been observed in blood cells. Confirmation studies in newborns should be done on placental tissues, skin, cord blood, and urine sediment, and, in abortuses, on kidney, skin, and placental tissues.

Other Mosaic Trisomies and Monosomies

Trisomy 22 mosaicism was reported in a collection of nine cases *(28)*. Of these, four continued and five terminated. Four of eight reported cases showed a normal outcome, and in the others, one fetal demise, one neonatal death with IUGR, one liveborn with IUGR, and one abortus with multiple congenital abnormalities were seen.

One to five cases each of mosaic trisomies 2, 3, 4, 5, 6, 7, 11, 14, 15, 16, 17, and 19 were reported in one series *(216a)*. Of these, only mosaic trisomies 2, 3, 4, 5, 7, 14, 15, and 16 were associated with an abnormal outcome, but the cases reported were few and interpretation must be done with caution.

In the *ref.* 28, it was noted that 13 cases of autosomal monosomy mosaicism had been prenatally diagnosed. These included five cases of monosomy 21 (–21), three of –22, two of –17 and one case each of –9, –19, and –20. Of seven cases with phenotypic information and four cases with confirmatory cytogenetic studies, only one case with –22 was reported to have multiple congenital abnormalities, including congenital heart disease. Another case of –21 was confirmed but was reported to be phenotypically normal. If autosomal mosaic monosomy is detected, particularly of chromosomes 21 or 22, further work-up, such as PUBS and ultrasound examination, is indicated.

Mosaicism of an Autosomal Structural Abnormality

In 78 reported cases of mosaicism for a balanced autosomal structural abnormality, phenotypic information was available in 16 cases, and all were associated with a normal phenotype *(28)*. However, for unbalanced autosomal structural abnormality mosaicism, 25/52 (48%) were reported to be phenotypically abnormal, and 28/48 (58.3%) were cytogenetically confirmed. Such a finding thus merits consideration of PUBS and ultrasound examination.

Culture Failure

Rates of culture failure vary from lab to lab, and guidelines for acceptable rates exist (see Chapter 6). In our laboratory's experience, the rate of amniotic cell culture failure is about 0.1%. Cell culture failure is more likely to occur in advanced-gestation amniocentesis specimens, as the number of nonviable cells in the fluid is very high, and they appear to slow the growth of the viable cells. The usual counseling provided in such cases is that the fetal outcome is not related to the lack of cell growth. However, there is a report describing 32 of 7852 (0.4%) amniocentesis specimens that were classified as unexplained growth failures, and, in this group, 10 women did not repeat the procedure while 22 did *(217)*. Of the 10 who did not, a fetal bladder-outlet obstruction, two stillbirths, and one acardiac twin resulted. Of the 22 who repeated, 18 had normal fetal karyotypes, but four were aneuploid. Of these, two had trisomy 21, one had trisomy 13, and one had Pallister–Killian syndrome, or tetrasomy 12p.

Maternal Cell Contamination

After cell culture and cytogenetic analysis of amniotic fluid specimens, maternal cell contamination is rarely found. Maternal cells were detected in 0.17% of 44,170 cultured amniotic fluid samples in one study *(212b)*. Because this would be expected to be detected in only male pregnancies (as a mixture of XX and XY cells), the frequency of maternal contamination was estimated at twice this rate, or 0.34%. If *in situ* hybridization techniques are used on uncultured cells, thus identifying both maternal and fetal nuclei, the proportion of maternal cell contamination (MCC) increases dramatically, being present at a level of 20% in half of amniotic fluid specimens. This was found to be strongly associated with the sampling technique in a survey of 36 amniotic fluid specimens *(218)*. Maternal cell contamination of less than 20% was seen in 19 specimens in which the placenta was posterior, and in two others, which were bloody specimens in pregnancies with posterior placentae, more than 20% MCC was seen. In cases in which the placenta was anterior, less than 20% MCC was seen in two cases, and more than 20% in 13 cases. It was thought that the maternal cells were introduced into the amniotic fluid specimen as a result of placental bleeding during amniocentesis. The authors stated that molecular cytogenetic analysis, or FISH, should not be performed on uncultured amniotic fluid cells without preselecting fetal cells. The preselection could consist of simultaneous analysis of the morphology of the nuclei and of the *in situ* hybridization findings.

REFERENCES

1. Menees, T.O., Miller, J.D., and Holly, L.E. (1930) Amniography: preliminary report. *Am. J. Roentgen.* **24**, 363–366.

2. Bevis, D.C.A. and Manc, M.B. (1952) The antenatal prediction of hemolytic disease of the newborn. *Lancet* **21(4)**, 395–398.

3. Marberger, E., Boccaabella, R.A., and Nelson, W.O. (1955) Oral smear as a method of chromosomal sex detection. *Proc. Soc. Exp. Biol. Med.* **89**, 488.

4. Barr, M.L. (1955) The skin biopsy test of chromosomal sex in clinical practice. *Anat. Rec.* **121**, 387.

5. James, F. (Jan. 28, 1956) Letter to the editor, *Lancet* 202–203.

6. Fuchs, F. and Riis, P. (1956) Antenatal sex determination. *Nature* **177**, 330.

7. Makowski, E.L., Prem. K.A., and Karer, J.H. (1956) Detection of sex of fetuses by the incidence of sex chromatin body in nuclei of cells in amniotic fluid. Letter to editor. *Science* **124**, 542–543.

8. Shettles, L.B. (1956) Nuclear morphology of cells in human amniotic fluid in relation to sex of infant. *Am. J. Obstet. Gynecol.* **71**, 834–838.

9. Steele, M.W. and Breg, W.R. (1966) Chromosome analysis of human amniotic-fluid cells. *Lancet* **i**, 383–385.

10. Jacobson, C.B. and Barter, R.G. (1967) Intrauterine diagnosis and management of genetic defects. *Am. J. Obstet. Gynecol.* **99**, 796–806.

11. Nadler, H.L. and Gerbie A.B. (1968) Role of amniocentesis in the intrauterine detection of genetic disorders. *N. Engl. J. Med.* **282**, 596–599.

12. Milunsky, A. (1979) The prenatal diagnosis of chromosomal disorders. In *Genetic Disorders and the Fetus* (Milunsky, A., ed.), Plenum Press, New York, p. 93.

13. Acosta-Sison, H. (1958) Diagnosis of hydatidiform mole. *Obstet. Gynecol.* **12**, 205.

14. Alvarez, H. (1966) Diagnosis of hydatidiform mole by transabdominal placental biopsy. *Am. J. Obstet. Gynecol.* **95**, 538.

15. Mohr, J. (1968) Foetal genetic diagnosis: development of techniques for early sampling of foetal cells. *Acta Pathol. Microbiol. Scand.* **73**, 73–77.

16. Kullander, S. and Sandahl, B. (1973) Fetal chromosome analysis after transcervical placental biopsies during early pregnancy. *Acta Obstet. Gynecol. Scand.* **52**, 355–359.

17. Hahnemann, N. (1974) Early prenatal diagnosis; a study of biopsy techniques and cell culturing from extraembryonic membranes. *Clin. Genet.* **6**, 294–306.

18. Dept. of Obstetrics and Gynecology (1975) Fetal sex prediction by sex chromatin of chorionic villi cells during early pregnancy. *Chinese Med. J.* **1**, 117–126.

19. Horwell, D.H., Loeffler, F.E., and Coleman, D.V. (1983) Assessment of a transcervical aspiration technique for chorionic villus biopsy in the first trimester of pregnancy. *Br. J. Obstet. Gynaecol.* **90**, 196–98.

20. Liu, D.T.Y., Mitchell, J., Johnson, J., and Wass D.M. (1983) Trophoblast sampling by blind transcervical aspiration. *Br. J. Obstet. Gynaecol.* **90**, 1119–1123.

21. Niazi, M., Coleman, D.V., and Loeffler, F.E. (1981) Trophoblast sampling in early pregnancy. Culture of rapidly dividing cells from immature placental villi. *Br. J. Obstet. Gynaecol.* **88**, 1081–1085.

22. Kazy, Z., Rozovsky, I.S., and Bakharev, V.A. (1982) Chorion biopsy in early pregnancy: a method of early prenatal diagnosis for inherited disorders. *Prenat. Diagn.* **2**, 39–45.

23. Kuliev, A., Jackson, L., Froster, U., Brambati, B., Simpson, J.L., Verlinsky, Y., Ginsberg, N., Smidt-Jensen, S., and Zakut, H. (1996) Chorionic villus sampling safety. Report of World Health Organization/EURO meeting in association with the Seventh International Conference on Early Prenatal Diagnosis of Genetic Disorders. *Am. J. Obstet. Gynecol.* **174**, 807–811.

24. Valenti, C. (1973) Antenatal detection of hemoglobinopathies. *Am. J. Obstet. Gynecol.* **115**, 851–853.

25. Hobbins, J.C. and Mahoney, M.J. (1974) In utero diagnosis of hemoglobinopathies. *N. Engl. J. Med.* **290**, 1065–1067.

26. Daffos, F., Capella-Pavlovsky, M., and Forestier, F. (1983) A new procedure for fetal blood sampling in utero: preliminary results of fifty-three cases. *Am. J. Obstet. Gynecol.* **146,** 985–987.

27. Daffos, F., Capella-Pavloksky, M., and Forestier, F. (1985) Fetal blood sampling during pregnancy with use of a needle guided by ultrasound: a study of 606 consecutive cases. *Am. J. Obstet. Gynecol.* **153,** 655–660.

28. Hsu, L. (1992) Prenatal diagnosis of chromosomal abnormalities through amniocentesis. In *Genetic Disorders and the Fetus,* 3rd edition. (Milunsky, A., ed.), Johns Hopkins University Press, Baltimore.

29. Buckton, K.E., O'Riordan, M.L., Ratcliffe, S., Slight, J., Mitchell, M., and McBeath, S. (1980) A G-band study of chromosomes in liveborn infants. *Ann. Hum. Genet.* **43,** 227–239.

30. Hook, E.B. (May 13, 1978) Rates of Down's syndrome in live births and at midtrimester amniocentesis. *Lancet* **i,** 1053–1054.

31. Schreinemachers, D.M., Cross, P.K., and Hook, E.B. (1981) Rates of trisomies 21, 18, 13 and other chromosome abnormalities in about 20,000 prenatal studies compared with estimated rates in live births. *Hum. Genet.* **61,** 318–324.

32. Hook, E.B. (1983) Chromosome abnormalities and spontaneous fetal death following amniocentesis: further data and associations with maternal age. *Am. J. Hum. Genet.* **35,** 110–116.

32a. Halliday, J.L., Watson, L.F., Lumley, J., Danks, D.M., and Sheffield, L.J. (1995) New estimates of Down syndrome risks at chorionic villus sampling, amniocentesis, and livebirth in women of advanced maternal age from a uniquely defined population. *Prenat. Diagn.* **15,** 455–465.

33. Warburton, D. (1991) De novo balanced chromosome rearrangements and extra marker chromosomes identified at prenatal diagnosis: clinical significance and distribution of break points. *Am. J. Hum. Genet.* **45,** 995–1013.

33a. Warburton, D., Byrne, J., and Canki, N. (1991) *Chromosome Anomalies and Prenatal Development: An Atlas.* Oxford University Press, New York.

34. Simpson, J.L. (1990) Incidence and timing of pregnancy losses: relevance to evaluating safety of early prenatal diagnosis. *Am. J. Med. Genet.* **35,** 165–173.

35. Warburton, D., Kline, J., Stein, Z., and Susser, M. (1980) Monosomy X: a chromosomal anomaly associated with young maternal age. *Lancet* **i,** 167–169.

36. The NICHD National Registry for Amniocentesis Study Group (1976) Midtrimester amniocentesis for prenatal diagnosis: safety and accuracy. *JAMA* **236,** 1471–1476.

37. Rapp, R. (1993) Amniocentesis in sociocultural perspective. *J. Genet. Counsel.* **2,** 183–196.

38. Tkachuk, D.C., Pinkel, D., Kuo, W.-L., Weier, H.-U., and Gray, J.W. (1991) Clinical applications of fluorescence in situ hybridization. *GATA,* **8,** 67–74.

39. Klinger, K., Landes, G., Shook, D., Harvey, R., Lopez, L., Locke, P., Lerner, T., Osathanondh, R., Leverone, B., Houseal, T., Pavelka, K., and Dackowski, W. (1992) Rapid detection of chromosome aneuploidies in uncultured amniocytes by using fluorescence in situ hybridization (FISH). *Am. J. Hum. Genet.* **51,** 55–65.

40. Lebo, R.V., Flandermeyer, R.R., Diukman, R., Lynch, E.D., Lepercq, J.A., and Golbus, M.S. (1992) Prenatal diagnosis with repetitive in situ hybridization probes. *Am. J. Med. Genet.* **43,** 848–854.

41. Callen, D.F., Freemantle, C.J., Ringenbergs, M.L., Baker, E., Eyre, H.J., Romain, D., and Haan, E.A. (1990) The isochromosome 18p syndrome: confirmation of cytogenetic diagnosis in nine cases by in situ hybridization. *Am. J. Hum. Genet.* **47,** 493–498.

42. Callen, D.F., Eyre, H., Yip, M-Y., Freemantle, J., and Haan, E.A. (1992) Molecular cytogenetic and clinical studies of 42 patients with marker chromosomes. *Am. J. Med. Genet.* **43,** 709–715.

43. Blennow, E., Bui, T-H., Kristoffersson, U., Vujic, M., Anneren, G., Holmberg, E., and Nordenskjöld, M. (1994) Swedish survey on extra structurally abnormal chromosomes in 39,105 consecutive prenatal diagnoses: prevalence and characterization by fluorescence in situ hybridization. *Prenat. Diagn.* **14,** 1019–1028.

44. Brondum-Nielsen, K. and Mikkelsen, M. (1995) A 10-year survey, 1980–1990, of prenatally diagnosed small supernumerary marker chromosomes, identified by FISH analysis. Outcome and follow-up of 14 cases diagnosed in a series of 12,699 prenatal samples. *Prenat. Diagn.* **15,** 615–619.

45. Bettio, D., Rizzi, N., Giardino, D., Gurrieri, F., Silvestri, G., Grugni, G., and Larizza, L. (1997) FISH characterization of small supernumerary marker chromosomes in two Prader–Willi patients. *Am. J. Med. Genet.* **68,** 99–104.

45a. Gosden, C.M. (1983) Amniotic fluid cell types and culture. *Br. Med. Bull.* **39,** 348–354.

46. Eiben, B., Goebel, R., Hansen, S., and Hammans, W. (1994) Early amniocentesis—A cytogenetic evaluation of over 1500 cases. *Prenat. Diagn.* **14,** 497–501.

47. Leschot, N.J., Verjaal, M., and Treffers, P.E. (1985) Risks of midtrimester amniocentesis; assessment in 3000 pregnancies. *Br. J. Obstet. Gynaecol.* **92,** 804–807.

48. Tabor, A., Madsen, M., Obel, E.B., Philip, J., Bang, J., and Norgaard-Pedersen, B. (June 7, 1986). Randomised controlled trial of genetic amniocentesis in 4606 low-risk women. *Lancet* **i,** 1287–1293.

49. Simpson, J.L., Mills, J.L., Holmes, L.B., Ober, C.L., Aarons, J., Jovanovic, L., and Knopp, R.H. (1987) Low fetal loss rates after ultrasound-proved viability in early pregnancy. *JAMA* **258,** 2555–2557.

50. Pauker, S.P. and Pauker, S.G. (1979) The amniocentesis decision: an explicit guide for parents. *Birth Defects Orig. Article Ser.* **XV,** 289–324.

51. Evans, M.I., Drugan, A., Koppitch, F.C., Zador, I.E., Sacks, A.J., and Sokol, R. (1989) Genetic diagnosis in the first trimester: the norm for the 1990s. *Am. J. Obstet Gynecol.* **160,** 1332–1339.

52. Evans, M.I., Johnson, M.P., and Holzgreve, W. (1994) Early amniocentesis—what exactly does it mean? *J. Reprod. Med.* **2,** 77–78.

53. Burrows, P.E., Lyons, E.A., Phillips, H.J., and Oates, I. (1982) Intrauterine membranes: sonographic findings and clinical significance. *J. Clin. Ultrasond* **10,** 1–8.

54. Hanson, F.W., Zorn, E.M., Tennant, F.R., Marianos, S., and Samuels, S. (1987) Amniocentesis before 15 weeks' gestation: outcome, risks, and technical problems. *Am. J. Obstet. Gynecol.* **156,** 1524–1531.

55. Johnson, A. and Godmilow, L. (1998) Genetic amniocentesis at 14 weeks or less. *Clin. Obstet. Gynecol.* **31,** 345–352.

56. Wald, N.J., Terzian, E., and Vickers, P.A. (1983) Congenital talipes and hip malformation in relation to amniocentesis: a case-control study. *Lancet* **ii,** 246–249.

57. Stripparo, L., Buscaglia, M., Longatti, L., Ghisoni, L., Dambrosio, F., Guerner, S. Rosella, F., Lituania, M., Cordone, M., De Biasio, P., Passamonti, U., Gimelli, G., and Cuoco, C. (1990) Genetic amniocentesis: 505 cases performed before the sixteenth week of gestation. *Prenat. Diagn.* **10,** 359–365.

58. Elejalde, B.R., de Elejalde, M.M., Acuna, J.M., Thelen, D., Trujillo, C., and Marrmann, M. (1990) Prospective study of amniocentesis performed between weeks 9 and 16 of gestation: its feasibility, risks, complications and use in early genetic prenatal diagnosis. *Am. J. Med. Genet.* **35,** 188–196.

59. Penso, C.A., Sandstrom, M.M., Garber, M.-F., Ladoulis, M., Stryker, J.M., and Benacerraf, B.B. (1990) Early amniocentesis: report of 407 cases with neonatal follow-up. *Obstet. Gynecol.* **76,** 1032–1036.

60. Hanson, F.W., Happ, R.L., Tennant, F.R., Hune, S., and Peterson, A.G. (1990) Ultrasonography-guided early amniocentesis in singleton pregnancies. *Am. J. Obstet. Gynecol.* **162,** 1376–1383.

61. Hackett, G.A., Smith, J.H., Rebello, M.T., Gray, C.T.H., Rooney, D.E., Beard, R.W., Loeffler, F.E., and Coleman, D.V. (1991) Early amniocentesis at 11–14 weeks' gestation for the diagnosis of fetal chromosomal abnormality—a clinical evaluation. *Prenat. Diagn.* **11**, 311–315.

62. Crandall, B.F., Kulch, P., and Tabsh, K. (1994) Risk assessment of amniocentesis between 11 and 15 weeks: comparison to later amniocentesis controls. *Prenat. Diagn.* **14**, 913–919.

63. Kerber, S. and Held, K.R. (1993) Early genetic amniocentesis—4 years' experience. *Prenat. Diagn.* **13**, 21–27.

64. Lockwood, D.H. and Neu, R.L. (1993) Cytogenetic analysis of 1375 amniotic fluid specimens from pregnancies with gestational age less than 14 weeks. *Prenat. Diagn.* **13**, 801–805.

65. Diaz Vega, M., de la Cueva, P., Leal, C., and Aisa, F. (1996) Early amniocentesis at 10–12 weeks' gestation. *Prenat. Diagn.* **16**, 307–312.

66. Brumfield, C.G., Lin, S., Conner, W., Cosper, P., Davis, R.O., and Owen, J. (1996) Pregnancy outcome following genetic amniocentesis at 11–14 versus 16–19 weeks' gestation. *Obstet. Gynecol.* **88**, 114–118.

66a. Bravo, R.R., Shulman, L.P., Phillips, O.P., Grevengood, C., and Martens, P.R. (1995) Transplacental needle passage in early amniocentesis and pregnancy loss. *Obstet. Gynecol.* **86**, 437–440.

67. Wilson, R.D. (1995) Early amniocentesis: a clinical review. *Prenat. Diagn.* **15**, 1259–1273.

68. Shulman, L.P., Elias, S., Phillips, O.P., Grevengood, C., Dungan, J.S., and Simpson, J.L. (1994) Amniocentesis performed at 14 weeks' gestation or earlier: comparison with first-trimester transabdominal chorionic villus sampling. *Obstet. Gynecol.* **83**, 543–548.

69. Nicolaides, K., de Lourdes Brizot, M., Patel, F., and Snijders, R. (1994) Comparison of chorionic villus sampling and amniocentesis for fetal karyotyping at 10–13 weeks' gestation. *Lancet* **344**, 435–439.

69a. Saura, R., Roux, D., Taine, L., Maugey, B., Laulon, D., Laplace, J.P., and Horovitz, J. (1994) Early amniocentesis veresus chorionic villus sampling for fetal karyotyping. *Lancet* **344**, 825–826.

69b. Bombard, A.T., Carter, S.M., and Nitowsky, H.M. (1994) Early amniocentesis versus chorionic villus sampling for fetal karyotyping. *Lancet* **344**, 826.

70. Vandenbussche, F.P.H.A., Kanhai, H.H.H., and Keirse, M.J.N.C. (1994) Safety of early amniocentesis. Letter. *Lancet* **344**, 1032.

71. Eiben, B., Osthelder, B., Hamman, W., and Goebel, R. (1994) Safety of early amniocentesis versus CVS. Letter. *Lancet* **344**, 1303–1304.

72. Crandall, B.F., Hanson, F.W., Tennant, F., and Perdue, S.T. (1989) α-Fetoprotein levels in amniotic fluid between 11 and 15 weeks. *Am. J. Obstet. Gynecol.* **160**, 1204–1206.

73. Nevin, J., Nevin, N.C., Dornan, J.C., Sim, D., and Armstrong, M.J. (1990) Early amniocentesis: experience of 222 consecutive patients, 1987–1988. *Prenat. Diagn.* **10**, 79–83.

74. Boyd, P.A., Keeling, J.W., Selinger, M., and MacKenzie, I.Z. (1990) Limb reduction and chorion villus sampling. *Prenat. Diagn.* **10**, 437–441.

75. Firth, H.V., Boyd, P.A., Chamberlain, P., MacKenzie, I.Z., Lindenbaum, R.H., and Huson, S.M. (1991) Severe limb abnormalities after chorion villus sampling at 56–66 days' gestation. *Lancet* **337**, 762–763.

76. Mastroiacovo, P. and Cavalcanti, D.P. (1991) Letter to the editor: Limb-reduction defects and chorion villus sampling. *Lancet* **337**, 1091.

77. Hsieh, F-J., Chen, D., Tseng, L-H., Lee, C-N., Ko, T-M., Chuang, S-M., and Chen, H-Y. (1991) Letter to the editor: Limb-reduction defects and chorion villus sampling. *Lancet* **337**, 1091–1092.

78. Monni, G., Ibba, R.M., Lai, R., Giovanni, O., and Cao, A. (1991) Letter to the editor: Limb-reduction defects and chorion villus sampling. *Lancet* **337,** 1091.

79. Mahoney, M.J. (1991) Letter to the editor: Limb abnormalities and chorionic villus sampling. *Lancet* **337,** 1422–1423.

80. Jackson, L.G., Wapner, R.J., and Brambati, B. (1991) Letter to the editor: Limb abnormalities and chorionic villus sampling. *Lancet* **337,** 1423.

81. Jahoda, M.G.J., Brandenburg, H., Cohen-Overbeek, T., Los, F.J., Sachs, E.S., and Waldimiroff, J.W. (1993) Terminal transverse limb defects and early chorionic villus sampling: evaluation of 4,300 cases with completed follow-up. *Am. J. Med. Genet.* **46,** 483–485.

82. Holmes, L.B. (1993) Report of National Institute of Child Health and Human Development Workshop on Chorionic Villus Sampling and Limb and Other Defects, October 20, 1992. *Teratology* **48,** 7–13.

83. Mastroiacovo, P., Tozzi, A.E., Agosti, S., Bocchino, G., Bovicelli, L., Dalpra, L., Carbone, L.D.L., Lituania, M., Luttichaum, A., Mantegazza, F., Nocera, G., Pachi, A., Passamonti, U., Piombo, G., and Vasta, A.F. (1993) Transverse limb reduction defects after chorion villus sampling: a retrospective cohort study. *Prenat. Diagn.* **13,** 1051–1056.

84. Hsieh, F-J., Shyu, M-K., Sheu, B-C., Lin, S-P., Chen, C-P., and Huang, F-Y. (1995) Limb defects after chorionic villus sampling. *Obstet. Gynecol.* **85,** 84–88.

85. Olney, R.S., Khoury, M.J., Alo, C.J., Costa, P., Edmonds, L.D., Flood, T.J., Harris, J.A., Howe, H.L., Moore, C.A., Olsen, C.L., Panny, S.R., and Shaw, G.M. (1995) Increased risk for transverse digital deficiency after chorionic villus sampling: results of the United States Multistate Case-Control Study, 1988–1992. *Teratology* **51,** 20–29.

86. Froster, U.G. and Jackson, L. (1996) Limb defects and chorionic villus sampling: results from an international registry, 1992–94. *Lancet* **347,** 489–494.

87. Kuliev, A., Jackson, L., Froster, U., Brambati, B., Simpson, J.L., Verlinsky, Y., Ginsberg, N., Smidt-Jensen, S., and Zakut, H. (1996) Chorionic villus sampling safety. *Am. J. Obstet. Gynecol.* **174,** 807–811.

88. Halliday, J., Lumley, J., Sheffield, L.J., and Lancaster, P.A.L. (1993) Limb deficiencies, chorion villus sampling, and advanced maternal age. *Am. J. Med. Genet.* **47,** 1096–1098.

89. Mastroiacovo, P. and Botto, L.D. (1994) Chorionic villus sampling and transverse limb deficiencies: maternal age is not a confounder. *Am. J. Med. Genet.* **53,** 182–186.

90. Cutillo, D.M., Hammond, E.A., Reeser, S.L., Kershner, M.A., Lukin, B., Godmilow, L., and Donnenfeld, A.E. (1994) Chorionic villus sampling utilization following reports of a possible association with fetal limb defects. *Prenat. Diagn.* **14,** 327–332.

91. California Genetic Disease Branch. Unpublished data.

92. Rhoads, G.G., Jackson, L.G., Schlesselman, S.E., De La Cruz, F.F., Desnick, R.J., Golbus, M.S., Ledbetter, D.H., Lubs, H.A., Mahoney, M.J., Pergament, E., Simpson, J.L., Carpenter, R.J., Elias, S., Ginsberg, N.A., Goldberg, J.D., Hobbins, J.C., Lynch, L., Shiono, P.H., Wapner, R.J., and Zachary, J.M. (1989) The safety and efficacy of chorionic villus sampling for early prenatal diagnosis of cytogenetic abnormalities. *N. Engl. J. Med.* **320,** 609–617.

93. Silver, R.K., MacGregor, S.N., Sholl, J.S., Hobart, E.D., and Waldee, J.K. (1990) An evaluation of the chorionic villus sampling learning curve. *Am. J. Obstet. Gynecol.* **163,** 917–922.

94. Meade, T.W., Ämmälä, P., Aynsley-Green, A., Bobrow, M., Chalmers, I., Coleman, D.V., Elias, J.A., Ferguson-Smith, M.A., Gosden, C., Grant, A., Hahnemann, N., Liu, D.T.Y., MacKenzie, I.Z., McPherson, K., Milner, R.D.G., Modell, B., O'Toole, O., Rodeck, C.H., Stott, P.C., Terzian, E., Ward, R.H.T., Weatherall, D., Ayers, S.E., Cloake, E., Thomson, M.A.R., Hally, M., Hennigan, M., and Modle, J. (1991) Medical Research Council European trial of chorion villus sampling. *Lancet* **337,** 1491–1499.

95. Centers for Disease Control and Prevention (1995) Chorionic villus sampling and amniocentesis: recommendations for prenatal counseling. *Morb. Mortal. Wkly. Rep.* **44**(RR-9), 1–12.

96. Brambati, B., Lanzani, A., and Tului, L. (1990) Transabdominal and transcervical chorionic villus sampling: efficiency and risk evaluation of 2,411 cases. *Am. J. Med. Genet.* **35**, 160–164.

97. Brambati, B., Trzian, E., and Tognoni, G. (1991) Randomized clinical trial of transabdominal versus transcervical chorionic villus sampling methods. *Prenat. Diagn.* **11**, 285–293.

98. Smidt-Jensen, S. and Hahnemann, N. (1988) Transabdominal chorionic villus sampling for fetal genetic diagnosis. Technical and obstetrical evaluation of 100 cases. *Prenat. Diagn.* **8**, 7–17.

99. Kalousek, D.K. and Dill, F.J. (1983) Chromosomal mosaicism confined to the placenta in human conceptions. *Science* **221**, 665–667.

100. Verjaal, M., Leschot, N.J., Wolf, H., and Treffers, P.E. (1987) Karyotypic differences between cells from placenta and other fetal tissues. *Prenat. Diagn.* **7**, 343–348.

101. Hogge, W.A., Schonberg, S.A., and Golbus, M.S. (1986) Chorionic villus sampling: experience of the first 1000 cases. *Am. J. Obstet. Gynecol.* **154**, 1249–1252.

102. Harrison, K., Barrett, I.J., Lomax, B.L., Kuchinka, B.D., and Kalousek, D.K. (1993) Detection of confined placental mosaicism in trisomy 18 conceptions using interphase cytogenetic analysis. *Hum. Genet.* **92**, 353–358.

102a. Goldberg, J.D. and Wohlferd, M.M. (1997) Incidence and outcome of chromosomal mosaicism found at the time of chorionic villus sampling. *Am. J. Obstet. Gynecol.* **176**, 1349–1353.

103. Kalousek, D.K., Dill, F.J., Pantzar, T., McGillivray, B.C., Yong, S.L., and Wilson, R.D. (1987) Confined chorionic mosaicism in prenatal diagnosis. *Hum. Genet.* **77**, 163–167.

104. Ledbetter, D.H., Zachary, J.M., Simpson, J.L., Golbus, M.S., Pergament, E., Jackson, L., Mahoney, M.J., Desnick, R.J., Schulman, J., Copeland, K.L., Verlinsky, Y., Yang-Feng, T., Schonberg, S.A., Babu, A., Tharapel, A., Dorfmann, A., Lubs, H.A., Rhoads, G.G., Fowler, S.E., and De La Cruz, F. (1992) Cytogenetic results from the U.S. collaborative study on CVS. *Prenat. Diagn.* **12**, 317–355.

105. Wang, B.T., Rubin, C.H., and Williams, J. (1993) Mosaicism in chorionic villus sampling: an analysis of incidence and chromosomes involved in 2612 consecutive cases. *Prenat. Diagn.* **13**, 179–190.

106. Wang, B.T., Peng, W., Cheng, K.T., Chiu, S-F., Ho, W., Khan, Y., Wittman, M., and Williams, J. (1994) Chorionic villi sampling: laboratory experience with 4,000 consecutive cases. *Am. J. Med. Genet.* **53**, 307–316.

107. Smith, K., Gregson, N.M., Howell, R.T., Pearson, J., and Wolstenholme, J. (1994) Cytogenetic analysis of chorionic villi for prenatal diagnosis: an ACC collaborative study of U.K. data. *Prenat. Diagn.* **14**, 363–379.

108. Wolstenholme, J., Rooney, D.E., and Davison, E.V. (1994) Confined placental mosaicism, IUGR, and adverse pregnancy outcome: a controlled retrospective U.K. collaborative survey. *Prenat. Diagn.* **14**, 345–361.

109. Kalousek, D.K., Howard-Peebles, P.N., Olson, S.B., Barrett, I.J., Dorfmann, A., Black, S.H., Schulman, J.D., and Wilson, R.D. (1991) Confirmation of CVS mosaicism in term placentae and high frequency of intrauterine growth retardation association with confined placental mosaicism. *Prenat. Diagn.* **11**, 743–750.

110. Hahnemann, J.M. and Vejerslev, L.O. (1997) European collaborative research on mosaicism in CVS (EUCROMIC)—fetal and extrafetal cell lineages in 192 gestations with CVS mosaicism involving single autosomal trisomy. *Am. J. Med. Genet.* **70**, 179–187.

111. Schuring-Blom, G.H., Keijzer, M., Jakobs, M.E., Van Den Brande, D.M., Visser, H.M., Wiegant, J., Hoovers, J.M.N., and Leschot, N.J. (1993) Molecular cytogenetic analysis

of term placentae suspected of mosaicism using fluorescence *in situ* hybridization. *Prenat. Diagn.* **13,** 671–679.

112. Henderson, K.G., Shaw, T.E., Barrett, I.J., Telenius, A.H.P., Wilson, R.D., and Kalousek, D.K. (1996) Distribution of mosaicism in human placentae. *Hum. Genet.* **97,** 650–654.

112a. Schubert, R., Raff, R., and Schwanitz, G. (1996) Molecular-cytogenetic investigations of ten term placentae in cases of prenatally diagnosed mosaicism. *Prenat. Diagn.* **16,** 907–913.

113. Klein, J., Graham, J.M., Platt, L.D., and Schreck, R. (1994) Trisomy 8 mosaicism in chorionic villus sampling: case report and counselling issues. *Prenat. Diagn.* **14,** 451–454.

114. Weiner, C.P. and Okamura, K. (1996) Diagnostic fetal blood sampling—technique related losses. *Fetal Diagn. Ther.* **11,** 169–175.

115. Orlandi, F., Damiani, G., Jakil, C., Lauricella, S., Bertolino, O., and Maggio, A. (1990) The risks of early cordocentesis (12–21 weeks): analysis of 500 procedures. *Prenat. Diagn.* **10,** 425–428.

116. Weiner, C.P. (1988) Cordocentesis. *Obstet. Gynecol. Clin. North Am.* **15,** 283–301.

117. Boulot, P., Deschamps, F., Lefort, G., Sarda, P., Mares, P., Hedon, B., Laffargue, F., and Viala, J.L. (1990) Pure fetal blood samples obtained by cordocentesis: technical aspects of 322 cases. *Prenat. Diagn.* **10,** 93–100.

118. Maxwell, D.J., Johnson, P., Hurley, P., Neales, K., Allan, L., and Knott, P. (1991) Fetal blood sampling and pregnancy loss in relation to indication. *Br. J. Obstet. Gynaecol.* **98,** 892–897.

119. Watson, M.S., Breg, W.R., Hobbins, J.C., and Mahoney, M.J. (1984) Cytogenetic diagnosis using midtrimester fetal blood samples: application to suspected mosaicism and other diagnostic problems. *Am. J. Med. Genet.* **19,** 805–813.

120. Shalev, E., Zalel, Y., Weiner, E., Cohen, H., and Shneur, Y. (1994) The role of cordocentesis in assessment of mosaicism found in amniotic fluid cell culture. *Acta Obstet. Gynecol. Scand.* **73,** 119–122.

121. Liou, J.D., Chen, C-P., Breg, W.R., Hobbins, J.C., Mahoney, M.J., and Yang-Feng, T.L. (1993) Fetal blood sampling and cytogenetic abnormalities. *Prenat. Diagn.* **13,** 1–8.

122. Gosden, C., Nicolaides, K.H., and Rodeck, C.H. (1988) Fetal blood sampling in investigation of chromosome mosaicism in amniotic fluid cell culture. *Lancet* **i,** 613–617.

123. Kaffe, S., Benn, P., and Hsu, L.Y.F. (1988) Fetal blood sampling in investigation of chromosome mosaicism in amniotic fluid cell culture. *Lancet* **ii,** 284.

123a. Hook, E.B. and Crosse, P.K. (1979) Risk of chromosomally normal women to deliver chromosomally abnormal offspring, by maternal age. *Am. J. Hum. Genet.* **31,** 137A.

124. Paladini, D., Calabro, R., Palmieri, S., and D'Andrea, T. (1993) Prenatal diagnosis of congenital heart disease and fetal karyotyping. *Obstet. Gynecol.* **81,** 679–682.

125. Bronshtein, M., Zimmer, E.Z., Gerlis, L.M., Lorber, A., and Drugan, A. (1993) Early ultrasound diagnosis of fetal congenital heart defects in high-risk and low-risk pregnancies. *Obstet. Gynecol.* **82,** 225–229.

126. Copel, J.A., Cullen, M., Green, J.J., Mahoney, M.J., Hobbins, J.C., and Kleinman, C.S. (1988) The frequency of aneuploidy in prenatally diagnosed congenital heart disease: an indication for fetal karyotyping. *Am. J. Obstet. Gynecol.* **158,** 409–412.

127. Raymond, F.L., Simpson, J.M., Mackie, C.M., and Sharland, G.K. (1997) Prenatal diagnosis of 22q11 deletions: a series of five cases with congenital heart defects. *J. Med. Genet.* **34,** 679–682.

128. Johnson, M.C., Hing, A., Wood, M.K., and Watson, M.S. (1997) Chromosome abnormalities in congenital heart disease. *Am. J. Med. Genet.* **70,** 292–298.

129. Schechter, A.G., Fakhry, J., Shapiro, L.R., and Gewitz, M.H. (1987) *In utero* thickening of the chordae tendinae: a cause of intracardiac echogenic foci. *J. Ultrasound Med.* **6,** 691–695.

130. Levy, D.W. and Minitz, M.C. (1988) The left ventricular echogenic focus: a normal finding. *Am. J. Roentgenol.* **150,** 85–86.
131. Twining, P. (1993) Echogenic foci in the fetal heart: incidence and association with chromosomal disease (abstract). *Ultrasound Obstet. Gynecol.* **190,** 175.
132. Petrikovsky, B.M., Challenger, M., and Wyse, L.J. (1995) Natural history of echogenic foci within ventricles of the fetal heart. *Ultrasound Obstet. Gynecol.* **5,** 92–94.
133. Bronshtein, M., Jakobi, P., and Ofir, C. (1996) Multiple fetal intracardiac echogenic foci: not always a benign sonographic finding. *Prenat. Diagn.* **16,** 131–135.
134. Bromley, B., Lieberman, E., Laboda, L., and Benacerraf, B.R. (1995) Echogenic intracardiac focus: a sonographic sign for fetal Down syndrome. *Obstet. Gynecol.* **86,** 998–1001.
135. Norton, M.E., Brown, P., and Ashour, A.M. (1997) Echogenic focus in the fetal heart as a risk factor for Down syndrome. *Am. Coll Med Genet Fourth Annual Meeting* **A80,** 159.
136. Benacerraf, B.R., Mandell, J., Estroff, J.A., Harlow, B.L., and Frigoletto, F.D. (1990) Fetal pyelectasis: a possible association with Down syndrome. *Obstet. Gynecol.* **76,** 58–60.
137. Wickstrom, E.A., Thangavelu, M., Parilla, B.V., Tamura, R.K., and Sabbagha, R.E. (1996) A prospective study of the association between isolated fetal pyelectasis and chromosomal abnormality. *Obstet. Gynecol.* **88,** 379–382.
138. Corteville, J.E., Dicke, J.M., and Crane, J.P. (1992) Fetal pyelectasis and Down syndrome: is genetic amniocentesis warranted? *Obstet. Gynecol.* **79,** 770–772.
139. Vintzileos, A.M. and Egan, J.F. (1995) Adjusting the risk for trisomy 21 on the basis of second-trimester ultrasonography. *Am. J. Obstet. Gynecol.* **172,** 837–844.
140. Degani, S., Leibovitz, Z., Shapiro, I., Gonen, R., and Ohel, G. (1997) Fetal pyelectasis in consecutive pregnancies: a possible genetic predisposition. *Ultrasound Obstet. Gynecol.* **10,** 19–21.
141. Johnson, C.E., Elder, J.S., Judge, N.E., Adeeb, F.N., Grisoni, E.R., and Fattlar, D.C. (1992) The accuracy of antenatal ultrasonography in identifying renal abnormalities. *Am. J. Dis. Child.* **146,** 1181–1184.
142. Chudleigh, P., Pearce, M., and Campbell, S. (1984) The prenatal diagnosis of transient cysts of the fetal choroid plexus. *Prenat. Diagn.* **4,** 135–137.
143. Thorpe-Beeston, J.G., Gosden, C.M., and Nicolaides, K.H. (1990) Choroid plexus cysts and chromosomal defects. *Br. J. Radiol.* **63,** 783–786.
144. Montemagno, R., Soothill, P.W., Scarcelli, M., and Rodeck, C.H. (1995) Disappearance of fetal choroid plexus cysts during the second trimester in cases of chromosomal abnormality. *Br. J. Obstet. Gynaecol.* **102,** 752–753.
145. Nicolaides, K.H., Rodeck, C.H., and Gosden, C.M. (1986) Rapid karyotyping in nonfetal malformations. *Lancet* **i,** 283–287.
146. Shields, L.E., Uhrich, S.B., Easterling, T.R., Cyr, D.R., and Mack, L.A. (1996) Isolated fetal choroid plexus cysts and karyotype analysis: is it necessary? *J. Ultrasound Med.* **15,** 389–394.
147. Morcos, C.L., Carlson, D.E., and Platt, L.D. (1997) Choroid plexus cysts and the risk of aneuploidy. *Am. J. Obstet. Gynecol.* (abstract) **213,** S70.
148. Porto, M., Murata, Y., Warneke, L.A., and Keegan, K.A. (1993) Fetal choroid plexus cysts: an independent risk factor for chromosomal anomalies. *J. Clin. Ultrasound* **21,** 103–108.
149. Kupferminc, M.J., Tamura, R.K., Sabbagha, R.E., Parilla, B.V., Cohen, L.S., and Pergament, E. (1994) Isolated choroid plexus cyst(s): an indication for amniocentesis. *Am. J. Obstet. Gynecol.* **171,** 1068–1071.
150. Walkinshaw, S., Pilling, D., and Spriggs, A. (1994) Isolated choroid plexus cysts—the need for routine offer of karyotyping. *Prenat. Diag.* **14,** 663–667.

151. Gupta, J.K., Cave, M., Lilford, R.J., Farrell, T.A., Irving, H.C., Mason, G., and Hau, C.M. (1995) Clinical significance of fetal choroid plexus cysts. *Lancet* **346,** 724–729.

152. Gross, S.J., Shulman, L.P., Tolley, E.A., Emerson, D.S., Felker, R.E., Simpson, J.L., and Elias, S. (1995) Isolated fetal choroid plexus cysts and trisomy 18: a review and meta-analysis. *Am. J. Obstet. Gynecol.* **172,** 83–87.

153. Hook, E.B. (1993) Choroid plexus cysts diagnosed prenatally as an independent risk factor for cytogenetic abnormality. *Hum. Genet.* **91,** 514–518.

154. Nyberg, D.A., Kramer, D., Resta, R.G. Kapur, R., Mahony, B.S., Luthy, D.A., and Hickok, D. (1993) Prenatal sonographic findings of trisomy 18: review of 7 cases. *J. Ultrasound Med.* **12,** 103–113.

155. Sarno, A.P., Polzin, W.J., and Kalish, V.B. (1993) Fetal choroid plexus cysts in association with cri du chat (5p–) syndrome. *Am. J. Obstet. Gynecol.* **169,** 1614–1615.

156. Meyer, P., Chitkara, U., Holbrook, R.H., El-Sayed, Y., Druzin, M., and Tung, R. (1997) Complex choroid plexus cysts and the risk of aneuploidy. *Am. J. Obstet. Gynecol.* (abstract) **215,** S70.

157. DiGiovanni, L.M., Quinlan, M.P., and Verp, M.S. (1997) Choroid plexus cysts: infant and early childhood developmental outcome. *Obstet. Gynecol.* **90,** 191–194.

158. Landwehr, Jr., J.B., Johnson, M.P., Hume, R.F., Yaron, Y., Sokol, R.J., and Evans, M.I. (1996) Abnormal nuchal findings on screening ultrasonography: aneuploidy stratification on the basis of ultrasonographic anomaly and gestational age at detection. *Am. J. Obstet. Gynecol.* **175,** 995–999.

159. Grandjean, H. and Sarramon, M.F. (1995) Sonographic measurement of nuchal skinfold thickness for detection of Down syndrome in the second-trimester fetus: a multicenter prospective study. *Obstet. Gynecol.* **85,** 103–106.

160. Nadel, A., Bromley, B., and Benacerraf, B.R. (1993) Nuchal thickening or cystic hygromas in first- and early second-trimester fetuses: prognosis and outcome. *Obstet. Gynecol.* **82,** 43–48.

161. van Vugt, J.M.G., van Zalen-Sprock, R.M., and Kostense, P.J. (1996) First-trimester nuchal translucency: a risk analysis on fetal chromosome abnormality. *Radiology* **200,** 537–540.

162. Pandya, P.P., Kondylios, A., Hilbert, L., Snijders, R.J.M., and Nicolaides, K.H. (1995) Chromosomal defects and outcome in 1015 fetuses with increased nuchal translucency. *Ultrasound Obstet. Gynecol.* **5,** 15–19.

163. Comas, C., Martinez, J.M., Ojuel, J., Casals, E., Puerto, B., Borrell, A., and Fortuny, A. (1995) First-trimester nuchal edema as a marker of aneuploidy. *Ultrasound Obstet. Gynecol.* **5,** 26–29.

164. Szabó, J., Gellén, J., and Szemere, G. (1995) First-trimester ultrasound screening for fetal aneuploidies in women over 35 and under 35 years of age. *Ultrasound Obstet. Gynecol.* **5,** 161–163.

165. Brizot, M.L., Snijders, R.J.M., Butler, J., Bersinger, N.A., and Nicolaides, K.H. (1995) Maternal serum hCG and fetal nuchal translucency thickness for the prediction of fetal trisomies in the first trimester of pregnancy. *Br. J. Obstet. Gynaecol.* **102,** 127–132.

166. Chen, C-P., Liu, F-F., Jan, S-W., Lee, C-C, Town, D-D., and Lan, C-C. (1996) Cytogenetic evaluation of cystic hygroma associated with hydrops fetalis, oligohydramnios or intrauterine fetal death: the roles of amniocentesis, postmortem chorionic villus sampling and cystic hygroma paracentesis. *Acta Obstet. Gynecol. Scand.* **75,** 454–458.

167. Vintzileos, A.M., Egan, J.F.X., Smulian, J.C., Campbell, W.A., Guzman, E.R., and Rodis, J.F. (1996) Adjusting the risk for trisomy 21 by a simple ultrasound method using fetal long-bone biometry. *Obstet. Gynecol.* **87,** 953–958.

168. Shah, Y.G., Eckl, C.J., Stinson, S.K., and Woods, J.R. (1990) Biparietal diameter/femur length ratio, cephalic index, and femur length measurements: not reliable screening techniques for Down syndrome. *Obstet. Gynecol.* **75,** 186–188.

169. Benacerraf, B.R., Nadel, A., and Bromley, B. (1994) Identification of second-trimester fetuses with autosomal trisomy by use of a sonographic scoring index. *Radiology* **193,** 135–140.

170. Lai, F.M. and Yeo, G.S. (1995) Reference charts of foetal biometry in Asians. *Singapore Med. J.* **36,** 628–636.

171. Scioscia, A.L., Pretorius, D.H., Budorick, N.E., Cahill, T.C., Axelrod, F.T., and Leopold, G.R. (1992) Second-trimester echogenic bowel and chromosomal abnormalities. *Am. J. Obstet. Gynecol.* **167,** 889–894.

172. Nyberg, D.A., Dubinsky, T., Resta, R.G,. Mahony, B.S., Hickok, D.E., and Luthy, D.A. (1993) Echogenic fetal bowel during the second trimester: clinical importance. *Radiology* **188,** 527–531.

173. Sepulveda, W., Hollingsworth, J., Bower, S., Vaughan, J.I., and Fisk, N.M. (1994) Fetal hyperechogenic bowel following intra-amniotic bleeding. *Obstet. Gynecol.* **83,** 947–950.

174. Bromley, B., Doubilet, P. Frigoletto, F.D., Krauss, C., Estroff, J.A., and Benacerraf, B.R. (1994) Is fetal hyperechoic bowel on second-trimester sonogram an indication for amniocentesis? *Obstet. Gynecol.* **83,** 647–651.

175. Yaron, Y., Hassan, S., Kramer, R.L. Zador, I., Ebrahim, S.A.D., Johnson, M.P., and Evans, M.I. (1997) Fetal echogenic bowel in the second trimester—prognostic implication. *Am. J. Obstet. Gynecol.* **176,** 209.

176. Muller, F., Dommergues, M., Aubry, M.-C., Simon-Bouy, B., Gautier, E., Qury, J.-F., and Narcy, F. (1995) Hyperechogenic fetal bowel: an ultrasonographic marker for adverse fetal and neonatal outcome. *Am. J. Obstet. Gynecol.* **173,** 508–513.

177. MacGregor, S.N., Tamura, R., Sabbagha, R., Brenhofer, J.K., Kambich, M.P., and Pergament, E. (1995) Isolated hyperechoic fetal bowel: significance and implications for management. *Am. J. Obstet. Gynecol.* **173,** 1254–1258.

178. Pletcher, B.A., Williams, M.K., Mulivor, R.A., Barth, D., Linder, C., and Rawlinson, K. (1991) Intrauterine cytomegalovirus infection presenting as fetal meconium peritonitis. *Obstet. Gynecol.* **78,** 903–905.

179. Forouzan I. (1992) Fetal abdominal echogenic mass: an early sign of intrauterine cytomegalovirus infection. *Obstet. Gynecol.* **80,** 535–537.

180. Peters, M.T., Lowe, T.W., Carpenter, A., and Kole, S. (1995) Prenatal diagnosis of congenital cytomegalovirus infection with abnormal triple-screen results and hyperechoic fetal bowel. *Am. J. Obstet. Gynecol.* **173,** 953–954.

181. Slotnick, R.N. and Abuhamad, A.Z. (1996) Prognostic implications of fetal echogenic bowel. *Lancet* **347,** 85–87.

182. Achiron, R., Seidman, D.S., Horowitz, A., Mashiach, S., Goldman, B., and Lipitz, S. (1996) Hyperechogenic bowel and elevated serum alpha-fetoprotein: a poor fetal prognosis. *Obstet. Gynecol.* **88,** 368–371.

183. Sepulveda, W., Bower, S., and Fisk, N.M. (1995) Third-trimester hyperechogenic bowel in Down syndrome. *Am. J. Obstet. Gynecol.* **172,** 210–211.

184. Sepulveda, W., Reid, R., Nicolaidis, P., Prendiville, O., Chapman, R.S., and Fisk, N.M. (1996) Second-trimester echogenic bowel and intraamniotic bleeding: association between fetal bowel echogenicity and amniotic fluid spectrophotometry at 410 nm. *Am. J. Obstet. Gynecol.* **174,** 839–842.

185. DeVore G. and Alfi, O. (1995) The use of color Doppler ultrasound to identify fetuses at increased risk for trisomy 21: an alternative for high risk patients who decline genetic amniocentesis. *Obstet. Gynecol.* **85,** 378–386.

186. Bahado-Singh, R.O., Deren, Ö., Tan, A., D'Ancona, R.L., Hunter, D., Copel, J.A., and Mahoney, M.J. (1996) Ultrasonographically adjusted midtrimester risk of trisomy 21 and significant chromosomal defects in advanced maternal age. *Am. J. Obstet. Gynecol.* **175,** 1563–1568.

187. Vintzileos, A.M., Campbell, W.A., Guzman, E.R., Smulian, J.C., McLean, D.A., and

Anath, C.V. (1997) Second-trimester ultrasound markers for detection of trisomy 21: which markers are best? *Obstet. Gynecol.* **89,** 941–944.

188. Feuchtbaum, L.B., Cunningham, G., Waller, D.K., Lustig, L.S., Tompkinson, D.G., and Hook, E.B. (1995) Fetal karyotyping for chromosome abnormalities after an unexplained elevated maternal serum alpha-fetoprotein screening. *Obstet. Gynecol.* **86,** 248–254.

189. Merkatz, I.R., Nitowsky, H.M., Macri, J.N., and Johnson, W.E. (1984) An association between low maternal serum alpha-fetoprotein and fetal chromosomal abnormalities. *Am. J. Obstet. Gynecol.* **148,** 886–894.

190. Bogart, M.H., Pandian, M.R., and Jones, O.W. (1987) Abnormal maternal serum chorionic gonadotropin levels in pregnancies with fetal chromosome abnormalities. *Prenat. Diagn.* **7,** 623–630.

191. Canick, J.A., Knight, G.J., Palomaki, G.E., Haddow, J.E., Chuckle, H.S., and Wald, N.J. (1988) Low second trimester maternal serum unconjugated oestriol in pregnancies with Down's syndrome. *Br. J. Obstet. Gynaecol.* **95,** 330–333.

192. Reynolds, T.M., Nix, A.B., Dunstan, F.D., and Dawson, A.J. (1993) Age-specific detection and false-positive rates: an aid to counseling in Down syndrome risk screening. *Obstet. Gynecol.* **81,** 447–450.

193. Palomaki, G.E., Haddow, J.E., Knight, G.J., Wald, N.J., Kennard, A., Canick, J.A., Saller, D.N., Blitzer, M.G., Dickerman, L.H., Fisher, R., Hansmann, D., Hansmann, M., Luthy, D.A., Summers, A.M., and Wyatt, P. (1995) Risk-based prenatal screening for trisomy 18 using alpha-fetoprotein, unconjugated oestriol and human chorionic gonadotropin. *Prenat. Diagn.* **15,** 713–723.

194. Benn, P.A., Horne, D., Briganti, S., and Greenstein, R.M. (1995) Prenatal diagnosis of diverse chromosome abnormalities in a population of patients identified by triple-marker testing as screen positive for Down syndrome. *Am. J. Obstet. Gynecol.* **173,** 496–501.

195. Palomaki, G.E., Knight, G.J., McCarthy, J.E., Haddow, J.E., and Donhowe, J.M. (1997) Maternal serum screening for Down syndrome in the United States: a 1995 survey. *Am. J. Obstet. Gynecol.* **176,** 1046–1051.

196. Krantz, D.A., Larsen, J.W., Buchanan, P.D., and Macri, J.N. (1996) First-trimester Down syndrome screening: free β-human chorionic gonadotropin and pregnancy-associated plasma protein A. *Am. J. Obstet. Gynecol.* **174,** 612–616.

197. Wald, N.J., Kennard, A., and Hackshaw, A.K. (1995) First trimester serum screening for Down's syndrome. *Prenat. Diagn.* **15,** 1227–1240.

198. Gardner, R.J.M. and Sutherland, G.R. (1996) *Chromosome Abnormalities and Genetic Counselling,* 2nd edition, Oxford University Press, New York.

199. Schwarz, R. and Johnston, R.B. (1996) Folic acid supplementation—when and how. *Obstet. Gynecol.* **88,** 886–887.

200. Seller, M.J. (1995) Neural tube defects, chromosome abnormalities and multiple closure sites for the human neural tube. *Clin. Dysmorphol.* **4,** 202–207.

201. Daniel, A., Hook, E.B., and Wulf, G. (1989) Risks of unbalanced progeny at amniocentesis to carriers of chromosome rearrangements: data from United States and Canadian laboratories. *Am. J. Med. Genet.* **33,** 14–53.

202. Kaiser, P. (1988) Pericentric inversions: their problems and clinical significance. In *The Cytogenetics of Mammalian Autosomal Rearrangements.* Alan R. Liss, New York, pp. 163–210.

203. Madan, K. (1995) Paracentric inversions: a review. *Hum. Genet.* **96,** 503–515.

204. Grass, F., McCombs, J., Scott, C.I., Young, R.S., and Moore, C.M. (1984) Reproduction in XYY males: two new cases and implications for genetic counseling. *Am. J. Med. Genet.* **19,** 553–560.

205. Shurtleff, D.B. and Lemire, R.J. (1995) Epidemiology, etiologic factors, and prenatal diagnosis of open spinal dysraphism. *Neurosurg. Clin. North Am.* **6,** 183–193.

206. Lindhout, D., Omtzigt, J.G., and Cornel, M.C. (1992) Spectrum of neural-tube defects in 34 infants prenatally exposed to antiepileptic drugs. *Neurology* **42,** 111–118.

207. Omtzigt, J.G., Los, F.J., Hagenaars, A.M., Stewart, P.A., Sachs, E.S., and Lindhout, D. (1992) Prenatal diagnosis of spina bifida aperta after first-trimester valproate exposure. *Prenat. Diagn.* **12,** 893–897.

208. Ferguson-Smith, M.A. and Yates, J.R.W. (1984) Maternal age specific rates for chromosome aberrations and factors influencing them: report of a collaborative European study on 52965 amniocenteses. *Prenat. Diagn.* **4,** 5–44.

209. Roecker, G.O. and Huether, C.A. (1983) An analysis for paternal-age effect in Ohio's Down syndrome births, 1970–1980. *Am. J. Hum. Genet.* **35,** 1297–1306.

210. Hook, E.B. and Regal, R.R. (1984) A search for a paternal-age effect upon cases of 47,+21 in which the extra chromosome is of paternal origin. *Am. J. Hum. Genet.* **36,** 413–421.

211. Roth, M.-P. and Stoll, C. (1983) Paternal age and Down's syndrome diagnosed prenatally: no association in French data. *Prenat. Diagn.* **3,** 327–335.

212. Orioli, I.M., Castilla, E.E., Scarano, G., and Mastroiacova, P. (1995) Effect of paternal age in achondroplasia, thanatrophoric dysplasia, and osteogenesis imperfecta. *Am. J. Med. Genet.* **59,** 209–217.

212a. Worton, R.G. and Stern, R. (1984) A Canadian collaborative study of mosaicism in amniotic fluid cell cultures. *Prenat. Diagn.* **4,** 131–144.

212b. Bui, T.-H., Iselius, L., and Lindsten, J. (1984) European collaborative study on prenatal diagnosis: mosaicism, pseudomosaicism and single abnormal cells in amniotic fluid cell cultures. *Prenat. Diagn.* **4,** 145–162.

212c. Hsu, L.Y.F. (1984) United States survey on chromosome mosaicism and pseudomosaicism in prenatal diagnosis. *Prenat. Diagn.* **4,** 97–130.

213. Robinson, W.P., Binkert, F., Bernasconi, F., Lorda-Sanchez, I., Werder, E.A., and Schinzel, A.A. (1995) Molecular studies of chromosomal mosaicism: relative frequency of chromosome gain or loss and possible role of cell selection. *Am. J. Hum. Genet.* **56,** 444–451.

214. Stoll, C., Chognot, D., Halb, A., and Luckel, J.C. (1992) Trisomy 9 mosaicism in two girls with multiple congenital malformations and mental retardation. *J. Med. Genet.* **30,** 433–435.

215. Guichet, A., Briault, S., Toutain, A., Paillet, C., Descamps, P., Pierres, F., Body, G., and Moraine, C.L. Prenatal diagnosis of trisomy 8 mosaicism in CVS after abnormal ultrasound findings at 12 weeks. *Prenat. Diagn.* **15,** 769–772.

216. Brosens, J., Overton, C., Lavery, S.A., and Thornton, S. (1996) Trisomy 12 mosaicism diagnosed by amniocentesis. *Acta Obstet. Gynecol. Scand.* **75,** 79–81.

216a. Hsu, L.Y.F., Yu, M.-F., Neu, R.L., Van Dyke, D.L., Benn, P.A., Bradshaw, C.L., Shaffer, L.G., Higgins, R.R., Khodr, G.S., Morton, C.C., Wang, H., Brothman, A.R., Chadwick, D., Disteche, C.M., Jenkins, L.G., Kalousek, D.K., Pantzar, T.J., and Wyatt P. (1997) Rare trisomy mosaicism diagnosed in amniocytes, involving an autosome other than chromosomes 13, 18, 20, and 21: karyotype/phenotype correlations. *Prenat. Diagn.* **17,** 201–242.

217. Persutte, W.H. and Lenke, R.R. (1995) Failure of amniotic-fluid-cell growth: is it related to fetal aneuploidy? *Lancet* **14,** 96–97.

218. Nuss, S., Brebaum, D., and Grond-Ginsbach, C. (1994) Maternal cell contamination in amniotic fluid samples as a consequence of the sampling technique. *Hum. Genet.* **93,** 121–124.

219. Elias, S. and Simpson, J.L. (1994) Techniques and safety of genetic amniocentesis and chorionic villus sampling. In *Diagnostic Ultrasound Applied to Obstetrics and Gynecology,* 3rd edition (Sabbagha, R.A., ed.), Raven-Lippincott, Philadelphia, p. 113.

12

Cytogenetics of Spontaneous Abortion

Solveig M.V. Pflueger, Ph.D., M.D.

INTRODUCTION

Pregnancy loss is quite common, with 15–20% of recognized pregnancies resulting in failure. The majority of these occur early in gestation, though losses in the second and third trimester are not rare. Approximately 2–5% of women will experience two or more losses. The majority of pregnancy failures are associated with cytogenetic abnormalities, with more than 50% of early miscarriages and as many as 5% of stillbirths exhibiting abnormal karyotypes.

Loss of a wanted pregnancy is always stressful for both the patient and her partner. A number of questions and concerns may be raised regarding the loss, including: What happened and why? How likely is it to happen again? What can be done to improve the chances of a successful future pregnancy? Is this even possible? Answering such concerns is important in helping the patient through the grieving process and in facilitating resolution. The answers that are provided may ultimately impact family planning and management of any future pregnancies the couple may undertake.

Unfortunately, early pregnancy losses are often given less attention than they merit, both by medical care providers and by society. The patient who loses an older child or who experiences a stillbirth at term can expect an attempt at explanation for the loss from her health care provider. She will also be offered sympathy and support from family and friends. Rituals associated with mourning and disposition of the remains help bring closure. However, the patient who experiences an early loss often feels isolated and alone. Her friends may be uncomfortable with discussing the loss, if they are even aware of it, and so may avoid the issue altogether. She may have been told by her caregiver that such early losses are common and that there is no reason she cannot have a successful pregnancy in the future, but this does not explain why the loss happened to her and usually does little to alleviate her sense of guilt and failure. These feelings of inadequacy are often amplified in the patient with recurrent losses *(1–5)*. Answering the patient's questions, whether verbalized or not, will help bring about closure to the loss and may open dialog with the patient and her partner about their specific concerns for the future. This in turn may have significant impact on management of future pregnancies. Thus, anyone caring for women of childbearing potential should be familiar with the causes and recurrence risks for pregnancy loss.

From: *The Principles of Clinical Cytogenetics*
Edited by: S. Gersen and M. Keagle © Humana Press Inc., Totowa, NJ

Table 1
Intrauterine Mortality per 100 Ova Exposed to Fertilization

Week after ovulation	Embryonic demise	Survivors
—	16 (not fertilized)	100
0	15	84
1	27	69
2	5.0	42
6	2.9	37
10	1.7	34.1
14	0.5	32.4
18	0.3	31.9
22	0.1	31.6
26	0.1	31.5
30	0.1	31.4
34	0.1	31.3
38	0.2	31.32

| Live births | 31 |
| Natural wastage | 69 |

Data from ref. *12*.

THE SCOPE OF THE PROBLEM

When examining the chances for success of a given conceptus, the results of human reproduction are quite poor. Approximately 78% of all conceptions fail to go to term *(6)*. Combined data from three studies of women attempting pregnancy revealed a postimplantation loss rate of 42% in documented conceptions confirmed by positive human chorionic gonadotrophin (hCG) levels *(7–9)*. A 4-year follow-up of 3084 pregnancies demonstrated a 23.7% loss rate following the first missed period *(10)*. The net overall fecundity for patients 20–30 years of age has been estimated at 21–28% *(11)*, a level that is quite low compared with most other mammalian species. Leridon *(12)* provides a useful summary table of pregnancy survival from fertilization to term, with only 31 survivors per 100 ova exposed to fertilization (**Table 1**). Although most of the losses occur vary early in gestation, losses continue to occur throughout the second and third trimesters of pregnancy with a slight increase in mortality at term.

RELATIONSHIP BETWEEN CYTOGENETIC
ABNORMALITIES AND GESTATIONAL AGE

Multiple studies have suggested that approximately 50% of early pregnancy losses are associated with cytogenetic abnormalities. Evaluation of 1205 pregnancy losses of varying gestational ages submitted to the author's laboratory between 1992 and 1996 revealed 539 (45%) cases with identified cytogenetic abnormalities *(13)*. The likelihood of a cytogenetic abnormality varies with the gestational age and morphology of the abortus. In evaluating products of conception, the development age at which growth arrest occurred is a more useful parameter than gestational age at the time of miscar-

Table 2
Chromosomal Abnormalities and Gestational Age

Gestational age (weeks)	Number of cases	Abnormal cases	Percent abnormal
2	23	18	78.0%
3	374	258	69.0%
4	203	125	61.6%
5	139	85	62.2%
6	302	211	69.9%
7	56	27	48.2%
Total weeks 2–7	**1097**	**724**	**66.0%**
8	36	8	22.2%
9	42	6	14.3%
10	14	7	50.0%
11	8	1	12.5%
12	8	3	37.5%
Total weeks 8–12	**108**	**25**	**23%**
Total	**1205**	**749**	**62.2%**

Adapted from Boué et al. *(14)*.

Table 3
Abnormal Development and Gestational Age

	4 Weeks or less		5–8 Weeks		9–12 Weeks	
Study	Normal	Abnormal	Normal	Abnormal	Normal	Abnormal
Milamo	0	48	21	40	71	10
Miller and Poland	10	73	51	71	121	56
Total	10 (8%)	121 (92%)	72 (39%)	111 (61%)	192 (74%)	66 (26%)

Adapted from Boué et al. *(14)*.

riage, as products of conception are often retained *in utero* for several weeks following embryonic demise.

Overall, the earlier the developmental age, the greater the likelihood of an abnormal karyotype in a spontaneous pregnancy loss. Boué and colleagues *(14)* found that approximately two thirds of losses under 8 weeks and nearly one fourth of those between 8 and 12 weeks had abnormal karyotypes (**Table 2**).

It is also of interest to note that the earlier the pregnancy undergoes growth arrest, the more likely it is for there to be anomalous development and that there will be an abnormal karyotype (**Table 3**).

Gestational Age

Examination of induced abortuses confirms the greater incidence of karyotypic abnormalities earlier in pregnancy *(15)* (**Table 4**). A total of 1197 pregnancies were examined. The rate of chromosomal abnormality varied with gestational age; 9.3% of cases were abnormal at 3–4 weeks, falling to 5.4% at 9–10 weeks.

Table 4
Chromosomal Abnormalities in Induced Abortuses

Developmental age	Number of cases	Abnormal cases (percent)
3–4 Weeks	108	10 (9.3%)
5–6 Weeks	570	37 (6.5%)
7–8 Weeks	389	25 (6.4%)
9–10 Weeks	130	7 (5.4%)

Data from ref. *15.*

The likelihood of detecting congenital anomalies in therapeutic terminations is variable and may be a reflection of the thoroughness of the examination and the skill of the examiner. However, identification of anomalous development may have considerable impact upon future reproduction, and it is the opinion of the author that careful anatomic evaluation of aborted products of conception should be considered regardless of whether the pregnancy is aborted spontaneously or induced.

In the second trimester, ascending infection becomes more frequent as a cause of spontaneous pregnancy loss. Abnormal karyotypes become less prevalent as pregnancy progresses since many of the less viable abnormal gestations have already undergone growth arrest and miscarriage. Gaillard et al. *(16)* studied 422 consecutive second trimester losses. Of these, 78.6% were recent demises without extensive maceration. Ascending infection could explain 85% of these. Structural anomalies were seen in 7/6% of fetuses. Cytogenetic abnormalities were confirmed or suspected in half of these. The majority of abnormal fetuses showed maceration consistent with longstanding intrauterine fetal demise. This again confirms the observation that cytogenetic abnormalities are associated with early demise, but that there is also frequent retention of the products of conception for some time prior to spontaneous abortion. The macerated fetus is at significant risk for chromosomal abnormality whereas the fresh fetal demise without gross congenital anomalies is more often due to other etiologies including but not limited to infection, endocrine disorders, abnormal uterine anatomy, and immunologic factors.

Although cytogenetic abnormalities are frequent in early pregnancy, they are much less common at term. Approximately one in 200 live newborns exhibits readily identified aneuploid karyotypes, and one study estimates that with moderate levels of banding, 0.061% of infants will show unbalanced structural abnormalities and 0.522% will harbor balanced rearrangements *(17)*. The rate of unbalanced karyotypes showing numerical or structural abnormality is much higher in stillbirths, approximating 5–7% overall. Here, too, the risk is greatest for macerated stillbirths, especially in the presence of congenital anomalies. Cytogenetic abnormalities and associated congenital anomalies are also a significant factor in neonatal deaths.

The likelihood of survival for a pregnancy with an aberrant karyotype is a reflection of the particular cytogenetic abnormality and the extent of its deleterious effects on embryonic growth and development. Davison and Burn *(18)* examined the likelihood of loss for various chromosomal abnormalities, confirming virtual 100% loss for autosomal monosomies and tetraploids. Autosomal trisomies resulted in a 96.5% loss rate. Although greater than 99% of monosomy X pregnancies failed, only 11% of sex chro-

Table 5
Prenatal Loss of Chromosomally Abnormal Fetuses

Autosomal monosomy	100.0%
Tetraploid	100.0%
Triploid	99.9%
Monosomy X	99.8%
Autosomal trisomy	96.5%
Mosaics	68.8%
Structural rearrangements	53.4%
Sex chromosomal trisomy	11.0%

Adapted from Davison and Burn *(18)*.

Table 6
Percent of Chromosomal Anomalies
Among Spontaneous Abortions and Live Births

Anomaly	*Spontaneous abortions*	*Live births*
Autosomal trisomies		
13	1.10%	0.01%
16	5.58%	0.00%
18	0.84%	0.02%
21	2.00%	0.11%
Other	11.81%	0.00%
total trisomies	21.33%	1.34%
Monosomy X	0.84%	0.01%
Sex chromosome trisomies	0.33%	0.15%
Triploids	5.79%	0.00%
Tetraploids	2.39%	0.00%
Total abnormal	41.52%	0.60%
Number karyotyped	3353	31521

Adapted from Kline and Stein *(19)*.

mosome trisomies were lost spontaneously. Mosaic and structurally rearranged karyotypes show intermediate loss rates of 68.8% and 53.4% respectively (**Table 5**).

Although there have been several reports of tetraploid conceptuses and near complete autosomal monosomies surviving into the third trimester, these are exceptionally rare.

Summarizing data from several series, Kline and Stein *(19)* compared the frequency of chromosomal anomalies of spontaneous abortions and live births (**Table 6**).

Summary

These data indicate that the majority of chromosomally abnormal pregnancies fail, that the losses are selective rather than random, and that the differing survival potential is dependent upon the particular cytogenetic abnormality involved.

Cytogenetic abnormalities are a significant factor in human pregnancy wastage at all stages of gestation, as well as into the neonatal period. However, the incidence of karyotypic abnormalities is greatest during early pregnancy, with the majority of aberrant gestations resulting in early spontaneous loss. Very early pregnancy loss is most likely to be the result of chromosomal abnormalities, especially when there is evidence of marked embryonic growth arrest at the time of delivery. The clinical significance of the loss and the potential impact on future reproductive risks for the couple is dependent upon the type of chromosomal error.

TYPES OF ERRORS LEADING TO CHROMOSOMALLY ABNORMAL CONCEPTUSES

Although most chromosomal abnormalities are associated with poor outcome early in pregnancy, the underlying mechanisms leading to an aberrant karyotype and the risk for recurrence vary considerably depending on the particular abnormal chromosomal complement. Generally speaking, most karyotypic abnormalities fall into one of four classes: errors in meiosis (gametogenesis), errors in mitosis leading to mosaicism, errors in fertilization, and structural abnormalities and rearrangements.

A classic study of 1498 abortuses by Boué and colleagues *(20,21)* revealed 921 abnormal karyotypes (61.5%). Among the chromosomally abnormal losses were 636 nondisjunctional events: 141 monosomies (15.3%), 479 trisomies (52.0%), and 16 double trisomies (1.7%). There were 183 triploids (19.9%), 57 tetraploids (6.2%), and 10 cases of mosaicism (1.1%). Structural abnormalities were identified in 35 abortuses (3.8%). With improved cytogenetic and molecular methods being used today the incidence of detectable abnormalities may have been even higher. However, the study clearly shows that cytogenetic abnormalities are present in the majority of early spontaneous losses, and the data provide a useful breakdown of the types of abnormalities that are observed. Normal karyotypes were seen in 577 abortuses (38%), although there may have been a few undetected underlying abnormalities such as subtle rearrangements, uniparental disomy, or tissue specific mosaicism that could have gone undetected in this sample.

Analysis of 1205 products of conception of varying gestational ages received in our laboratory between 1992 and 1996 revealed 539 (47.2%) abnormal karyotypes. Of these 50.6% were trisomies, 11.3% were monosomies, 4.2% were tetraploid, and 14.8% were triploid *(13)*. Although the total percentage of abnormal karyotypes is lower in our series than that of Boué and colleagues, this can be explained by a higher proportion of cases from later in gestation in our population, giving a greater number of losses due to nonchromosomal etiologies. The distribution of types of abnormalities among the aberrant karyotypes is similar, however.

Errors in Meiosis

During meiosis, the usual parental diploid chromosome complement of 46 is reduced to the haploid number of 23. Nondisjunctional events during meiosis I or II of either oögenesis or spermatogenesis can result in monosomic or trisomic conceptuses due to formation of a gamete with fewer or greater than the usual number of chromosomes (see Chapter 2). With the exception of monosomy X, complete, apparently nonmosaic monosomies are almost invariably lethal early in gestation and are not usually identified in recognized pregnancies. Gene dosage effects or imprinting failure may be factors contributing to the high embryonic lethality of the autosomal monosomies.

Trisomies, on the other hand, are relatively common and represent the most frequently encountered group of abnormalities leading to spontaneous pregnancy loss. Approximately 25% of karyotyped spontaneous abortions will be trisomic *(21–23)*. All autosomal trisomies have been reported in multiple studies with the unique exception of chromosome 1, trisomy for which appears to be lethal prior to implantation, and thus the conceptus would be unlikely to survive long enough to be seen in routine series of spontaneous abortions. The majority of trisomic conceptuses, even those with karyotypes that may be viable in the neonate, result in miscarriage.

Trisomies

The frequency of particular autosomal trisomies varies with gestational age. At term, trisomy 21 (Down syndrome) is the most common and is seen in approximately 1 in 700 live births. Trisomy 18 (Edward syndrome) and trisomy 13 (Patau syndrome) are seen in approximately 1 in 6000–8000 and 1 in 12,000 births, respectively. Trisomy 8 is much less frequent and most cases are mosaic. Although individual case reports indicate that other unusual autosomal trisomies and rare autosomal monosomies do occasionally occur in the neonate, these aneuploidies are typically seen only in a mosaic state and generally appear to be lethal when a normal cell line is absent.

The distribution of trisomies in spontaneous abortions is quite different from that seen at term. The most common trisomy observed in spontaneous abortuses is trisomy 16, accounting for 31.0% of trisomic conceptuses and 7.27% of all spontaneous abortions. This is followed by trisomy 22, seen in 11.4% of trisomies and 2.26% of spontaneous abortions. Trisomy 21 is third most frequent, accounting for 10.5% of trisomies and 2.11% of spontaneous abortions (**Table 7**) *(18,23)*.

Double trisomies also occur and show a strong association with advancing maternal age, even more so than the age effect seen with the viable trisomies such as Down syndrome *(14)*.

Identification of trisomic conceptuses is of clinical importance because of the question of possible increased risk for aneuploidy in subsequent pregnancies. The recurrence risk for a couple with a previous trisomic infant is often cited as approaching 1% *(24,25)*. Verp and Simpson *(26)* combined data from several smaller studies to suggest that the risk for an aneuploid liveborn following a trisomic abortus may also approximate 1%. Connor and Ferguson-Smith offer an empirical risk of 1.5% for a trisomy (not necessarily a viable trisomy) in any subsequent pregnancy following a trisomic abortus *(27)*. This raises the issue as to whether prenatal diagnosis should be offered to couples who have experienced a previous trisomic abortus.

There may be increased susceptibility to trisomic conceptuses in some patients with a history of trisomic pregnancies. The risk for these patients would be for nondisjunction in general, not for a specific trisomy. Our laboratory has seen several patients with three or more consecutive trisomies, each involving different chromosomes, suggesting a population of couples who are at significant risk for recurrence. At this time, however, it is difficult to determine which women with a trisomic abortus are more likely to experience recurrent nondisjunctional events. Thus, couples may benefit from genetic counseling following a trisomic or any other chromosomally abnormal pregnancy.

Table 7
Distribution of Individual Trisomies
Among Trisomic Spontaneous Abortions

Chromosome number	Percent of trisomies	Percent of abortuses
1	Single case report	0
2	4.0	1.11
3	0.9	0.25
4	2.4	0.64
5	0.3	0.04
6	0.9	0.14
7	4.0	0.89
8	4.6	0.79
9	2.3	0.72
10	2.0	0.36
11	0.3	0.04
12	1.2	0.18
13	4.1	1.07
14	4.8	0.82
15	7.4	1.68
16	31.0	7.27
17	0.3	0.18
18	4.6	1.15
19	0.2	0.01
20	2.2	0.61
21	10.5	2.11
22	11.4	2.26

Data from refs. *18, 23.*

The majority of autosomal trisomies are maternal in origin, with errors in meiosis I being more frequent than meiosis II, although there appears to be some variability depending upon the chromosome involved. Of 436 informative cases reviewed by Hassold and colleagues, 407 trisomies were maternal in origin *(28)*. All cases of trisomy 16 and trisomy 22 were also maternal in origin, 19% of trisomies involving chromosomes 2 through 12 were paternal in origin, and 27% of trisomies of chromosomes 13 through 15 were paternally derived. Paternal nondisjunction was also associated with sex chromosome aneuploidies, being responsible for 44% of XXY and 6% of 47 XXX conceptions.

Examination of oöcytes reveals a significant percentage of cytogenetic abnormalities. Kamiguchi and colleagues found abnormal chromosomal complements in 24.3% of unfertilized oöcytes *(29)*. Aneuploidy was most commonly observed, followed by diploidy and structural abnormalities. A review of 1559 published cases revealed chromosomal abnormalities in 24% of mature oöcyte karyotypes *(30)*. The majority were aneuploid (22.8%); fewer had structural aberrations (1.2%). The particular chromosomes involved showed an unequal distribution with an excess of D and G group aneuploidies and less than expected A and C group examples (see Chapter 3 for description of chromosome groups). It is of interest to note that only one oöcyte with an extra

Table 8
Effect of Maternal Age as Seen in Abortuses

Maternal age (years)	Number karyotyped	Percent abnormal	Percent trisomic	Percent nontrisomic
20	104	18.3	4.8	13.5
20–24	256	28.5	12.1	16.4
25–29	339	26.3	10.6	15.6
30–34	161	32.3	19.3	13.0
35–39	99	34.3	25.3	9.0
40+	32	65.6	50.0	15.6

Adapted from Creasy *(36)*.

chromosome 16 was identified, although this is the most common trisomy in spontaneous abortions. The difference in distribution of trisomies suggests that postmeiotic viability may be as significant as meiotic error in determining the incidence of particular trisomies in the human species.

Cytogenetic studies of human spermatocytes also reveal abnormalities in paternal gametogenesis. The reported studies have used several different methods for karyotype preparation. In 1987, Martin et al. reported that 3–4% of sperm exhibited aneuploidy due to nondisjunctional events, and 10% had structural abnormalities *(31)*. More recently, FISH techniques have been used, allowing for examination of far greater numbers of sperm. FISH does have inherent limitations based on the particular chromosome specific probes utilized; only the specific aneuploidies being probed for will be detected. Using FISH techniques, Miharu and colleagues analyzed 450,580 sperm from 9 fertile and 12 infertile men *(32)*. Disomy for chromosomes 1, 16, X, and Y ranged from 0.34% to 0.84% in infertile subjects and from 0.32% to 0.61% in fertile subjects. Guttenbach and colleagues examined 16,127 sperm from eight healthy donors for disomy of chromosome 18 and found a range of 0.25–0.5% *(33)*. Examination of 76,253 sperm from seven donors revealed a range of 0.31—0.34% of disomy for chromosomes 3, 7, 10, 11, 17, and X *(34)*. Although FISH studies have limitations, the data suggest that the rate of paternal meiotic nondisjunction appears relatively constant for the various chromosomes studied.

Over all, maternal age is the best known predictor of risk for nondisjunctional events, in particular those resulting from errors in meiosis I. The association between maternal age and risk for Down syndrome has long been established, and risk for trisomic abortuses also increases with advancing maternal age *(19,35,36)* (**Table 8**).

Not all chromosomal trisomies appear to have the same association with maternal age. Warburton et al. found that age-associated nondisjunction appeared to have a greater effect on the smaller chromosomes, with mean maternal age increasing with decreasing size of the trisomic chromosome *(22)*. Susceptibility to nondisjunction may not be the same for all chromosomes, and recurrence risks may be dependent upon the particular chromosome involved in the trisomy, the parent contributing the extra chromosome, and the background risk associated with maternal age. Regardless of the exact risk, many couples who have experienced a trisomic conceptus find the availability of prenatal diagnosis reassuring in planning subsequent gestations.

Sex Chromosome Aneuploidy

Sex chromosome aneuploidies are among the most common chromosomal abnormalities, both in spontaneous pregnancy loss and in liveborn infants. By far the most frequent sex chromosome aneuploidy at conception is 45,X, accounting for approximately 1–2% of clinically recognized pregnancies. It is the single most frequent abnormal karyotype seen in spontaneous abortions. The vast majority of monosomy X conceptuses terminate in miscarriage, less than 1% of affected pregnancies surviving to term *(18,37)*. The incidence of Turner syndrome in surviving pregnancies is approximately 1 in 1000 female live births. No 45,Y karyotypes have been reported. This is not an unexpected finding, considering the important contributions of genes located on the X chromosome.

The three sex chromosome trisomies, 47,XXX, 47,XXY, and 47,XYY, are much less frequent than monosomy X in spontaneous pregnancy loss, but are similar to monosomy X in frequency at term, each affecting approximately 1 in 1000 infants of the appropriate sex. Affected infants with sex chromosome trisomies are not usually markedly dysmorphic, and are often not identified unless cytogenetic studies are performed for other reasons. These conditions are frequently not recognized until later in life when behavioral changes or, in the case of 47,XXY, infertility, may cause the patient to present for evaluation. Some affected individuals may never be identified. The mild phenotypic expression at birth appears to reflect an absence of markedly deleterious effects during embryogenesis. This would explain the relatively low frequency of sex chromosome trisomies of 0.2% among spontaneous abortuses *(23)*.

Monosomy X gestations vary considerably in phenotype and may exhibit marked dysmorphism. The majority undergo early embryonic growth arrest and present as an empty gestational sac or as an umbilical cord ending with a small nodule of necrotic embryonic tissue (See **Fig. 1**). A lesser number survive into the second trimester, at which time the phenotype is often that of an hydropic fetus with massive cystic hygroma (see **Fig. 2**). Renal and cardiac anomalies are frequently seen as well. During the third trimester, the appearance may be similar to that seen in the second trimester, with cystic hygroma and dorsal edema over the hands and feet, the classic Turner syndrome phenotype. There are also 45,X infants who appear minimally affected and may not be recognized at birth, presenting later in childhood or adolescence with hypogonadism and short stature.

Several explanations have been proposed for the wide variability in phenotype. Although the majority of 45,X conceptuses surviving to term appear to have a maternally derived X, parental origin of the monosomy does not appear to affect phenotype or viability *(38,39)*. Rather, survival of the early pregnancy may be dependent upon presence, in some tissues, of a second sex chromosome, either another X or a Y. The nonmosaic 45,X conceptus appears unlikely to survive, whereas a mosaic gestation with a second sex chromosome, regardless of whether it is an X or a Y, has a better chance of undergoing orderly morphogenesis early in gestation and of surviving to term *(39,40)*. This second cell line may be absent from many tissues and is often difficult to detect with routine cytogenetic studies, but can sometimes be identified using multiple sampling sites or FISH techniques. Although extensive efforts at identification of a second cell line may not be justified in routine evaluation of a monosomy X

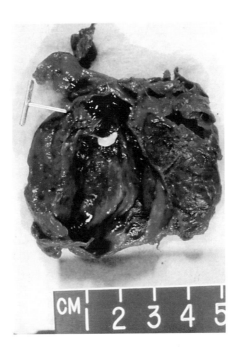

Fig. 1. Gestational sac with very small embryo, consistent with an underlying cytogenetic abnormality, often a nonviable trisomy, or, as in this case, monosomy X.

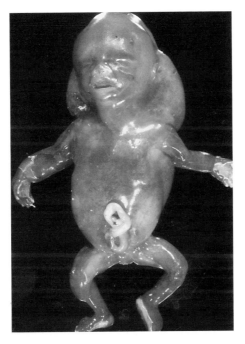

Fig. 2. 45,X spontaneous loss at mid-gestation. Note marked cystic hygroma and generalized edema.

abortus, such techniques are often helpful in evaluation of Turner syndrome patients with suspected low-level Y chromosome mosaicism. The presence of genes originating on the Y chromosome may place the patient at increased risk for gonadoblastoma.

Whereas the mean maternal age for most trisomic conceptuses is increased over the normal population, this is not the case with monosomy X. Rather, the mean maternal age for monosomy X is the same or lower than expected for the reproductive age population as a whole *(14)*. The evidence that many cases of monosomy X are the result of postzygotic nondisjunction may possibly explain the difference in maternal age between aneuploid pregnancies with monosomy X and those with autosomal trisomies. Mitotic nondisjunction during embryogenesis appears to be a different process, which may not exhibit the same maternal age effect; hence the maternal age for monosomy X would not be expected to be increased over the mean reproductive age of the population. Although patients who have experienced a pregnancy with monosomy X often choose to have prenatal cytogenetic evaluation in subsequent gestations, the recurrence risk for postzygotic/mitotic nondisjunctional events has not been established.

Errors in Mitosis

Malsegregation in the first mitotic division can give rise to tetraploidy. Tetraploid conceptions are usually lost relatively early in gestation, although there are rare exceptions.

Mitotic nondisjunction often results in mosaicism—the presence of two or more cell lines with a different genetic make-up. As has been suggested for Turner syndrome, mosaic aneuploidy may be better tolerated by the developing conceptus than complete aneuploidy, and there is evidence that survival of a trisomic fetus to term may be more likely if there is a normal cell line present within the placenta.

The question of tissue specific mosaicism has long been an issue in prenatal diagnosis, especially with the advent of chorionic villus sampling (CVS). Early nondisjunction can result in a generalized pattern of mosaicism, whereas divergence later in gestation can lead to mosaicism confined to either the fetus or the placenta. Mosaicism confined to the amnion may present a dilemma in interpretation of amniotic fluid karyotype, yet may not pose a problem for the fetus *(41)*. Within the placental chorionic villous tissue, there may be karyotypic differences between the direct preparation and long-term culture methods. This is a reflection of the different origins of the trophoblast cells and the extraembryonic mesodermal cells.

Confined placental mosaicism is a potential concern even in the fetus with a normal karyotype. The presence of confined placental mosaicism has been associated with abnormal mid-trimester hCG levels *(42)*, and with increased risk for adverse pregnancy outcome, including growth retardation and fetal demise *(43)*. Confined placental mosaicism may also be a factor leading to spontaneous abortion. A normal fetal karyotype does not rule out a cytogenetic abnormality in the placenta as a factor leading to pregnancy failure, suggesting the need for karyotype analysis of both fetal and placental tissues in unexplained stillbirths *(43)*. Although the incidence of mosaicism in CVS series is often cited in the 1–2% range, Kalousek and colleagues detected confined placental mosaicism in 11 of 54 spontaneous abortions studied and have suggested that the frequency may be especially high in growth-disorganized embryos *(44)*. The cytogenetic contribution to human pregnancy failure may thus be even higher than estimates based on early series, as those cases were often examined using only single tissue source, and some morphologically aberrant conceptuses classified as euploid may actually have been the result of undiagnosed mosaicism.

Recent molecular studies have shown that mosaic autosomal trisomies can arise either from errors in meiosis, with subsequent loss of one of the chromosomes leading to production of a euploid cell line, or from the postzygotic duplication of one of the chromosomes in an originally euploid cell line. The likelihood of one or the other mechanism may vary depending on the particular chromosome involved. Robinson and colleagues suggest that the mosaic trisomies involving chromosomes 13, 18, 21, and X most often result from somatic loss of a supernumerary chromosome that arose from meiotic nondisjunction *(45)*. Mosaic trisomy 8, on the other hand, may be more likely to survive when the aneuploid line is derived later, as a result of a postzygotic error in mitosis in a conceptus that was originally chromosomally normal.

Mosaicism in the placenta may be a significant determining factor in survival of the trisomic conceptus. Those cases of trisomies 13 and 18 that survive to term appear to have a diploid cell line in the cytotrophoblasts, whereas those lost early in gestation are less likely to show a normal cell line *(46,47)*. Mosaicism does not appear to be necessary for survival in trisomy 21, possible due to a less deleterious effect of this trisomy on placental function *(46)*.

The presence of a euploid cell line in the fetus does not necessarily imply a genetically normal fetus. If the mosaicism is the result of "rescue" of a trisomic cell line, the possibility of both remaining chromosomes of the pair originating from a single parent becomes a concern. This condition, uniparental disomy, can often have severe consequences in the affected fetus due to potential loss of heterozygosity with expression of recessive traits only carried by one parent, or due to effects of inappropriate imprinting (see Chapter 16). Thus, multiple sampling sites should be evaluated in cases where a cytogenetic abnormality is strongly suspected, even if a normal karyotype is identified on initial evaluation. Molecular studies may be indicated to rule out uniparental disomy in ongoing pregnancies that have been identified as mosaic. More study regarding the effects of uniparental disomy on embryogenesis is clearly needed.

Chimerism

Another possible cause for the presence of more than one cell line in a fetus is chimerism. The chimera of classical mythology was a creature with the head of a lion, the body of a goat, and the tail of a serpent. Although the mythical chimera composed of several unrelated species is purely fanciful, individuals with cells derived from two separate fertilized eggs are known to exist in humans and other mammals. Postzygotic fusion of dizygotic twin zygotes results in a single chimeric individual.

Chimerism can explain the presence of two cell lines, in a single individual, where neither can be derived from the other. This is the most likely mechanism underlying 46,XX/46,XY hermaphroditism, and may also explain a 45,X/69,XXY fetus described by Betts and colleagues *(48)*. A number of diploid/triploid mosaics have also been reported *(49)*. Some of these are probably chimeras, although another possible mechanism here is dispermy, in which a single maternal haploid pronucleus is fertilized by a haploid sperm in the usual manner, resulting in the diploid line. A second fertilization event then occurs in one of the daughter cells after the first cell division, leading to the triploid cell line *(50)*.

Errors in Fertilization

Errors in fertilization can lead to pregnancies with an extra complete set of chromosomes (triploidy), and also abnormal diploid pregnancies in which both sets of chromosomes come from one parent (hydatidiform or complete molar pregnancies). Because paternal triploids may exhibit changes in the villi that resemble hydatidiform moles, these are sometime referred to as partial moles. Both partial and complete molar pregnancies have been instrumental in advancing our understanding of imprinting and the role imprinting plays in fetal development and carcinogenesis. Imprinting may have functions not only in gene expression early in embryogenesis, but may also play a significant role in surveillance for chromosome loss later in life, and thus may help reduce the risk of cancer *(51)*.

An extra haploid set of chromosomes from either the mother (digyny) or the father (diandry) can result in a triploid conceptus. A 69,XYY karyotype is indicative of a paternal origin for the extra chromosomal set, whereas a 69,XXX or 69,XXY karyotype could represent either digyny or diandry. A variety of events can lead to the presence of an extra set of chromosomes.

The paternally derived triploid usually results from either fertilization of a normal haploid egg by two separate sperm (dispermy), or from fertilization of the egg by a diploid sperm. Fertilization by a haploid sperm with subsequent endoreduplication of the paternal chromosomal complement is also a possible mechanism. The latter process would result in isodisomy for all paternal chromosomes, as would an error in meiosis II *(52)*. The maternally derived triploid, on the other hand, most often originates from an error during maternal meiosis I or II, resulting in a diploid egg, although other mechanisms including fertilization of a primary oöcyte have also been postulated *(53)*. Together, triploidy accounts for 1–3% of all recognized pregnancies, and 15–20% of all chromosomally abnormal miscarriages, placing the triploidy among the most frequent chromosomal aberrations in human conception *(54,55)*.

Although the net result of either diandry or digyny is a pregnancy with 69 chromosomes, the phenotype of the paternal triploid is quite different from that of the maternal. On microscopic section paternal triploids will often show a mixture of hydropic villi together with smaller, more normal appearing villi, a phenotype sometimes referred to as a "partial mole." Most present as a "blighted ovum" with an empty gestational sac in the first trimester. Those that survive into the second trimester exhibit an abnormal fetal-to-placental weight ratio with a very large placenta showing grossly hydropic villi (see **Fig. 3**). Alpha-fetoprotein (AFP) and hCG levels are characteristically elevated.

The maternal triploid fetus is growth retarded with a disproportionately large cranium. The placenta is small and fibrotic in appearance, with none of the hydropic degeneration seen in the paternal triploid (see **Fig. 4**). In contrast with the paternal triploid, AFP and hCG levels are low.

The risk for triploid gestations appears to decrease with advancing maternal age. A decline in survival of aberrant conceptuses, in older women, to the stage of recognized

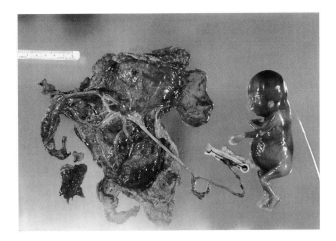

Fig. 3. Paternal triploid, 69,XXY karyotype. Patient presented with markedly elevated β-hCG at 16 weeks. Note very large placenta in relation to the size of the fetus.

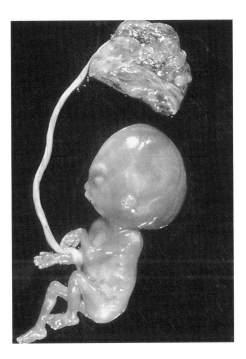

Fig. 4. Maternal triploid. Note very small placenta in relation to fetal size, and fetus with micrognathia, syndactyly, and disproportionately large cranium in relation to body.

pregnancy is one possible explanation. Younger patients appear more likely to present with paternal triploids, whereas maternal triploids are more frequent in older patients.

The complete mole is a pregnancy characterized by marked placental overgrowth with large, cystic–hydropic villi. The fetus is absent and the villi do not exhibit fetal vascularization. The trophoblastic layers on the surface of the villi show varying degrees of proliferation. Patients usually exhibit markedly elevated hCG levels, although a method-dependent artifact can result in falsely low levels *(56)*. Despite the markedly abnormal phenotype, molar pregnancies usually exhibit a diploid karyotype of 46,XX in approximately 90% of cases, and 46,XY in 6–10% of cases *(57)*. Both haploid sets of chromosomes are of paternal origin, however. Mechanisms are probably similar to paternal triploids, but with fertilization of an "empty" egg. Duplication of the chromosomes of a haploid sperm appears frequent and may explain the preponderance of 46,XX karyotypes, whereas fertilization by a diploid sperm could result in either a 46,XY or a 46,XX karyotype. The 46,YY karyotype appears to be nonviable *(55)*.

Hydatidiform moles pose a risk of undergoing malignant transformation, becoming choriocarcinomas. Because of this the diagnosis is critical for patient management. The triploid conceptus does not appear to have the same malignant potential *(58,59)*. The mechanism for malignant transformation in the complete mole appears to be relaxation of imprinting with expression of genes that would normally be repressed *(60,61)*. Imprinting has also been suggested as an explanation for the difference in phenotype between the maternal and paternal triploids *(62)*.

Although the experienced perinatal pathologist should have little difficulty in recognizing the true hydatidiform mole based on the histologic appearance of the villi, cytogenetic evaluation should be considered whenever there is a question of the diagnosis, as follow-up with serial hCG levels is crucial to prevent a malignancy in cases of complete mole.

Both complete hydatidiform moles and most triploids appear to represent random errors at the time of fertilization. As such, a significant impact on the risk for other chromosomal abnormalities in subsequent pregnancies is not expected. Berkowitz et al. studied 1205 pregnancies following a complete molar pregnancy and found no increase in risk for stillbirth, prematurity, ectopic gestations, malformations, or spontaneous abortion *(63)*. However, there appears to be a recurrence risk of about 1% for a future mole following a molar pregnancy *(63–65)*. Early sonographic surveillance is suggested for future gestations to rule out recurrent mole, and postnatal hCG determinations are recommended to rule out persistent trophoblastic disease *(63)*. Several pedigrees suggesting familial predisposition to molar pregnancies have also been reported *(66–69)*, although the significance of family history on risk has not yet been established with certainty.

Structural Rearrangements

Structural rearrangements are less common than the other types of chromosomal abnormalities in pregnancy losses. Approximately 1–2% of spontaneous abortions show structural rearrangements. Jacobs summarized 5726 karyotyped spontaneous abortions, revealing 0.28% balanced and 1.54% unbalanced rearrangements *(70)*. Balanced rearrangements include Robertsonian translocations, reciprocal translocations, and inversions (see Chapter 9). A survey of the literature by Dewald and Michels revealed translocations in 2.1% of couples with recurrent miscarriage *(71)*. Translocations were found in 1.7% of male patients and 2.6% of female patients. This compares with an incidence of 1 in 500 (0.2%) in newborns *(72)*. The frequency of balanced rearrangements in spontaneous abortions is not markedly increased over that seen in live births. This is not unexpected, as balanced rearrangements are typically not associated with significant phenotypic alterations and are usually compatible with embryonic and fetal life.

The most frequent unbalanced rearrangements result from Robertsonian translocations. These may occur *de novo* or be familial in origin. The incidence of unbalanced Robertsonian translocations is much higher in spontaneous abortuses than live births, reflecting the uterine mortality of trisomic conceptuses.

Other unbalanced structural rearrangements seen in spontaneous abortions include abnormal chromosomes with extra or missing material, ring chromosomes, and small supernumerary chromosomes. *De novo* rearrangements are more frequently paternal in origin *(73)*. Analysis of human sperm revealed considerable variability among donors (0–17.8%) with a median of 9.3% abnormal sperm, consisting primarily of breaks, fragments, and small deletions. Increased susceptibility of sperm to chromosomal damage could explain the paternal origin of the majority of rearrangements.

Although many structural rearrangements are *de novo*, the majority appear to be familial. Numerous studies of patients experiencing recurrent pregnancy loss have shown that these individuals are at increased risk of carrying a balanced chromosomal rearrangement. Cytogenetic analysis to rule out structural rearrangements and genetic counseling is indicated for couples who have experienced two or more losses. Because most balanced rearrangement carriers can produce both balanced and unbalanced gametes, a combination of normal and abnormal conceptions is frequently seen in such couples, and rearrangements may be more likely in those who have experienced both miscarriages and live births than in those with only miscarriages *(74,77)*.

Campana and colleagues note that the chromosomes and breakpoints involved in structural rearrangements do not appear to be random *(75)*. Survival of pregnancies with unbalanced chromosomal complements appears to be dependent upon the particular chromosome and segment(s) involved.

Structural rearrangements appear to occur with greater frequency in females than in males. Braekeleer and Dao found translocations or inversions in 2.6% of females with a history of reproductive failure compared with 1.4% in males, and Gadow and colleagues found that 3.5% of women and 1.7% of men with recurrent loss had balanced translocations *(76,77)*. Both reports suggest increased risk for sterility in male carriers as a possible explanation. Chromosomal rearrangement appears to be associated with increased risk for infertility as well as for pregnancy loss.

The risk for poor pregnancy outcome when one member of a couple carries a structural rearrangement varies considerably depending upon the particular type of rearrangement and the chromosome(s) involved. Counseling must be individualized for each family with attention given to potential viability of any unbalanced meiotic products.

The risk figures that are used in counseling are often based on pooled data from translocations involving various chromosomes and breakpoints. Generally, it has been suggested that a male carrier is at lower risk for abnormal offspring than a female carrier. However, such generalizations may not be applicable in all cases and more specific risks figures based on the particular chromosomes involved may be beneficial in evaluating reproductive options for a family in which a balanced translocation has been identified *(78)*.

The cause of reproductive failure in patients with balanced translocations is most likely the production of unbalanced gametes as a result of abnormal segregation during meiosis. Inversions can also lead to unbalanced gametes through crossover events involving the inverted segment. A discussion of the implications of specific rearrangements with regard to abnormal segregation products can be found in Chapter 9; see also ref *(25)*.

Chromosomally Normal Pregnancy Loss

Identification of the cytogenetically normal spontaneous abortion may be more important clinically than identification of the aberrant gestation. The risk of repeat miscarriage is higher when the prior loss is chromosomally normal *(79)*. Boué and colleagues found a risk of repeat loss of 23% after a chromosomally normal miscarriage compared with 16.5% following a chromosomally abnormal loss *(14)*. Morton and colleagues found that in women under 30, the risk for miscarriage was 22.7% following a chromosomally normal loss, 15.4% following a trisomy, and 17% following other chromosome abnormalities. In women over 30, these risks were 25.1%, 24.7%, and 20.3%, respectively *(80)*. Cytogenetic study of repeated spontaneous abortions suggests that those patients who experience a chromosomally normal pregnancy loss are more likely to show normal karyotypes in subsequent losses *(81,82)*.

Women with recurrent pregnancy losses and normal fetal karyotypes may be more likely to have underlying uterine abnormalities or endocrine dysfunction. Menstrual irregularities and elevated lutenizing hormone levels are more common in women with normal fetal karyotypes than in women with abnormal fetal karyotypes *(83)*.

Immunologic disorders have also been linked with recurrent normal pregnancy loss *(84)*. Systemic lupus erythematosus is perhaps the best known, but other autoimmune conditions have also been implicated *(85)*. Because patients with antiphospholipid antibodies and pregnancy failure frequently respond to treatment with prednisone and low-dose aspirin or heparin, it is important to recognize autoimmune disease as a frequent cause of recurrent chromosomally normal pregnancy losses *(86–88)*. Alloimmune disorders are less well understood, but also appear to play a role in recurrent pregnancy failure. Several therapies including immunization with paternal while cells have been suggested *(89)*.

Mutations that are lethal in the embryo are known from animal models and may also be a factor in recurrent euploid abortion in man *(90)*. Mutations in genes responsible for early organization of the embryo can have devastating effects on embryogenesis, with resultant pregnancy failure. Parental sharing of HLA antigens may also increase risk for spontaneous abortion, although the mechanism is not yet clearly understood *(91)*. More study of such genes and their effects on embryonic development is needed to determine the frequency of their contribution to poor pregnancy outcome.

SPECIMENS FOR CYTOGENETIC STUDIES

Although cytogenetic studies may be very helpful in managing patients with recurrent pregnancy loss, fetal karyotypes are infrequently performed. Cowchock and colleagues reported a success rate of 84% in a series of 100 samples, showing that chromosome analysis is indeed feasible in most specimens *(79)*. Chorionic villi are often the tissue of choice, as skin biopsies from deceased fetal tissue can be associated with a higher failure rate. As previously mentioned, with spontaneous pregnancy loss it is frequently the case that the tissue is retained *in utero* for several days or even weeks following embryonic or fetal demise. Because of this, fetal tissue is often autolyzed and is unlikely to respond to standard culture methods. Placental tissue, on the other hand, often remains viable for a much longer period of time, as necessary substrates for survival are provided by contact with the maternal blood supply. Ideally, both fetal and placental sources should be utilized. The advantage of the fetal tissue is that there is little risk for maternal cell contamination. If the fetus appears macerated, however, a high success rate is not to be expected. Placental tissue usually grows well but adds the risk of maternal cell contamination. This risk is reduced if the technologist processing the sample is experienced in the identification of membrane and chorionic villi.

Direct preparations using the *in situ* method of tissue culture work well with cells derived from spontaneously aborted tissues, and have the advantage of rapid results with a high success rate and minimal risk for maternal cell contamination *(92)*. However, if maternal cells are present in the original sample, trypsinization of slow growing cultures to increase cell yield appears to increase the risk for maternal cell overgrowth.

Flow cytometry can sometimes provide useful information in cases that are not amenable to cell culture. Fluorescence *in situ* hybridization (FISH) using either tissue sections or disaggregated cells may be used in cases in which the tissue was accidentally fixed in formalin prior to receipt in the cytogenetics laboratory. It must be remembered, however, that this will detect only those chromosome abnormalities for which probes are available, and is typically used only to diagnose aneuploidy.

Peripheral blood cytogenetic studies should be considered for any couple experiencing recurrent pregnancy loss. In examining parental chromosomes, structural rearrangements including translocations and inversions are the obvious focus. Such rearrangements may have significant impact on the couple's risk for miscarriage or infants with anomalies.

Cytogenetic abnormalities that are less clear in terms of their implications for future reproduction may also be seen. It is not uncommon to find mosaic aneuploidy in couples with recurrent pregnancy loss. Low-level sex chromosome aneuploidy is sometimes seen in lymphocytes, but is not usually found in cultured fibroblasts. The risk appears to increase with age, but does not appear to be correlated with reproductive history *(93)*. Discussion with a cytogeneticist can be invaluable in interpreting whether unexpected findings are of potential clinical significance or artifact unrelated to the reproductive history.

EVALUATION OF PREGNANCY LOSSES

Although a complete evaluation of a pregnancy loss requires extensive specialized testing, including cytogenetic studies, such tests are costly and labor intensive. With increasing emphasis on delivery of cost-effective health care, cytogenetic studies simply cannot be justified for every unsuccessful pregnancy. However, a careful examination by a pathologist can often go a long way toward answering the patient's questions about the loss, and a more thorough evaluation by a pathologist with training and interest in developmental anatomy can often provide considerable information without significant increase in cost.

Such an examination can establish how far the pregnancy proceeded prior to developmental arrest, and whether the pregnancy appears to have been developing normally.

Table 9
Phenotype of Abortus and Incidence of Abnormal Karyotypes

Appearance of abortus	Percent chromosomally abnormal
Incomplete specimen, no embryo	47.3
Incomplete embryo/fetus	40.0
Intact empty sac	64.3
Severely disorganized embryo	68.6
Normal embryo	54.1
Embryo with focal abnormalities	55.0
Normal fetus	3.3
Fetus with malformations	18.2

Data from ref. *36.*

The developmental age is especially helpful because the earlier the growth arrest, the more likely it is that the conceptus will exhibit an abnormal karyotype.

Any embryo or fetus should be examined closely for evidence of congenital anomalies. Embryos with malformations and growth-retarded embryos are more likely to exhibit abnormal karyotypes. Some isolated anomalies, such as cleft palate or neural tube defects, may be associated with significant recurrence risks yet may have normal karyotypes. Specific anomalies may also be indicative of an underlying syndromal process, with or without an abnormal karyotype. Single gene defects with significant recurrence risk can sometimes be identified from a careful fetal examination. Evidence of infectious processes or teratogen exposure may also be present, with their own implications for future pregnancy management. Anatomic evaluation can therefore play a useful role when traditional cytogenetic studies either are not indicated, as in a first loss with no other risk factors, or are not possible, for example, a formalin-fixed or otherwise nonviable specimen.

Although most chromosomal abnormalities are not associated with distinct phenotypes, especially in very early losses, there does appear to be some correlation between specimen morphology and the likelihood of an abnormal karyotype. Creasy studied the prevalence of chromosomal abnormalities and phenotype *(36).* The results are summarized in **Table 9**.

Even though the degree of correlation between specimen types and risk for chromosomal abnormalities is far from ideal, some information regarding the likelihood of a karyotype abnormality can be gained from the embryonic pathology. Absence of abnormalities in a pregnancy that has progressed to the fetal stage is a good predictor for a normal chromosomal complement.

Although morphology can help predict the likelihood of a chromosomal etiology for the loss, it cannot be expected to identify the particular karyotype abnormality involved. However, even distinguishing probable chromosomal from nonchromosomal losses is of considerable benefit to the patient, as it can help in determining need for further studies and in predicting risk for recurrence.

There is a strong correlation between the chromosomal constitutions of first and subsequent abortions. The patient with a chromosomally abnormal abortus is more likely to experience abnormal karyotypes in subsequent losses, whereas a patient with a normal karyotype in one loss is more likely to show normal karyotypes in any future pregnancy losses *(81,82,94)*.

Chromosome studies are especially useful for stillbirths suspected of having cytogenetic abnormalities, such as infants with congenital anomalies or intrauterine growth retardation. There may also be increased risk in the presence of fetal hydrops, maceration, or a history of prior losses *(95,96)*. Cytogenetic studies should probably be performed in any case in which a pathophysiologic explanation for the demise is not identified *(97)*. A careful anatomic evaluation of both fetus and placenta is indicated in all stillbirths, as are photographs and radiographic studies to document morphology when there is question of a skeletal dysplasia or other anomalies. These can be reviewed later by a specialist in fetal dysmorphology if there is any question of anomalous development. Additional special studies for congenital infection, hematologic disorders, or metabolic disease may also be indicated in some cases. Overall, a cause of death can be assigned in approximately 80% of cases *(95,97)*.

A wide range of problems can result in decreased fertility or pregnancy failure, and the work-up for an infertile couple can be extensive and costly *(98)*. Identifying those couples whose losses are explained as being due to karyotypic abnormalities may be a cost-effective alternative. Cowchock and colleagues argue that if cytogenetic studies cost $500, with an 84% chance of culture success and a 40% chance of detecting a chromosomal abnormality that would explain the loss, one of every three women with multiple miscarriage would be spared further costly and invasive evaluations for recurrent pregnancy loss *(81)*. This would save approximately $2000 in expenses for testing that would otherwise be done as part of a multiple miscarriage protocol. Given the availability of therapy for many patients with nonchromosomal causes of pregnancy loss, the cost–benefit ratio may actually be even better.

Although recurrent spontaneous abortion is often defined as three consecutive losses, today many couples find that three miscarriages are more than they are willing to accept before looking for answers. There may indeed be justification for initiating further evaluation after the second failed pregnancy. Coulam compared 214 couples with a history of two or more consecutive abortions with 179 couples with a history of three or more abortions *(99)*. Both groups showed 6% of losses that were chromosomal, 1% that were anatomic, and 5% that were hormonal. Sixty-five percent of the group with two losses and 66% of the group with three losses had immunologic causes. Twenty-three percent of the group with two losses and 22% of the group with three losses were unexplained. The absence of any significant difference in prevalence between the two groups suggests that there is little to be gained by delaying evaluation until after the third pregnancy loss.

Tharapel and colleagues reviewed published surveys of couples with two or more pregnancy losses (8208 women and 7834 men) and found an overall prevalence of major chromosome abnormalities of 2.9% *(100)*. They go on to suggest that even with normal parental chromosomes, prenatal diagnosis should be offered because of the high incidence of chromosomal abnormalities in spontaneous pregnancy loss. Drugan and colleagues identified five anomalous fetuses, including one trisomy 18, two trisomy 21, one trisomy 13, and one monosomy X fetus among 305 couples with recurrent pregnancy loss *(101)*. This 1.6% risk is greater than the risk usually cited for amniocentesis. A control group of 979 patients revealed only three abnormalities,

(0.3%), all sex chromosome aneuploidies. This would suggest an increased risk for nondisjunction among couples experiencing repeated pregnancy failure. Their conclusion is that prenatal diagnosis is sufficiently safe and the risk for an abnormal result is sufficiently high to justify offering prenatal diagnosis to couples with a history of two losses. Although this conclusion is based on a relatively small sample size and not all obstetrical care givers would agree, a discussion of risks and benefits of prenatal diagnosis would appear to be justified in this patient population.

Although considerable advances have been made in understanding the causes underlying pregnancy failure and there is considerable hope for more specific therapies for couples experiencing nonchromosomal losses, there is unfortunately little to offer the couple who may be at increased risk for cytogenetically abnormal pregnancies. When a rearrangement is incompatible with normal pregnancy outcome (such as an isochromosome 21), use of donor ova or sperm may be an option. The issues are not so clear for the couple with recurrent aneuploidy or polyploidy.

Preimplantation assessment of the fetal karyotype using FISH may be a consideration for some patients undergoing in vitro fertilization for other reasons. Simultaneous use of probes for chromosomes 13, 18, 21, X, and Y can enhance the likelihood of transfer of normal embryos; however some mosaic aneuploid conceptions and aneuploidy for other chromosomes would still be missed *(102)*. It is important to remember that the majority of embryos with cytogenetic abnormalities will be lost spontaneously, thus the unknowing transfer of cytogenetically abnormal embryos potentially contributes to the less than optimal success rate for IVF procedures. Better methods for identifying chromosomally normal embryos for transfer are needed *(103)*.

SUMMARY

Humans experience a wide range of chromosomal abnormalities at conception. The incidence is surprisingly high when compared with other mammals, such as the mouse. When considering pregnancy loss in this context, spontaneous abortion can be seen as a means of "quality control" in an otherwise inefficient reproduction system *(14)*. Our understanding of the mechanisms involved in meiosis, fertilization, and mitosis is still quite limited, and the factors affecting survival of the embryo are not yet fully understood. Maternal age appears to increase the incidence of abnormal conceptions, but may also decrease the efficiency of this control process.

Although our understanding of pregnancy loss is limited and we cannot fully predict risks, we can attempt to offer patients some explanation as to why a given pregnancy has failed and whether there is any treatment that might improve chances for future success. We can also make prenatal diagnosis available in those cases in which there is increased risk for cytogenetic abnormalities or when additional reassurance of a normal fetal karyotype is needed. It is important to keep in mind that even with a history of a chromosomally abnormal pregnancy, most couples have a good chance for a subsequent successful outcome.

REFERENCES

1. Kay, J. (1987) Pregnancy loss and the grief process. In *Pregnancy Loss: Medical Therapeutics and Practical Considerations* (Woods, J.R. and Esposito, J.L., eds.), Williams & Wilkins, Baltimore, pp. 5–20.

2. Hager, A. (1987) Early pregnancy loss: miscarriage and ectopic pregnancy. In *Pregnancy Loss: Medical Therapeutics and Practical Considerations* (Woods, J.R. and Esposito, J.L., eds.), Williams & Wilkins, Baltimore, pp. 23–50.

3. Friedman, R. and Gradstein, B. (1992) *Surviving Pregnancy Loss.* Little, Brown, Boston.

4. Seibel, M. and Graves, W.L. (1980) The psychological implications of spontaneous abortions. *J. Reprod. Med.* **25,** 161–165.

5. Stack, J.M. (1990) Psychological aspects of early pregnancy loss. In *Early Pregnancy Failure* (Huisjes, H.J. and Lind, T., eds.), Churchill Livingstone, New York, pp. 212–239.

6. Roberts, C.J. and Lowe, C.R. (1975) Where have all the conceptions gone? *Lancet* **i,** 498–499.

7. Edmonds, D.K., Lindsay, K.S., Miller, J.F., Williamson, E., and Wood, P.J. (1982) Early embryonic mortality in women. *Fertil. Steril.* **38,** 447–453.

8. Miller, J.F., Williamson, E., Glue, J., Gordon, Y.B., Grudzinskas, J.G., and Sykes, A. (1980) Fetal loss after implantation: a prospective study. *Lancet* **ii,** 554–556.

9. Wilcox, A.J., Weinberg, C.R., O'Connor, J.F., Baird, D.D., Schlatterer, J.P., Canfield, R.E., Armstrong, E.G., and Nisula, B.C. (1988) Incidence of early loss of pregnancy. *N. Engl. J. Med.* **319,** 189–194.

10. French, F.E. and Bierman, J.M. (1962). Probabilities of fetal mortality. *Pub. Health Rep.* **77,** 835–847.

11. Short, R.V. (1979) When a conception fails to become a pregnancy. In *Maternal Recognition of Pregnancy.* Excerpta Medica, Amsterdam, pp. 377–387.

12. Leridon, H. (1977). *Human Fertility.* University of Chicago Press, Chicago.

13. Yusuf, R., and Naeem, R. (1997) Personal communication.

14. Boué, A., Boué, J., and Gropp, A. (1985). Cytogenetics of pregnancy wastage. *Adv. Hum. Genet.* **14,** 1–57.

15. Yamamoto, M. and Wantanabe, G. (1979) Epidemiology of gross chromosome anomalies at the early stage of pregnancy. *Contrib. Epidemiol.* **1,** 101–106.

16. Gaillard, D.A., Paradis, P., Lallemand, A.V., Vernet, V.M., Carquin, J.S., Chippaux, C.G., and Visseaux-Coletto, B.J. (1993) Spontaneous abortions during the second trimester of gestation. *Arch. Pathol. Lab. Med.* **117,** 1022–1026.

17. Jacobs, P.A., Browne, C., Gregson, N., Joyce, C., and White, H. (1992) Estimates of the frequency of chromosome abnormalities detectable in unselected newborns using moderate levels of banding. *J. Med. Genet.* **29,** 103–108.

18. Davison, E.V. and Burn, J. (1990) Genetic causes of early pregnancy loss. In *Early Pregnancy Failure* (Huisjes, H.J. and Lind, T., eds.), Churchill Livingstone, New York, pp. 55–78.

19. Kline, J. and Stein, Z. (1986) The epidemiology of spontaneous abortion. In *Early Pregnancy Failure* (Huisjes, H.J. and Lind, T., eds.), Churchill Livingstone, New York, pp. 240–256.

20. Boué, J., Philippe, E., Giroud, A., and Boué, A. (1976) Phenotypic expression of lethal chromosomal anomalies in human abortuses. *Teratology* **14,** 3–20.

21. Boué, J., Boué, A., and Lazar, P. (1975) Retrospective and prospective epidemiological studies of 1500 karyotyped spontaneous human abortions. *Teratology* **12,** 11–26.

22. Warburton, D., Byrne, J., and Canki, N. (1991) *Chromosome Anomalies and Prenatal Development: An Atlas. Oxford Monographs on Medical Genetics No. 20.* Oxford University Press, New York.

23. Simpson, J.L. (1986) *Genetics. CREOG Basic Science Monograph in Obstetrics and Gynecology.* Council on Resident Education in Obstetrics and Fynecology, Washington D.C.

24. Robinson, A. and Linden, M.G. (1993) *Clinical Genetics Handbook,* 2nd edition. Blackwell Scientific Publications, Oxford.

25. Gardner, R.J.M. and Sutherland, G.R. (1996) *Chromosome Abnormalities and Genetic Counseling.* Oxford University Press, New York.

26. Verp, M.S. and Simpson, J.L. (1985) Amniocentesis for cytogenetic studies. In *Human Parental Diagnosis* (Filkins, K. and Russo, J.F., eds.), Marcel Dekker, New York, pp. 13–48.

27. Connor, J.M. and Ferguson-Smith, M.A. (1987) *Essential Medical Genetics,* 2nd edition. Blackwell, Oxford.

28. Hassold, T., Hunt, P.A., and Sherman, S. (1993) Trisomy in humans: incidence, origin and etiology. *Curr. Opin. Genet. Dev.* **3,** 398–403.

29. Kamiguchi, Y., Rosenbusch, B., Sterzik, K., and Mikamo, K. (1993) Chromosomal analysis of unfertilized human oöcytes prepared by a gradual fixation–air drying method. *Hum. Genet.* **90,** 533–541.

30. Pellestor, F. (1991) Frequency and distribution of aneuploidy in human female gametes. *Hum. Genet.* **86,** 283–288.

31. Martin, R.H., Rademaker, A.W., Hildebrand, K., Long-Simpson, L., Peterson, D., and Yamamato, J. (1987) Variation in the frequency and type of sperm chromosomal abnormalities among normal men. *Hum. Genet.* **77,** 108–114.

32. Miharu, N., Best, R.G., and Young, S.R. (1994). Numerical chromosome abnormalities in spermatozoa of fertile and infertile men detected by fluorescence in situ hybridization. *Hum. Genet.* **93,** 502–506.

33. Guttenbach, M., Schakowski, R., and Schmid, M. (1994a) Incidence of chromosome 18 disomy in human sperm nuclei as detected by nonisotopic in situ hybridization. *Hum. Genet.* **93,** 421–423.

34. Guttenbach, M., Schakowski, R., and Schmid, M. (1994b) Incidence of chromosome 3, 7, 10, 11, 17 and X disomy in mature human sperm nuclei as determined by nonradioactive in situ hybridization. *Hum. Genet.* **93,** 7–12.

35. Hassold, T. (1986) Chromosome abnormalities in human reproductive wastage. *Trends Genet.* **2,** 105–110.

36. Creasy, M.R., Crolla, J.A., and Alberman, E.D. (1976) A cytogenetic study of human spontaneous abortions using banding techniques. *Hum. Genet.* **31,** 177–196.

37. Hook, E.B. and Warburton, D. (1983) The distribution of chromosomal genotypes associated with Turner's syndrome: livebirth prevalence rates and severity in genotypes associated with structural X abnormalities of mosaicism. *Hum. Genet.* **64,** 24–27.

38. Hassold, T., Benham, F., and Leppert, M. (1988) Cytogenetic and molecular analysis of sex-chromosome monosomy. *Am. J. Hum. Genet.* **42,** 534–541.

39. Hassold, T., Pettay, D., Robinson, A., and Uchida, I. (1992) Molecular studies of parental origin and mosaicism in 45,X conceptuses. *Hum. Genet.* **89,** 647–652.

40. Held, K.R., Kerber, S., Kaminsky, E., Singh, S., Goetz, P., Seemanova, E., and Goedde, H.W. (1992) Mosaicism in 45,X Turner syndrome: does survival in early pregnancy depend on the presence of two sex chromosomes? *Hum. Genet.* **88,** 288–294.

41. Baldinger, S., Millard, C., Schmeling, D., and Bendel, R.P. (1987) Prenatal diagnosis of trisomy 20 mosaicism indicating an extra embryponic origin. *Prenat. Diag.* **7,** 273–276.

42. Morssink, L.P., Sikkema-Raddatz, B., Beekhuis, J.R., deWolf, B.T.H.M., and Mantingh, A. (1996) Placental mosaicism is associated with unexplained second-trimester elevation of MshCG levels, but not with elevation of MSAFP levels. *Prenat. Diag.* **16,** 845–851.

43. Kalousek, D.K. and Barrett, I. (1994) Confined placental mosaicism and stillbirth. *Pediatr. Pathol.* **14,** 151–159.

44. Kalousek, D.K., Barrett, I.J., and Gärtner, A.B. (1992) Spontaneous abortion and confined placental mosaicism. *Hum. Genet.* **88,** 642–646.

45. Robinson, W.P., Binkert, F., Bernasconi, F., Lorda-Sanchez, I., Werder, E.A., and

Schinzel, A.A. (1995) Molecular studies of chromosomal mosaicism: relative frequency of chromosome gain or loss and possible role of cell selection. *Am. J. Hum. Genet.* **56,** 444–451.

46. Kalousek, D.K., Barrett, I.J., and McGillivray, B.C. (1989) Placental mosaicism and intrauterine survival of trisomies 13 and 18. *Am. J. Hum. Genet.* **44,** 338–343

47. Harrison, K.J., Barrett, I.J., Lomax, B.L., Kuchinka, B.D., and Kalousek, D.K. (1993) Detection of confined placental mosaicism in trisomy 18 conceptions using interphase cytogenetic analysis. *Hum. Genet.* **92,** 353–358.

48. Betts, D.R., Fear, C.N., Barry, T., and Seller, M.J. (1989) A 45,X/69,XXY fetus. *Clin. Genet.* **35,** 285–288.

49. Blackburn, W.R., Miller, W.P., Superneau, D.W., Cooley, N.R., Zellweger, H., and Wertelecki, W. (1982) Comparative studies of infants with mosaic and complete triploidy: an analysis of 55 cases. *Birth Defects Orig. Article Ser.* **18,** 251–274.

50. Dewald, G., Alvarez, M.N., Cloutier, M.D., Kelalis, P.P., and Gordon, H. (1975) A diploid-triploid human mosaic with cytogenetic evidence of double fertilization. *Clin. Genet.* **8,** 149–160.

51. Thomas, J.H. (1995) Genomic imprinting proposed as a surveillance mechanism for chromosome loss. *Proc. Natl. Acad. Sci. USA* **92,** 480–482.

52. Neuber, M., Rehder, H., Zuther, C., Lettau, R., and Schwinger, E. (1993) Polyploidies in abortion material decrease with maternal age. *Hum. Genet.* **91,** 563–566.

53. O'Neill, G.T. and Kaufmann, M.H. (1987) Ovulation and fertilization of primary and secondary oöcytes in LT/Sv strain mice. *Gamete Res.* **18,** 27–36.

54. Jacobs, P.A., Hassold, T.J., Matsuyama, A.M., and Newlands, I.M. (1978) Chromosome constitution of gestational trophoblastic disease. *Lancet* **ii,** 49.

55. Lindor, N.M., Ney, J.A., Gaffey, T.A., Jenkins, R.B., Thibodeau, S.N., and Dewald, G.W. (1992) A genetic review of complete and partial hydatidiform moles and nonmolar triploidy. *Mayo Clin. Proc.* **67,** 791–799.

56. Levavi, H., Neri, A., Bar, J., Regev, D., Nordenberg, J., and Ovadia, J. (1993) "Hook effect" in complete hydatidiform molar pregnancy: a falsely low level of β-hCG. *Obstet. Gynecol.* **82,** 720–721.

57. Berkowitz, R.S., Goldstein, D.P., and Bernstein, M.R. (1991) Evolving concepts of molar pregnancy. *J. Reprod. Med.* **36,** 40–44.

58. Benirschke, K. and Kaufmann, P. (1995) *Pathology of the Human Placenta,* 3rd edition. Springer-Verlag, New York.

59. Szulman, A.E. (1984) Syndromes of hydatidiform moles: partial vs. complete. *J. Reprod. Med.* **11,** 788–791.

60. Mutter, G.L., Stewart, C.L., Chaponot, M.L., and Pomponio, R.J. (1993) Oppositely imprinted genes H19 and insulin-like growth factor 2 are coexpressed in human androgenic trophoblast. *Am. J. Hum. Genet.* **53,** 1096–1102.

61. Ariel, I., Lustin, O., Oyer, C.E., Eikin, M., Gonik, B., Rachmilewitz, J., Biram, H., Goshen. R., deGroot, N., and Hochberg, A. (1994) Relaxation of imprinting in trophoblastic disease. *Gynecol. Oncol.* **53,** 212–219.

62. McFadden, D.E., Kwong, L.C., Yam, I.Y.L., and Langlois, S. (1993) Parental origin of triploidy in human fetuses: evidence for genomic imprinting. *Hum. Genet.* **92,** 465–469.

63. Berkowitz, R.S., Bernstrin, M.R., Laborde, O., and Goldstein, D.P. (1994) Subsequent pregnancy experience in patients with gestational trophoblastic disease. New England Trophoblastic Disease Center, 1965–1992. *J. Reprod. Med.* **39,** 228–232.

64. Matalon, M. and Modan, B. (1972). Epidemiologic aspects of hydatidiform mole in Israel. *Am. J. Obstet. Gynecol.* **112,** 107–112.

65. Rice, L.W., Lage, J.M., Berkowitz, R.S., Goldstein, D.P., and Bernstein, M.R. (1989) Repetitive complete and partial hydatidiform mole. *Obstet. Gynecol.* **74,** 217–219.

66. LaVecchia, C., Franceschi, S., Fasoli, M., and Mangioni, C. (1982) Gestational tropho-blastic neoplasms in homozygous twins. *Obstet. Gynecol.* **60,** 250–252.
67. Ambrani, L.M., Vaidya, R.A., Rao, C.S., Daftary, S.D., and Motashaw, N.D. (1980) Familial occurrence of trophoblastic disease—report of recurrent molar pregnancies in sisters in three families. *Clin. Genet.* **18,** 27–29.
68. Parazzini, F., LaVecchia, C., Franceschi, S., and Mangili, G. (1984) Familial trophoblas-tic disease: case report. *Am. J. Obstet. Gynecol.* **149,** 382–383.
69. Kircheisen, R. and Schroeder-Kurth, T. (1991) Familiäres blasenmolen-syndrom und genetische aspekte dieser gestörten trophoblastentwicklung. *Geburtshilfe Frauenheilkd* **51,** 560–571.
70. Jacobs, P.A. (1981). Mutation rates of structural chromosome rearrangements in man. *Am. J. Hum. Genet.* **33,** 44–54.
71. Dewald, G. and Michels, V.V. (1986) Recurrent miscarriages: cytogenetic causes and genetic counseling of affected families. *Clin. Obstet. Gynecol.* **29,** 865–885.
72. Jacobs, P.A. (1977). Epidemiology of chromosome abnormalities in man. *Am. J. Epidemiol.* **105,** 180–191.
73. Olson, S.B. and Magenis, R.D. (1988) Preferential paternal origin of de novo structural chromosome arrangements. In *The Cytogenetics of Mammalian Autosomal Rearrange-ments* (Daniel, A., ed.), Alan R. Liss, New York, pp. 583–599.
74. Simpson, J.L., Elias, S., Meyers, C.M., Ober, C., and Martin, A.O. (1989) Translocations are infrequent among couples having repeated spontaneous abortions but no other abnor-mal pregnancies. *Fertil. Steril.* **51,** 811–814.
75. Campana, M., Serra, A., and Neri, G. (1986) Role of chromosome aberrations in recur-rent abortion: a study of 269 balanced translocations. *Am. J. Med. Genet.* **24,** 341–356.
76. Braekeleer, M. De and Dao, T.-N. (1990) Cytogenetic studies in couples experiencing repeated pregnancy losses. *Hum. Reprod.* **5,** 519–528.
77. Gadow, E.C., Lippold, S., Otano, L., Serafin, E., Scarpati, R., and Metayoshi, T. (1991) Chromosome rearrangements among couples with pregnancy losses and other adverse reproductive outcomes. *Am. J. Med. Genet.* **41,** 279–281.
78. Petrosky, D.L. and Borgaonkar, D.S. (1984) Segregation analysis in reciprocal transloca-tion carriers. *Am. J. Hum. Genet.* **19,** 137–159.
79. Cowchock, F.S., Gibas, Z., and Jackson, L.G. (1993) Chromosomal errors as a cause of spontaneous abortion: the relative importance of maternal age and obstetrical history. *Fertil. Steril.* **59,** 1011–1014.
80. Morton, N.E., Chiu, D., Holland, C., Jacobs, P.A., and Pettay, D. (1987) Chromosome anomalies as predictors of recurrence risk for spontaneous abortion. *Am. J. Med. Genet.* **28,** 353–360.
81. Hassold, T. (1980) A cytogenetic study of repeated spontaneous abortions. *Hum. Genet.* **32,** 723–730.
82. Warburton, D., Kline, J., Stein, Z., Hutzler, M., Chin, A., and Hassold, T. (1987) Does the karyotype of a spontaneous abortion predict the karyotype of a subsequent abortion?— Evidence from 272 women with two karyotyped spontaneous abortions. *Am. J. Hum. Genet.* **41,** 465–483.
83. Hasegawa, I., Imai, T., Tanaka, K., Fujimori, R., and Sanada, H. (1996) Studies on the cytogenetic and endocrinologic background of spontaneous abortion. *Fertil. Steril.* **65,** 52–54.
84. Scott, J.R., Rote, N.S., and Branch, D.W. (1987) Immunologic aspects of recurrent abor-tion and fetal death. *Obstet. Gynecol.* **70,** 645–656.
85. Gimovsky, M.L. and Montoro, M. (1991) Systemic lupus erythematosus and other con-nective tissue diseases in pregnancy. *Clin. Obstet. Gynecol.* **34,** 35–50.
86. Lubbe, W.F. and Liggins, G.C. (1988) Role of lupus anticoagulant and autoimmunity in recurrent pregnancy loss. *Semin. Reprod. Endocrinol.* **6,** 181–190.

87. Branch, P.W. and Ward, K. (1989) Autoimmunity and pregnancy loss. *Semin. Reprod. Endocrinol.* **7,** 168–177.
88. Brown, H.L. (1991) Antiphospholipid antibodies and recurrent pregnancy loss. *Clin. Obstet. Gynecol.* **34,** 17–26.
89. Mowbry, J.F., Gibbings, C., Liddell, H. Reginald, P.W., Underwood, J.L., and Beard, R.W. (1985) Control trial of treatment of recurrent spontaneous abortion by immunization with paternal cells. *Lancet* **i,** 941–943.
90. McDonough, P.G. (1988) The role of molecular mutation in recurrent euploidic abortion. *Semin. Reprod. Endocrinol.* **6,** 155–161.
91. Gill, T.J. (1992). Influence of MHC and MHC-linked genes on reproduction. *Am. J. Hum. Genet.* **50,** 1–5.
92. Ohno, M., Maeda, T., and Matsunobu, A. (1991) A cytogenetic study of spontaneous abortions with direct analysis of chorionic villi. *Obstet. Gynecol.* **77,** 394–398.
93. Nowinski, G.P., VanDyke, D.L., Tilley, B.C., Jacobsen, G., Babu, V.R., Worsham, M.J., Wilson, G.N., and Weiss, L. (1990) The frequency of aneuploidy in cultured lymphocytes is correlated with age and gender but not reproductive history. *Am. J. Hum. Genet.* **46,** 1101–1111.
94. Simpson, J.L. (1980) Genes, chromosomes, and reproductive failure. *Fertil. Steril.* **33,** 107–116.
95. Curry, C.J.R. (1992) Pregnancy loss, stillbirth, and neonatal death: a guide for the pediatrician. *Pediatr. Clin. North Am.* **39,** 157–192.
96. Mueller, R.F., Sybert, V.P., Johnson, J., Brown, Z.A., and Chen, W.-J. (1983) Evaluation of a protocol for post-mortem examination of stillbirths. *N. Engl. J. Med.* **309,** 586–590.
97. Schauer, G.M., Kalousek, D.K., and Magee, J.F. (1992) Genetic causes of stillbirth. *Semin. Perinatol.* **16,** 341–351.
98. Harger, J. H. (1993) Recurrent spontaneous abortion and pregnancy loss. In *Gynecology and Obstetrics, A Longitudinal Approach* (Moore, T.R., Reiter, R.C., Rebar, R.W., and Baker, V.V., eds.), Churchill Livingston, New York, pp. 247–261.
99. Coulam, C.B. (1991) Epidemiology of recurrent spontaneous abortion. *Am. J. Reprod. Immunol.* **26,** 23–27.
100. Tharapel, A.T., Tharapel, S.A., and Bannerman, R.M. (1985) Recurrent pregnancy losses and parental chromosome abnormalities: *Br. J. Obstet. Gynecol.* **92,** 899–914.
101. Drugan, A., Koppitch, F.C., Williams, J.C., Johnson, M.P., Moghissi, K.S., and Evans, M.I. (1990) Prenatal genetic diagnosis following recurrent early pregnancy loss. *Obstet. Gynecol.* **75,** 381–384.
102. Munné, S., Grifo, J., Cohen, J., and Weier, H-U.G. (1994) Chromosome abnormalities in human arrested preimplantation embryos: a multiple-probe FISH study. *Am. J. Hum. Genet.* **55,** 150–159.
103. Zenzes, M.T. and Casper, R.F. (1992) Cytogenetics of human oöcytes, zygotes, and embryos after in vitro fertilization. *Hum. Genet.* **88,** 367–375.

13

Cancer Cytogenetics

AnneMarie W. Block, Ph.D.

INTRODUCTION

Chromosome abnormalities in neoplasia are acquired during the process of tumorigenesis. These aberrations are restricted to the tumor tissue and are not present in other cells of the body. The distribution of abnormalities is highly nonrandom and clonal in nature. Both benign and malignant tumors have been shown to exhibit chromosome abnormalities.

Two major classes of mitotic events affect the chromosomes in neoplastic cells: abnormal segregation and chromosome breakage. The former (nondisjunction, anaphase lag) results in changes in chromosome numbers that range from monosomy and trisomy to haploidy and polyploidy. The latter, caused by exogenous factors (ionizing radiation, viruses, chemical mutagenic agents) and endogenous factors (enzyme systems comprising DNA replication, transcription, and recombination) results in deletions, duplications, inversions, translocations, insertions, and amplifications. The consequences of such events are (1) gain of whole chromosomes or chromosomal segments, (2) loss of whole chromosomes or segments, and (3) relocation of chromosomal segments within a chromosome or between chromosomes leading to recombination of genes located at breaks.

Chromosome aberrations in neoplasia have been classified into two main categories: primary and secondary abnormalities. Primary abnormalities are strongly (nonrandomly) correlated with a specific tumor type, may be the only cytogenetic abnormality, and may play an important role in the earliest stages of tumor initiation. Secondary abnormalities are also nonrandom, but less disease specific, and are postulated to be later events contributing to the process of tumor progression.

In the clinical setting, cytogenetic analysis has proven to be an invaluable tool in the diagnosis, prognosis, and management of hematologic malignancies while aiding in the differential diagnosis between solid tumor types with common features.

AN HISTORICAL LOOK

The significance of mitotic and chromosomal abnormalities observed in cancer cells to the development of tumors was first appreciated during the late 19th and early 20th centuries. The first paper on the subject of human malignant tumors was that of Arnold *(1)* in 1879. In 1890, von Hansemann *(2)* postulated that all carcinomas are character-

From: *The Principles of Clinical Cytogenetics*
Edited by: S. Gersen and M. Keagle © Humana Press Inc., Totowa, NJ

ized by asymmetric karyokinesis (mitosis) leading to an unequal chromatin distribution, resulting in cancer. Boveri *(3,4)* working with sea urchin eggs and *Ascaris,* correlated the Mendelian theory of segregation of specific characteristics with the reduction phenomenon of the chromosomes at meiosis. The stemline concept was first proposed by Winge *(5)* and subsequently extended to tumor development at the pathophysiologic level by Foulds *(6).* Cytogenetic analysis of cultured mammalian cells by Levan *(7)* and Hauschka *(8)* provided further evidence of this new view of tumor cells. Makino *(9),* Hsu *(10),* and Sandberg *(11)* furnish fascinating historical perspectives of this period for the interested reader.

A new era in cancer cytogenetics was ushered in by the technical innovation in the late 1950s of hypotonic treatment of cells to obtain better spreading of metaphase chromosome preparations. Related to this discovery was the description by Nowell and Hungerford *(12)* in 1960 of a specific chromosome abnormality consistently associated with chronic myelogenous leukemia (CML) (see Chapter 1). This abnormality, an apparent deletion of one of the "G group" chromosomes, was designated the Philadelphia or Ph chromosome. The introduction of banding techniques in the early 1970s showed the Ph chromosome in CML and the 14q+ marker chromosome in Burkitt's lymphoma (BL) to be derived from specific translocations *(13,14).* During the past 20 years, a vast literature has accumulated describing in ever-increasing detail the pattern of chromosome changes occurring in neoplasia *(11,15–17).* Paralleling this compilation is the application of molecular genetic techniques that have increased our understanding of the pathogenetic mechanisms underlying the neoplastic process.

To date nearly 27,000 cytogenetically abnormal neoplasms have been reported in the literature and catalogued in the Cancer Chromosome Data Bank *(15).* Of these cases, 215 balanced and 1588 unbalanced neoplasia-associated rearrangements have been identified among 75 different neoplastic disorders *(17)* (see **Figs. 1 and 2**).

HEMATOLOGIC NEOPLASMS

Chronic Myeloid Leukemia

Accounting for approximately 15–20% of all leukemia cases, CML is most common in the 40- to 50-year age group. Bone marrow morphology is characterized by granulocytic and megakaryocytic hyperplasia without maturation arrest. Eosinophilia, basophilia, and myelofibrosis are common.

Because all hematopoietic cell lineages are involved in the disease process, the leukemogenic event is thought to take place at the level of the pluripotent stem cell *(18).* Later accelerated and blast crisis phases are characterized by an increase in immature cells in the bone marrow and peripheral blood, anemia, thrombocytopenia, extramedullary accumulation of blast cells, and an associated reduced response to therapy. Cells in the terminal phase may morphologically resemble myeloblasts or lymphoblasts.

CML was the first neoplastic disease to be associated with a specific recurrent chromosome abnormality *(12).* More than 90% of patients with CML exhibit a (9;22) translocation, t(9;22)(q34.1;q11.2) (see **Fig. 1t**). In honor of the city of discovery, the derivative chromosome 22 formed by this reciprocal rearrangement has been described as the Philadelphia (Ph) chromosome.

Fig. 1. Representative examples of translocations seen in hematologic disorders. [From ref. (*371*), with permission.]

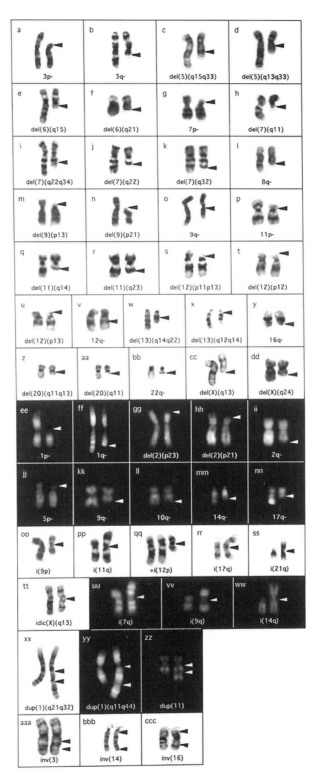

Fig. 2. Representative examples of inversions, isochromosomes, duplications, and deletions seen in hematologic disorders. [From ref. (*371*), with permission.]

Two subgroups of variant Ph-producing translocations have been observed in 5–10% of CML cases. In simple variant translocations, the segment lost from chromosome 22 is translocated to a chromosome other than 9. Complex variant translocations involve one or more chromosomes in addition to 9 and 22. All chromosomes, with the exception of Y, have been reported as variant translocation partners with a nonrandom clustering to Giemsa-negative breakpoints *(15)*. High-resolution banding studies have, however, demonstrated simple variant translocations to cryptically involve band 9q34 *(19)*. The classic and variant Ph-positive CML types do not differ in hematology or prognosis *(11)*.

A molecular result of the t(9;22) rearrangement is the translocation and subsequent fusion of the *ABL* oncogene on chromosome 9 to the breakpoint cluster region *(BCR)* gene on chromosome 22 *(20)*. The break in 9q34 occurs 5′ to or within the *ABL* oncogene. Most breaks occur in the long intron between *ABL* exons 1a and 1b. Breaks in *BCR* occur in the centrally located major *bcr* (M-*bcr*) between exons b2 and b3 or b3 and b4. The product of *BCR/ABL* transcription is an aberrant 8.5-kb mRNA, which translates into a chimeric 210-kD protein with tyrosine-specific kinase activity.

Ph-Negative CML

Between 5% and 10% of traditional CML cases have no observable Ph chromosome. These patients differ from Ph-positive patients in a number of ways (older, male, lower platelet and WBC counts) and frequently display nonclonal chromosome abnormalities. Molecular cytogenetic studies using fluorescence *in situ* hybridization (FISH) have demonstrated submicroscopic rearrangements of chromosomes 9 and 22 *(22)*. Trisomy 8 (20%), monosomy 7 (20%), i(17)(q10) (10%, **Fig. 2rr**), and structural rearrangements of chromosome 3 (>10%) are the most common aberrations reported in karyotypically abnormal Ph-negative CML.

Secondary Changes

Secondary chromosome aberrations occur in 75–80% of patients when disease progression occurs. **Table 1** lists the major (70%) routes of cytogenetic evolution in CML accelerated phase and blast crisis. Gains of chromosome 8, an extra Ph chromosome [der(22)t(9;22)] and i(17)(q10) (**Fig. 2rr**) are the most common major route solitary changes. Combinations of these changes along with gain of chromosome 19 usually occur later in karyotype evolution. More than 50% of minor route cases (30%) include –Y, –7, –17, +17, +21 or t(3;21)(q26;q22) (**Fig. 1k**) *(15)*. Near-haploid leukemia stemlines have also been reported *(23,24)*. The presence of these additional chromosome abnormalities at diagnosis has been shown to be associated with a poor prognosis *(25)*. The type of secondary chromosome aberration observed has also been correlated with the phenotype of blast crisis *(26,27)*. Hyperdiploidy, trisomies of chromosomes 8, 19, 21, and i(17)(q10) are more common in myeloid blast crisis. An extra Ph chromosome is seen in both lymphoid and myeloid blast crises. The translocation t(3;21) (**Fig. 1k**) is associated with megakaryocytic lineage involvement *(28)*. Other structural chromosome abnormalities associated with acute myeloid leukemia, inv(3)(q21q27) (**Fig. 2aaa**), t(15;17)(q22;q12) (**Fig. 1bb**), and inv(16)(p13q22) (**Fig. 2ccc**), have also been sporadically reported in patients whose bone morphology showed megakaryoblastic, promyelocytic, and myelomonocytic transformation with eosinophilia, respectively.

Table 1
Major Routes of Cytogenetic Evolution
in CML Blast Crisis

Additional change	Frequency (%)
+Ph[a]	15
i(17q)	12
+8	11
+Ph, +8	8
+8, i(17q)	7
+Ph, +8, +19	5
+Ph, +19	4
+8, +19	2
+Ph, +8, i(17q)	2
+19	1
i(17q), +Ph	1
+8, i(17q), +19	1
+Ph, +8, i(17q), +19	1
i(17q), +19	>1
i(17q), +19, +Ph	>1

[a]+der(22)t(9;22)(q34.1;q11.2)

Acute Myeloid Leukemia

Acute myeloid leukemia (AML) is characterized by an excessive accumulation of immature nonlymphatic bone marrow precursor cells due to a block in cell maturation. The incidence of AML is 2.5 cases/100,000 per year with a male predominance. Incidence rises sharply after 55–60 years of age. AML historically has been classified on the basis of similarities between leukemia cells and the precursor cells that are associated with the developmental stages in normal hematopoiesis *(29–31)*. The French–American–British (FAB) Cooperative Group classifies AML by morphologic and nonmorphologic criteria (*29,30;* see **Table 2**).

The Cancer Chromosome Data Bank reports more than 6000 cytogenetically characterized cases of AML with 41 recurrent balanced (39 sole) and 178 (118 sole) unbalanced abnormalities *(17)*. Clonal chromosome abnormalities are present in more than three fourths of successfully karyotyped patients with AML *(16,32–34)*. The distribution of chromosome aberrations is nonrandom; specific rearrangements are associated with hematologic subgroups: t(8;21) in M2 (**Fig. 1r**), t(15;17) in M3 (**Fig. 1bb**), inv(16) in M4$_{EO}$ (**Fig. 2ccc**), and t(9;11) in M5 (**Fig. 1s**) (see **Table 3**). A significant proportion of cytogenetically abnormal AML patients (>55%) display single numerical or structural chromosome abnormalities (15–20% have gain/loss of a single chromosome as the sole change); the remaining 45% have two or more abnormalities *(7,35)*. Cytogenetically normal patients may display submicroscopic changes detectable only by molecular genetic techniques *(36,37)*. The diagnostic karyotype has been shown to be an independent prognostic parameter in AML *(35)*.

Additional chromosome aberrations may be present at diagnosis or appear for the first time at relapse. These secondary chromosome abnormalities are believed to reflect

Table 2
FAB Classification of AML

FAB type	Incidence (%)	Description
M0	<5	Acute myeloblastic leukemia without maturation
M1	20	Acute myeloblastic leukemia with minimal differentiation
M2	30	Acute myeloblastic leukemia with maturation
M3; M3 variant	10	Acute promyelocytic leukemia (hypergranular); acute promyelocytic leukemia (hypo- or microgranular)
M4; M4$_{Eo}$	25	Acute myelomonocytic leukemia; M4 with eosinophilia
M5a; M5b	10	Acute monoblastic leukemia; acute promonocytic–monocytic leukemia
M6	<5	Acute erythroblastic leukemia
M7	<5	Acute megakaryoblastic leukemia

later genetic events contributing to disease progression *(17)*. Even though secondary abnormalities are less specific than primary changes, these aberrations are nonrandom and may be dependent upon the type of primary chromosome rearrangement or type of leukemia *(38)*.

Secondary AML

Two different classes of therapy related AML (tAML) have been reported. One class [–5/del(5q) and –7/del(7q); see **Fig. 2c, d, h, i, j, k**] is causally related to medical exposure to alkylating agents, and is preceded by a myelodysplastic syndrome (MDS) phase and poor response to chemotherapy *(39,40)*. The second class develops as a late effect after DNA-topoisomerase II inhibitors combined with alkylating agents and radiation *(41)*. This leukemia, which is characterized by balanced rearrangements involving 11q23 and 21q22, begins earlier, without an initial MDS, and has a more favorable response to chemotherapy.

Molecular cloning of leukemia-specific breakpoints involved in recurrent translocations, inversions, and deletions has led to the discovery of genes that become leukemogenic by the mechanisms of either gene activation or gene fusion (**Table 3**). Gene activation occurs when the transcriptional regulatory elements of one gene become juxtaposed to the complete reading frame of another gene. Under the control of a new promoter and/or enhancer region, the expression pattern of the affected gene becomes disrupted. Gene fusions occur when segments from two different genes are joined to give rise to a chimeric structure consisting of the 5′ end of one gene and the 3′ end of another. The resultant chimeric proteins contain structural and functional domains that differ from the wild-type proteins and it is generally accepted that it is the expression of these fusion products that leads to the formation of leukemia. However, the mechanism of leukemogenesis in most fusions is still unknown *(42)*.

Myelodysplastic Syndromes

Myelodysplastic syndromes (MDS), often termed "preleukemias," are a heterogeneous group of clonal hematologic disorders characterized by bone marrow dysfunc-

(*text continued on page 359*)

Table 3

Association of Recurrent Primary Chromosome Aberrations in AML with Disease Subtype, Molecular Abnormality, Clinical Characteristics and Prognosis

Cytogenetic abnormality	Genes involved	Protein	FAB type(s)	Clinical characteristics/prognosis
+1				
del(1)(q21)				
i(1)(q10)				
t(1;3)(p36;q21)			M1, M4	Preceded by MDS; dysmegakaryocytopoiesis
t(1;11)(p32;q23)[b]	AF1p (1p32)	Murine eps 15 homolog	M0, M5	
	MLL (11q23)	Drosophila trithorax homolog		
t(1;11)(q21;q23)[b]	AF1q (1q21)	No homolog to any known protein	M4, M5	
	MLL (11q23)	Drosophila trithorax homolog		
t(1;17)(q36;q21)				
t(1;22)(p13;q13)			M7	Thrombocytopenia, hepatosplenomegaly, bone marrow fibrosis; 2° assoc. with der(1)t(1;22)
del(2)(p21)				
del(2)(p23)				
t(2;3)(p23;q25)				
t(2;11)(p21;q23)				
+3				
del(3)(q21)				
inv(3)(q21q26)[a]	Ribophorin (3q21)	RER transmembrane glycoprotein	M0, M1, M2, M4, M5, M6, M7[d]	Poor prognosis
	EVI1 (3q26)	Multiple zinc fingers		
t(3;3)(q21;q26)[a,e]	Ribophorin (3q21)	RER transmembrane glycoprotein	M1, M2, M4, M6[d]	
	EVI1 (3q26)	Multiple zinc fingers		
t(3;5)(q25;q34)			M6	Megakaryocytosis; young age at diagnosis; Sweet's syndrome
t(3;11)(q21;q13)				
t(3;12)(q26;p13)				
t(3;21)(p14;q22)				

(continued)

Table 3 (Continued)

Cytogenetic abnormality	Genes involved	Protein	FAB type(s)	Clinical characteristics/prognosis
t(3;21)(q26;q22)[b]	EVI1 (3q26) AML1 (21q22)	Multiple zinc fingers CBFα, Drosophila runt homolog	[d]	Genotoxic exposure
t(3;21)(q26;q22)[b]	EAP (3q26) AML1 (21q22)	Ribosomal protein L22 CBFα, Drosophila runt homolog	[d]	
t(3;21)(q26;q22)[b]	MDS1 (3q26) AML1 (21q22)	40% Homolog to RIZ CBFα, Drosophila runt homolog	[d]	
+4			M2, M4	
del(4)(p15)				
t(4;12)(q11–13;p13)				
+5				
−5				Genotoxic exposure; poor prognosis
del(5)(q13q33)[f]				Genotoxic exposure; 2° assoc. with t(1;3), inv(3), −7/del(7q), poor prognosis
t(5;16)(q33;q22)				
+6				
del(6)(q15)				
del(6)(q21-22)				
i(6)(q10)				
t(6;9)(p23;q34)[b]	DEK (6p23) CAN (9q34)	nuclear protein nucleoporin	M1, M2, M4[d]	Preceded by MDS; bone marrow basophilia; young age at diagnosis; Auer rods
t(6;11)(q27;q23)[b]	AF6 (6q27) MLL (11q23)	GLGF motif Drosophila trithorax homolog	M4, M5	Localized infections
+7			Multiple	2° assoc. with t(1;22)
−7				Genotoxic exposure; 2° assoc. with inv(3); hepatosplenomegaly; poor prognosis
del(7)(p12p21)			Multiple	Genotoxic exposure; poor prognosis
del(7)(q22)[g]			M1, M2, M4	Preceded by MDS; poor prognosis; genotoxic exposure
der(1;7)(q10;p10)				

353

(continued)

Table 3 (*Continued***)**

Cytogenetic abnormality	Genes involved	Protein	FAB type(s)	Clinical characteristics/prognosis
i(7)(q10)				
t(7;11)(p15;p15)[b]	HOXA9 (7p15) NUP98 (11p15)	Class 1 homeobox Nucleoporin	M2, M4[d]	Auer rods
t(7;11)(p13;p15) t(7;12)(q21;q13) t(7;21)(q31;q22) +8[l]			M2, M4, M5	Preceded by MDS; 2° assoc. with der(1;7), t(6;9), t(9;11), t(9;22), t(15;17), inv(16); sole change in older patients; leukemia cutis; intermediate prognosis
del(8)(q22) t(8;16)(p11;p13)			M4, M5 (M5b)	Mainly infants/children; erythrophagocytosis; DIC; poor prognosis
t(8;21)(q22;q22)[b]	ETO (8q22) AML1 (21q22)	Zinc finger CBFα, *Drosophila* runt homolog	M2[d]	Auer rods and eosinophilia; older children and adults; granulocytic sarcomas; intermediate prognosis
t(8;22)(p11;q13) +9 −9			M2, M4, M5	
del(9)(p21) del(9)(q22)[h]			Multiple	2° assoc. with t(8;21); agranular blasts; Auer rods
t(9;11)(p21−22;(q23)[b]	AF9 (9p22) MLL (11q23)	Nuclear protein, ENL homology *Drosophila* trithorax homolog	M4, M5 (M5a)	
t(9;22)(q34.1;q11.2)	ABL (9q34) BCR (22q11)	Tyrosine kinase Serine kinase	M1, M2	2° assoc. with der(22)t(9;22)
+10 ins(10;11)(p11;q23q24) t(10;11)(p11−15;q13−23)[b]	AF10 (p12) MLL (11q23)	Leucine zipper, zinc finger *Drosophila* trithorax homolog	M0, M4, M5	

(continued)

Table 3 (Continued)

Cytogenetic abnormality	Genes involved	Protein	FAB type(s)	Clinical characteristics/prognosis
t(10;17)(p13;q12–21) +11[b,c]	MLL (11q23)	Drosophila trithorax homolog	M1, M2, M4	Sole change in older patients; Auer rods
del(11)(p11–12p14–15) del(11)(q14–23) del(11)(q23;q25) i(11)(q10) t(11;11)(q23q25) t(11;15)(q23;q14–15)				
t(11;17)(q23;q21)[b]	MLL (11q23) AF17 (17q21)	Drosophila trithorax homolog Leucine zipper, zinc finger	M5	
t(11;17)(q23;q21)[b]	PLZF (11q23) RARα (17q12)	Zinc finger Retinoic acid receptor α	M3	Poor prognosis
t(11;17)(q23;q23) t(11;17)(q23;q25)			M2, M4, M5	
t(11;19)(q23;p13.3)[b]	MLL (11q23) ENL (19p13.3)	Drosophila trithorax homolog Transcription factor	M4, M5	Mixed myeloid/lymphoid phenotype; mostly in male infants
t(11;19)(q23;p13.1)	MLL (11q23) ELL (19p13.1)	Drosophila trithorax homolog Transcription enhancer	M4, M5	
t(11;20)(p15;q11) t(11;22)(q23;q11) +12				
del(12)(p12)[i]			M2, M4 Baso	2° assoc. with −7; genotoxic exposure 2° AML; poor prognosis
+i(12)(p10) t(12;13)(p13;q14) t(12;14)(q13;q32) t(12;14)(q24;q32)				Primarily in males; mediastinal germ cell tumors

(continued)

Table 3 (Continued)

Cytogenetic abnormality	Genes involved	Protein	FAB type(s)	Clinical characteristics/prognosis
t(12;17)(q13;p11–12)				
t(12;19)(q13;q13.3)				
t(12;22)(p12–13;q11–13)[b]	TEL (12p13) MN1 (22q11)	ETS-related transcription factor Nuclear protein	M1, M4, M7[d]	
+13			Multiple	2° assoc. with t(6;9); seen as a sole abnormality in males; sole change in older patients; undifferentiated/biphenotypic
−13				
del(13)(q12q14)[j]				
i(13)(q10)				Sole change in older patients
+14				
−14				
i(14)(q10)				
+15				
t(15;17)(q22;q12)[b]	PML (15q21) RARα (17q21)	Zinc finger Retinoic acid receptor α	M3, M3v	Older children or adults; absence of extramedullary leukemia; DIC; Auer rods; good prognosis
−16				
del(16)(q22)				
der(16)t(1;16)[k]				
inv(16)(p13q22)[b]	MYH11 (16p13) CBFβ (16q22)	Smooth muscle myosin heavy chain Heterodimerizes with AML1	M4$_{EO}$	Lymphadenopathy, hepatomegaly; good prognosis
t(16;16)(p13;q22)[b]	MYH11 (16p13) CBFβ (16q22)	Smooth muscle myosin heavy chain Heterodimerizes with AML1	M4$_{EO}$	Lymphadenopathy, hepatomegaly; good prognosis
t(16;21)(p11;q22)[b]	FUS (16p11) ERG (21q22)	RNA-binding protein ETS-related transcription factor	M1, M2, M4, M5, M7	Lymphadenopathy, hepatomegaly

(continued)

Table 3 (Continued)

Cytogenetic abnormality	Genes involved	Protein	FAB type(s)	Clinical characteristics/prognosis
+17				2° assoc. with +21
−17			Multiple	MDS features; seen in older patients; poor prognosis
i(17)(q10)				
del(17)(q22)				
+18				2° assoc. with t(10;11)
−18				2° assoc. with t(1;22), t(9;22); seen as sole change in young patients
+19				
−19				
+20			Multiple	
−20			Multiple	MDS features
del(20)(q11q13)				
+21				
−21				
i(21)(q10)			M4	MDS features; 2° assoc. with inv(16)
+22				2° assoc. with t(1;22)
−22				
del(22)(q11−13)				2° assoc. with t(8;21)
−X				
del(X)(q24)				Found exclusively in women;seen in older patients
idic(X)(q13)				

(continued)

357

Table 3 (Continued)

Cytogenetic abnormality	Genes involved	Protein	FAB type(s)	Clinical characteristics/prognosis
t(X;11)(q13;q23)				
t(X;11)(q24–25;q23)				
+Y				2° assoc. with der(1;17)
–Y			Multiple	Age related phenomenon in elderly patients; 2° assoc. with t(8;21)

[a]Gene activation.
[b]Gene fusion.
[c]Partial tandem gene duplication.
[d]Also observed in MDS.
[e]Also interpreted as ins(3;3)(q21;q21q26).
[f]Other breakpoints (between bands 5q11 and 5q35) also reported.
[g]Other breakpoints (between bands 7q11 and 7q36) also reported.
[h]Other breakpoints (between bands 9q11 and 9q34) also reported.
[i]Breakpoints in 12p11 and 12p13 also reported.
[j]Other breakpoints (between bands 13q11 and 13q34) also reported.
[k]Also observed in ALL, solid tumors.
[l]Also observed in solid tumors.

Table 4
Recurrent Primary Chromosome
Aberrations in Myelodysplastic Syndromes

del(1)(p22)	+9
t(1;3)(p36;q21)	t(9;11)(p22;q23)
der(1;7)(q10;p10)	t(9;22)(q34.1;q11.2)
t(2;11)(p21–22;q23)	+11
t(3;21)(q26;q22)	del(11q)[c]
3q26 abnormalities[a]	del(12)(p11–13)
–5	del(13q)[d]
del(5q)[b]	16q22 abnormalities
t(5;12)(q33;p13)	i(17q)
+6	+19
t(6;9)(p23;q34)	del(20q)[e]
–7	+21
del(7q)	–X
+8	–Y
t(8;21)(q22;q22)	

[a]Includes inv(3)(q21q26), t(3;3)(q21;q26), ins(3;3)(q21;q21q26).
[b]del(5)(q13q33) observed in 5q– syndrome.
[c]Includes breakpoints at 11q14 and 11q23.
[d]Includes breakpoints 13q11–22.
[e]Includes breakpoints 20q11–13.

tion involving all lineages. Myelodysplastic syndromes have an incidence of 1 per 100,000 persons per year in the United States, and affect mostly the elderly. Approximately 20–40% of patients progress to acute myeloid leukemia.

The FAB Study Group recognizes five different myelodysplastic subgroups: refractory anemia (RA), refractory anemia with ringed sideroblasts [RARS; also called acquired idiopathic sideroblastic anemia (AISA)], chronic myelomonocytic leukemia (CMMoL), refractory anemia with excess of blasts (RAEB), and refractory anemia with excess of blasts in transformation (RAEBT) *(43,44)*.

One third to one half of successfully karyotyped cases are abnormal *(45)*. More than 2200 MDS cases with 13 recurrent balanced and 95 unbalanced abnormalities have been reported *(17)*. The karyotypic abnormalities found in MDS are also observed in AML (**Table 4**) *(11,16,17,46)*. This apparent karyotypic similarity suggests a neoplastic progression between the two diseases. It should be noted, however, that subgroup specific AML-associated abnormalities t(8;21)/M2, t(15;17)/M3, and inv(16)/M4$_{EO}$, are rarely seen in MDS. The most common chromosome changes in MDS are del(5q) (30%), –7 (20–25%), +8 (20%), del(20q) (5–10%), –Y (5–10%), and del(7q) (5–10%) *(46)* (see **Fig. 2c, d, z, aa, h, i, j, k**).

The most common changes in RA are del(5q), –7, and +8 *(46)*. Deletion 5q is seen in 50% of RA patients, with a specific del(5)(q13q33) associated with the "5q– syndrome" described by van den Berghe et al. *(47)*.

Trisomy 8 is present in one fourth to one third of all abnormal RARS cases. Also observed in RARS are del(5q) (25%), –7 (10%), del(11q) (10%), and del(20q) *(48–50)*.

Table 5
Recurrent Primary Chromosome Aberrations in Chronic Myeloproliferative Disorders

Diagnosis	Cytogenetic abnormalities[a]
Polycythemia vera	del(20q), +8, +9, 1q gain, del(13q)
Idiopathic myelofibrosis[b]	der(1q)[c], –7, +8, +9, del(5q), del(13q), del(20q), i(17q), t(9;22)(q34.1;q11)
Essential (idiopathic) thrombocytopenia	t(9;22)(q34;q11.2), +9

[a]Abnormalities listed in order of frequency.

[b]Also known as myelofibrosis with myeloid metaplasia, agnogenic myeloid metaplasia, myelosclerosis, osteosclerosis, aleukemic myelosis.

[c]t/dup/del(1)(q21–32) resulting in partial trisomy 1q, t(1;5)(q23;q23), t(1;6)(q23;p23).

Monosomy 7 (20%), +8 (20%), and –Y (10–15%) are the most common changes in CMMoL *(51)*. Deletion of the long arm of chromosome 5 (25%) and –7 are common in RAEBT, but monosomy 5 (10%) is less frequent as a sole abnormality *(46)*.

Secondary MDS

Chromosome abnormalities in therapy-related MDS (tMDS) are more frequent than *de novo* MDS. Common changes in tMDS are –7 (40%), del(5q) (25%, **Fig. 2c, d, e**), and –5 (10%), followed by der(21q), +8, del(7q), der(12p) (**Fig. 2s, t, u**), der(1;7) (**Fig. 1b**), –12, –17, der(17p), der(3p), der(6p), and –18 *(52)*. Like tAML, the incidence of cytogenetic abnormalities correlates with the type of previous therapy *(53–56)*. Secondary MDS and AML may also develop after occupational exposure to genotoxins *(57)*. Patients with –7 or complex karyotypes have a higher risk to progression to acute leukemia and shorter survival *(16,58)*.

Chronic Myeloproliferative Disorders

Chronic myeloproliferative disorders (CMD or MPD) are characterized by clonal proliferation of one or more lineages of the myeloid series. Maturation is less affected, with no maturation arrest or block.

The various MPD subtypes differ by hematopoietic cell type affected. Overproduction of erythrocytes is common in polycythemia vera (PV), overproduction of platelets is seen in essential thrombocythemia (ET), while myeloid metaplasia and reactive fibrosis are common in idiopathic myelofibrosis (IMF). Shift from one subtype to another can occur with progression to AML.

A single recurrent balanced abnormality and 20 unbalanced abnormalities have been observed in over 600 reported cases *(17)*. While no karyotypic abnormality is specific for CMD, certain numerical and structural abnormalities (also found in other myeloid disorders) have been reported (**Table 5**).

Two thirds of cytogenetically abnormal PV cases have either del(20q) (25%, **Fig. 2z, aa**), +8 (15–20%), +9 (15–20%), 1q gain (10%) or del(13q) (10%, **Fig. 2w, x**) *(50,59)*. IMF carries a 10–15% risk of progression to AML and may represent a variant of the M7 subtype. Monosomy 7 syndrome, which often resembles juvenile CML, is classified as IMF *(60,61)*.

Table 6
FAB Classification of ALL

FAB type	Description
L1	Small cells with homogeneous nuclear chromatin, regular nuclear shape, indistinct nucleoli. No T- and B-cell antigen expression. CD10 [common ALL antigen (CALLA)] positive. Most common type of childhood ALL.
L2	Larger cells with more heterogeneity in size and distribution of nuclear chromatin than in L1. Irregular nuclear shape; one or more large nucleoli. CALLA positive. Less common in childhood.
L3	Large and homogeneous cells with finely stippled nuclear chromatin, regular nucleus, prominent nucleoli, and often prominent vacuolation of basophilic cytoplasm. Morphologic counterpart of Burkitt-type leukemia.

Juvenile myelomonocytic leukemia (JMMoL) has been recently recognized as a separate diagnostic entity. Absence of t(9;22), a questionable association with –7, and an apparently normal karyotype have been reported in this disorder *(62)*. The CML-specific t(9;22)(q34;q11) is present in one third of ET patients, while +9 is observed in 10% of patients with abnormal karyotypes *(63,64)*.

Acute Lymphocytic Leukemia

Acute lymphocytic leukemia (ALL) is characterized by the accumulation of immature lymphoid cells in the bone marrow and peripheral blood. ALL is more common in children than in adults, with a median occurrence at 3–5 years of age in children and under 30 years of age in adults. The incidence of ALL in childhood is 3/100,000 per year. Both childhood and adult ALL are characterized by a male predominance.

As lymphoid cells differentiate and mature, the structural changes that they undergo are accompanied by functional changes *(65,67)*. The principal change occurring in the maturation of immature lymphoblasts to B lymphocytes is somatic recombination of immunoglobulin gene germline DNA in a highly ordered sequence. These loci are on three different chromosomes: heavy-chain locus (*IgH*) at 14q32, kappa light chain locus (*Igκ*) at 2p12, and lambda light chain locus (*Igλ*) at 22q11. The current concept of T-cell maturation follows a similar schema utilizing germline DNA recombination at the T-cell receptor locus. Three loci are thought to be essential in the production of T-cell receptor molecules: alpha chain locus (*TCRα*) at 14q11, beta chain locus (*TCRβ*) at 7q35, and the gamma chain locus (*TCRγ*) at 7p13. The delta chain receptor (*TCRδ*) lies within the α-chain locus at 14q11.

The FAB and MIC (Morphology, Immunology, and Cytogenetics) classification proposals take into account characteristic morphology, functional aspects of cell maturation, and differentiation in an attempt to subdivide the disease into meaningful subgroups *(68,69;* **Tables 6 and 7**). The MIC system defines four major immunologic subclasses of B-lineage ALL:

(text continued on page 366)

Table 7
Association of Recurrent Chromosome Aberrations in ALL with Disease, Lineage/Subtype, Molecular Abnormality, Clinical Characteristics and Prognosis

Cytogenetic abnormality	Genes involved	Protein	Phenotype	FAB type	Clinical characteristics/prognosis
t(1;7)(p32;q34)[a]	TAL1 (1p32) TCRβ (7q34)	βHLH transcription factor[c] T-cell receptor promoter	T lineage		
t(1;7)(p34;q34)[a]	LCK (1p34)	Lymphocyte-specific protein tyrosine kinase	T lineage		
t(1;11)(p32:q23)[b,d]	TCRβ (7q34) AF1p (1p32) MLL (11q23)	T-cell receptor promoter Murine eps 15 homolog Drosophila trithorax homolog		L1	Sole karyotypic abnormality in majority of reported cases
t(1;14)(p33;q11)[a]	TAL1 (1p32) TCRδ (14q11)	βHLH transcription factor T-cell receptor promoter	T lineage	L1	
t(1;19)(p23;p13)[b]	PBX1 (1p23) E2A (19p13)	Homeobox domain βHLH transcription factor	Pre B-cell	L1	Common rearrangement, 6% of reported cases
dup(1)(q12–21q31–32)			Pre B-cell, B-cell	L2, L3	2° abnormality; associated with t(8;14) and massive hyperdiploidy
t(2;8)(p12:q24)[a]	IGK (2p12) MYC (8q24)	Igκ promoter βHLH transcription factor	B-cell	L3	Burkitt's lymphoma
t(4;11)(q21:q23)[b,d]	AF4 (4q21) MLL (11q23)	Serine and proline-rich Drosophila trithorax homolog	Early B precursor, biphenotypic, mixed lineage	L1, L2	Associated with high WBC, increased blasts, hepatosplenomegaly and/or lymphadenopathy; mostly female and very young. Associated with genotoxic exposure in older patients. >10% of ALL cases, poor prognosis
t(5;14)(q31:q32)[a]	IL3 (5q31) IGH (14q32)	Hematopoietic growth factor Ig heavy chain promoter	Pre B with eosinophilia		

(continued)

362

Table 7 (Continued)

Cytogenetic abnormality	Genes involved	Protein	Phenotype	FAB type	Clinical characteristics/prognosis
del(6q)[e]			B or T lineage	L1, L2	5–10% of karyotypically abnormal cases, only change in two thirds
i(6p)					Rare, 2° abnormality
dic(7;9)(p11–13;p11)			Pre B-cell	L1	Rare
i(7q)					Rare, 2° abnormality often associated with massive hyperdiploidy
t(7;9)(q34–36;q34.3)[a]	TCRβ (7q35) TAN1 (9q34.3)	T-cell receptor promoter Notch homolog	T lineage		
t(7;9)(q34;q32)[a]	TCRβ (7q34) TAN2 (9q32)	T-cell receptor promoter βHLH transcription factor	T lineage		
t(7;10)(q34–36;q24)[a]	TCRβ (7q35) HOX11 (10q24)	T-cell receptor promoter Homeobox	T lineage		Variant of t(10;14)
t(7;11)(q34–36;p13)[a]	TCRβ (7q35) LMO2 (11p13)	T-cell receptor promoter Zinc finger	T lineage		
t(7;14)(p15;q32) t(7;19)(q35;p13)[a]	TCRβ (7q35) LYL1 (19p13)	T-cell receptor promoter βHLH transcription factor	T lineage		
+8					Rarely seen as sole change (1–2% of cases)
t(8;14)(q24;q11)[a]	MYC (8q24) TCRα (14q11)	βHLH transcription factor T-cell receptor promoter	T lineage		High WBC, bulky extramedullary leukemia, CNS involvement (one third of cases), mostly male
t(8;14)(q24;q32)[a]	MYC (8q24) IGH (14q32)	βHLH transcription factor Ig heavy chain promoter	B-cell	L3	Burkitt's lymphoma
t(8;22)(q24;q11)[a]	MYC (8q24) IGL (22q11)	βHLH transcription factor Ig light chain promoter	B-cell	L3	Burkitt's lymphoma
del(a)(p21-22)			B or T lineage	L1, L2	Sole abnormality in 50% of patients, mostly male; good prognosis
t/dic(9;12)(p11–12;p11–13)			Pre B or early B precursor		

(continued)

Table 7 (Continued)

Cytogenetic abnormality	Genes involved	Protein	Phenotype	FAB type	Clinical characteristics/prognosis
i(9q)			Pre-B cell	L2	Rare, 2° abnormality; associated with t(1;19); older children
t(9;22)(q34.1;q11.2)[b,f]	ABL (9q34) BCR (22q11)	Tyrosine kinase Serine kinase	B lineage, mixed lineage	L1, L2	Children: 5% of cases, older age, CNS leukemia, high leukocyte count, increase in circulating blasts, L2 morphology Adult: 20–25% of cases, coarse, pink peroxidase-negative granules in Ph-positive cells, poor prognosis
t(10;11)(p13–14;q14–21)			T lineage	L2	Seen also in AML Seen in older children, CD10-positive
t(10;14)(q24;q11)[a]	HOX11 (10q24) TCRδ (14q11)	Homeobox T-cell receptor promoter	T lineage		Sole abnormality in 50% of cases
t(11;14)(p15;q11)[a]	LMO1 (11p15) TCRα/δ (14q11)	Zinc finger T-cell receptor promoter	T lineage		Seen in 5–10% of T-ALL cases
t(11;14)(p13;q11)[a]	LMO2 (11p13) TCRδ (14q11)	Zinc finger T-cell receptor promoter	T lineage		Seen in 5–10% of T-ALL cases; most common breakpoint on 11p
t(11;19)(q23;p13)[b,d]	MLL (11q23) ENL (19p13)	Drosophila trithorax homolog Transcription factor	Biphenotypic mixed lineage	L1, L2	Sole change in most cases. Most frequent in infants, high WBC, organomegaly, early CNS involvement, poor prognosis. Lineage infidelity
t/del(12p)			B or T lineage	L1	Seen in 10% of childhood B precursor ALL. See t(12;21) below
t(12;17)(p13;q21)			Early B precursor		Common 12p rearrangement
t(12;21)(p13;q22)[b]	TEL (12p13) AML1 (21q22)	ETS family Drosophila runt homolog	B precursor		Often cryptic rearrangement, >20% of B precursor ALL

(continued)

Table 7 (Continued)

Cytogenetic abnormality	Genes involved	Protein	Phenotype	FAB type	Clinical characteristics/prognosis
inv(14)(q11q32.3)	$TCR\alpha\beta$ (14q11) IGH (14q32)	Chimeric Ig–TCR gene Ig heavy chain promoter			
inv(14)(q11q32.1)	$TCR\alpha\beta$ (14q11) $TCL1$ (14q32)	Chimeric Ig–TCR gene MTCP1 homology			
t(14;14)(q11;q32.1)	$TCR\alpha\beta$ (14q11) $TCL1$ (14q32)	Chimeric Ig–TCR gene MTCP1 homology			
t(14;18)(q32;q21)	IGH (14q32) $BCL2$ (18q21)	Ig heavy chain promoter Apoptosis regulator	B lineage	L2, L3	Associated with Burkitt's translocations t(8;14), t(2;8) and t(8;22). Poor prognosis
t(14;22)(q32;q11)	IGH (14q32) BCR (22q11)	Ig heavy chain promoter Serine kinase			May be variant Ph translocation
i(17q)					Most common isochromosome in ALL; 2° abnormality associated with massive hyperdiploidy.
t(17;19)(q22;p13)[b]	HLF (17q22) $E2A$ (19p13)	Leucine zipper βHLH transcription factor	Pre B	L1	Variant of t(1;19)
-20			B lineage	L1	Rarely seen as sole change
+21			B lineage	L1, L2	Rarely seen as sole change (>2%)
t(X;11)(q13;q23)[b]	AFX (Xq13) MLL (11q23)	Serine and proline-rich *Drosophila* trithorax homolog			

[a] Gene activation.
[b] Gene fusion.
[c] βHLH family of transcription factors, named for a conserved protein motif consisting of a basic domain, amphipathic α helix loop and a second amphipathic α helix.
[d] Also observed in AML.
[e] 6q21 always deleted.
[f] Ph chromosome results from fusion of second exon of ABL (e1–a2 junction) and first exon of BCR (minor breakpoint cluster region, in-bcr). 7.5-kD mRNA with 190-kD kinase whose substrate specificity does not differ from 210 kD CML enzyme. Seen in 80% of children and 50% of adults.

365

- early B-precursor ALL (formerly known as null ALL), cells committed to B lineage differentiation—L1 or L2
- common ALL (cells express CALLA)—L1 or L2
- pre-B-ALL (cells express immunoglobulin in the cytoplasm)—L1 or occasionally L3
- B-ALL (cells express immunoglobulin on cell surface)—L3.

T-cells are classified into two immunologic categories:

- early T-precursor ALL (immature, erythrocyte-rosette receptor negative, L1 or L2)
- T-cell ALL (mature, receptor positive, L1 or L2).

The majority (85%) of ALL cases are T-lineage specific, with only 2–4% of adults having a B-cell phenotype.

Most ALL patients (>85% adult, >90% children) have clonal chromosome abnormalities *(70,71)*. More than 3500 cytogenetically characterized ALL cases have been reported with 32 balanced and 108 unbalanced recurrent abnormalities *(17)*. Karyotypic aberrations are more commonly detected in B-lineage leukemia, while only about three fourths of T-cell cases have such findings *(72)*.

Recognition of ploidy as a distinctive cytogenetic feature in childhood ALL has greatly enhanced the ability to predict clinical outcome and devise risk-specific therapy *(71)*. ALL can be classified into five subtypes based on the modal number of chromosomes:

- Hyperdiploidy (47–50 chromosomes) is found in 25–30% of childhood ALL and 7% of adult ALL with B-lineage phenotype. Hyperdiploidy accounts for 15–20% of childhood ALL. Gains of almost every chromosome have been observed, with chromosomes 8, 10, and 21 seen most often.
- Massive hyperdiploidy (>50 chromosomes) has long been recognized as conferring the most durable responses to treatment *(71,72)*. The modal distribution of chromosomes peaks at 55 chromosomes.

The karyotypes of patients with more than 50 chromosomes have certain features in common: extra copies of chromosomes X, 4, 6, 10, 14, 17, 18, 20, and 21, dup(1q), and i(17)(q10). Gain of chromosome 6 and the combination of trisomy 4 and trisomy 10 have been strongly associated with favorable outcome *(73,74)*.

The simultaneous presence of a structural abnormality (especially i(17)(q10), **Fig. 2rr**) in massively hyperdiploid cases has been shown to unfavorably influence prognosis *(75)*. Massive hyperdiploidy with a standard pattern of chromosome gains is seen in a lower frequency (<10%) in adult ALL and is associated with the Ph chromosome in one third of reported cases. Tetraploidy (2%) and triploidy (3%) are more frequent in adult than childhood ALL *(70)*.

- Patients whose leukemia cell chromosomes contain structural rearrangements in the context of a diploid net chromosome number (pseudodiploid) comprise the largest ploidy group in childhood ALL (40%).
- The incidence of cases that lack apparent cytogenetic abnormalities in ALL varies widely in part because of the indistinct chromosome morphology that is often encountered. A reported 10–15% of cases in some series have lacked detectable chromosome abnormalities by standard cytogenetic evaluation. The incidence in T-cell cases may be as high as 30%. Whether these cytogenetically "normal" cases represent mitotically inactive clones or clones with submicroscopic genetic changes is unresolved *(71)*.

• Hypodiploidy (<46 chromosomes) is found in 7–8% of cases of childhood ALL. Most hypodiploid cases (80%) have a modal number of 45 and arise from the loss of a whole chromosome, unbalanced translocations, or the formation of dicentric chromosomes. Hypodiploidy with 30–39 chromosomes is seen in 2% of adult ALL. Although very rare (<1%), near haploid cases have been well studied *(76)*. The main clone in near haploid cases contains at least one copy of each chromosome, with a second copy of the sex chromosomes and chromosome 21 found in most cases (90%). Other chromosomes found with two copies include chromosomes 10, 14, and 18. The morphologic features of these cases are usually poorly defined and few or no structural abnormalities can be distinguished. In many near-haploid cases, there is a second abnormal line with a hyperdiploid karyotype. This hyperdiploid line usually contains exactly two copies of all chromosomes in the near-haploid line. Several of the reported cases have included structural chromosome abnormalities in both cell lines, suggesting a common precursor cell with a near-haploid karyotype *(77,78)*. Near-haploid leukemia is associated with poor prognosis *(79)*.

Translocations occur in up to 50% of childhood ALL cases *(80;* **Table 7**). Chromosome translocations affect T-cell receptor genes (TCR) in approximately one half of T-cell ALL cases and involve immunoglobulin genes (Ig) in virtually all B-cell ALL cases, suggesting that *Ig/TCR* recombinase and class-switch enzymes are instrumental in facilitating interchromosomal translocations in these cases *(81–83)*.

B-cell ALL is associated with one of three specific chromosomal translocations: t(8;14)(q24;q32) (**Fig. 1q**) in approximately 80% of cases and t(2;8)(p12;q24) (**Fig. 1e**) or t(8;22)(q24;q11) (**Fig. 1mm**) in the remainder *(82)*. These rearrangements bring the *MYC* proto-oncogene on chromosome 8q24 near the transcriptionally active gene for the Ig heavy chain on 14q32, the κ chain on 2p12, or the λ light chain on 22q11, leading to aberrant activation of *MYC (84)*. *MYC* is a transcription factor that is involved in the control of cell proliferation and its binding to DNA and activation is enhanced by the MAX and MAZ proteins *(85,86)*.

Approximately 25% of cases of pre-B-ALL have a specific chromosome translocation, t(1;19)(q23;p13) (**Fig. 1ff**). Overall, the t(1;19) is found in 5–6% of childhood ALL cases, making it the most common chromosomal translocation in this disease *(87)*. The translocation breakpoint on chromosome 19 occurs within a single intron of the *E2A* gene, which encodes a transcription factor containing a helix–loop–helix DNA binding motif and dimerization domain. The breakpoint on chromosome 1q23 interrupts a homeobox gene *PBX1*. As a result, chimeric transcripts with oncogenic potential are formed that fuse coding sequences of the *E2A* gene in frame with the *PBX* sequences *(88)*. Approximately 1% of early pre-B cases also have a t(1;19) that is cytogenetically indistinguishable from that found in pre-B-ALL. They have different clinical and molecular findings, however. These cases are associated with hyperdiploidy and lack of E2A–PBX1 fusion protein, and have a favorable prognosis *(89)*.

The t(9;22)(q34.1;q11.2) is the most common of all ALL-associated chromosome abnormalities (15–20% of all cases), but is rare in childhood ALL (2–5%) and is associated with poor prognosis in all age groups. Monosomy 7 has been associated with Ph-positive ALL in adults and children *(90,91)*.

Most of the translocations involving the 11q23 region (*MLL* gene) result from exchanges with chromosome 4; however, chromosomes 1, 10, and, as seen previously, 19 can also contribute to this process *(17,* see **Fig. 1c, 1, u, pp**). Whereas most chromosome abnormalities are specific for a given hematologic disorder, abnormalities involving 11q23 are indiscriminately found in both AML and some cases of lymphoma. Reciprocal translocations involving 11q23 in acute leukemia have been found to involve more than 25 partner chromosome regions, and on occasion, there are identified partner chromosomes (*92*, **Fig. 1o, s**). Translocations involving chromosome 11q23 occur in 5–10% of adult leukemia cases and approximately 60% of infant acute leukemia cases, and are associated with poor prognosis *(93)*.

Approximately 30–40% of the abnormal karyotypes in T-cell leukemia have nonrandom breakpoints within the 14q11, 7q35, or 7p15 regions *(94)*. The prominent recurring chromosomal abnormalities in T-cell ALL are listed in **Table 7**. The most frequent childhood T-ALL abnormality is t(11;14)(p13;q11) (**Fig. 1w**), detected in 7% of all cases. The translocation t(10;14)(q24;q11) (**Fig. 1nn**) is reported in 5–10% of T-cell ALL patients, and t(11;14)(p15;q11) is seen in 1% of T-cell neoplasias.

Abnormalities of nonspecific lineage involve the long arm of chromosome 6 and the short arms of chromosomes 9 and 12. Deletions of 6q, mostly localized to 6q15 and 6q21, have been reported in 4–13% of childhood ALL *(95)*. No conclusive molecular evidence for loss of heterozygosity has been reported for 6q deletions. Abnormalities of 9p (**Fig. 2m, n**) have been reported in 7–12% of childhood ALL *(96)*. The critical region involved is 9p21–22, which encompasses the interferon-β1 gene (*INFB1*). Abnormalities of the short arm of chromosome 12 (**Fig. 2s, t, u**) appear in approximately 10% of cases of ALL with breakpoints most commonly reported at p12 as well as p11 and p13 *(97)*. Most cases with a 12p abnormality have a B-lineage immunophenotype, although some T-cell cases have been reported. Reported abnormalities have included deletions, balanced translocations, unbalanced translocations, dicentric chromosomes, and inversions *(98)*. It should be noted that the t(12;21)(p13;q22) that leads to the *TEL–AML1* fusion is essentially undetectable cytogenetically when only standard banding techniques are used, due to the marked similarity of chromosomal segments involved in the translocation. It has recently been shown that this cryptic event is one of the most common genetic abnormalities in pediatric ALL, occurring in approximately 22% of patients and associated with a B-lineage phenotype and generally good prognosis *(99–101)*.

Clonal karyotypic evolution has been reported to occur in almost 50% of ALL cases. Most frequent of the secondary numerical changes were +8 (10%), +21 (9%), –7 (9%), +X (7%), and +4 (5%). The most common secondary structural rearrangements were del(22q) (9%, **Fig. 2bb**), i(7q) (6%, **Fig. 2uu**), dup(1q) (6%), and del(9p) (5%) *(38)*.

Chronic Lymphoproliferative Disorders

Chronic lymphoproliferative disorders (CLDs) are characterized by clonal proliferation and accumulation in the bone marrow and peripheral blood of relatively mature cells of B and T lineage. These diseases are associated with the various cell types of the B- and T-cell differentiation pathways. The major chronic lymphoproliferative disorders include:

- B lineage: chronic lymphoproliferative leukemia (CLL), prolymphocytic leukemia (PLL), hairy cell leukemia (HCL), Waldenström's macroglobulinemia (WM), plasma cell leukemia (PCL), and multiple myeloma (MM)

- T lineage: chronic lymphocytic leukemia, adult T-cell leukemia/lymphoma (ATL), prolymphocytic leukemia, hairy cell leukemia, cutaneous T-cell lymphoma, Sezary syndrome, and mycosis fungoides *(102)*.

More than 1500 CLD cases with recurrent chromosome aberrations have been reported, with 14 balanced and 108 unbalanced abnormalities described *(103–116,* **Tables 8 and 9**). Approximately 90% of all reported cases are of the B-cell type. Because the spontaneous mitotic activity of these leukemic cells is low, mitogenic stimulation has improved cytogenetic analyses *(117)*. Clonal chromosome abnormalities reflect the developmental differences between the two cell types with B-cell neoplasias involving breakpoints at the immunoglobulin gene loci, in particular, the heavy chain locus at 14q32. T-cell neoplasias are characterized by rearrangements involving the T-cell receptor loci, especially the α-chain locus at 14q11. Other associated changes include +12 and del(13q) in CLL, del(6q) (**Fig. 2e, f**) in HCL and ATL, and 2p rearrangements in Sezary syndrome. Trisomy 12 and add(14q) have been associated with unfavorable prognosis in B-cell CLL, while karyotypic abnormalities have been linked to shorter survival in MM *(114,118–121)*.

Malignant Lymphomas

Lymphomas are caused by localized proliferation of lymphatic cells. Lymphomas originate mostly in lymph nodes but also develop in organs such as the bone marrow, spleen, intestinal tract, skin, and liver, where lymphatic cells are abundant. Based on histopathologic and clinical criteria, malignant lymphomas are divided into non-Hodgkin's lymphoma (NHL) and Hodgkin's disease (HD).

Non-Hodgkin's Lymphomas

Historically, lymphomas have been categorized using various classification systems *(122–125)*. Unfortunately, incompatibility exists among the systems. A Working Formulation for NHL nomenclature is currently in place that recognizes histopathologic subtypes within three main categories: high-, intermediate-, and low-grade malignant lymphomas *(126)*. Most recently, the "Revised European American Lymphoma" scheme has entered the evaluation and testing stage *(127)*. This classification system takes advantage of current immunophenotypic, molecular biologic, and genetic techniques *(128)*. **Table 10** *(127)* lists the probable equivalents for each disease entity in the Rappaport, Lukes–Collins and Kiel Classifications and in the Working Formulation.

Successful chromosome analysis has been reported in more than 2300 NHL cases, with 14 balanced and 95 unbalanced recurrent abnormalities *(17)*. More than 80% of patients with NHL have clonal chromosome abnormalities *(129–138)*, with low-grade lymphomas more difficult to karyotype than high-grade tumors *(135,137)*. The highest frequency (>90%) of clonal aberrations have been reported in small, noncleaved cell lymphomas (SNC), diffuse, large-cell lymphoma (DLC), and follicular, predominantly large cell lymphoma (FL). In contrast, the highest frequencies of tumors with normal metaphases have been reported in diffuse, small cleaved cell lymphoma (DSC, >30%); lymphoblastic lymphoma (LBL, >30%); follicular, predominantly small, cleaved cell lymphoma (FSC, 25%); and diffuse, mixed small and large cell lymphoma (DM, 25%) *(135,139)*. These normal karyotypic findings may reflect the presence of reactive, non-

Table 8
Chromosome Changes in B-Lineage Chronic Lymphoproliferative Disorders

Diagnosis	Cytogenetic abnormalities	Frequency (%)[a]
Chronic lymphocytic leukemia	+12	30
	add(14)(q32)[b]	20
	del/t(13q)	20
	del(6q)[c]	10
	del(11q)	>10
	+18	>5
	+3	>5
	del(12p)	>5
	del(14q)	>5
Richter's syndrome	11 abnormalities	
	14 abnormalities	
	t(5;12)(q21–22;q23–24)	
Prolymphocytic leukemia[e]	add(14)(q32)[d]	50
Hairy cell leukemia[g]	add(14)(q32)[f]	20
	del(6)(q23)	10
	del(14)(q22q32)	10
Waldenström's macroglobulinemia	del(3)(p25)	5
	i(6p)	5
Multiple myeloma[j]	1q abnormalities[h]	40
	add(14)(q32)[i]	25
Plasma cell leukemia	1p/1q abnormalities[k]	70
	add(14)(q32)[l]	50
	del(6q)[m]	10

[a]Percentage of cytogenetically abnormal cases.

[b]Multiple translocation partners including chromosomes 1, 2, 5, 6, 7, 8, 9, 11, 12, 13, 14, 17, 18, 19, 22. Translocations include t(2;14)(p13;q32), t(11;14)(q13;q32), t(14;17)(q32;q23), t(14;18)(q32;q21), t(14;19)(q32;q13).

[c]Breakpoints at 6q13, 6q15 and 6q21; sole abnormality in one third of cases.

[d]Most common rearrangement (50% of cases) is t(11;14)(q13;q32), t(14;17)(q32;q11) also reported.

[e]+12, t(6;12)(q15;p13) also reported. Common secondary changes are del(1)(q32)(20%) and del(6)(q21)(10%).

[f]Only reported cases have chromosomes 1, 9, 14, 17, 18, 22 as translocation partners.

[g]Reported changes also include 1q42 abnormalities, +5, chromosome 2 and 5 rearrangements.

[h]Multiple translocation partners (50%), deletions of 1p (30%), rarely observed as sole change. t(1;20)(q12;p13) also reported.

[i]t(11;14)(q13;q32) observed in one third of cases.

[j]Also reported are add/del(17p); del(6q) with breakpoints at bands q21, q23 and q25, del(7q) some with breakpoint at q36, der(1;7)(q10;p10) and numerical abnormalities –7, –9, +12, +14, +15, –18, +19, –21, –22, –X,–Y.

[k]Probably secondary change, 50% also have add(14q) marker.

[l]t(11;14)(q13;q32) and t(8;14)(q24;q32) reported.

[m]Breakpoints at 6q11, 6q14 and 6q21; t(11;14)(q13;q32) and t(9;22)(q34.1;q11.2) also present.

neoplastic cells that preferentially divide in vitro. Alternatively, the genomic alteration may be submicroscopic, occurring at the gene level *(140)*.

Approximately one third of karyotypically abnormal cases are pseudodiploid. Hypodiploidy is seen in less than 8% of tumors, while hyperdiploidy is most common (>50–65%) *(129–133,139)*. One study correlated ploidy with histologic subtype and

Table 9
Chromosome Changes in T-Lineage Chronic Lymphoproliferative Disorders

Diagnosis	Cytogenetic abnormality	Frequency (%)[a]
Chronic lymphocytic leukemia	14q11 rearrangements[b]	40
Prolymphocytic leukemia	14q11 rearrangements	50
	i(8q)[c]	30
	del/t(7q34–36)	15
Adult T-cell leukemia/lymphoma[g]	del(6q)[d]	25
	add(14q)[e]	20
	14q11 rearrangements[f]	15
Cutaneus T-cell lymphoma	t(2p)	10
	del(6q)	10
Mycosis fungoides[h]	del(6q)	20
	add(14)(q32)	20
Sezary syndrome	del/t(2p)	30
	del(6q)	20
	add(14)(q32)	20

[a]Percentage of cytogenetically abnormal cases.
[b]Includes inv(14)(q11q32), t(14;14)(q11;q32), del/t(14)(q11).
[c]Most common secondary change.
[d]Breakpoints at 6q15 and 6q21, often associated with changes of 14q11 and/or 14q32.
[e]Multiple translocation partners include 1p/q, 3q, 5p/q, 9q, 10p/q, 11q, 12q, 18q, Y. t(14;14)(q11;q32) most commonly identified rearrangement.
[f]Includes inv(14)(q11q32) (30%), t(14;14)(q11;q32) and del(14)(q11q32). Multiple partners include Xq, 1p, 1q, 3p, 8q, 10p, 11p, 12q, 18p.
[g]Numerical changes involving gains of chromosomes 3, 7, 21 and loss of X and Y seen as sole changes.
[h]Usually complex karyotype. Double minute (dmin) chromosomes reported.

found that small lymphocytic lymphoma (SLL), LBL, and SNC were most likely to be pseudodiploid; FSC and DM hyperdiploid (47–48 chromosomes); and DLC and immunoblastic lymphoma (IBL) hyperdiploid (>49 chromosomes) *(138)*.

In comparison to acute leukemias, single clonal aberrations in NHL are rare, with complex numerical and structural changes occurring in most cases. Structural abnormalities (reciprocal and unbalanced translocations, insertions, deletions, duplications, para- and pericentric inversions, ring chromosomes, isochromosomes) are more frequent than numerical changes. Gains of whole chromosomes are more common than losses *(15,130)*. The most frequent recurring cytogenetic abnormalities seen among five series were add(14)(q32) (40–65% of abnormal cases), t(14;18) (0–41%, **Fig. 1y**), t(8;14) (0–15%, **Fig. 1q**), add(3)(q27–29) (5–9%), del(6q) (15–21%, **Fig. 2e, f**), i(17)(q10) (5–10%, **Fig. 2rr**), +3 (9–14%), +7 (7–19%), +12 (5–19%), +18 (8–16%), and +21 (4–11%) *(140)*.

Abnormalities strongly associated with histologic type and often found as the sole karyotypic change are assumed to be early changes that may play a crucial role in tumor initiation. A list of recurrent structural and numerical chromosome abnormalities found as single changes in NHL are given in **Table 11** *(140)*. Recurring abnormalities that have been predominantly observed in NHL are given in **Table 12** *(15,140–144)*. These changes are less specific and are believed to reflect later events in tumor progression.

Table 10
Comparison of the Proposed Classification with the Kiel Classification and Working Formulation[a]

Kiel Classification	Revised European-American Lymphoma Classification	Working Formulation
B-lymphoblastic	Precursor B-lymphoma/leukemia	Lymphoblastic
B-lymphocytic, CLL[b]	B-cell chronic lymphocytic leukemia/ prolymphocytic leukemia/small lymphocytic lymphoma	**Small lymphocytic, consistent with CLL**
B-lymphocytic, prolymphocytic leukemia		Small lymphocytic, plasmacytoid
Lymphoplasmacytoid immunocytoma		
Lymphoplasmacytic immunocytoma	Lymphoplasmacytoid lymphoma	**Small lymphocytic, plasmacytoid**
		Diffuse, mixed small and large cell
		Small lymphocytic
Centrocytic	Mantle cell lymphoma	**Diffuse, small cleaved cell**
Centroblastic, centrocytoid subtype		Follicular, small cleaved cell
		Diffuse, mixed small and large cell
		Diffuse, large cleaved cell
Centroblastic–centrocytic, follicular	Follicular center lymphoma, follicular	
	Grade I	**Follicular, predominantly small cleaved cell**
	Grade II	**Follicular, mixed small and large cell**
	Grade III	Follicular, predominantly large cell
Centroblastic, follicular	Follicular center lymphoma, diffuse small cell [provisional]	**Diffuse, small cleaved cell**
Centroblastic–centrocytic, diffuse		Diffuse, mixed small and large cell
—	Extranodal marginal zone B-cell lymphoma (low grade B-cell lymphoma of MALT type)	**Small lymphocytic**
		Diffuse, small cleaved cell
		Diffuse, mixed small and large cell
Monocytoid, including marginal zone	Nodal marginal zone B-cell lymphoma [provisional]	**Small lymphocytic**
Immunocytoma		Diffuse, small cleaved cell
		Diffuse, mixed small and large cell
		Unclassifiable
—	Splenic marginal zone B-cell lymphoma [provisional]	**Small lymphocytic**
Hairy cell leukemia	Hairy cell leukemia	Diffuse small cleaved cell
Plasmacytic	Plasmacytoma/myeloma	Extramedullary plasmacytoma

(continued)

Table 10 (Continued)

Kiel Classification	Revised European-American Lymphoma Classification	Working Formulation
Centroblastic (monomorphic, polymorphic, and multilobated subtypes)	Diffuse large B-cell lymphoma	**Diffuse, large cell**
B-Immunoblastic		Large cell immunoblastic
B-large cell anaplastic (Ki-1$^+$)		Diffuse, mixed small and large cell
—		
	Primary mediastinal large B-cell lymphoma	**Diffuse, large cell**
		Large cell immunoblastic
Burkitt's lymphoma	Burkitt's lymphoma	Small noncleaved cell, Burkitt's
—	High-grade B-cell lymphoma, Burkitt-like [provisional]	**Small noncleaved cell, non-Burkitt's**
? Some cases of centroblastic and immunoblastic		Diffuse, large cell
		Large cell immunoblastic
T-lymphoblastic	Precursor T-lymphoblastic lymphoma/leukemia	Lymphoblastic
T-lymphocytic, CLL type	T-cell chronic lymphocytic leukemia/prolymphocytic leukemia	**Small lymphocytic**
T-lymphocytic, promyelocytic leukemia		Diffuse small cleaved cell
T-lymphocytic, CLL type	Large granular lymphocytic leukemia	**Small lymphocytic**
—	–T-cell type	Small lymphocytic
	–NK-cell type	Diffuse, small cleaved cell
Small cell cerebriform (mycosis fungoides, Sezary syndrome)	Mycosis fungoides/Sezary syndrome	Mycosis fungoides
T-zone	Peripheral T-cell lymphomas, unspecified (including provisional subtype: subcutaneous panniculitic T-cell lymphoma)	Diffuse, small cleaved cell
Lymphoepithelioid		**Diffuse, mixed small and large cell**
Pleomorphic, small T-cell		Diffuse, large cell
Pleomorphic, medium-sized and large T-cell		**Large cell immunoblastic**
T-immunoblastic	Hepatosplenic γ-δ T cell lymphoma [provisional]	—
—		

(continued)

Table 10 (*Continued*)

Kiel Classification	Revised European-American Lymphoma Classification	Working Formulation
Angioimmunoblastic (AILD, LgX)	Angioimmunoblastic T-cell lymphoma	**Diffuse, mixed small and large cell**
		Diffuse, large cell
		Large cell immunoblastic
—	Angiocentric lymphoma	Diffuse, small cleaved cell
		Diffuse, mixed small and large cell
		Diffuse, large cell
		Large cell immunoblastic
—	Intestinal T-cell lymphoma	Diffuse, small cleaved cell
		Diffuse, mixed small and large cell
		Diffuse, large cell
		Large cell immunoblastic
Pleomorphic small T-cell, HTLV1[+]	Adult T-cell lymphoma/leukemia	Diffuse, small cleaved cell
Pleomorphic medium-sized and large T-cell, HTLV1[+]		**Diffuse, mixed small and large cell**
		Diffuse, large cell
		Large cell immunoblastic
T-large cell anaplastic (Ki-1[+])	Anaplastic large cell lymphoma, T- and null-cell types	Large cell immunoblastic

[a]Taken from ref. *127*.
[b]When more than one Kiel or Working Formulation category is listed, those in boldface type comprise the majority of the cases.

Table 11
Structural and Numerical Chromosome Abnormalities
Observed as the Sole Change in Non-Hodgkin's Lymphoma[a,b]

del(1)(q32)
t(2;5)(p23;q35), t(2;8)(p12;q24)
+3, del(3)(p21–25)
del(6)(q13–24–q21–27)
–7, +7, i(7)(q10)
t(8;13)(p11;q11–12), t(8;14)(q24;q32), t(8;22)(q24;q11)
t(9;14)(p13;q32), i(9)(q10), t(9;17)(q34;q23), t(9;22)(q34.1;q11.2)
t(10;14)(q24;q11)
t(11;14)(q13;q32), dup(11)(q13q23-25), del(11)(q14–23), t(11;18)(q21;q21)
+12
–14, +14, inv(14)(q11q32), del(14)(q11–24–q22–32), t(14;18)(q32;q21), t(14;19)(q32;q13)
del(16)(q22)
–17
t(18;22)(q21;q11)
+20
+22
–X, +X, –Y

[a]Reported in at least two cases karyotyped in two or more laboratories.
[b]Data from ref. *140.*

The associations of recurrent chromosome aberrations in NHL with disease, histologic group, molecular abnormality, and clinical characteristics/prognosis are given in **Table 13**. Selected recurring abnormalities within histologic subgroups are given in **Table 14**.

Tables 11–14 list 14 reciprocal translocations that appear to be primary aberrations in NHL by virtue of their appearance as solitary changes. The most specific changes are the t(8;14) (**Fig. 1q**), t(2;8) (**Fig. 1e**), and t(8;22) (**Fig. 1mm**) associated with Burkitt's lymphoma (BL). In t(8;14) the proto-oncogene *MYC,* located at 8q24, is juxtaposed to the enhancer elements of the immunoglobulin heavy chain gene (*IgH*) at 14q32. In the variant translocations, the kappa light chain gene (*Igκ*) at 2p12 or the lambda light chain gene (*Igλ*) at 22q11 is translocated to a telomeric region of the *MYC* oncogene. As a result of these rearrangements, *MYC* is activated and constitutively produced at high levels *(145).* A similar transposition of genes from chromosomes 11, 18, and 19 onto the *IgH* gene locus in B-cell-specific translocations has led to the identification of the genes *BCL1* at 11q13 (**Fig. 1v**), *BCL2* at 18q21 (**Fig. 1y**), and *BCL3* at 19q13 (**Fig. 12z**) *(145).* In the T-lineage associated with rearrangements, recurrent translocations involve T-cell receptor loci at bands 14q11 (*TCRα, TCRλ*), 7p14–15 (*TCRγ*), and 7q35 (*TCRβ*). Most genes activated by juxtaposition to TCR loci are putative transcription factors involved in cell differentiation and not expressed in normal T-cells *(146).*

The protein products of most genes activated by juxtaposition to the *Ig* or *TCR* loci remain unchanged and are identical to their nontranslocated counterparts *(145).* By contrast, NHL translocations that do not involve *Ig* or *TCR* loci create fusion genes

Table 12
Recurring Abnormalities That Have Been Observed Predominantly in Addition to Other Abnormalities in Non-Hodgkin's Lymphoma[a,b]

inv(1)(p22p36), del(1)(p32–36), del(1)(p22), del(1)(p13), del(1)(q11–12), del(1)(q21), del(1)(q42), dup(1)(q11–21q31–32), dup(1)(q21–25q44), i(1)(p10), i(1)(q10), t(1;3)(q21–25;q27–29), t(1;5)(q32;q31), t(1;6)(p34–36;q23–25), t(1;7)(q21–23;q23), t(1;14)(p36;q22), t(1;14)(p35–36;q32), t(1;14)(p31–32;q32), t(1;14)(p22;q32), t(1;14)(q21–23;q32), t(1;17)(p36;q21), t(1;18)(p35–36;q21), der(1)t(1;1)(p36;q12–21), der(1)t(1;2)(p34–36;q31), der(1)t(1;6)(q21–23;p21)

del(2)(p22–23), del(2)(p21), del(2)(p11–12), del(2)(q21–22), del(2)(q31–32), del(2)(q33), dup(2)(p11–13p21), t(2;2)(p25;q31), t(2;2)(q33;q37), t(2;3)(p11;q11–12), t(2;3)(p12;q27), t(2;3)(q21–23;q27), t(2;14)(p11–13;q32), t(2;14)(q21–23;q32), t(2;18)(p12;q21)

del(3)((p25), del(3)(p21), del(3)(q11–12), del(3)(q21–23), del(3)(q25), del(3)(q27), dup(3)(q13–21q26–29), i(3)(p10), i(3)(q10), t(3;6)(q27–29;p21), t(3;7)(q27–29;p13), t(3;10)(p14;q22–23), t(3;11)(q21;p13), t(3;14)(p21;q32), t(3;14)(p11;q12), t(3;14)(q27;q32), t(3;17)(p11;q12), t(3;22)(q27–28;q11)

del(4)(p13–14), del(4)(q21–22), del(4)(q32–33), t(4;11)(q21;q23), der(4)t(1;4)(q11–12;q34–35)

del(5)(p13), del(5)(q12–15), del(5)(q12–14q31–32), del(5)(q31), i(5)(p10), t(5;7)(q13;q35–36), t(5;14)(q11;q32), t(5;14)(q23;q32)

del(6)(p21–23), del(6)(p11–12), i(6)(p10), t(6;7)(p23;q32), t(6;12)(q21–22;p13), t(6;14)(p21;q32), t(6;14)(q21;q32), der(6)t(1;6)(q21;q13–14)

del(7)(p15), del(7)(p13–14), del(7)(q22), del(7)(q22q32–34), del(7)(q32), del(7)(q34–35), dup(7)(q11q32–36), t(7;11)(q11;q24–25), t(7;14)(p15;q32)

del(8)(p21), del(8)(q22), del(8)(q24), i(8)(q10), t(8;9)(q10), t(8;9)(q23;p21–22), t(8;9)(q24;p13), t(8;11)(p21;q13), t(8;14)(q22;q32)

del(9)(p21–22), del(9)(p13), del(9)(p11–12), del(9)(q11–13), del(9)(q11–13q21–34), del(9)(q22), del(9)(q32–33), i(9)(p10), t(9;22)(q24;q11)

del(10)(p13), del(10)(p11), del(10)(q22), del(10)(q23–24), t(10;14)(q24;q32), t(10;15)(p11;q11)

del(11)(p13), del(11)(p11–12), del(11)(q14q23), del(11)(q21), del(11)(q23), i(11)(q10), t(11;13)(q13;q14), (11;14)(p13;q11), t(11;14)(q21;q32), t(11;14)(q23;q32)

(continued)

376

Table 12 (Continued)

del(12)(p11–12), del(12)(q12–13), del(12)(q22), del(12)(q24), dup(12)(q13q21–22), dup(12)(q13q24), i(12)(q10), t(12;14)(p11;q32), der(12)t(1;12)(q21–23;q24), der(12)t(9;12)(q13;q24)

del(13)(q12q14), del(13)(q14), del(13)(q14q22), del(13)(q22), del(13)(q31), i(13)(q10), t(13;14)(q11;q32), der(13)t(1;13)(q12–22;p13), der(13)t(1;13)(q21–22;q34)

t(14;14)(q11;q32), t(14;15)(q32;q15), t(14;15)(q32;q21–24), der(14)t(1;14)(q21–25;q32)

del(15)(q22), del(15)(q24), i(15)(q10), t(15;22)(p11;q11), der(15)t(1;15)(q12–25;q22–24), der(15)t(9;15)(p11;p11)

del(16)(p13)

del(17)(p13), del(17)(p12), del(17)(p11), del(17)(q21), del(17)(q23), dup(17)(q21–22q24–25), i(17)(q10), der(17)t(1;17)(q21–25;q25)

del(18)(q21–22), i(18)(q10), der(18)t(1;18)(q25–31;q23), der(18)t(14;18)(q32;q21)

del(19)(q13), der(19)t(1;19)(q21–25;p13), der(19)t(1;19)(q25;q13), der(19)t(14;19)(q11;p13), der(19)t(17;19)(q11–12;p11)

del(20)(q11)

i(21)(q10), t(21;22)(q22;q11)

del(22)(q11–12), der(22)t(1;22)(q11–12;p11–13), der(22)t(17;22)(q11;p11)

del(X)(p11), del(X)(q24), del(X)(q26–27), der(X)t(X;1)(p21–22;q11–21), der(X)t(X;1)(p11;q23–25), der(Y)t(Y;1)(q12;q21–25)

[a]Reported in at least two cases karyotyped in two or more different laboratories.
[b]Data from ref. *140*.

377

encoding hybrid proteins with excessive tyrosine kinase activity. The t(2;5)(p23;q35) (**Fig. 1hh**) fuses the nuclear phosphoprotein gene (*NPM*) at 5q35 to the protein tyrosine kinase gene *ALK* (anaplastic lymphoma kinase) at 2p23 while the t(9;22)(q34;q11) in NHL creates a *BCR–ABL* fusion protein similar to that observed in Ph-positive ALL *(145,147)*.

A novel, zinc-finger encoding oncogene *BCL6/LAZ3* that shares homology with several transcription factors is rearranged in the recurring translocations t(3;14)(q27;q32), t(3;22)(q27–28;q11), t(2;3)(p12;q27), t(2;3)(q21–23;q27), t(1;3)(q21–25;q27–29), t(3;6)(q27–29;p21), and t(3;7)(q27–29;p12–13) *(143,148–150)*. A variant of t(14;18)(q32;q21), t(2;18)(p12;q21), leads to activation of *BCL2 (151)*.

Deletions involving virtually all chromosome arms have been documented in NHL and are the second most common type of structural abnormality in NHL. Deletion is a mechanism by which tumor suppressor genes may be inactivated, resulting in disturbances in the cell's growth control. Loss and inactivation of p53 mapped at 17p13 have been reported in greater than 30% of BL, and p53 mutations have been observed in later stages of diffuse and follicular NHL *(152,153)*. Loss of material from 9p21, resulting in loss of *MTS1* (CDKN2/p16), is associated with progression of high-grade NHL *(154)*. Retinoblastoma *(RB1)* gene abnormalities/aberrant expression have been observed in a small number of low grade B-lineage malignancies and greater than 50% of high-grade lymphomas *(155,156)*.

Isochromosomes are fairly common in NHL, occurring in 10% of karyotypically abnormal tumors *(157)*. The most frequently occurring isochromosomes include i(17)(q10) (**Fig. 2rr**), i(6)(p10), i(1)(q10), i(18)(q10), and i(7)(q10) (**Fig. 2uu**).

As demonstrated in **Table 12,** 17 unbalanced translocations involve an extra copy of chromosome 1q, with breakpoints in or near the heterochromatic region, leading to partial trisomy 1q. These unbalanced rearrangements, as well as duplications of 1q, are common in NHL and are regarded as nonspecific secondary abnormalities contributing to tumor progression through a possible proliferative advantage *(11,15)*.

Numerical chromosome abnormalities are usually observed in complex NHL karyotypes and are considered secondary changes reflecting the karyotypic instability characteristic of advanced lymphomas. However, trisomies 3, 7, 12, 14, 20, and 22, +X; monosomies 7, 14, and 17; as well as –X/–Y, have been observed as solitary changes and are considered primary abnormalities *(15)*.

The prognostic relevance of chromosome abnormalities in NHL is not as clear cut as in the acute leukemias, often complicated by small numbers of patients, lack of uniformity with regard to histology and tumor grade and treatment protocols *(134,158–162)*. Some series show association with a number of different aberrations and shorter survival times, but the findings are still preliminary and require confirmation *(140)*.

Hodgkin's Disease

Successful chromosome analysis has been reported in more than 178 HD cases with 1 balanced and 11 unbalanced recurrent abnormalities *(17)*. Karyotypic abnormalities have been reported in all subtypes of HD, but no pathologic correlations have been made *(163,164)*. The number of cases with reported clonal abnormalities in HD is approximately 10 times lower than that found in NHL *(15)*. Specific problems in cytogenetic analysis of HD include lower mitotic index, overrepresentation of cells with a

Table 13

Association of Recurrent Chromosome Aberrations in NHL with Disease, Histologic Group, Molecular Abnormality, Clinical Characteristics, and Prognosis

Cytogenetic abnormality	Genes involved	Frequency (%)	Histologic group	Clinical characteristics/prognosis
1p/1q abnormalities[a]		25–50	Variable, B and T lineage	Shorter duration of complete remission and shorter survival
t(2;3)(p12;q27)[b]	IGK (2p12), LAZ3 (3q27)	<5	DLC; follicular B cell	Longer survival and freedom from disease progression (?)
t(2;5)(p23;q35)[c]	ALK (2p23), NPM (5q35)	5–10	Ki-1+ anaplastic large cell	Shorter survival
t(2;8)(p12;q24)[d]	IGK (2p12), MYC (8q24)	5–10	Burkitt's lymphoma	
t(2;18)(p12;q21)	IGK (2p12), FVT1 (18q21)	5–10	Follicular B cell	
+3[e]/dup(3p)		>10	FM, FL, DM, PL	Longer survival, no remission
del(3)(p21–25)		<5		No correlation in DM, DLC, IBL
t(3;14)(q27;q32)	LAZ3 (3q27), IGH (14q32)	>20	DLC; follicular B cell	Longer survival and freedom from disease progression (?)
t(3;22)(q27;q11)[f]	LAZ3 (3q27), IGL	15–20	DLC; follicular B cell	Longer survival and freedom from disease progression (?)
6p rearrangements[g]			T lineage	
del(6q)[h]		15	Variable; mostly B lineage	Shorter survival[i]; more frequent in high grade T cell lymphomas
−7/7p abnormality		<5		Poor response to chemotherapy, shorter disease-free and overall survival, no correlation between −7 and survival in DM, DLC and IBL
+7[j]		10–15	Variable, B and T lineage	Shorter survival; more frequent in high-grade T cell lymphomas
t(8;14)(q24;q32)	MYC (8q24), IGH (14q32)	>10	Burkitt's lymphoma	
t(8;22)(q24;q11)[k]	MYC (8q24), IGL (22q11)	5–10	Burkitt's lymphoma	

(continued)

Table 13 (Continued)

Cytogenetic abnormality	Genes involved	Frequency (%)	Histologic group	Clinical characteristics/prognosis
9q rearrangements[l]		10–15	Variable, B and T lineage	
t(9;14)(p13;q32)	? IGH (14q32)	<5	SLL B cell type	Plasmacytoid SLL
t(10;14)(q24;q32)	HOX11 (10q24) IGH (14q32)	<5	Variable, B lineage	
del(11q)[m]		>10	SLL	
t(11;14)(q13;q32)	?NCAM (11q23) BCL1/PRAD1 (11q13) IGH (14q32)	>10	Small lymphocytic or centrocytic B lineage	Originate from mantle cells in lymph nodes
t(11;18)(q21;q21)[n]		5–10	SLL B lineage	Indolent course, preferential involvement of mucosal sites, mature B-cell phenotype with CD5⁻ and λ light chain expression
+12[o]		10–15	SLL or DLC B lineage	Shorter survival in low-grade NHL
14q11 rearrangements[p]	TCRA (14q11) TCRD (14q11)		T lineage	Shorter survival
del(14q)[q]		<5	Variable, B and T lineage	
14q32 rearrangements[r]	IGH (14q32)	40–50	Variable B lineage	
t(14;18)(q32;q21)[s]	IGH (14q32) BCL2 (18q21)	>10 (sole)	Follicular or DLC B lineage	Geographic and racial differences in t(14;18) frequency may exist; shorter disease-free survival
+18[t]		10	Variable, B and T lineage	Shorter survival; longer survival in patients with t(14;18)
t(18;22)(q21;q11)[u]	BCL2/?FVT1 (18q21) IGL (22q11)	<5	Follicular B lineage	
–X/+X/–Y[v]		10	Variable, B and T lineage	

[a] Most frequent breakpoints are 1p32–36 and 1q21–23. Usually secondary change. del(1)(q32) as sole change.
[b] Variant of t(3;14)(q27;q32).
[c] Variant translocations include t(1;5)(p32;q35) and t(5;6)(q35;p12).
[d] Variant of t(8;14)(q24;q32).

[e] +3 in t(14;18)-positive NHL has longer survival than other patients with t(14;18); +3 or +3p seen in 45% of karyotypically abnormal MALT lymphomas, usually associated with other abnormalities.

[f] Variant of t(3;14)(q27;q32).

[g] Association with break in 6p21.

[h] Deletions affecting 6q25–27, 6q23, and 6q21. Usually secondary change, often found together with 11q+.

[i] Shorter survival, increased risk of transformation to DLC lymphoma if break at 6q23–26 in follicular NHL. Shorter survival and diminished probability of achieving complete remission or no correlation between del(6q) and survival in DM, DLC and IBL if break at 6q21–25 in DM, DLC, and IBL. Shorter survival than patients without del(6q) or than patients with the del(6q) and t(14;18) in patients with del(6q) in the absence of t(14;18).

[j] Usually secondary change. Reported as primary change in MALT (mucosa-associated lymphoid tissue) lymphomas.

[k] Variant of t(8;14)(q24;q32). Occurs twice as frequently as t(2;8).

[l] Includes breaks in 9q11–13 associated with diffuse lymphomas with large cell component and typical response to chemotherapy. See also commonly deleted 9q31–32 segment with diffuse B-cell histology, young age, and a poor clinical outcome. Breaks in 9q34 associated with high-grade T-cell lymphomas.

[m] Deletions interpreted as interstitial as well as terminal.

[n] Seen only as sole change.

[o] Often found together with other lymphoma-associated aberrations, especially t(14;18) in follicular lymphomas. Present in 25% of NHL in which del(14q) accompanied by other aberrations.

[p] Most common rearrangements are inv(14)(q11q32) and t(14;14)(q11;q32).

[q] Most breaks assigned to 14q22 when del(14q) was the only change.

[r] Recurrent translocation partners include 1p22, 1p31, 1q21-23, 1q42, 2p13, 2q21, 3p21, 3q27, 5q23, 7p15, 8q24, 9p13, 10q22-24, 11q12-14, 11q22-23, 12p11, 14q11, 15q15, 18q21, 19q13.

[s] Associated secondary changes include trisomy 3, 5, 7, 8, 9, 12, 17, 18, 20, 21, +x; loss of sex chromosomes, del(6q), i(17)(q10), i(18)(q10), +der(18)t(14;18), dup(1q). Also seen are 1p/q, 3q21-27, 7q32, 10q23–25 and Xp22 abnormalities. +7, +12, +17 and i(17)(q10) more common in intermediate and high-grade tumors. +3, del(6q), +7, del(13q), +12, +18 associated with transformation of initially low-grade t(14;18) NHL into intermediate or high-grade lymphoma.

[t] Usually secondary change; nonrandom association between +3 and +18 noted.

[u] Variant of t(14;18).

[v] Secondary change. –X associated with T-cell lymphomas.

Table 14
Selected Recurring Cytogenetic Abnormalities
Within Histologic Subgroups of Non-Hodgkin's Lymphoma[a]

Histologic group	Associated abnormalities
Small lymphocytic	del(6q),del(11)(q14–23)[b], del(14)(q22–24), +3, +12[b]
Small lymphocytic extranodal	t(11;18)(q21;q21)
Small lymphocytic plasmacytoid	t(9;14)(p13;q32)
Mucosa-associated lymphoid tissue	–3
Follicular, predominantly small cleaved cell	t(14;18)(q32;q21), del(6q), t(6p) or i(6)(p10)
Follicular, mixed small cleaved and large cell	t(14;18)(q32;q21), del(2q), +3/3q[c], +8
Follicular, predominantly large cell	t(14;18)(q32;q21), +7
Diffuse, small cleaved cell	t(11;14)(q13;q32), del(8p), del(20q)
Intermediate lymphocytic/mantle cell	t(11;14)(q13;q32)
Diffuse, mixed small and large cell	+3, +5, +14
Diffuse, large cell	t(3;14)(q27;q32), t(3;22)(q27;q11), +X, +4, +7, +9, +12,+21
Ki-1[+] anaplastic large cell	t(2;5)(p23;q35)
Large cell immunoblastic	del(3p), del(5q), del(6q), +X, +3, +5, +7, +18
Lymphoblastic	t(9;17)(q34;q23), t(8;13)(p11;q11–12)
Small noncleaved cell	t(8;14)(q24;q32), t(2;8)(p12;q24), t(8;22)(q24;q11)

[a]Data from ref. *140.*
[b]As a sole abnormality.
[c]Trisomy 3 or structural changes of 3q.

normal karyotype, and paucity of Reed–Sternberg cells in lymph node specimens. Recent series show a culture success rate of 75% (14–92%); however, only 44% of these cases have clonal abnormalities *(163–168).*

Recurrent chromosomal abnormalities seen in HD are given in **Table 15**. Structural rearrangements of chromosome 1 (deletions, translocations, duplications, isochromosome 1q) are found in one third of cases. Rearrangements resulting in "14q+," mostly by addition to 14q32, are seen in 20% of cases. The translocations t(11;14) and t(14;18) are rarely observed. Variable-sized deletions of part of the long arm of chromosome 6 are found in 15% of cases. In comparison to NHL, abnormalities of 3q, 7q, 12p (mostly deletions), and 13p (addition of unknown material) occur at higher frequencies. The most common numerical changes are trisomies 1, 3, 7, 8, and 21.

Solid Tumors

Karyotypic information relating to solid tumors has lagged behind our knowledge of hematologic malignancies. Solid tumors represent less than 30% of the almost 27,000 cytogenetically catalogued abnormal neoplasms to date *(15),* yet make up almost 95% of all human tumors, a disproportionate value in relation to their involvement in cancer morbidity and mortality.

Original solid tumor cytogenetic studies were hampered by technical difficulties relating to cell culture, and interpretation was hindered by poor banding quality. A histologically defined list of recurrent structural and numerical chromosome aberrations observed in solid tumors is given in **Table 16** *(169–299).* **Table 17** lists solid

Table 15
Recurrent Chromosome Aberrations Detected in Hodgkin's Disease[a,b]

+1, add(1)(p36), del(1)(p12–13), i(1)(q10), del(1)(q21)	add(13)(p11–13)
add(2)(p24–25), i(2)(q10), add(2)(q36–37)	add(14)(p11–p13), i(14)(q10),
+3, del(3)(p12–13), del(3)(q21), t/add(3)(q27–29)	t(14;15)(q10;q10), del(14)(q22–24),
add(4)(q34–35)	t/add(14)(q32), t(14;18)(q32;q21)
add(5)(p14–15)	–15, add(15)(p11–13), t(15;22)(q10;q10)
i(6)(p10), del(6)(q11–16), del(6)(q21–22)	add(16)(p13), add(16)(q22)
+7, i(7)(p10), i(7)(q10), del(7)(q21–22)	add(17)(p13), i(17)(q10)
+8, i(8)(q10), add(8)(q24)	+18, add(18)(p11–12), i(18)(q10)
add(9)(p21), add(9)(p24), i(9)(p10)	add(19)(p13), add(19)(q13)
t(10;14)(q22–23;q32)	add(20)(q13)
add(11)(p15), del(11)(p12–13), del(11)(q13–14)	+21, add(21)(p11–13), i(21)(q10)
del(12)(p11–13), add(12)(q24)	–22, add(22)(p11), i(22)(q10)
	add(X)(p22)

[a]Reported in at least two cases karyotyped in two or more different laboratories.
[b]Data from ref. *140*.

tumor breakpoints associated with inherited cancer syndromes *(300)*. Although rare, studies of specific mutations responsible for these syndromes have provided insight into the pathogenesis of cancer. Upon examination of the list in **Table 16** two karyotypic patterns emerge, one characterized by unbalanced multiple and unspecific abnormalities, the other characterized by simple, disease-specific balanced changes *(301)*. Most epithelial tumor types (notably breast, colon, lung, prostate, and ovarian carcinoma) and a few mesenchymal neoplasms (osteosarcoma and malignant fibrous histiocytoma) display the former pattern. Hematologic malignancies that are secondary in nature also fit into this classification. A disease-specific pattern, observed most commonly in leukemias and lymphomas, is seen in mesenchymal tumors. These disease-specific balanced structural rearrangements usually involve translocations and inversions.

Traditionally, the development and progression of cancer is thought to be a multistep process in which sequential genetic abnormalities in individual tumors confer incremental proliferative advantages upon successful clonal overgrowth *(302)*. The striking cytogenetic differences observed between human leukemias and solid tumors may reflect early vs late stages in a common genetic evolutionary sequence, or may reflect fundamentally different mechanisms of genetic evolution *(303)*. One mechanism proposed for breast cancer evolution begins with unbalanced rearrangements, followed by chromosomal loss and endoreduplication events leading to formation of hyperdiploid clones. Near-diploid cells are then eliminated, followed by additional cycles of chromosomal loss with an occasional endoreduplication event leading to selection of a polyploid tumor population. The rate by which chromosomes are lost and rearranged is postulated to be an indicator of tumor progression *(304)*.

Because the development of solid tumors is dependent upon a combination of deletions and amplifications of multiple chromosome segments, these recurrently lost and

(text continued on page 394)

Table 16
Recurrent Structural and Numerical Chromosome Aberrations in Solid Tumors

Benign epithelial neoplasms

Adenoma

Adrenocortical [169]	−Y
Breast [170]	del(1)(q10;p10), del(1)(q12), del(3)(p12–14), r(9)(p24q34), rearrangements involving 12p11–13, 12q13–15
Colorectal [171, 172]	loss 1p36 through del/unbalanced t, +7, +8, del(8p), del(12q), +13, i(13q), −14, −18
Kidney [173]	+7, +7, +17, −Y
Parathyroid [174]	t(1;5)(p22;q32)
Pituitary gland [175]	t(1;21)(q32;q22), +9
Salivary gland [176]	t(3;8)(p21;q12), rearrangements of 3p21, 8q12, 12q15, t/inv 12q13–15, inv(12)(p13q13)
Thyroid [177, 178]	+5, +7, +12, +22, t(2;3)(q12–13;q24–25), 19q13
Barrett's esophagus [179]	−Y

Oncocytoma

Kidney [180]	−1, −Y

Papilloma

Basal cell [181]	t(2;6;11)(q21;q27;p13)[a]
Basosquamous [182]	+7, polyclonal changes

Thecoma

Ovary [183]	+12

Warthin's tumor (adenolymphoma)

Salivary gland [184]	t(11;19)(q21;p13), +12

Germ cell and gonadal stromal cell neoplasms

Seminoma/dysgerminoma [240, 241]	i(12p), triploid–tetraploid with +3, +7, +8, −11, +12, −13, −18, +X
Teratoma (mature and immature) [240, 241]	i(12p), hypotetraploid with +7, +8, −11, +12, +i(12p), −15, −22, +X
Combined tumors [240, 241]	i(12p)

Gonadal stromal cell tumors

Sertoli cell tumor, ovary [242]	+i(1q)[a]

(continued)

384

Table 16 (Continued)

Malignant epithelial neoplasms

Adenocarcinoma

Adrenal cortical [185]

Breast [186–195] — t(4;11)(q35;p13)[a]
del(1p), del(1)(p35–36), 1q gain, del(1)(q21–31), gain 1q41–44, del(3)(p11–14), del(3)(p14–23), del(3)(p24–26), del(6)(q13–27), +7, del(7)(q31), +8/8q gain/abnormalities of 8q24, del(8)(p21–22), del(11)(p15.5), del(11)(q13), del(11)(q23–24), del(16)(q12.1), del(16)(q21), del(16)(q22–23), del(16)(q24.2), del(16)(q24.3), der(16)t(1;16), del(17p), abnormalities of 17p13.1, +18, +20, –X

Colorectal [196] — del(1)(p36), i(1)(q10), +7, i(8)(q10), +13, i(13)(q10), –14, –17, del(17)(p11), i(17)(q10), –18, +20, –22, –Y

Esophagus [197] — –Y

Gall bladder [198]

Kidney [199–201] — Rearrangements of 1, 3; del(18q), del(21q)
–3, del(3)(p14–21), del(3)(p11–14), t(3;5)(p13;q22), t(3;8)(p14;q24), t(3;11)(p11–14;p15), i(5p), del(6)(q21–23), +7, –8, +10, t(X;1)(p11;q21), –Y

Lung [202] — Rearrangements of 1p, 1q, 6q, 13q, 19p; losses 1p, 5q, 11p, 17p; i(8q)

Ovary [183, 203, 204] — Rearrangements of 1p, 1q; del(1)(q21–41), rearrangements of 3p, 3q, 6q; del(6)(q15–25), rearrangements of 7p, –8, rearrangements of 9q, 11p, 11q; del(11)(p11–15), +12, –13, –14, –17, rearrangements of 17q, 19q; add(19)(p13), –22, +22, –Y

Pancreas [205] — del(1p), del(1q), rearrangements of 1p, 1cen; del(3p), del(6q), rearrangements of 6q, +7, rearrangements of 7p, del(7q), del(8p), rearrangements of 8p, 9cen, 17p; –18, rearrangements of 19q, +20, dmin

Prostate [206] — Rearrangements of 1q, +7, del(7q), del(8p), del(10)(q24), –Y, dmin

Salivary gland [207] — del(6)(q22–25), +8, –Y

Stomach [208] — Rearrangements of 3p21, i(5p), +8, i(8q), +9, i(9q), add(9p), –Y

Thymus [209] — t(15;19)(q15;p13)[a]

Thyroid [210–212] — del(3)(p25p26), +7, +12, –Y, inv(19)(q11q21), rearrangements of 10q11, 10q21

Uterus [183]

Endometrium — i(1)(q10), +2, del(6)(q21–25), +7, +10, rearrangements of 11q21–25, +12,–22

Stroma — t(7;17)(p15–21;q11.2–21), del(11)(q13–21q21–23)

Uterine cervix [183] — Rearrangements of 1p11–15, i(4p), i(5p), rearrangements of 11p11–15

(continued)

385

Table 16 (Continued)

Adenoid cystic carcinoma	
Bartholin's gland [213]	Rearrangements of 1, 4, 6, 11, 14, 22
Respiratory tract [214]	Rearrangements involving 19p13
Salivary gland [215]	t(6;9)(q23;p21), rearrangements of 9p21, 17p, 17p
Adenosquamous carcinoma	
Lung [216]	Hyperdiploid with rearrangements
Basal cell carcinoma	
Skin [217, 218]	Polyclonal changes, +6, del(9)(q22), +12, +18, −X, −Y
Carcinoid tumor	
Lung [219]	+7
Chordoma	
Bone [220]	−3, −4, −10, −13
Hepatoblastoma	
Liver [221]	+2, 2q gain
Hepatocellular carcinoma	
Liver [222]	Rearrangements of 1 with 1p loss, rearrangements of 6 with 6q loss, −16
Merkel cell carcinoma	
Skin [223]	del(1p), partial trisomy 11p
Mucoepidermoid carcinoma	
Lung [224]	t(11;19)(q14−21;p12)
Salivary gland [225]	t(11;19)(q14−21;p11−12)
Squamous cell carcinoma	
Anal canal [226]	Rearrangements with 3p and 11q loss
Esophagus [227]	Rearrangements of 1, 3, 9, 11, all leading to deletion with breakpoints at 1p34−pter, 3p13, 9q11−12, 11q11−12, rearrangements at acrocentric centromeres
Head and neck (oropharynx, larynx, pyriform sinus) [228, 229]	Polyclonal changes, rearrangements of 1p22, 1q25, del(3p), rearrangements of 5q13, i(8q), rearrangements of 11q13, hsr11q13, i(13q), i(14q), i(15q), del(18q), del(Xp), rearrangements of Y, −Y
Lung [230, 231]	rearrangements/gain of 1q, 5p; rearrangements/loss 3p, 5q, 8p, +7, −Y
Penis [232]	del(2)(q33q36), rearrangements of 4p, der(5;15)(q10;q10), rearrangements of 8p[a]

(continued)

Table 16 (*Continued*)

Skin [233]	Polyclonal changes
Vulva [234]	losses 3cen–p14, 8p, 10q23–25, 18q22–23, 22q13, Xp; gains 3q25–qter, 11q21–qter, rearrangements of 5q
Transitional cell carcinoma	
Bladder [235]	del/t 1q21, i(5p), +7, –9, del(10)(q22–24), i(11p), del/t 11p11–q11, del/t/dup 13q14, del(21)(q22), –Y
Kidney [199]	i(5p), +7, –9
Undifferentiated carcinoma, large cell	
Lung [236]	rearrangements with loss of 1p, 1q, 3p, 6q, 7q, 17p; rearrangemetns with gain of 5q
Undifferentiated carcinoma, small cell	
Lung [237, 238]	del(3)(p14p23), rearrangements with loss of 5q, 6, 9p, 13q, 17p
Wilms tumor [239]	i(1q), del(11)(p13), del(11)(p15), der(16)t(1;16)
Benign mesenchymal neoplasms	
Chondroma/chondroblastoma/chondromyxoid fibroma [243]	t with 12q13–15
Endometrial polyp [244]	t with 6p21–22, t with 12q14–15
Fibroma	
Bone [245]	del(4)(p14)
Ovary [246]	+12
Fibromatosis/desmoid tumor [247, 248]	5q abnormalities, +7,+8,+20,–Y
Hamartoma	
Liver [249]	t(11;19)(q13;q13.4)[a]
Pulmonary chondroid [250, 251]	t/inv with 6p21–22, t with 12q14–15, t with 14q24
Skeletal (osteocartilaginous exostoses) [252]	del/t leading to loss of 8q24.1
Hemangioma [253]	+5,–Y[a]
Hibernoma [254]	t with 11q13

387

(*continued*)

Table 16 (*Continued*)

Leiomyoma	
Small intestine [255]	del(1p), −15[a]
Uterus [256]	t(1;2)(p36;p24), del(3q), del(7q), t with 6p21−22, +12, t with 12q14−15, del(13q)
Lipoblastoma [257]	t with 8q11−13
Lipoma (ordinary) [258]	t with 6p21−22, t with 12q15, del(13q)
Lymphangioma [259]	i(7q),−X[a]
Meningioma [260]	−22, del(22)(q11−13), idic(22)(q11), −Y
Mesoblastic nephroma [261]	+8,+11,+17
Myxoma [262]	−Y, telomere associations/t with 12p12
Neurofibroma [263, 264]	+18,−22[a]
Neurolemmoma (benign Schwannoma) [265]	+5,+7, −12, −15, +20, −22, −X, −Y
Spindle cell and pleomorphic lipoma [258]	loss of 16q13−qter
Tenosynovial giant cell tumors [266]	t with 1p11-13, with 16q24
Malignant mesenchymal neoplasms	
Borderline/locally aggressive neoplasms	
Aggressive angiomyxoma [267]	t with 12q14−15
Atypical lipomatous tumor	+ ring or long marker (sequences 12q13−q15)
(Atypical lipoma/well-diff. liposarcoma) [268]	
Congenital fibrosarcoma [269]	Combination of trisomies (8, 11, 17, 20)
Dermatofibrosarcoma protuberans [270]	+r (sequences 17q and 22q)
Desmoid tumors [248]	+8, +20
Giant cell fibroblastoma [271]	t(17;22)(q21;q13)

(*continued*)

Table 16 (*Continued*)

Malignant neoplasms

Alveolar soft part sarcoma [272]	t with 17q25
Clear cell sarcoma [273]	+8, t(12;22)(q13;q12)
Desmoplastic small round cell [274]	t(11;22)(p13;q12), abnormalities of 11p13, 22q12
Epitheliod cell sarcoma [275]	del(1p)[a], +2[a]
Extraskeletal myxoid chondrosarcoma [276]	t(9;22)(q22–31;12)
Hemangiopericytoma [277]	Abnormalities of 3p12–21, 7p15, 11p11–13, 12q13–15, 19q13, t(12;19)(q13;q13)
Leiomyosarcoma (GI) [278]	Monosomy 1p12→1pter with hypodiploid chromosome number, –22
Liposarcoma (myxoid/round cell) [279, 280]	t(12;16)(q13;p11), t(12;22)(q13;q12)
Mesothelioma [281]	del(3p), –4, +5, +7, del(9p), +20, –22
Osteosarcoma [282]	+1, rearrangements of 1p11–13, 1q10–12, 1q21–22; –6/del(6q), –9, –10, rearrangements of 11p15, 12p13; –13, rearrangements of 17p12–13, –17, rearrangements of 19q13, 22q11–13
Parosteal osteosarcoma [283]	+r
Rhabdomyosarcoma [284, 285]	
Alveolar	t(2;13)(q35;q14), t(1;3)(p36;q14), abnormalities of 13q14
Embryonal	+2, +8, +11, rearrangements of 12q13
Synovial sarcoma [286]	t(X;18)(p11.2;q11.2)

Neuroglial neoplasms

Gliomas

Astrocytoma [287]	+7, –10, –22, –Y
Ependymoma [288]	+7, +12, –17, –22
Oligodendroglioma [289]	+7, –22, –X/–Y

(continued)

389

Table 16 (*Continued*)

Neuronal neoplasms	
Neuroblastoma [290]	Rearrangements/del leading to loss of 1p32–36, t(1;17)(p36;q12), dmin, hsr
Primitive neuroectodermal tumors/medulloblastoma [291]	del(6q), +7, del(7)(q32), del(7)(q33), t(8;11)(q11;p11), −10, t(11;22)(q24;q12), −17, i(17q), −22, −X, −Y, dmin/hsr (2p)
Retinoblastoma [292]	Rearrangements involving 1q, i(6p), del(13)(q14)
Peripheral primitive neuroectodermal tumors	
Askin tumor [293]	t(11;22)(q24;q12)
Esthesioneuroblastoma [294]	t(11;22)(q24;q12)
Ewing's sarcoma [295, 296]	+8, t(11;22)(q24;q12), der(16)t(1;16)
Peripheral neuroepithelioma [297]	t(11;22)(q24;q12)
Melanocytic neoplasms	
Malignant melanoma	
Cutaneous [298]	t(1;6)(q11–21;q11–13),t(1;19)(q11;q13); t of 1q11–q12 leading to 1q gain; del/t 1p11–q12; i(6p),t of 6p11–q11 leading to 6p gain; other t of 6q11–13, t of 7q11,+7,del/t 9p, −10
Uveal [299]	−3, partial del of 3,+8, i(8q)

[a]Single case report.

Table 17
Solid Tumor Breakpoints Associated with Inherited Cancer Syndromes

Locus	Gene involved	Proposed gene product function	Syndrome/genetic condition	Primary tumor	Associated cancer/trait
1q25	Unknown	Unknown	Hereditary prostate cancer[a]	Prostate cancer	Unknown
2p16	MSH2	DNA mismatch repair	Hereditary nonpolyposis colorectal cancer (HNPCC)[a]	Colorectal cancer	Endometrial, ovarian, hepatobiliary and urinary tract cancer; glioblastoma (Turcot syndrome)
2q32	PMS1	DNA mismatch repair	Hereditary nonpolyposis colorectal cancer (HNPCC)[a]	Colorectal cancer	Endometrial, ovarian, hepatobiliary and urinary tract cancer; glioblastoma (Turcot syndrome)
3p25	VHL	?Regulates transcriptional elongation by RNA polymerase II	von Hippel–Lindau (VHL) syndrome[a]	Renal cancer (clear cell)	Pheochromocytomas, retinal angiomas, hemangioblastomas
3p21	MLH1	DNA mismatch repair	Hereditary nonpolyposis colorectal cancer (HNPCC)[a]	Colorectal cancer	Endometrial, ovarian, hepatobiliary and urinary tract cancer; glioblastoma (Turcot syndrome)
5q21	APC	Regulation of β-catenin; microtubule binding	Familial adenomatous polyposis (FAP)[a]	Colorectal cancer	Colorectal adenomas, duodenal and gastric tumors, CHRPE[b], jaw osteomas and desmoid tumors (Gardner syndrome), medulloblastoma (Turcot syndrome)
7p22	PMS2	DNA mismatch repair	Hereditary nonpolyposis colorectal cancer (HNPCC)[a]	Colorectal cancer	Endometrial, ovarian, hepatobiliary and urinary tract cancer; glioblastoma (Turcot syndrome)
7q31	MET	Transmembrane receptor for HGF[c]	Hereditary papillary renal cancer (HPRC)[a]	Renal cancer (papillary type)	?Other cancers

(continued)

Table 17 (Continued)

Locus	Gene involved	Proposed gene product function	Syndrome/genetic condition	Primary tumor	Associated cancer/trait
8q24.1	EXT1	Unknown	Multiple exostoses[a]	Exostoses	Chondrosarcoma
9p21	p16 (CDKN2)	Inhibitor of CDK4 and CDK6 cyclin-dependent kinases	Familial melanoma[a]	Melanoma	Pancreatic cancer, dysplastic nevi, atypical moles
9q22.3	PTCH	Transmembrane receptor for hedgehog signaling molecule	Nevoid basal cell carcinoma syndrome (NBCCS)[a]	Basal cell skin cancer	Jaw cysts, palmar and plantar pits, medulloblastomas, ovarian fibromas
10q11.2	RET	Transmembrane receptor tyrosine kinase for GDNF[d]	Multiple endocrine neoplasia type 2 (MEN2)[a]	Medullary thyroid cancer	Type 2A: pheochromocytoma, parathyroid hyperplasia; Type 2B: pheochromocytoma, mucosal hamartoma; familial medullary thyroid cancer
10q23	PTEN (MMAC1)	Dual-specificity phosphatase with similarity to tensin	Cowden disease[a]	Breast cancer, thyroid cancer (follicular type)	Intestinal hamartomous polyps, skin lesions
11p15	?p57KIP2 ?others-contiguous gene disorder	Cell cycle regulator	Wiedemann–Beckwith syndrome[a]	Wilms tumor	Organomegaly, hemi-hypertrophy, hepatoblastoma, adrenocortical cancer
11p13	WT1	Transcriptional repressor	Wilms tumor[a]	Wilms tumor	WAGR (Wilms, aniridia, genitourinary abnormalities, mental retardation)
11p11-13	EXT2	Unknown	Multiple exostoses[a]	Exostoses	Chondrosarcoma
11q13	MEN1	Unknown	Multiple endocrine neoplasia type 1 (MEN1)[a]	Pancreatic inlet cell	Parathyroid hyperplasia ?p57KIP2 pituitary adenomas
12q13	CDK4	Cyclin-dependent kinase	Familial melanoma[a]	Melanoma	Pancreatic cancer, dysplastic nevi, atypical moles
13q12	BRCA2	Interacts with Rad51 protein; ?repair of double strand breaks	Familial breast cancer 2[a]	Breast cancer	Male breast cancer, pancreatic cancer, ?others (?ovarian)

(continued)

Table 17 (Continued)

Locus	Gene involved	Proposed gene product function	Syndrome/genetic condition	Primary tumor	Associated cancer/trait
13q14.3	RB1	Cell cycle and transcriptional regulation; E2F binding	Familial retinoblastoma[a]	Retinoblastoma	Osteosarcoma
15q26.1	BLM	?DNA helicase	Bloom syndrome[e]	Solid tumors	Immunodeficiency, short stature
17p13.1	p53 (TP53)	Transcriptional factor; response to DNA damage and stress; apoptosis	Li–Fraumeni syndrome[a]	Sarcomas, breast cancer	Brain tumors, leukemia
17q11.2	NF1	GAP[f] for p21 ras proteins; microtubule binding?	Neurofibromatosis type 1 (NF1)[a]	Neurofibromas	Neurofibrosarcoma, AML, brain tumors
17q21	BRCA1	Interacts with Rad51 protein; ?repair of double strand breaks	Familial breast cancer 1[a]	Breast cancer	Ovarian cancer
17q25	Unknown	Unknown	Palmoplantar keratoderma[a]	Esophageal cancer	Leukoplakia
19p	EXT3	Unknown	Multiple exostoses	Exostoses	Chondrosarcoma
22q12.2	NF2	Links membrane proteins to cytoskeleton?	Neurofibromatosis type 2 (NF2)[a]	Acoustic neuromas, meningiomas	Gliomas, ependymomas

[a]Dominant inheritance.
[b]Congenital hypertrophy of the retinal pigment epithelium.
[c]Hepatocyte growth factor.
[d]Glial derived neurotrophic factor.
[e]Recessive inheritance.
[f]GTPase-activating protein.
Adapted from Fearon (300).

gained segments have been postulated to be candidate loci for tumor suppressor genes and dominantly acting growth regulatory genes *(305–307)*. Mertens et al. *(308)* profiled chromosomal gains and losses in 11 different tumor types representing 3185 neoplasms. The profiles revealed unique combinations of imbalances among the tumor types. However, enough considerable overlap existed to speculate that similar molecular mechanism may be operative. Deletions were more common than gains in all tumor types, with chromosomes X, Y, 4, 10, 13–15, 18, and 22 and chromosome segments 1p22-pter, 3p13-pter, 6q14-qter, 8p, 9p, and 11p deleted in the majority of tumors. Distinct candidate tumor suppressor gene loci were indicated at 1p12–13, 1p22, 1p34, 1p36, 7q22, 7q32, and 7q36. The only chromosomes or chromosome segments more often gained than deleted were chromosomes 7 and 20 and the long arms of chromosomes 1 and 12, suggesting the presence of dominantly acting growth regulatory genes at those loci.

Nonneoplastic conditions have traditionally been distinguished from neoplastic conditions by the observation of a normal karyotype. A major exception to this rule has been the observation of trisomy 7, a previously described common clonal change in malignant tumors, in normal brain *(309,310)*, kidney *(311–313)*, and lung *(314)* as well as in non-neoplastic lesions such as atherosclerotic plaque *(315,316)*, chronic pyelonephritis *(311,312)*, Dupuytren's contracture *(317)*, liver steatosis *(318)*, osteoarthritis *(319)*, Peyronie's disease *(320)*, prostatic hyperplasia *(321)*, rheumatoid synovitis *(322,323)*, thyroid hyperplasia *(178,324)*, and varicose vein *(325)*. Clonal loss and gain of a sex chromosome, usually –X/+X or –Y, has been seen in both benign and malignant tumors, as well as in normal cells and nonneoplastic lesions *(11)*. The significance of these findings is not known.

Chromosome abnormalities in benign tumors have also been well documented *(169–184,243–266)*, with cytogenetic complexity distinguishing malignant tumors from benign ones *(326–335)*.

The same characteristic chromosomal change may be found in different tumor types or may reflect pathogenetic similarities between tumors. Adenoid cystic carcinomas of the respiratory tract and salivary gland share 19p abnormalities *(214,215)* while a t(11;19)(q14–21;p11–12) is observed in mucoepidermoid tumors of the respiratory tract and adenolymphoma (Warthin tumors) of the salivary gland *(184,224,225)*. Involvement of the 14q24 breakpoint is observed in pulmonary hamartomas and uterine myomas *(251,256)*, while rearrangements of chromosome 22 are seen in malignant rhabdoid tumors of the kidney and brain *(336,337)*.

Germ cell tumors may have the same pathogenetic mechanisms regardless of whether they develop in males or females. An isochromosome for the short arm of chromosome 12 is seen in dysgerminomas, yolk sac tumors, and mature teratomas of the female as well as 80% of testicular germ cell tumors.

Trisomy 12, a common numerical change in CLL *(103)*, is reported in benign sex cord stromal tumors of the ovary (fibromas, thecomas, granulosa cell tumors), fibrothecomas, adenomas, uterine leiomyomas, ovarian carcinomas, and thyroid carcinoma *(177,183,203,204,210–212,246,256)*. Genetic events in pancreatic carcinogenesis overlap with other gastrointestinal tract adenocarcinomas *(196,205)*. Hepatoblastoma, an embryonal proliferation of the liver, and embryonal rhabdomyosarcoma share gains of chromosome 2 or 2q, supporting the hypothesis that neoplastic proliferations of these cells may result from mutations of the same gene *(338)*.

Gains of chromosomes 7 and 8 and loss of Y have been observed in progressive fibroses involving hyperproliferation (carpal tunnel syndrome, Dupuytren's contracture of the palm of the hand and Peyronie's disease of the penis) as well as in desmoid tumors, an aggressive fibromatosis *(247,248,317,320,339)*.

An unbalanced translocation between chromosomes 1 and 16, der(16)t(1;16), resulting in trisomy 1q and loss of genetic material from 16q, may represent a general pathway of clonal evolution in several different tumor types including Ewing's sarcoma and peripheral primitive neurectodermal tumor (PNET), Ph-positive ALL, myxoid liposarcoma, rhabdomyosarcoma, breast cancer, endometrial adenocarcinoma, myelodysplastic syndromes, acute myeloid leukemia, multiple myeloma, retinoblastoma, and Wilms tumor *(190–192,340–343)*. While the acquisition of extra material from the long arm of chromosome 1 is a very common and apparently nonspecific clonal phenomenon that contributes to the clonal evolution of many tumor types *(11,344)*, the loss of genes from the long arm of chromosome 16 is speculated to represent a later step in the neoplastic process *(343)*.

Multiple unrelated clonal cell populations, reflecting genetic heterogeneity, have been reported in squamous cell and basal cell tumors of the head and neck, breast, and lung *(217,218,228,229,345)*. This apparent polyclonality may be the result of selection differences, making the results heavily dependent upon sample processing and disaggregation techniques, as well as length in culture and subjectivity of analysis *(346)*.

An important mechanism for activation of oncogene overexpression is DNA sequence amplification, visualized cytogenetically as a homogeneously staining region (hsr) or double minutes (dmin). Dmin have been reported in neuroblastoma and breast, colon, pancreatic, and ovarian carcinomas. Hsr's have been identified for the regions 8p1 in breast and 19q13.1–13.2 in ovarian carcinomas as well as 12q13–14 in gliomas *(347)*.

Recurrent translocation events in soft tissue tumors of mesenchymal origin show striking similarities (**Table 18**). In a manner analogous to some leukemias, all result in the production of a tumor-specific chimeric RNA transcript that is predicted to encode a novel oncogene transcription factor. Genes having RNA binding domains (*EWS, FUS*) are fused with transcription factors (*FLI1, ERG, ETV, ATF, CHOP,* and *WT*) *(348)*.

As previously discussed, the existence of a recurrent translocation in a disorder suggests that this rearrangement may involve genes that are directly related to its pathogenesis. Ewing's sarcoma, peripheral neuroepithelioma, Askin tumor, and esthesioneuroblastoma all share t(11;22)(q24;q12) and are considered closely related, if not identical *(293–297)*. The t(11;22) leads to generation of a chimeric gene in which the 5′ portion of *EWS* on chromosome 22 is fused in frame to 3′ sequences from *FLI1* on chromosome 11. As a result, the RNA binding domain of *EWS* is replaced by the DNA binding domain of *FLI1* and the fusion portion has properties of a transcription factor with a transcriptional activation domain contributed by *FLI1*.

Variant translocation partners for *EWS* in Ewing's sarcoma have also been identified. These include *ETV1* (7p22), *ETV4* (17q21), and *ERG* (21q22) *(349–351)*. All four translocation partners are members of the *ETS* family of transcription factors that contain a characteristic DNA binding motif, the *ETS* domain.

EWS also promiscuously fuses with transcription factors in other sarcomas in a manner reminiscent of *MLL* in acute leukemia. *EWS* forms chimeric proteins with *ATF1*

Table 18
Molecularly Characterized Rearrangements in Soft Tissue Tumors

Tumor	Cytogenetic abnormality	Gene(s) involved (locus)	Protein
Benign			
Aggressive angiomyxoma	t(12q13–15;v)	*HMGIC*(12q14–15)	HMG DNA binding protein
Chondroma	t(12q13–15;v)	*HMGIC*(12q14–15)	HMG DNA binding protein
Endometrial polyp	t(6p21;v)	*HMGI(Y)*(6p21)	HMG DNA binding protein
	t(12q13–15;v)	*HMGIC*(12q14–15)	HMG DNA binding protein
Fibroadenoma	t(12q13–15;v)	*HMGIC*(12q14–15)	HMG DNA binding protein
Hibernoma	t(11q13;v)	Unknown	
Lipoblastoma	Rearrangements involving 8q	Unknown	Unknown
Plemorphic adenoma	t(3;8)(p21;q12)	*PLAG1*(8q12)	HMG DNA binding domain
	t(12q13–15;v)	*HMGIC*(12q14–15)	HMG DNA binding domain
Pulmonary chondroid hamartoma	t(12q13–15;v)	*HMGIC*(12q14–15)	HMG DNA binding domain
Sporadic lipoma	t(6p21;v)	*HMGI(Y)*(6p21)	HMG DNA binding domain
	t(12q13–15;v)	*HMGIC*(12q14–15)	HMG DNA binding domain
Uterine leiomyoma	t(6p21;v)	*HMGI(Y)*(6p21)	HMG DNA binding domain
	t(12q13–15;v)	*HMGIC*(12q14–15)	HMG DNA binding domain
Malignant			
Chondrosarcoma (extraskeletal myxoid)	t(9;22)(q22–31;q12)	*CHN*(9q31)	Steroid/thyroid receptor gene super family
		EWS(22q12)	RNA binding
Clear cell sarcoma	t(12;22)(q13;q12)	*ATF1*(12q13)	Transcription factor (bZIP family)
		EWS(22q12)	RNA binding protein
Desmoplastic small round cell tumor	t(11;22)(p13;q12)	*WT1*(11p13)	Wilms tumor gene
		EWS(22q12)	RNA binding protein

(continued)

396

Table 18 (*Continued*)

Tumor	Cytogenetic abnormality	Gene(s) involved (locus)	Protein
Malignant (continued)			
Ewing's sarcoma/PNET	t(7;22)(p22;q12)	*ETV1*(7p22)	ETS transcription factor family
		EWS(22q12)	RNA binding protein
	t(11;22)(q24;q12)	*FLI1*(11q24)	ETS transcription factor family
		EWS(22q12)	RNA binding protein
	t(17;22)(q21;q12)	*ETV4*(17q12)	ETS transcription factor family
		EWS(22q12)	RNA binding protein
	t(21;22)(q22;q12)	*ERG*(21q22)	ETS transcription factor family
		EWS(22q12)	RNA binding protein
Liposarcoma (myxoid/round cell)	t(12;16)(q13;p11)	*CHOP*(12q13)	Transcription factor (C/EBPβ family)
		FUS(16p11)	RNA binding protein
	t(12;22)(q13;q12)	*CHOP*(12q13)	Transcription factor (C/EBPβ family)
		EWS(22q12)	RNA binding protein
Rhabdomyosarcoma (alveolar)	t(1;13)(p36;q14)	*PAX7*(1p36)	Homeobox homolog
		FKHR(13q14)	Forkhead domain
	t(2;13)(q35;q14)	*PAX3*(2q35)	Homeobox homolog
		FKHR(13q14)	Forkhead domain
Synovial sarcoma	t(X;18)(p11.2;q11.2)	*SYT*(18q11.2)	Unknown
		SSX1, SSX2 (Xp11.2)	Homology to Krüppel-associated box transcriptional repressor region

v = variable translocation partner

397

(12q13) in clear cell sarcoma (malignant melanoma of the soft parts) and with *WT1* (11p13) in desmoplastic small round cell tumors *(273,274). CHN* (9q22), a member of the steroid hormone receptor superfamily, is the *EWS* fusion partner in myxoid chondrosarcoma *(276). FUS* (16p11), a homologue of *EWS,* fuses with the *CHOP* gene (12q13), another transcription factor, in myxoid liposarcoma *(279).* A cytogenetic variant of myxoid liposarcoma t(12;22)(q13;q12) fuses *CHOP* with *EWS (352).*

Benign mesenchymal tumors are also characterized by specific chromosome abnormalities (**Tables 16 and 17**), with 12q13–15, 6p21–23, 13q12–22, 1pter, 2p11, 7q21–22, 8q12, and 21q12 the regions most frequently involved *(243–268).* Members of the high-mobility group (HMG) protein family have recently been identified as target genes in these tumors *(353–355).*

HMG proteins are a family of nonhistone chromatin-associated proteins that regulate chromatin architecture as well as gene expression, and are divided into three families: the HMG box containing *HMG1/HMG2,* the active chromatin associated *HMG14/ HMG17,* and the HMGI proteins, the latter family consisting of *HMGI, HMGI(Y),* and *HMGIC. HMGIC,* mapped to 12q14–15 is frequently rearranged or overexpressed in lipomas, uterine leiomyomas, leiomyomas, fibroadenomas, pleomorphic adenomas, aggressive angiomyxomas, endometrial polyps, chondromas, and pulmonary hamartomas. *HMGI(Y),* mapped to 6p21, is rearranged in lipomas, endometrial polyps and uterine leiomyomas. *HMG17* (1p36.1–35), *HMG14* (21q22), *HMGIL* (13q12), and *HMG2* (4q31) are all mapped to regions involved in mesenchymal benign tumors or in tumors with a mesenchymal component.

Benign tumors with *HMGI(Y)* or *HMGIC* rearrangements rarely undergo malignant transformation, sharply contrasting with the highly malignant properties of tumors with *EWS* translocations. The reason why so many different translocation partners participate in translocations involving *HMGI* loci is unknown.

Like the *EWS* gene family, the *HMGI* family translocations involve chimeric DNA-binding proteins which most likely exert their oncogenic effects by perturbing normal gene expression. The critical domains in *HMGI* loci are DNA binding domains, while in the case of *EWS* a transcriptional activating domain is the critical domain.

In 1973, Sakurai and Sandberg published the first clinical–cytogenetic correlation that showed that patients with a normal karyotype had a significantly better prognosis than those with chromosome abnormalities *(356).* Twenty-five years later, we are at the same level of understanding with information relating to the prognostic impact of cytogenetics in solid tumors. The presence of an abnormal clone has been demonstrated to be a poor prognostic sign in bladder cancer *(357),* malignant gliomas *(358),* ovarian cancer *(359),* and prostate cancer *(360).* The degree of cytogenetic complexity, rather than the presence of an abnormal karyotype, was associated with better survival outcome in a series of ovarian *(359),* colon *(361),* pancreatic *(334),* renal *(362),* and head and neck carcinomas *(363).* Mitelman et al., however, raise the issue of interpretation of normal karyotypic findings in solid tumors, warning of submicroscopic changes and failure to analyze neoplastic cells in culture *(364).* Citing specific examples, Trent et al. correlated significantly shorter survival in metastatic malignant melanoma patients with structural rearrangements of chromosomes 7 or 11 *(365).* A derivative "19p+" chromosome was associated with relapse in malignant fiberous histiocytoma *(366),* while 11q13 rearrangements in primary squamous cell carcinomas of the head and neck were associated with poor prognosis *(363)*. *MYCN* amplifications identified by hsr/dmin and 1p deletions correlated with poor survival/prognosis in neuroblastoma *(367–370)*.

ACKNOWLEDGMENT

The author wishes to thank Mrs. Diane Tymorek for her assistance in preparation of this manuscript.

REFERENCES

1. Arnold, J. (1879) Über feinere struktur der zellen unter normalen und pathologischen bedingungen. *Virchows Arch. Path. Anat.* **77,** 181–206.
2. von Hansemann, D. (1890) Ueber asymmetrische Zelltheilung in Epithelkrebsen und deren biologische Bedeutung. *Virchows Arch. A. Pathol. Anat.* **119,** 299–326.
3. Boveri, T. (1902) Über mehrpolige Mitosen als Mittel zur Analyse des Zellkerns. *Verh. Phys. Med. Ges. Wurzb.*
4. Boveri, T. (1914) *Zur Frage der Entstehung malinger Tumoren.* Verlag von Gustav Fischer, Jena.
5. Winge, Ö. (1930) Zytologische Untersuchungen über die natur malinger tumoren. II. Teerkarzinomen bei Mäusen. *Z. Zellforsch. Miktosk. Anat.* **10,** 683–735.
6. Foulds, L. (1958) The natural history of cancer. *J. Chronic Dis.* **8,** 2–37.
7. Levan. A. (1959) Relation of chromosome status to the origin and progression of tumors. The incidence of chromosome numbers. In *Genetics and Cancer,* University of Texas Press, Austin, 151 pp.
8. Hauschka, T.S. (1961) The chromosomes in ontogeny and oncology. *Cancer Res.* **21,** 957–974.
9. Makino, S. (1975) *Human Chromosomes.* Igaku Shoin, Tokoyo.
10. Hsu, T.C. (1979) *Human and Mammalian Cytogenetics. An Historical Perspective.* Springer-Verlag, Heidelberg.
11. Sandberg, A.A. (1990) *The Chromosomes in Human Cancer and Leukemia,* 2nd edition. Elsevier, New York.
12. Nowell, P.C. and Hungerford, D.A. (1960) A minute chromosome in human chronic granulocytic leukemia. *Science* **132,** 1497.
13. Rowley, J.D. (1973) A new consistent chromosomal abnormality in chronic myelogenous leukaemia identified by quinacrine fluorescence and Giemsa staining. *Nature* **243,** 290–293.
14. Zech, L., Haglund, U., Nilsson, K., and Klein, G. (1976) Characteristic chromosomal abnormalities in biopsies and lymphoid-cell lines from patients with Burkitt and non-Burkitt lymphomas. *Int. J. Cancer* **17,** 47–56.
15. Mitelman, F. (1994) *Catalog of Chromosome Aberrations in Cancer,* 5th edition. Wiley-Liss, New York.
16. Heim, S. and Mitelman, F. (1995) *Cancer Cytogenetics,* 2nd edition. Wiley-Liss, New York.
17. Mitelman, F., Mertens, F., and Johansson, B. (1997) A breakpoint map of recurrent chromosome rearrangements in human neoplasia. *Nature* **15** (Suppl), 417–474.
18. Schuh, A.C., Sutherland, D.R., Horsfall, W., Mills, G.B., Dubé, I., Baker, M.A., Siminovich, K., Bailey, D., and Keating, A. (1990) Chronic myeloid leukemia arising in a progenitor common to T cells and myeloid cells. *Leukemia* **4,** 631–636.
19. Hagemeijer, A., Bartram, C.R., Smit, E.M.E., van Agthoven, A.J., and Bootsma, D. (1984) Is the chromosomal region 9q34 always involved in variants of the Ph[1] translocation? *Cancer Genet. Cytogenet.* **13,** 1–16.

20. Heisterkamp, N., Voncken, J.-W., van Schaick, H., and Groffen, J. (1993) Ph-positive leukemia. In *The Causes and Consequences of Chromosomal Aberrations* (Kirsch, I.R., ed), CRC Press, Boca Raton, pp. 359–376.

21. Kantarjian, H.M., Keating, M.J., Walters, R.S., McCredie, K.B., Smith, T.L., Talpaz, M., Beran, M., Cork, A., Trujillo, J.M., and Freireich, E.J. (1986) Clinical and prognostic features of Philadelphia chromosome negative chronic myelogenous leukemia. *Cancer* **58,** 2023–2030.

22. Lazaridou, A., Chase, A., Melo, J., Garicochea, B., Diamond, J., and Goldman, J. (1994) Lack of reciprocal translocation in BCR–ABL positive Ph-negative chronic myeloid leukemia. *Leukemia* **8,** 454–457.

23. Andersson, B.S., Beran, M., Pathak, S., Goodacre, A., Barlogie, B., and McCredie, K.B. (1987) Ph-positive chronic myeloid leukemia with near-haploid conversion in vivo and establishment of a continuously growing cell line with similar cytogenetic pattern. *Cancer Genet. Cytogenet.* **24,** 335–343.

24. Mayne, K.M. and Maher, E.J. (1989) Near-haploid cell line in megakaryoblastic transformation of Philadelphia-positive chronic myeloid leukemia. *Cancer Genet. Cytogenet.* **39,** 133–136.

25. Sokal, J.E., Gomez, G.A., Baccarani, M., Tura, S., Clarkson, B.D., Cervantes, F., Rozman, C., Carbonell, F., Anger, B., Heimpel, H., Nissen, N.I., and Robertson, J.E. (1988) Prognostic significance of additional cytogenetic abnormalities at diagnosis of Philadelphia chromosome-positive chronic granulocytic leukemia. *Blood* **72,** 294–298.

26. Alimena, G., DeCuia, M.R., Diverio, D., Gastaldi, R., and Nanni, M. (1987) The karyotype of blast crisis. *Cancer Genet. Cytogenet.* **6,** 39–50.

27. Bernstein, R. and Gale, R.P. (1990) Do chromosome abnormalities determine the type of acute leukemia that develops in CML? *Leukemia* **4,** 65–68.

28. Lafage-Pochitaloff-Huvalé, M., Sainty, D., Adriaanssen, H.J., Lopez, M., Maraninchi, D., Simonetti, J., Mannoni, P., Carcassonne, Y., and Hagemeijer, A. (1989) Translocation (3;21) in Philadelphia positive chronic myeloid leukemia: high resolution chromosomal analysis and immunological study on five new cases. *Leukemia* **3,** 554–559.

29. Bennett, J.M., Catovsky, D., Daniel, M-T., Flandrin, G., Galton, D.A.G., Gralnick, H.R., and Sultan, C. (1985) Proposed revised criteria for the classification of acute myeloid leukemia. *Ann. Intern. Med.* **103,** 626–629.

30. Bennett, J.M., Catovsky, D., Daniel, M-T., Flandrin, G., Galton, D.A.G., Gralnick, H.R., and Sultan, C. (1991) Proposal for the recognition of minimally differentiated acute myeloid leukaemia (AML-M0). *Br. J. Haematol.* **78,** 325–328.

31. Cheson, B.D., Cassileth, P.A., Head, D.R., Schiffer, C.A., Bennett, J.M., Bloomfield, C.D., Brunning, R., Gale, R.P., Grever, M.R., Keating, M.J., Sawitsky, A., Stass, S., Weinstein, H., and Woods, W.G. (1990) Report of the National Cancer Institute-sponsored workshop on defintions of diagnosis and response in acute myeloid leukemia. *J. Clin. Oncol.* **8,** 813–819.

32. Stasi, R., Del Poeta, G., Masi, M., Tribalto, M., Venditti, A., Papa, G., Nicoletti, B., Vernole, P., Tedeschi, B., Delaroche, I., Mingarelli, R., and Dallapiccola, B. (1993) Incidence of chromosome abnormalities and clinical significance of karyotype in de novo acute myeloid leukemia. *Cancer Genet. Cytogenet.* **67,** 28–34.

33. Kalwinsky, D.K., Raimondi, S.C., Schell, M.J., Mirro, Jr. J., Santana, V.M., Behm, F., Dahl, G.V., and Williams, D. (1990) Prognostic importance of cytogenetic subgroups in de novo pediatric nonlymphocytic leukemia. *J. Clin. Oncol.* **8,** 75–83.

34. Martinez-Climent, J.A., Lane, N.J., Rubin, C.M., Morgan, E., Johnstone, H.S., Mick, R., Murphy, S.B., Vardiman, J.W., Larson, R.A., Le Beau, M.M., and Rowley, J.D. (1995) Clinical and prognostic significance of chromosomal abnormalities in childhood acute myeloid leukemia de novo. *Leukemia* **9,** 95–101.

35. Mrózek, K., Heinonen, K., de la Chapelle, A., and Bloomfield, C.D. (1997) Clinical significance of cytogenetics in acute myeloid leukemia. *Semin. Oncol.* **24,** 17–31.
36. Caligiuri, M.A., Schichman, S.A., Strout, M.P., Mrózek, K., Baer, M.R., Frankel, S.R., Barcos, M., Herzig, G.P., Croce, C.M., and Bloomfield, C.D. (1994) Molecular rearrangement of the *ALL-1* gene in acute myeloid leukemia without cytogenetic evidence of 11q23 chromosomal translocations. *Cancer Res.* **54,** 370–373.
37. Caligiuri, M.A., Strout, M.A., Arthur, D.C., Baer, M.R., Shah, D., Block, A.W., Mrózek, K., Yu, F., Oberkircher, A., Lawrence, D., Knuutila, S., Ruutu, T., Herzig, G.P., Schiffer, C.A., and Bloomfield, C.D. (1996) Rearrangement of *ALL1* is a recurrent molecular defect in adult myeloid leukemia (AML) with normal cytogenetics that predicts a short complete remission (CR) duration. *Proc. Am. Soc. Clin. Oncol.* **15,** 360.
38. Johansson, B., Mertens, F., and Mitelman, F. (1994) Secondary chromosomal abnormalities in acute leukemias. *Leukemia* **8,** 953–962.
39. Pedersen-Bjergaard, J. and Philip, P. (1991) Two different classes of therapy-related and de novo acute myeloid leukemia? *Cancer Genet. Cytogenet.* **55,** 119–124.
40. Pedersen-Bjergaard, J. and Rowley, J.D. (1994) The balanced and unbalanced chromosome aberrations in acute myeloid leukemia may develop in different ways and may contribute differently to malignant transformation. *Blood* **83,** 2780–2786.
41. Sandoval, C., Pui, C-H., Bowman, L.C., Heaton, D., Hurwitz, C.A., Raimondi, S.C., Behn, F.G., and Head, D.R. (1993) Secondary acute myeloid leukemia in children previously treated with alkylating agents, intercalating topoisomerase II inhibitors, and irradiation. *J. Clin. Oncol.* **11,** 1039–1045.
42. Caligiuri, M.A., Strout, M.P., and Gilliland, D.G. (1997) Molecular biology of acute myeloid leukemia. *Semin. Oncol.* **24,** 32–44.
43. Bennett, J.M., Catovsky, D., Daniel, M.T., Flandrin, G., Galton, D.A.G., Gralnick, H.R., and Sultan, C. (1982) Proposals for the classification of the myelodysplastic syndromes. *Br. J. Haematol.* **51,** 189–199.
44. Bennett, J.M. (1992) FAB and MIC classification of myelodysplastic syndromes: points of controversy. In *The Myelodysplastic Syndromes* (Mufti, G.J. and Galton, D.A.G., eds.), Churchill Livingstone, Edinburgh, pp. 15–121.
45. Second International Workshop on Chromosomes in Leukemia (1980) Chromosomes in preleukemia. *Cancer Genet. Cytogenet.* **2,** 108–113.
46. Heim, S. (1992) Cytogenetic findings in primary and secondary MDS. *Leuk. Res.* **16,** 43–46.
47. van den Berghe, H., Vermaelen, K., Mecucci, C., Barbieri, D., and Tricot, G. (1985) The 5q-anomaly. *Cancer Genet. Cytogenet.* **17,** 189–255.
48. Pierre, R.V., Catovsky, D., Mufti, G.J., Swansbury, G.J., Mecucci, C., Dewald, G.W., Ruutu, T., Van den Berghe, H., Rowley, J.D., Mitelman, F., Reeves, B.R., Alimena, G., Garson, O.M., Lawler, S.D., and de la Chapelle, A. (1989) Clinical-cytogenetic correlations in myelodysplasia (preleukemia). *Cancer Genet. Cytogenet.* **40,** 149–161.
49. Suciu, S., Kuse, R., Weh, H.J., and Hossfeld, D.K. (1990) Results of chromosome studies and their relation to morphology, course, and prognosis in 120 patients with de novo myelodysplastic syndrome. *Cancer Genet. Cytogenet.* **44,** 15–26.
50. Aatola, M., Armstrong, E., Teerenhovi, L., and Borgström, G.H. (1992) Clinical significance of the del(20q) chromosome in hematologic disorders. *Cancer Genet. Cytogenet.* **62,** 75–80.
51. Michaux, J-L. and Martiat, P. (1993) Chronic myelomonocytic leukaemia (CMML)—a myelodysplastic or myeloproliferative syndrome? *Leuk. Lymphoma* **9,** 35–41.
52. Johansson, B., Mertens, F., Heim, S., Kristoffersson, U., and Mitelman, F. (1991) Cytogenetics of secondary myelodysplasia (sMDS) and acute nonlymphocytic leukemia (sANLL). *Eur. J Haematol.* **47,** 17–27.

53. Pui, C-H., Behm, F.G., Raimondi, S.C., Dodge, R.K., George, S.L., Rivera, G.K., Mirro, Jr. J., Kalwinsky, D.K., Dahl, G.V., Murphy, S.B., Crist, W.M., and Williams, D.L. (1989) Secondary acute myeloid leukemia in children treated for acute lymphoid leukemia. *N. Engl. J. Med.* **321,** 136–142.

54. Rubin, C.M., Arthur, D.C., Woods, W.G., Lange, B.J., Nowell, P.C., Rowley, J.D., Nachman, J., Bostrom, B., Baum, E.S., Suarez, C.R., Shah, N.R., Morgan, E., Maurer, H.S., McKenzie, S.E., Larson, R.A., and Le Beau, M.M. (1991) Therapy-related myelodysplastic syndrome and acute myeloid leukemia in children: correlation between chromosomal abnormalities and prior therapy. *Blood* **78,** 2982–2988.

55. Berger, R. and Flandrin, G. (1992) Chromosomal abnormalities in secondary acute myeloid leukemia and the myelodysplastic syndromes. In *The Myelodysplastic Syndromes* (Mufti, G.J. and Galton, D.A.G., eds.), Churchill Livingstone, Edinburgh, pp. 129–139.

56. Gill Super, H.J., McCabe, N.R., Thirman, M.J., Larson, R.A., Le Beau, M.M., Pedersen-Bjergaard, J., Philip, P., Diaz, M.O., and Rowley, J.D. (1993) Rearrangements of the *MLL* gene in therapy-related acute myeloid leukemia in patients previously treated with agents targeting DNA-topoisomerase II. *Blood* **82,** 3705–3711.

57. Levine, E.G. and Bloomfield, C.D. (1992) Leukemias and myelodysplastic syndromes secondary to drug, radiation, and environmental exposure. *Semin. Oncol.* **19,** 47–84.

58. Yunis, J.J., Lobell, M., Arnesen, M.A., Oken, M.M., Mayer, M.G., Rydell, R.E., and Brunning, R.D. (1988) Refined chromosome study helps define prognostic subgroups in most patients with primary myelodysplastic syndrome and acute myelogenous leukaemia. *Br. J. Haematol.* **68,** 189–194.

59. Diez-Martin, J.L., Graham, D.L., Petitt, R.M., and Dewald, G.W. (1991) Chromosome studies in 104 patients with polycythemia vera. *Mayo Clin. Proc.* **66,** 287–299.

60. Mertens, F., Johansson, B., Heim, S., Kristoffersson, U., and Mitelman, F. (1991) Karyotypic patterns in chronic myeloproliferative disorders: report on 74 cases and review of the literature. *Leukemia* **5,** 214–220.

61. Baranger, L., Baruchel, A., Leverger, G., Schaison, G., and Berger, R. (1990) Monosomy-7 in childhood hemopoietic disorders. *Leukemia* **4,** 345–349.

62. Aricó, M., Biondi, A., and Pui, C-H. (1997) Juvenile myelomonocytic leukemia. *Blood* **90,** 479–488.

63. Sessarego, M., Defferrari, R., Dejana, A.M., Rebuttato, A.M., Fugazza, G., Salvidio, E., and Ajmar, F. (1989) Cytogenetic analysis in essential thrombocythemia at diagnosis and at transformation. A 12-year study. *Cancer Genet. Cytogenet.* **43,** 57–65.

64. Cournoyer, D., Noël, P., Schmidt, M.A., and Dewald, G.W. (1987) Trisomy 9 in hematologic disorders: possible association with primary thrombocytosis. *Cancer Genet. Cytogenet.* **27,** 73–78.

65. Clark, E.A. and Ledbetter, J.A. (1989) Structure, function, and genetics of human B cell-associated surface molecules. *Adv. Cancer Res.* **52,** 81–149.

66. Chan, A. and Mak, T.W. (1990) Genomic organization of the T cell receptor. *Cancer Detect. Prevent.* **14,** 261–267.

67. Jennings, C.D. and Foon, K.A. (1997) Recent advances in flow cytometry: application to the diagnosis of hematologic malignancy. *Blood* **90,** 2863–2892.

68. Bennett, J.M., Catovsky, D., Daniel, M-T., Flandrin, G., Galton, D.A.G., Gralnick, H.R., and Sultan, C. (1976) Proposals for the classification of the acute leukaemias. *Br. J. Haematol.* **33,** 451–458.

69. First MIC Cooperative Study Group (1986) Morphologic, immunologic, and cytogenetic (MIC) working classification of acute lymphoblastic leukemia. *Cancer Genet. Cytogenet.* **23,** 189–197.

70. The Groupe Français de Cytogénétique Hématologique (1996) Cytogenetic abnormalities in adult lymphoblastic leukemia: correlations with hematologic findings and out-

come. A collaborative study of the Groupe Français de Cytogénétique Hématologique. *Blood* **87**, 3135–3142.

71. Raimondi, S.C. (1993) Current status of cytogenetic research in childhood acute lymphoblastic leukemia. *Blood* **81**, 2237–2251.

72. Pui, C-H., Williams, D.L., Roberson, P.K., Raimondi, S.C., Behm, F.G., Lewis, S.H., Rivera, G.K., Kalwinsky, D.K., Abromowitch, M., Crist, W.M., and Murphy, S.B. (1988) Correlation of karyotype and immunophenotype in childhood acute lymphoblastic leukemia. *J. Clin. Oncol.* **6**, 56–61.

73. Jackson, J.F., Boyett, J., Pullen, J., Brock, B., Patterson, R., Land, V., Borowitz, M., Head, D., and Crist, W. (1990) Favorable prognosis associated with hyperdiploidy in children with acute lymphocytic leukemia correlates with extra chromosome 6. A Pediatric Oncology Group study. *Cancer* **66**, 1183–1189.

74. Harris, M.B., Shuster, J.J., Carroll, A., Look, A.T., Borowitz, M.J., Crist, W.M., Nitschke, R., Pullen, J., Steuber, C.P., and Land, V.J. (1992) Trisomy of leukemic cell chromosomes 4 and 10 identifies children with B-progenitor cell acute lymphoblastic leukemia with a very low risk of treatment failure. A Pediatric Oncology Group study. *Blood* **79**, 3316–3324.

75. Pui, C-H., Raimondi, S.C., Dodge, R.K., Rivera, G.K., Fuchs, L.A.H., Abromowitch, M., Look, A.T., Furman, W.L., Crist, W.M., and Williams, D.L. (1989) Prognostic importance of structural chromosomal abnormalities in children with hyperdiploid (>50 chromosomes) acute lymphoblastic leukemia. *Blood* **73**, 1963–1967.

76. Pui, C-H., Carroll, A.J., Raimondi, S.C., Land, V.J., Crist, W.M., Shuster, J.J., Williams, D.L., Pullen, D.J., Borowitz, M.J., Behm, F.G., and Look, A.T. (1990) Clinical presentation, karyotypic characterization, and treatment outcome of childhood acute lymphoblastic leukemia with a near-haploid or hypodiploid <45 line. *Blood* **75**, 1170–1177.

77. Oshimura, M., Freeman, A.I., and Sandberg, A.A. (1977) Chromosomes and causation of human cancer and leukemia XXIII. Near-haploidy in acute leukemia. *Cancer* **40**, 1143–1148.

78. Tallents, S., Forster, D.C., Garson, O.M., Michael, P.M., Briggs, P., Brodie, G.N., Pilkington, G., and Januszewicz, E. (1987) Hybrid biphenotypic acute leukemia with extreme hypodiploidy. *Pathology* **19**, 197–200.

79. Pui, C-H. and Crist, W.M. (1992) Cytogenetic abnormalities in childhood acute lymphoblastic leukemia correlates with clinical features and treatment outcome. *Leuk. Lymphoma* **7**, 259–274.

80. Pui, C-H., Crist, W.M., and Look, A.T. (1990) Biology and clinical significance of cytogenetic abnormalities in childhood acute lymphoblastic leukemia. *Blood* **76**, 1449–1463.

81. Raimondi, S.C., Behm, F.G., Roberson, P.K., Pui, C-H., Rivera, G.K., Murphy, S.B., and Williams, D.L. (1988) Cytogenetics of childhood T-cell leukemia. *Blood* **72**, 1560–1566.

82. Croce, C.M. and Nowell, P.C. (1985) Molecular basis of human B cell neoplasia. *Blood* **65**, 1–7.

83. Cleary, M.L. (1991) Oncogenic conversion of transcription factors by chromosome translocations. *Cell* **66**, 619–622.

84. Taub, R., Moulding, C., Battey, J., Murphy, W., Vasicek, T., Lenoir, G.M., and Leder, P. (1984) Activation and somatic mutation of the translocated c-*myc* gene in Burkitt lymphoma cells. *Cell* **36**, 339–348.

85. Blackwood, E.M. and Eisenman, R.N. (1991) Max: a helix-loop-helix zipper protein that forms a sequence specific DNA-binding complex with Myc. *Science* **251**, 1211–1217.

86. Bossone, S.A., Asselin, C., Patel, A.J., and Marcu, K.B. (1992) MAZ, a zinc finger protein, binds to c-MYC and C2 gene sequences regulating transcriptional initiation and termination. *Proc. Natl. Acad. Sci. USA* **89**, 7452–7456.

87. Williams, D.L., Look, A.T., Melvin, S.L., Roberson, P.K., Dahl, G., Flake, T., and Stass,

S. (1984) New chromosomal translocations correlate with specific immunophenotypes of childhood acute lymphoblastic leukemia. *Cell* **36,** 101–109.

88. Nourse, J., Mellentin, J.D., Galili, N., Wilkinson, J., Stanbridge, E., Smith, S.D., and Cleary, M.L. (1990) Chromosomal translocation t(1;19) results in synthesis of a homeobox fusion mRNA that codes for a potential chimeric transcription factor. *Cell* **60,** 535–545.

89. Privitera, E., Kamps, M.P., Hayashi, Y., Inaba, T., Shapiro, L.H., Raimondi, S.C., Behm, F., Hendershot, L., Carroll, A.J., Baltimore, D., and Look, A.T. (1992) Different molecular consequences of the 1;19 chromosomal translocation in childhood B-cell precursor acute lymphoblastic leukemia. *Blood* **79,** 1781–1788.

90. Crist, W., Carroll, A., Shuster, J., Jackson, J., Head, D., Borowitz, M., Behm, F., Link, M., Steuber, P., Ragab, A., Hirt, A., Brock, B., Land, V., and Pullen, J. (1990) Philadelphia chromosome positive childhood acute lymphoblastic leukemia: clinical and cytogenetic characteristics and treatment outcome. A Pediatric Oncology Group study. *Blood* **76,** 489–494.

91. Tuszynski, A., Dhut, S., Young, B.D., Lister, T.A., Rohatiner, A.Z., Amess, J.A.L., Chaplin, T., Dorey, E., and Gibbons, B. (1993) Detection and significance of *bcr–abl* mRNA transcripts and fusion proteins in Philadelphia-positive adult acute lymphoblastic leukemia. *Leukemia* **7,** 1504–1508.

92. Bernard, O.A. and Berger, R. (1995) Molecular basis of 11q23 rearrangements in hematopoietic malignant proliferations. *Genes Chromosom Cancer* **13,** 75–85.

93. Thirman, M.J., Gill, H.J., Burnett, R.C., Mbangkollo, D., McCabe, N.R., Kobayashi, H., Ziemin van der Poel, S., Kaneko, Y., Morgan, R., Sandberg, A.A., Chaganti, R.S.K., Larson, R.A., Le Beau, M.M., Diaz, M.O., and Rowley, J.D. (1993) Rearrangement of the *MLL* gene in acute lymphoblastic and acute myeloid leukemias with 11q23 chromosomal translocations. *N. Engl. J. Med.* **329,** 909–914.

94. Croce, C.M., Isobe, M., Palumbo, A., Puck, J., Ming, J., Tweardy, D., Crickson, J., David, M., and Rovera, G. (1985) Gene for alpha-chain of human T-cell receptor: location of chromosome 14 region involved in T-cell neoplasms. *Science* **227,** 1044–1047.

95. Hayashi, Y., Raimondi, S.C., Look, A.T., Behm, F.G., Kitchingman, G.R., Pui, C-H., Rivera, G.K., and Williams, D.L. (1990) Abnormalities of the long arm of chromosome 6 in childhood acute lymphoblastic leukemia. *Blood* **76,** 1626–1630.

96. Kowalczyk J. and Sandberg, A.A. (1983) A possible subgroup of ALL with 9p–. *Cancer Genet. Cytogenet.* **9,** 383–385.

97. Raimondi, S.C., Williams, D.L., Callihan, T., Peiper, S., Rivera, G.K., and Murphy, S.B. (1986) Nonrandom involvement of the 12p12 breakpoint in chromosome abnormalities of childhood acute lymphoblastic leukemia. *Blood* **68,** 69–75.

98. Raimondi, S.C., Rivera, G.K., Pui, C-H., Mahmoud, H., Look, A.T., and Williams, D.L. (1991) Frequency and heterogeneity of 12p structural abnormalities, including a new pericentric inversion, in childhood acute lymphoblastic leukemia. *Cytogenet. Cell Genet.* **58,** 27301a.

99. Romana, S.P., Poirel, H., LeConiat, M., Flexor, M.A., Mauchauffé, M., Jonveaux, P., Macintyre, E.A., Berger, R., and Bernard, O.A. (1995) High frequency of t(12;21) in childhood B lineage acute lymphoblastic leukemia. *Blood* **86,** 4263–4269.

100. Cayuela, J-M, Baruchel, A., Orange, C., Madani, A., Auclerc, M.F., Daniel, M-T., Schaison, G., and Sigaux, F. (1996) *TEL-AML1* fusion RNA as a new target to detect minimal residual disease in pediatric B-cell precursor acute lymphoblastic leukemia. *Blood* **88,** 302–308.

101. Stegmaier, K., Pendse, S., Barker, G.F., Bray-Ward, P., Ward, D.C., Montgomery, K.T., Krauter, K.S., Reynolds, C., Sklar, J., Donnelly, M., Bohlander, S.K., Rowley, J.D., Sallan, S.E., Gilliland, D.G., and Golub, T.R. (1995) Frequent loss of heterozygosity at the *TEL* gene locus in acute lymphoblastic leukemia in childhood. *Blood* **86,** 38–44.

102. Bennett, J.M., Catovsky, D., Daniel, M-T., Flandrin, G., Galton, D.A.G., Gralnick, H.R., and Sultan, C. (1989) Proposals for the classification of chronic (mature) B and T lymphoid leukaemias. French-American British (FAB) Cooperative Group. *J. Clin. Pathol.* **42**, 567–584.

103. Crossen, P.E. (1989) Cytogenetic and molecular changes in chronic B-cell leukemia. *Cancer Genet. Cytogenet* **43**, 143–150.

104. Juliusson, G. and Gahrton, G. (1990) Chromosome aberrations in B-cell chronic lymphocytic leukemia. Pathogenetic and clinical implications. *Cancer Genet. Cytogenet.* **45**, 143–160.

105. Escudier, S.M., Pereira-Leahy, J.M., Drach, J.W., Weier, H.U., Goodacre, A.M., Cork, M.A., Trujillo, J.M., Keating, M.J., and Andreeff, M. (1993) Fluorescent in situ hybridization and cytogenetic studies of trisomy 12 in chronic lymphocytic leukemia. *Blood* **81**, 2702–2707.

106. Zech, L., Godal, T., Hammarström, L., Mellstedt, H., Smith, C.I.E., Tötterman, T., and Went, M. (1986) Specific chromosome markers involved with chronic T lymphocyte tumors. *Cancer Genet. Cytogenet.* **21**, 67–77.

107. Brito-Babapulle, V., Pittman, S., Melo, J.V., Pomfret, M., and Catovsky, D. (1987) Cytogenetic studies on prolymphocytic leukemia. I. B-cell prolymphocytic leukemia. *Hematol. Pathol.* **1**, 27–33.

108. Brito-Babapulle, V. and Catovsky, D. (1991) Inversions and tandem translocations involving chromosome 14q11 and 14q32 in T-prolymphocytic leukemia and T-cell leukemias in patients with ataxia telangiectasia. *Cancer Genet. Cytogenet.* **55**, 1–9.

109. Brito-Babapulle, V., Pittman, S., Melo, J.V., Parreira, L., and Catovsky, D. (1986) The 14q+ marker in hairy cell leukaemia. A cytogenetic study of 15 cases. *Leuk. Res.* **10**, 131–138.

110. Matutes, E. and Catovsky, D. (1991) Mature T-cell leukemias and leukemia/lymphoma syndromes. Review of our experience in 175 cases. *Leuk. Lymphoma* **4**, 81–91.

111. Sadamori, N., Isobe, M., Shimizu, S., Yamamori, T., Itoyama, T., Ikeda, S., Yamama, Y., and Ichimaru, M. (1991) Relationship between chromosomal breakpoint and molecular rearrangement of T-cell antigen receptors in adult T-cell leukemia. *Acta Haematol.* **86**, 14–19.

112. Berger, R. and Bernheim, A. (1987) Cytogenetic studies of Sézary cells. *Cancer Genet. Cytogenet.* **27**, 79–87.

113. Heim, S. and Mitelman, F. (1989) Cytogenetically unrelated clones in hematological neoplasms. *Leukemia* **3**, 6–8.

114. Weh, H.J., Gutensohn, K., Selbach, J., Kruse, R., Wacker-Backhaus, G., Seeger, D., Fiedler, W., Fett, W., and Hossfeld, D.K. (1993) Karyotype in multiple myeloma and plasma cell leukaemia. *Eur. J. Cancer* **29A**, 1269–1273.

115. Manolova, Y., Manolov, G., Apostolov, P., and Levan, A. (1979) The same marker chromosome, mar17p+, in four consecutive cases of multiple myeloma. *Hereditas* **90**, 307–310.

116. Jonveaux, P. and Berger, R. (1992) Chromosome studies in plasma cell leukemia and multiple myeloma in transformation. *Genes Chromosom Cancer* **4**, 321–325.

117. Block, A.W. (1991) Cytogenetic analysis of B-cell chronic lymphoproliferative disorders: New technology, complexities and promises. *Prog. Clin. Biol. Res.* **368**, 145–173.

118. Juliusson, G., Oscier, D.G., Fitchett, M., Ross, F.M., Stockdill, G., Mackie, M.J., Parker, A.C., Castoldi, G.L., Cuneo, A., Knuutila, S., Elonen, E., and Gahrton, G. (1990) Prognostic subgroups in B-cell chronic lymphocytic leukemia defined by specific chromosomal abnormalities. *N. Engl. J. Med.* **323**, 720–724.

119. Han, T., Sadamori, N., Block, A.W., Xiao, H., Henderson, E.S., Emrich, L., and Sandberg, A.A. (1988) Cytogenetic studies in chronic lymphocytic leukemia, prolymphocytic leukemia, and hairy cell leukemia: a progress report. *Nouv. Rev. Fr. Hematol.* **30**, 393–395.

120. Dewald, G.W., Kyle, R.A., Hicks, G.A., and Greipp, P.R. (1985) The clinical signifi-

cance of cytogenetic studies in 100 patients with multiple myeloma, plasma cell leuke-mia, or amyloidosis. *Blood* **66**, 380–390.

121. Lisse, I.M., Drivsholm, A., and Christoffersen, P. (1988) Occurrence and type of chromo-somal abnormalities in consecutive malignant monoclonal grammopathies: correlation with survival. *Cancer Genet. Cytogenet.* **35**, 27–36.

122. Rappaport, H. (1966) *Tumors of the Hematopoietic System. Atlas of Tumor Pathology, Series I, Section III, Fascicle 8.* Armed Forces Institute of Pathology, Washington, D.C.

123. Lukes, R. J. and Collins, R. D. (1974) Immunologic characterization of human malignant lymphomas. *Cancer* **34**, 1488–1503.

124. Lennert, K. (1981) *Histopathology of Non-Hodgkin's Lymphomas: Based on the Kiel Classification.* Springer-Verlag, New York.

125. Stansfeld, A., Diebold, Noel, H., J., Kapanci, Y., Rilke, F., Kelényi, G., Sundstrom, C., Lennert, K., van Unnik, J.A.M., Mioduszewska, O., and Wright, D.H. (1988) Updated Kiel classification of lymphomas. *Lancet* **i**, 292–293.

126. The Non-Hodgkin's Lymphoma Pathologic Classification Project (1982) National Can-cer Institute sponsored study of classifications of non-Hodgkin's lymphomas: summary and description of a Working Formulation for clinical usage. *Cancer* **49**, 2112–2135.

127. Harris, N.L., Jaffe, E.S., Stein, H., Banks, P.M., Chan, J.K., Cleary, M.L., Delsol, G., De Wolf-Peeters, C., Falini, B., Gatter, K.C., Grogan, T.M., Isaacson, P.G., Knowles, D.M., Mason, D.Y., Muller-Hermelink, H-K., Pileri, S.A., Piris, M.A., Ralfkiaer, E., and Warnke, R.A. (1994) A revised European-American Classification of lymphoid neo-plasms: a proposal from the International Lymphoma Study Group. *Blood* **84**, 1361–1392.

128. Hiddemann, W., Longo, D.L., Coiffier, B., Fisher, R.I., Cabanillas, F., Cavalli, F., Nadler, L.M., DeVita, V.T., Lister, A., and Armitage, J.O. (1996) Lymphoma classification—the gap between biology and clinical management is closing. *Blood* **88**, 4085–4089.

129. Yunis, J.J., Oken, M.M., Kaplan, M.E., Ensrud, K.M., Howe, R.R., and Theologides, A. (1982) Distinctive chromosomal abnormalities in histologic subtypes of non-Hodgkin's lymphoma. *N. Engl. J. Med.* **307**, 1231–1236.

130. Bloomfield, C.D., Arthur, D.C., Frizzera, G., Levine, E.G., Peterson, B.A., and Gajl-Peczalska, K.J. (1983) Nonrandom chromosome abnormalites in lymphoma. *Cancer Res.* **43**, 2975–2984.

131. Yunis, J.J., Oken, M.M., Theologides, A., Howe, R.B., and Kaplan, M.E. (1984) Recur-rent chromosomal defects are found in most patients with non-Hodgkin's-lymphoma. *Cancer Genet. Cytogenet.* **13**, 17–28.

132. Kristoffersson, U., Heim, S., Olsson, H., Åkerman, M., and Mitelman, F. (1986) Cyto-genetic studies in non-Hodgkin lymphomas-results from surgical biopsies. *Hereditas* **104**, 1–13.

133. Juneja, S., Lukeis, R., Tan, L., Cooper, I., Szelag, G., Parkin, J.D., Ironside, P., and Garson, O.M. (1990) Cytogenetic analysis of 147 cases of non-Hodgkin's lymphoma: non-random chromosomal abnormalities and histological correlations. *Br. J. Haematol.* **76**, 231–237.

134. Schouten, H.C., Sanger, W.G., Weisenburger, D.D., Anderson, J., and Armitage, J.O. (1990) Chromosomal abnormalities in untreated patients with non-Hodgkin's lymphoma: associations with histology, clinical characteristics, and treatment outcome. *Blood* 75, 1841–1847.

135. Ofitt, K., Jhanwar, S.C., Ladanyi, M., Filippa, D.A., and Chaganti, R.S.K. (1991) Cyto-genetic analysis of 434 consecutively ascertained specimens of non-Hodgkin's lym-phoma: correlations between recurrent aberrations, histology, and exposure to cytotoxic treatment. *Genes Chromosom. Cancer* **3**, 189–201.

136. Bastard, C., Tilly, H., Lenormand, B., Bigorgne, C., Boulet, D., Kunlin, A., Monconduit, M., and Piguet, H. (1992) Translocations involving band 3q27 and Ig gene regions in non-Hodgkin's lymphoma. *Blood* **79**, 2527–2531.

137. Speaks, S.L., Sanger, W.G., Linder, J., Johnson, D.R., Armitage, J.O., Weisenburger, D., and Purtilo, D. (1987) Chromosomal abnormalities in indolent lymphoma. *Cancer Genet. Cytogenet.* **27,** 335–344.

138. Levine, E.G., Arthur, D.C., Frizzera, G., Peterson, B.A., Hurd, D.D., and Bloomfield, C.D. (1985) There are differences in cytogenetic abnormalities among histologic subtypes of the non-Hodgkin's lymphomas. *Blood* **66,** 1414–1422.

139. Fifth International Workshop on Chromosomes in Leukemia-Lymphoma (1987) Correlation of chromosome abnormalities with histologic and immunologic characteristics in non-Hodgkin's lymphoma and adult T cell leukemia-lymphoma. *Blood* **70,** 1554–1564.

140. Mrózek, K. and Bloomfield, C.D. (1996) Cytogenetics of non-Hodgkin's lymphoma and Hodgkin's disease. In *Neoplastic Diseases of the Blood,* 3rd edition (Wiernik, P.H., Canellos, G.P., Dutcher, J.P., and Kyle, R.A., eds.), Churchill Livingstone, New York, pp. 835–862.

141. Nashelsky, M.B., Hess, M.M., Weisenburger, D.D., Pierson, J.L., Bast, M.A., Armitage, J.O., and Sanger, W.G. (1994) Cytogenetic abnormalities in B-immunoblastic lymphoma. *Leuk. Lymphoma* **14,** 415–420.

142. Whang-Peng, J., Knutsen, T., Jaffe, E.S., Steinberg, S.M., Raffeld, M., Zhao, W.P., Duffey, P., Condron, K., Yano, T., and Longo, D.L. (1995) Sequential analysis of 43 patients with non-Hodgkin's lymphoma: clinical correlations with cytogenetic, histologic, immunophenotyping, and molecular studies. *Blood* **85,** 203–216.

143. Ohno, H., Kerckaert, J-P., Bastard, C., and Fukuhara, S. (1994) Heterogeneity in B-cell neoplasms associated with rearrangement of the *LAZ3* gene on chromosome band 3q27. *Jpn. J. Cancer Res.* **85,** 592–600.

144. Wlodarska, I., Delabie, J., deWolf-Peeters, C., Mecucci, C., Stul, M., Verhoef, G., Cassiman, J.J., and Van den Berghe, H. (1993) T-cell lymphoma developing in Hodgkin's disease: evidence for two clones. *J. Pathol.* **170,** 239–248.

145. Croce, C.M. (1993) Molecular biology of lymphomas. *Semin. Oncol.* (Suppl. 5) **20,** 31–46.

146. Rabbitts, T.H. (1994) Chromosomal translocations in human cancer. *Nature* **372,** 143–149.

147. Mitani, K., Sato, Y., Tojo, A., Ishikawa, F., Kobayashi, Y., Miura, Y., Miyazono, K., Urabe, A., and Takaku, F. (1990) Philadelphia positive chromosome positive B-cell type malignant lymphoma expressing an aberrant 190 kDa *bcr–abl* protein. *Br. J. Haematol.* **76,** 221–225.

148. Ye, B.H., Lista, F., Lo Coco, F., Knowles, D.M., Offit, K., Chaganti, R.S.K., and Dalla-Favera, R. (1993) Alterations of a zinc finger-encoding gene, *BCL-6,* in diffuse large-cell lymphoma. *Science* **262,** 747–750.

149. Miki, T., Kawamata, N., Hirosawa, S., and Aoki, N. (1994) Gene involved in the 3q27 translocation associated with B-cell lymphoma, *BCL5,* encodes a Krüppel-like zinc-finger protein. *Blood* **83,** 26–32.

150. Bastard, C., Deweindt, C., Kerckaert, J-P., Lenormand, B., Rossi, A., Pezzella, F., Fruchart, C., Duvall, C., Monconduit, M., and Tilly, H. (1994) *LAZ3* rearrangements in non-Hodgkin's lymphoma: correlation with histology, immunophenotype, karyotype, and clinical outcome in 217 patients. *Blood* **83,** 2423–2427.

151. Hillion, J., Mecucci, C., Aventin, A., Leroux, D., Wlodarska, I., van den Berghe, H., and Larsen, C.-J. (1991) A variant translocation t(2;18) in follicular lymphoma involved the 5′ end of *bcl-2* and *Ig*κ light chain gene. *Oncogene* **6,** 169–172.

152. Gaidano, G., Ballerini, P., Gong, J.Z., Inghirami, G., Neri, A., Newcomb, E.W., Magrath, I.T., Knowles, D.M., and Dalla-Favera, R. (1991) p53 mutations in human lymphoid malignancies: association with Burkitt lymphoma and chronic lymphocytic leukemia. *Proc. Natl. Acad. Sci. USA* **88,** 5413–5417.

153. Ichikawa, A., Hotta, T., Takagi, N., Tsushita, K., Kinoshita, T., Nagai, H., Murakami, Y., Hayashi, K., and Saito, H. (1992) Mutations of p53 gene and their relation to disease progression in B-cell lymphoma. *Blood* **79,** 2701–2707.

154. Stranks, G., Height, S.E., Mitchell, P., Jadayel, D., Yuille, M.A., De Lord, C., Clutterbuck, R.D., Treleaven, J.G., Powles, R.L., Nacheva, E., Oscier, D.G., Karpas, A., Lenoir, G.M., Smith, S.D., Millar, J.L., Catovsky, D., and Dyer, M.J.S. (1995) Deletions and rearrangement of *CDKN2* in lymphoid malignancy. *Blood* **85,** 893–901.

155. Ginsberg, A.M., Raffeld, M., and Cossman, J. (1991) Inactivation of the retinoblastoma gene in human lymphoid neoplasms. *Blood* **77,** 833–840.

156. Weide, R., Tiemann, M., Pflüger, K-H., Koppler, H.J., Parvizl, B., Wacker, H.H., Creipe, H.H., Havemann, K., and Parwaresch, M.R. (1994) Altered expression of the retinoblastoma gene product in human high grade non-Hodgkin's lymphomas. *Leukemia* **8,** 97–101.

157. Mertens, F., Johansson, B., and Mitelman, F. (1994) Isochromosomes in neoplasia. *Genes Chromosom. Cancer* **10,** 221–230.

158. Pirc-Danoewinata, H., Chott, A., Onderka, E., Drach, J., Schlogl, E., Jager, U., Thalhammer, F., Nowotny, H., Aryee, D., Steger, G.G., Locker, G., Michl, I., Djavanmard, M., Simonitsch, I., Radaszkiewicz, T., Hanak, H., Schneider, B., Heinz, R., and Marosi, C. (1994) Karyotype and prognosis in non-Hodgkin lymphoma. *Leukemia* **8,** 1929–1939.

159. Tilly, H., Rossi, A., Stamatoullas, A., Lenormand, B., Bigorgne, C., Kunlin, A., Monconduit, M., and Bastard, C. (1994) Prognostic value of chromosomal abnormalities in follicular lymphoma. *Blood* **84,** 1043–1049.

160. Levine, E.G., Arthur, D.C., Frizzera, G., Peterson, B.A., Hurd, D.D., and Bloomfield, C.D. (1988) Cytogenetic abnormalities predict clinical outcome in non-Hodgkin's lymphoma. *Ann. Intern. Med.* **108,** 14–20.

161. Offit, K., Wong, G., Filippa, D.A., Tao, Y., and Chaganti, R.S.K. (1991) Cytogenetic analysis of 434 consecutively ascertained specimens of non-Hodgkin's lymphoma: clinical correlations. *Blood* **77,** 1508–1515.

162. Kristoffersson, U., Heim, S., Mandahl, N., Olsson, H., Ranstam, J., Akerman, M., and Mitelman, F. (1987) Prognostic implication of cytogenetic findings in 106 patients with non-Hodgkin lymphoma. *Cancer Genet. Cytogenet.* **25,** 55–64.

163. Schouten, H.C., Sanger, W.G., Duggan, M., Weisenburger, D.D., MacLennan, K.A., and Armitage, J.O. (1989) Chromosomal abnormalities in Hodgkin's disease. *Blood* **73,** 2149–2154.

164. Tilly, H., Bastard, C., Delastre, T., Duval, C., Bizet, M., Lenormand, B., Daucé, J-P., Monconduit, M., and Piguet, H. (1991) Cytogenetic studies in untreated Hodgkin's disease. *Blood* **77,** 1298–1304.

165. Döhner, H., Bloomfield, C.D., Frizzera, G., Frestedt, J., and Arthur, D.C. (1992) Recurring chromosome abnormalities in Hodgkin's disease. *Genes Chrom. Cancer* **5,** 392–398.

166. Poppema, S., Kaleta, J., and Hepperle, B. (1992) Chromosomal abnormalities in patients with Hodgkin's disease: evidence for frequent involvement of the 14q chromosomal region but infrequent bcl-2 gene rearrangement in Reed–Sternberg cells. *J. Natl. Cancer Inst.* **84,** 1789–1793.

167. Koduru, P.R.K., Susin, M., Schulman, P., Catell, D., Goh, J.C., Karp, L., and Broome, J.D. (1993) Phenotypic and genotypic characterization of Hodgkin's disease. *Am. J. Hematol.* **44,** 117–124.

168. Schlegelberger, B., Weber-Matthiesen, K., Himmler, A., Bartels, H., Sonnen, R., Kuse, R., Feller, A.C., and Grote, W. (1994) Cytogenetic findings and results of combined immunophenotyping and karyotyping in Hodgkin's disease. *Leukemia* **8,** 72–80.

169. Gordon, R.D., Stowasser, M., Martin, N., Epping, A., Conic, S., Klemm, S.A., Tunny, T.J., and Rutherford, J.C. (1993) Karyotypic abnormalities in benign adrenocortical tumors producing aldosterone. *Cancer Genet. Cytogenet.* **68,** 78–81.

170. Dietrich, C.U., Pandis, N., Teixeira, M.R., Bardi, G., Gerdes, A.M., Andersen, J.A., and

Heim, S. (1995) Chromosome abnormalities in benign hyperproliferative disorders of epithelial and stromal breast tissue. *Int. J. Cancer* **60,** 49–53.

171. Mitelman, G., Mark, J., Nilsson, P.G., Dencker, H., Norryd, C., and Tranverg, K-G. (1974) Chromosome banding pattern in human colonic polyps. *Hereditas* **78,** 63–68.

172. Muleris, M., Zafrani, B., Validire, P., Girodet, J., Salmon, R-J., and Dutrillaux, B. (1994) Cytogenetic study of 30 colorectal adenomas. *Cancer Genet. Cytogenet.* **74,** 104–108.

173. Dal Cin, P., Gaeta, J., Huben, R., Li, F.P., Prout Jr., G.R., and Sandberg, A.A. (1989) Renal cortical tumors. Cytogenetic characterization. *Am. J. Clin. Pathol.* **92,** 408–414.

174. Örndal, C., Johansson, M., Heim, S., Mandahl, N., Månsson, B., Alumets, J., and Mitelman, F. (1990) Parathyroid adenoma with t(1;5)(p22;q32) as the sole clonal chromosome abnormality. *Cancer Genet. Cytogenet.* **48,** 225–228.

175. Papi, L., Baldassarri, G., Montali, E., Bigozzi, U., Ammannati, F., and Brandi, M.L. (1993) Cytogenetic studies in sporadic and multiple endocrine neoplasia type 1-associated pituitary adenomas. *Genes Chromosom. Cancer* **7,** 63–65.

176. Bullerdiek, J., Wobst, G., Meyer-Bolte, K., Chilla, R., Haubrich, J., Thode, B., and Bartnitzke, S. (1993) Cytogenetic subtyping of 220 salivary gland pleomorphic adenomas: correlation to occurrence, histological subtype, and in vitro cellular behavior. *Cancer Genet. Cytogenet.* **65,** 27–31.

177. Antonini, P., Lévy, N., Caillou, B. Vénuat, A-M., Schlumberger, M., Parmentier, C., and Bernheim, A. (1993) Numerical aberrations, including trisomy 22 as the sole anomaly, are recurrent in follicular thyroid adenomas. *Genes Chromosom. Cancer* **8,** 63–66.

178. Belge, G., Thode, B., Rippe, V., and Bartnitzke, J. (1994) A characteristic sequence of trisomies starting with trisomy 7 in benign thyroid tumors. *Hum. Genet.* **94,** 198–202.

179. Garewal, H.S., Sampliner, R., Liu, Y., and Trent, J.M. (1989) Chromosomal rearrangements in Barrett's esophagus. A premalignant lesion of esophageal adenocarcinoma. *Cancer Genet. Cytogenet.* **42,** 281–296.

180. Crotty, T.B., Lawrence, K.M., Moertel, C.A., Bartelt Jr, D.H., Batts, K.P., Dewald, G.W., Farrow, G.M., and Jenkins, R.B. (1992) Cytogenetic analysis of six renal oncocytomas and a chromophobe cell renal carcinoma. Evidence that –Y, –1 may be a characteristic anomaly in renal oncocytomas. *Cancer Genet. Cytogenet.* **61,** 61–66.

181. Mertens, F., Heim, S., Mandahl, N., Johansson, B., Rydholm, A., Biorklund, A., Wennerberg, J., Jonsson, N., and Mitelman, F. (1989) Clonal chromosome aberrations in a keratoacanthoma and a basal cell papilloma. *Cancer Genet. Cytogenet.* **39,** 227–232.

182. Mertens, F., Heim, S., Jin, Y., Johansson, B., Mandahl, N., Biörklund, A., Wennerberg, J., Jonsson, N., and Mitelman, F. (1989) Basosquamous papilloma. A benign epithelial skin tumor with multiple cytogenetic clones. *Cancer Genet. Cytogenet.* **37,** 235–239.

183. Rodriguez, E., Sreekantaiah, C., and Chaganti, R.S.K. (1994) Genetic changes in epithelial solid neoplasia. *Cancer Res.* **54,** 3398–3406.

184. Bullerdiek, J., Haubrich, J., Meyer, K., and Bartnitzke, S. (1988) Translocation t(11;19)(q21;p13.1) as the sole chromosome abnormality in a cystadenolymphoma (Warthin's tumor) of the parotid gland. *Cancer Genet. Cytogenet.* **35,** 129–132.

185. Limon, J., Dal Cin, P., Gaeta, J., and Sandberg, A.A. (1987) Translocation t(4;11)(q35;p13) in an adrenocortical carcinoma. *Cancer Genet. Cytogenet.* **28,** 343–348.

186. Dutrillaux, B., Gerbault-Seureau, M., and Zafrani, B. (1990) Characterization of chromosomal anomalies in human breast cancer: a comparison of 30 paradiploid cases with few chromosome changes. *Cancer Genet. Cytogenet.* **49,** 203–217.

187. Hainsworth, P.J., Raphael, K.L., Stillwell, R.G., Bennett, R.C., and Garson, O.M. (1991) Cytogenetic features of twenty-six primary breast cancers. *Cancer Genet. Cytogenet.* **52,** 205–218.

188. Thompson, F., Emerson, J., Dalton, W., Yang, J.-M., McGee, D., Villar, H., Knox, S., Massey, K., Weinstein, R., Bhattacharya, A., and Trent, J. (1993) Clonal chromosome

abnormalities in human breast carcinomas. I. Twenty-eight cases with primary disease. *Genes Chromosom. Cancer* **7,** 185–193.

189. Pandis, N., Heim, S., Bardi, G., Idvall, I., Mandahl, N., and Mitelman, F. (1993) Chromosome analysis of 20 breast carcinomas: cytogenetic multiclonality and karyotypic-pathologic correlations. *Genes Chromosom. Cancer* **6,** 51–57.

190. Pandis, N., Heim, S., Bardi, G., Idvall, I., Mandahl, N., and Mitelman, F. (1992) Whole-arm t(1;16) and i(1q) as sole anomalies identify gain of 1q as a primary chromosomal abnormality in breast cancer. *Genes Chromosom. Cancer* **5,** 235–238.

191. Pandis N., Yuesheng, J., Gorunova, L., Petersson, C., Bardi, G., Idvall, I., Johansson, B., Ingvar, C., Mandahl, N., Mitelman, F., and Heim, S. (1995) Chromosome analysis of 97 primary breast carcinomas: identification of eight karyotypic subgroups. *Genes Chromosom. Cancer* **12,** 173–185.

192. Pandis, N., Bardi, G., Jin, Y., Dietrich, C., Johansson, B., Andersen, J., Mandahl, N., Mitelman, F., and Heim, S. (1994) Unbalanced 1;16 translocation as the sole karyotypic abnormality in breast carcinoma and its lymph node metastasis. *Cancer Genet. Cytogenet.* **75,** 158–159.

193. Pandis, N., Jin, Y., Limon, J., Bardi, G., Idvall, I., Mandahl, N., Mitelman, F., and Heim, S. (1993) Interstitial deletion of the short arm of chromosome 3 as the primary chromosome abnormality in carcinomas of the breast. *Genes Chromosom. Cancer* **6,** 151–155.

194. Slovak, M.L., Ho, J., and Simpson, J.F. (1992) Cytogenetic studies of 46 human breast carcinomas. *Proc. Am. Assoc. Cancer Res.* **33,** 40.

195. Slovak, M.L. (1997) Breast tumor cytogenetic markers. In *Human Cytogenetic Cancer Markers* (Wolman, S.R. and Sell, S., eds.), Humana Press, Totowa, NJ, pp. 111–166.

196. Bardi, G., Sukhikh, T., Pandis, N., Fenger, C., Kronborg, O., and Heim, S. (1995) Karyotypic characterization of colorectal adenocarcinomas. *Genes Chromosom. Cancer* **12,** 97–109.

197. Hunter, S., Gramlich, T., and Varma, V. (1993) Y chromosome loss in esophageal carcinoma: an in situ hybridization study. *Genes Chromosom. Cancer* **8,** 172–177.

198. Bardi, G., Gorunova, L., Limon, J., Nedoszytko, B., Johansson, B., Pandis, N., Mandahl, N., Bak-Jensen, E., Andrén-Sandberg, Å., Rys, J., Niezabitowski, A., Mitelman, F., and Heim, S. (1994) Abnormal karyotypes in three carcinomas of the gallbladder. *Cancer Genet. Cytogenet.* **76,** 15–18.

199. Meloni, A.M., Bridge, J., and Sandberg, A.A. (1992) Reviews on chromosome studies in urological tumors. I. Renal tumors. *J. Urol.* **148,** 253–265.

200. Kovacs, G., Fuzesi, L., Emanuel, A., and Kung, H. (1991) Cytogenetics of papillary renal cell tumors. *Genes Chromosom. Cancer* **3,** 249–255.

201. Meloni, A.M., Dobbs, R.M., Pontes, J.E., and Sandberg, A.A. (1993) Translocation (X;1) in papillary renal cell carcinoma. A new cytogenetic subtype. *Cancer Genet. Cytogenet.* **65,** 1–6.

202. Johansson, M., Dietrich, C., Karaüzum, S., Mandahl, N., Hambraeus, G., Johansson, L., Praetorius Clausen, P., Mitelman, F., and Heim, S. (1994) Karyotypic abnormalities in adnocarcinomas of the lung. *Int. J. Oncol.* **5,** 17–26.

203. Jenkins, R.B., Bartelt, Jr., D., Stalboerger, P., Persons, D., Dahl, R.J., Podratz, K., Keeney, G., and Hartmann, L. (1993) Cytogenetic studies of epithelial ovarian carcinoma. *Cancer Genet. Cytogenet.* **71,** 76–86.

204. Thompson, F.H., Liu, Y., Emerson, J., Weinstein, R., Makar, R., Trent, J.M., Taetle, R., and Albert, D.S. (1994) Simple numeric abnormalities as primary karyotype changes in ovarian carcinoma. *Genes Chromosom. Cancer* **10,** 262–266.

205. Griffin, C.A. (1997) Pancreatic exocrine tumors. In *Human Cytogenetic Cancer Markers* (Wolman, S.R. and Sell, S., eds.), Humana Press, Totowa, NJ, pp. 403–423.

206. Brothman, A.R. and Williams, B.J. (1997) Prostate cancer. In *Human Cytogenetic Cancer Markers* (Wolman, S.R. and Sell, S., eds.), Humana Press, Totowa, NJ, pp. 223–246.

207. Sandros, J., Stenman, G., and Mark, J. (1990) Cytogenetic and molecular observations in human and experimental salivary gland tumors. *Cancer Genet. Cytogenet.* **44,** 153–167.
208. Ferti-Passantonopoulou, A.D., Panani, A.D., Vlachos, J.D., and Raptis, S.A. (1987) Common cytogenetic findings in gastric cancer. *Cancer Genet. Cytogenet.* **24,** 63–73.
209. Kubonishi, I., Takehara, N., Iwata, J., Sonobe, H., Ohtsuki, Y., Abe, T., and Miyoshi, I. (1991) Novel t(15;19)(q15;p13) chromosome abnormality in a thymic carcinoma. *Cancer Res.* **51,** 3327–3328.
210. Sozzi, G., Bongarzone, I., Miozzo, M., Cariani, C.T., Mondellini, P., Calderone, C., Pilotti, S., Pierotti, M.A., and Della Porta, G. (1992) Cytogenetic and molecular genetic characterization of papillary thyroid carcinomas. *Genes Chromosom. Cancer* **5,** 212–218.
211. Herrmann, M.A., Hay, I.D., Bartelt Jr., D.H., Ritland, S.R., Dahl, R.J., Grant, C.S., and Jenkins, R.B. (1991) Cytogenetic and molecular genetic studies of follicular and papillary thyroid cancers. *J. Clin. Invest.* **88,** 1596–1604.
212. Roque, L., Clode, A.L., Gomes, P., Rosa-Santos, J., Soares, J., and Castedo, S. (1995) Cytogenetic findings in 31 papillary thyroid carcinomas. *Genes Chromosom. Cancer* **13,** 157–162.
213. Kiechle-Schwartz, M., Kommoss, F., Schmidt, J., Lukovic, L., Walz, L., Bauknecht, T., and Pfeiderer, A. (1992) Cytogenetic analysis of an adenoid cystic carcinoma of the Bartholin's gland. A rare, semimalignant tumor of the female genitourinary tract. *Cancer Genet. Cytogenet.* **61,** 26–30.
214. Higashi, K., Jin, Y., Johansson, M., Heim, S., Mandahl, N., Biörklund, A., Wennerberg, J., Hambraeus, G., Johansson, L., and Mitelman, F. (1991) Rearrangement of 9p13 as the primary chromosomal aberration in adenoid cystic carcinoma of the respiratory tract. *Genes Chromosom. Cancer* **3,** 21–23.
215. Nordkvist, A., Mark, J., Gustafsson, H., Bang, H., and Stenman, G. (1994) Non-random chromosome rearrangements in adenoid cystic carcinoma of the salivary glands. *Genes Chromosom. Cancer* **10,** 115–121.
216. Flüry-Hérard, A., Viegas-Péquignot, E., De Cremoux, H., Chlecq, C., Bignon, J., and Dutrillaux, B. (1992) Cytogenetic study of five cases of lung adenosquamous carcinomas. *Cancer Genet. Cytogenet.* **59,** 1–8.
217. Kawasaki-Oyama, R.S., André, F.S., Caldeira, L.F., Castilho, W.H., Gasques, J.A.L., Bozola, A.R., Thomé, J.A., and Tajara, E.H. (1994) Cytogenetic findings in two basal cell carcinomas. *Cancer Genet. Cytogenet.* **73,** 152–156.
218. Nangia, R., Sait, S.N.J., Block, A.W., and Zhang, P.J. (1998) Trisomy 6 in basal cell carcinoma (BCC) correlates with metastatic potential: a dual color fluorescence in situ hybridization (FISH) study on paraffin tissue sections. *Modern Pathol.* **11,** 524.
219. Johansson, M., Heim, S., Mandahl, N., Hambraeus, G., Johansson, L., and Mitelman, F. (1993) Cytogenetic analysis of six bronchial carcinoids. *Cancer Genet. Cytogenet.* **66,** 33–38.
220. Mertens, F., Kreicbergs, A., Rydholm, A., Willén, H., Carlén, B., Mitelman, F., and Mandahl, N. (1994) Clonal chromosome aberrations in three sacral chordomas. *Cancer Genet. Cytogenet.* **73,** 147–151.
221. Bardi, B., Johansson, B., Pandis, N., Heim, S., Mandahl, N., Békássy, A., Hägerstrand, I., and Mitelman, F. (1992) Trisomy 2 as the sole chromosomal abnormality in a hepatoblastoma. *Genes Chromosom. Cancer* **4,** 78–80.
222. Bardi, G., Johansson, B., Pandis, N., Heim, S., Mandahl, N., Andrén-Sandberg, Å., Hägerstrand, I., and Mitelman, F. (1992) Cytogenetic findings in three primary hepatocellular carcinomas. *Cancer Genet. Cytogenet.* **58,** 191–195.
223. Gibas, Z., Weil, S., Chen, S-T., and McCue, P.A. (1994) Deletion of chromosome arm 1p in a Merkel cell carcinoma (MCC). *Genes Chromosom. Cancer* **9,** 216–220.
224. Johansson, M., Mandahl, N., Johansson, L., Hambraeus, G., Mitelman, F., and Heim, S. (1995) Translocation 11;19 in a mucopidermoid tumor of the lung. *Cancer Genet. Cytogenet.* **80,** 85–86.

225. Nordkvist, A., Gustafsson, H., Juberg-Ode, M., and Stenman, G. (1994) Recurrent rearrangements of 11q14-22 in mucopidermoid carcinoma. *Cancer Genet. Cytogenet.* **74,** 77–83.

226. Muleris, M., Salmon, R-J., Girodet, J., Zafrani, B., and Dutrillaux, B. (1987) Recurrent deletions of chromosomes 11q and 3p in anal canal carcinoma. *Int. J. Cancer* **39,** 595–598.

227. Whang-Peng, T., Banks-Schlegel, S.P., and Lee, E.C. (1990) Cytogenetic studies of esophageal carcinoma cell lines. *Cancer Genet. Cytogenet.* **45,** 101–120.

228. Jin, Y., Mertens, F., Mandahl, N., Heim, S., Olegård, C., Wennerberg, J., Biörklund, A., and Mitelman, F. (1993) Chromosome abnormalities in eighty-three head and neck squamous cell carcinomas: influence of culture conditions on karyotypic pattern. *Cancer Res.* **53,** 2140–2146.

229. Van Dyke, D.L., Worsham, M.J., Benninger, M.S., Krause, C.J., Baker, S.R., Wolf, G.T., Drumheller, T., Tilley, B.C., and Carey, T.E. (1994) Recurrent cytogenetic abnormalities in squamous cell carcinomas of the head and neck region. *Genes Chromosom. Cancer* **9,** 192–206.

230. Viegas-Péquignot, E., Flüry-Hérard, A., De Cremoux, H., Chlecq, C., Bignon, J., and Dutrillaux, B. (1990) Recurrent chromosome aberrations in human lung squamous cell carcinomas. *Cancer Genet. Cytogenet.* **49,** 37–49.

231. Johansson, M., Jin, Y., Mandahl, N., Hambraeus, G., Johansson, L., Mitelman, F., and Heim, S. (1995) Cytogenetic analysis of short-term cultured squamous cell carcinomas of the lung. *Cancer Genet. Cytogenet.* **81,** 46–55.

232. Xiao, S., Feng, X-L., Shi, Y.-H., Liu, Q-Z., and Lip, P. (1992) Cytogenetic abnormalities in a squamous cell carcinoma of the penis. *Cancer Genet. Cytogenet.* **64,** 139–141.

233. Pavarino, E.C., Antonio, J.R., Pozzeti, E.M., Larranaga, H.J., and Tajara, E.H. (1995) Cytogenetic study of neoplastic and nonneoplastic cells of the skin. *Cancer Genet. Cytogenet.* **85,** 16–19.

234. Worsham, M.J., Van Dyke, D.L., Grenman, S.E., Grenman, R., Hopkins, M.P., Roberts, J.A., Gasser, K.M., Schwartz, D.R., and Carey, T.E. (1991) Consistent chromosome abnormalities in squamous cell carcinoma of the vulva. *Genes Chromosom. Cancer* **3,** 420–432.

235. Sandberg, A.A. and Berger, C.S. (1994) Review of chromosome studies in urological tumors. II. Cytogenetics and molecular genetics of bladder cancer. *J. Urol.* **151,** 545–560.

236. Johansson, M., Dietrich, C., Mandahl, N., Hambraeus, G., Johansson, L., Praetorius Clausen, P., Mitelman, F., and Heim, S. (1994) Karyotypic characterization of bronchial large cell carcinomas. *Int. J. Cancer* **57,** 463–467.

237. Whang-Peng, J., Bunn Jr., P.A., Kao-Shan, C.S., Lee, E.C., Carney, D.N., Gazdar, A., and Minna J.D. (1982) A nonrandom chromosomal abnormality, del 3p(14-23), in human small cell lung cancer (SCLC). *Cancer Genet. Cytogenet.* **6,** 119–134.

238. Testa, J.R. and Graziano, S.L. (1993) Molecular implications of recurrent cytogenetic alterations in human small cell lung cancer. *Cancer Detect. Prevent.* **17,** 267–277.

239. Slater, R.M. and Mannens, M.M.A.M. (1992) Cytogenetics and molecular genetics of Wilms' tumor of childhood. *Cancer Genet. Cytogenet.* **61,** 111–121.

240. Surti, U. and Hoffner, L. (1997) Cytogenetic markers in selected gynecological malignancies. In *Human Cytogenetic Cancer Markers* (Wolman, S.R., Sell, S., eds.), Humana Press, Totowa, NJ, pp. 203–221.

241. Sandberg, A.A., Meloni, A.M., and Suijkerbuijk, R.F. (1996) Reviews of chromosome studies in urological tumors. III. Cytogenetics and genes in testicular tumors. *J. Urol.* **155,** 1531–1556.

242. Pejovic, T., Heim, S., Alm, P., Iosif, S., Himmelmann, A., Skaerris, J., and Mitelman, F. (1993) Isochromosome 1q as the sole karyotypic abnormality in a Sertoli cell tumor of the ovary. *Cancer Genet. Cytogenet.* **65,** 79–80.

243. Mandahl, N., Willén, H., Rydholm, A., Heim, S., and Mitelman, F. (1993) Rearrangement of band q13 on both chromosomes 12 in a periosteal chondroma. *Genes Chromosom. Cancer* **6,** 121–123.

244. Vanni, R., Dal Cin, P., Marras, S., Moerman, P., Andria, M., Valdes, E., Deprest, J., and Van Den Berghe, H. (1993) Endometrial polyp: another benign tumor characterized by 12q13–q15 changes. *Cancer Genet. Cytogenet.* **68,** 32–33.

245. Tarkkanen, M., Kaipainen, A., Karaharju, E., Böhling, T., Szymanska, J., Heliö, H., Kivioja, A., Elomaa, I., and Knutilla, S. (1993) Cytogenetic study of 249 consecutive patients examined for a bone tumor. *Cancer Genet. Cytogenet.* **68,** 1–21.

246. Leung, W.-Y., Schwartz, P.E., Ng, H-T., and Yang-Feng, T.L. (1990) Trisomy 12 in benign fibroma and granulosa cell tumor of the ovary. *Gynecol. Oncol.* **38,** 28–31.

247. Bridge, J.A., Sreekantaiah, C., Mouron, B., Neff, J.R., Sandberg, A.A., and Wolman, S.R. (1992) Clonal chromosomal abnormalities in desmoid tumors. Implications for histopathogenesis. *Cancer* **69,** 430–436.

248. Qi, H., Dal Cin, P., Hernández, J.M., Garcia, J.L., Sciot, R., Fletcher, C., Van Eyken, P., De Wever, I., and Van den Berghe, H. (1996) Trisomies 8 and 20 in desmoid tumors. *Cancer Genet. Cytogenet.* **92,** 147–149.

249. Mascarello, J.T., and Krous, H.F. (1992) Second report of a translocation involving 19q13.4 in a mesenchymal hamartoma of the liver. *Cancer Genet. Cytogenet.* **58,** 141–142.

250. Johansson, M., Dietrich, C., Mandahl, N., Hambraeus, G., Johansson, L., Claussen, P.P., Mitelman, F., and Heim, S. (1993) Recombinations of chromosomal bands 6p21 and 14q24 characterise pulmonary hamartomas. *Br. J. Cancer* **67,** 1236–1241.

251. Dal Cin, P., Kools, P., De Jonge, I., Moerman, P., Van de Ven, W., and Van Den Berghe, H. (1993) Rearrangement of 12q14–15 in pulmonary chondroid hamartoma. *Genes Chromosom. Cancer* **8,** 131–133.

252. Mertens, F., Rydholm, A., Kreicbergs, A., Willén, H., Jonsson, K., Heim, S., Mitelman, F., and Mandahl, N. (1994) Loss of chromosome band 8q24 in sporadic osteocartilaginous exostoses. *Genes Chromosom. Cancer* **9,** 8–12.

253. Mandahl, N., Jin, Y., Heim, S., Willén, H., Wennerberg, J., Biörklund, A., and Mitelman, F. (1990) Trisomy 5 and loss of the Y chromosome as the sole cytogenetic anomalies in a cavernous hemangioma/angiosarcoma. *Genes Chromosom. Cancer* **1,** 315–316.

254. Mrózek, K., Karakousis, C.P., and Bloomfield, C.D. (1994) Band 11q13 is nonrandomly rearranged in hibernomas. *Genes Chromosom. Cancer* **9,** 145–147.

255. Dal Cin, P., Aly, M.S., De Wever, I., Van Damme, B., and Van Den Berghe, H. (1992) Does chromosome investigation discriminate between benign and malignant gastrointestinal leiomyomatous tumors? *Diagn. Oncol.* **2,** 55–59.

256. Nilbert, M. and Heim, S. (1990) Uterine leiomyoma cytogenetics. *Genes Chromosom. Cancer* **2,** 3–13.

257. Fletcher, J.A., Kozakewich, H.P., Schoenberg, M.L., and Morton, C.C. (1993) Cytogenetic findings in pediatric adipose tumors: consistent rearrangement of chromosome 8 in lipoblastoma. *Genes Chromosom. Cancer* **6,** 24–29.

258. Fletcher, C.D.M., Akerman, M., Dal Cin, P., De Wever, I., Mandahl, N., Mertens, F., Mitelman, F., Rosai, J., Ryndholm, A., Sciot, R., Tallini, G., Van Den Berghe, H., Van De Ven, W., Vanni, R., and Willén, J. (1996) Correlation between clinicopathological features and karyotype in lipomatous tumors. *Am. J. Pathol.* **148,** 623–630.

259. Debiec-Rychter, M., Katuzewski, B., Saryusz-Wolska, H., and Jankowska, J. (1990) A case of renal lymphangioma with a karyotype 45,X,–X,i dic(7q) *Cancer Genet. Cytogenet.* **46,** 29–33.

260. Zang, K.D. (1982) Cytological and cytogenetical studies on human meningioma. *Cancer Genet. Cytogenet.* **4,** 249–274.

261. Schofield, D.E., Yunis, E.J., and Fletcher, J.A. (1993) Chromosome aberrations in mesoblastic nephroma. *Am. J. Pathol.* **143,** 714–724.

262. Dewald, G.W., Dahl, R.J., Spurbeck, J.L., Carney, J.A., and Gordon, H. (1987) Chromo-somally abnormal clones and nonrandom telomeric translocations in cardiac myxomas. *Mayo Clin. Proc.* **62,** 558–567.

263. Chadduck, W.M., Boop, F.A., and Sawyer, J.R. (1991) Cytogenetic studies of pediatric brain and spinal cord tumors. *Pediatr. Neurosurg.* **17,** 57–65.

264. Riccardi, V.M. and Elder, D.W. (1986) Multiple cytogenetic aberrations in neuro-fibrosarcomas complicating neurofibromatosis. *Cancer Genet. Cytogenet.* **23,** 199–209.

265. Bello, M.J., de Campos, J.M., Kusak, M.E., Vaquero, J., Sarasa, J.L., Pestana, A., and Rey, J.A. (1993) Clonal chromosome aberrations in neurinomas. *Genes Chromosom. Cancer* **6,** 206–211.

266. Dal Cin, P., Sciot, R., Samson, I., De Smet, L., De Wever, I., Van Damme, B., and Van den Berghe, H. (1994) Cytogenetic characterization of tenosynovial giant cell tumors (nodular tenosynovitis). *Cancer Res.* **54,** 3986–3987.

267. Kazmierczak, B., Wanschura, S., Meyer-Bolte, K., Caselitz, J., Meister, P., Barnitzke, S., Van De Ven, W., and Bullerdiek, J. (1995) Cytogenetic and molecular analysis of an aggressive angiomyxoma. *Am. J. Pathol.* **147,** 580–585.

268. Rosai, J., Akerman, M., Dal Cin, P., De Wever, I., Fletcher, C.D., Mandahl, N., Mertens, F., Mitelman, F., Ryndholm, A., Sciot, R., Tallini, G., Van den Berghe, H., Van Den Ven, W., Vanni, R., and Willén, H. (1996) Combined morphologic and karyotypic study of 59 atypical lipomatous tumors: evaluation of their relationship and differential diagnosis with other adipose tissue tumors. *Am. J. Surg. Pathol.* **20,** 1182–1189.

269. Schofield, D.E., Fletcher, J.A., Grier, H.E., and Yunis, E.J. (1994) Fibrosarcoma in in-fants and children. *Am. J. Surg. Pathol.* **18,** 14–24.

270. Naeem, R., Lux, M.L., Huang, S-F., Naber, S.P., Corson, J.M., and Fletcher, J.A. (1995) Ring chromosomes in dermatofibrosarcoma protuberans are composed of interspersed sequences from chromosomes 17 and 22. *Am. J. Pathol.* **147,** 1553–1558.

271. Dal Cin, P., Sciot, R., De Wever, I., Brock, P., Casteels-Van Daele, M., Van Damme, B., and Van Den Berghe, H. (1996) Cytogenetic and immunohistochemical evidence that giant cell fibroblastoma is related to dermatofibrosarcoma protuberans. *Genes Chromosom. Cancer* **15,** 73–75.

272. Sciot, R., Dal Cin, P., De Vos, R., Van Damme, B., De Wever, I., Van Den Berghe, H., and Desmet, V. (1993) Alveolar soft part sarcoma: evidence for its myogenic origin and for the involvement of 17q25. *Histopathology* **23,** 439–444.

273. Limon, J., Debiec-Rychter, M., Nedoszytko, B., Liberski, P.P., Babińska, M., and Szadowska, A. (1994) Aberrations of chromosome 22 and polysomy of chromosome 8 as nonrandom changes in clear cell sarcoma. *Cancer Genet. Cytogenet.* **72,** 141–145.

274. Rodriguez, E., Sreekantaiah, C., Gerald, W., Reuter, V.E., Motzer, R.J., and Chaganti, R.S.K. (1993) A recurring translocation, t(11;22)(p13;q11.2), characterizes intra-abdomi-nal desmoplastic small round-cell tumors. *Cancer Genet. Cytogenet.* **69,** 17–21.

275. Stenman, G., Kindblom, L-G., Willems, J., and Angervall, L. (1990) A cell culture, chro-mosomal and quantitative DNA analysis of a metastatic epitheliod sarcoma. Deletion 1p, a possible primary chromosomal abnormality in epithelioid sarcoma. *Cancer* **65,** 2006–2013.

276. Örndal, C., Carlén, B., Åkerman, M., Willén, H., Mandahl, N., Heim, S., Ryndholm, A., and Mitelman, F. (1991) Chromosomal abnormality t(9;22)(q22;q12) in an extraskeletal myxoid chondrosarcoma characterized by fine needle aspiration cytology, electron micro-scopy, immunohistochemistry and DNA cytometry. *Cytopathology* **2,** 261–270.

277. Henn, W., Wullich, B., Thönnes, M., Steudel, W-I., Feiden, W., and Zang, K.D. (1993) Recurrent t(12;19)(q13;q13.3) in intracranial and extracranial hemangiopericytoma. *Cancer Genet. Cytogenet.* **71,** 151–154.

278. Boghosian, L., Dal Cin, P., Turc-Carel, C., Karakousis, C., Rao, U., Sait, S.J., and Sandberg, A.A. (1989) Three possible cytogenetic subgroups of leiomyosarcomas. *Cancer Genet. Cytogenet.* **43,** 39–49.

279. Tallini, G., Akerman, M., Dal Cin, P., De Wever, I., Fletcher, C.D.M., Mandahl, N., Mertens, F., Mitelman, F., Rosai, J., Ryndholm, A., Sciot, R., Van Den Berghe, H., Van Den Ven, W., Vanni, R., and Willén, H. (1996) Combined morphologic and karyotypic study of 28 myxoid liposarcomas: implications for a revised morphologic typing. *Am. J. Pathol.* **20,** 1047–1055.

280. Dal, Cin, P. and Van Den Berghe, H. (1997) Ten years of the cytogenetics of soft tissue tumors. *Cancer Genet. Cytogenet.* **95,** 59–66.

281. Hagemeijer, A., Versnel, M.A., Van Drunen, E., Moret, M., Bouts, M.J., van der Kwast, T.H., and Hoogsteden, H.C. (1990) Cytogenetic analysis of malignant mesothelioma. *Cancer Genet. Cytogenet.* **47,** 1–28.

282. Bridge, J.A., Nelson, M., McComb, E., McGuire, M.H., Rosenthal, H., Vergara, G., Maale, G.E., Spanier, S., and Neff, J.R. (1997) Cytogenetic findings in 73 osteosarcoma specimens and a review of the literature. *Cancer Genet. Cytogenet.* **95,** 74–87.

283. Sinovic, J.F., Bridge, J.A., and Neff, J.R. (1992) Ring chromosome in parosteal osteosarcoma. Clinical and diagnostic significance. *Cancer Genet. Cytogenet.* **62,** 50–52.

284. Whang-Peng, J., Knutsen, T., Theil, K., Horowitz, M.E., and Triche, T. (1992) Cytogenetic studies in subgroups of rhabdomyosarcoma. *Genes Chromosom. Cancer* **5,** 299–310.

285. Dietrich, C.U., Brock Jacobsen, B., Starklint, H., and Heim, S. (1993) Clonal karyotypic evolution in an embryonal rhabdomyosarcoma with trisomy 8 as the primary chromosomal abnormality. *Genes Chromosom. Cancer* **7,** 240–244.

286. Limon, J., Mrózek, K., Mandahl, N., Nedoszytko, B., Verhest, A., Rys, J., Niezabitowski, A., Babinska, M., Nosek, H., Ochalek, T., Kopacz, A., Willén, H., Rydholm, A., Heim, S., and Mitelman, F. (1991) Cytogenetics of synovial sarcoma: presentation of ten new cases and review of the literature. *Genes Chromosom. Cancer* **3,** 338–345.

287. Thiel G., Losanowa, T., Kintzel, D., Nisch, G., Martin, H., Vorpahl, K. and Witkowski, R. (1992) Karyotypes in 90 human gliomas. *Cancer Genet. Cytogenet.* **58,** 109–120.

288. Griffin, C.A., Long, P.P., Carson, B.S., and Brem, H. (1992) Chromosome abnormalities in low-grade central nervous system tumors. *Cancer Genet. Cytogenet.* **60,** 67–73.

289. Ransom, D.T., Ritland, S.R., Kimmel, D.W., Moertel, C.A., Dahl, R.J., Scheithauer, B.W., Kelly, P.J., and Jenkins, R.B. (1992) Cytogenetic and loss of heterozygosity studies in ependymomas, pilocytic astrocytomas, and oligodendrogliomas. *Genes Chromosom. Cancer* **5,** 348–356.

290. Brodeur, G.M. (1990) Neuroblastoma: clinical significance of genetic abnormalities. *Cancer Surv.* **9,** 673–688.

291. Neumann, E., Kalousek, D.K., Norman, M.G., Steinbok, P., Cochrane, D.D., and Goddard, K. (1993) Cytogenetic analysis of 109 pediatric central nervous system tumors. *Cancer Genet. Cytogenet.* **71,** 40–49.

292. Cowell, J.K. and Hogg, A. (1992) Genetics and cytogenetics of retinoblastoma. *Cancer Genet. Cytogenet.* **64,** 1–11.

293. Füzesi, L., Heller, R., Schreiber, H., and Mertens, R. (1993) Cytogenetics of Askin's tumor. Case report and review of the literature. *Pathol. Res. Pract.* **189,** 235–241.

294. Whang-Peng, J., Freter, C.E., Knutsen, T., Nanfro, J.J., and Gazdar, A. (1987) Translocation t(11;22) in esthesioneuroblastoma. *Cancer Genet. Cytogenet.* **29,** 155–157.

295. Turc-Carel, C., Aurias, A., Mugneret, F., Lizard, S., Sidaner, I., Volk, C., Thiery, J.P., Olschwang, S., Philip, I., Berger, M.P., Philip, T., Lenoir, G.M., and Mazabraud, A. (1988) Chromosomes in Ewing's sarcoma. I. An evaluation of 85 cases and remarkable consistency of t(11;22)(q24;q12). *Cancer Genet. Cytogenet.* **32,** 229–238.

296. Douglass, E.C., Rowe, S.T., Valentine, M., Parham, D., Meyer, W.H., and Thompson, E.I. (1990) A second nonrandom translocation, der(16)t(1;16)(q21;q13), in Ewing sarcoma and peripheral neuroectodermal tumor. *Cytogenet Cell Cytogenet.* **53,** 87–90.

297. Miozzo, M., Sozzi, G., Calderone, C., Pilotti, S., Lombardi, L., Pierotti, M.A., and Della Porta, G. (1990) t(11;22) in three cases of peripheral neuroepithelioma. *Genes Chromosom. Cancer* **2,** 163–165.

298. Thompson, F.H., Emerson, J., Olson, S., Weinstein, R., Leavitt, S.A., Leong, S.P.L., Emerson, S., Trent, J.M., Nelson, M.A., Salmon, S.E., and Taetle, R. (1995) Cytogenetics in 158 patients with regional or disseminated melanoma: subset analysis of near diploid and simple karyotypes. *Cancer Genet. Cytogenet.* **83**, 93–104.

299. Horsman, D.E. and White, V.A. (1993) Cytogenetic analysis of uveal melanoma. Consistent occurrence of monosomy 3 and trisomy 8q. *Cancer* **71**, 811–819.

300. Fearon, E.R. (1997) Human cancer syndromes: Clues to the origin and nature of cancer. *Science* **278**, 1043–1050.

301. Nowell, P.C. (1976) The clonal evolution of tumor cell populations. *Science* **194**, 23–28.

302. Nowell, P.C. (1986) Mechanisms of tumor progression. *Cancer Res.* **46**, 2203–2207.

303. Shackney, S.E., Smith, C.A., Miller, B.W., Burholt, D.R., Murtha, K., Giles, H.R., Ketterer, D.M., and Pollice, A.A. (1989) Model for the genetic evolution of human solid tumors. *Cancer Res.* **49**, 3344–3354.

304. Magdelenat, H., Gerbault-Seureau, M., Laine-Bidron, C., Prieur, M., and Dutrillaux, B. (1992) Genetic evolution of breast cancer: II. Relationship with estrogen and progesterone receptor expression. *Breast Cancer Res. Treat.* **22**, 119–127.

305. Weinberg, R.A. (1989) Oncogenes, anti-oncogenes, and the molecular bases of multistep carcinogenesis. *Cancer Res.* **49**, 3713–3721.

306. Fearon, E.R. and Vogelstein, B. (1990) A genetic model for colorectal tumorigenesis. *Cell* **61**, 759–767.

307. Vogelstein, B. and Kinzler, K.W. (1993) The multistep nature of cancer. *Trends Genet.* **9**, 138–141.

308. Mertens, F., Johansson, B., Höglund, M., and Mitelman, F. (1997) Chromosomal imbalance maps of malignant solid tumors: a cytogenetic survey of 3185 neoplasms. *Cancer Res.* **57**, 2765–2780.

309. Heim, S., Mandahl, N., Jin, Y., Strömblad, S., Lindström, E., Salford, L.G., and Mitelman, F. (1989) Trisomy 7 and sex chromosome loss in human brain tissue. *Cytogenet. Cell Genet.* **52**, 136–138.

310. Moertel, C.A., Dahl, R.J., Stalboerger, P.G., Kimmel, D.W., Scheithauer, B.W., and Jenkins, R.B. (1993) Gliosis specimens contain clonal cytogenetic abnormalities. *Cancer Genet. Cytogenet.* **67**, 21–27.

311. Elfving, P., Cigudosa, J.C., Lundgren, R., Limon, J., Mandahl, N., Kristoffersson, U., Heim, S., and Mitelman, F. (1990) Trisomy 7, trisomy 10, and loss of the Y chromosome in short-term cultures of normal kidney tissue. *Cytogenet. Cell Genet.* **53**, 123–125.

312. Elfving, P., Åman, P., Mandahl, N., Lundgren, R., and Mitelman, F. (1995) Trisomy 7 in neoplastic epithelial kidney cells. *Cytogenet. Cell Genet.* **69**, 90–96.

313. Kovacs, C. and Brusa, P. (1989) Clonal chromosome aberrations in normal kidney tissue from patients with renal cell carcinoma. *Cancer Genet. Cytogenet.* **37**, 289–290.

314. Lee, J.S., Pathak, S., Hopwood, V., Tomasovic, B., Mullins, T.D., Baker, F.L., Spitzer, G., and Neidhart, J.A. (1987) Involvement of chromosome 7 in primary lung tumor and nonmalignant normal lung tissue. *Cancer Res.* **47**, 6349–6352.

315. Vanni, R., Cossu, L., and Licheri, S. (1990) Atherosclerotic plaque as a benign tumor? *Cancer Genet. Cytogenet.* **47**, 273–274.

316. Casalone, R., Granata, P., Minelli, E., Portentoso, P., Guidici, A., Righi, R., Castelli, P., Socrate, A., and Frigerio, B. (1991) Cytogenetic analysis reveals clonal proliferation of smooth muscle cells in atherosclerotic plaques. *Hum. Genet.* **87**, 139–143.

317. Wurster-Hill, D.H., Brown, F., Park, J.P., and Gibson, S.H. (1988) Cytogenetic studies in Dupuytren contracture. *Am. J. Hum. Genet.* **43**, 285–292.

318. Bardi, G., Johansson, B., Pandis, N., Heim, S., Mandahl, N., Hägerstrand, I., Holmin, T., Åndrén-Sandberg, Å., and Mitelman, F. (1992) Trisomy 7 in nonneoplastic focal steatosis of the liver. *Cancer Genet. Cytogenet.* **63**, 22–24.

319. Mertens, F., Pålsson, E., Lindstrand, A., Toksvig-Larsen, S., Knuutila, S., Larramendy,

M.L., El-Rifai, W., Limon, J., Mitelman, F., and Mandahl, N. (1996) Evidence of somatic mutations in osteoarthritis. *Hum. Genet.* **98**, 651–656.

320. Somers, K.D., Winters, B.A., Dawson, D.M., Leffell, M.S., Wright, G.L. Jr., Devine, C.J. Jr., Gilbert, D.A., and Horton, C.E. (1987) Chromosome abnormalities in Peyronie's disease. *J. Urol.* **137**, 672–675.

321. Aly, M.S., Dal Cin, P., Van de Voorde, W., van Poppel, H., Ameye, F., Baert, L., and Van den Berghe, H. (1994) Chromosome abnormalities in benign prostatic hyperplasia. *Genes Chromosom. Cancer* **9**, 227–233.

322. Ermis, A., Hopf, T., Hanselmann, R., Remberger, K., Welter, C., Dooley, S., Zang, K.D., and Henn, W. (1993) Clonal chromosome aberrations in cell cultures of synovial tissue from patients with rheumatoid arthritis. *Genes Chromosom. Cancer* **6**, 232–234.

323. Mertens, F., Örndal, C., Mandahl, N., Heim, S., Bauer, H.F.C., Rydholm, A., Tufvesson, A., Willén, H., and Mitelman, F. (1993) Chromosome aberrations in tenosynovial giant cell tumors and nontumorous synovial tissue. *Genes Chromosom. Cancer* **6**, 212–217.

324. Roque, L., Gomes, P., Correia, C., Soares, P., Soares, J., and Castedo, S. (1993) Thyroid nodular hyperplasia: Chromosomal studies in 14 cases. *Cancer Genet. Cytogenet.* **69**, 31–34.

325. Scappaticci, S., Capra, E., Cortinovis, M., Cortinovis, R., Arbustini, E., Diegoli, M., and Fraccaro, M. (1994) Cytogenetic studies in venous tissue from patients with varicose veins. *Cancer Genet. Cytogenet.* **75**, 26–30.

326. Al Saadi, A., Latimer, F., Madercic, M., and Robbins, T. (1987) Cytogenetic studies of human brain tumors and their clinical significance. II. Meningioma. *Cancer Genet. Cytogenet.* **26**, 127–141.

327. Casalone, R., Simi, P., Granata, P., Minelli, E., Giudici, A., Butti, G., and Solero, C.L. (1990) Correlation between cytogenetic and histopathological findings in 65 human meningiomas. *Cancer Genet. Cytogenet.* **45**, 237–243.

328. Doco-Fenzy, M., Cornillet, P., Scherpereel, B., Depernet, B., Bisiau-Leconte, S., Ferre, D., Pluot, M., Graftiaux, J-P., and Teyssier, J-R. (1993) Cytogenetic changes in 67 cranial and spinal meningiomas: relation to histopathological and clinical pattern. *Anticancer Res.* **13**, 845–850.

329. Vagner-Capodano, A.M., Grisoli, F., Gambarelli, D., Sedan, R., Pellet, W., and De Victor, B. (1993) Correlation between cytogenetic and histopathological findings in 75 human meningiomas. *Neurosurgery* **32**, 892–900.

330. Zang, K.D. (1982) Cytological and cytogenetical studies on human meningioma. *Cancer Genet. Cytogenet.* **6**, 249–274.

331. Pandis, N., Heim, S., Willén, H., Bardi, G., Flodérus, U-M., Mandahl, N., and Mitelman, F. (1991) Histologic-cytogenetic correlations in uterine leiomyomas. *Int. J. Gynecol. Cancer* **1**, 163–168.

332. Bomme, L., Bardi, G., Pandis, N., Fenger, C., Kronborg, O., and Heim, S. (1994) Clonal karyotypic abnormalities in colorectal adenomas: clues to the early genetic events in the adenoma-carcinoma sequence. *Genes Chromosom. Cancer* **10**, 190–196.

333. Bardi, G., Johansson, B., Pandis, N., Mandahl, N., Bak-Jensen, E., Lindström, C., Törnqvist, A., Frederiksen, H., Åndrén-Sandberg, Å., Mitelman, F., and Heim, S. (1993) Cytogenetic analysis of 52 colorectal carcinomas- non-random aberration pattern and correlation with pathologic parameters. *Int. J. Cancer* **55**, 422–428.

334. Johansson, B., Bardi, G., Pandis, N., Gorunova, L., Bäckman, P.L., Mandahl, N., Dawiskiba, S., Åndrén-Sandberg, Å., Heim, S., and Mitelman, F. (1994) Karyotypic pattern of pancreatic adenocarcinomas correlates with survival and tumour grade. *Int. J. Cancer* **58**, 8–13.

335. Pandis, N., Idvall, I., Bardi, G., Jin, Y., Gorunova, L., Mertens, F., Olsson, H., Ingvar, C., Beroukas, K., Mitelman, F., and Heim, S. (1996) Correlation between karyotypic pattern and clinicopathlogic features in 125 breast cancer cases. *Int. J. Cancer* **66**, 191–196.

336. Biegel, J.A., Burk, C.D., Parmiter, A.H., and Emanuel, B.S. (1992) Molecular analysis of a partial deletion of 22q in a central nervous system rhabdoid tumor. *Genes Chromosom. Cancer* **5,** 104–108.

337. Shashi, V., Lovell, M.A., von Kap-herr, C., Waldron, P., and Golden, W.L. (1994) Malignant rhabdoid tumor of the kidney: involvement of chromosome 22. *Genes Chromosom. Cancer* **10,** 49–54.

338. Fletcher, J.A., Kozakewich, H.P., Pavelka, K., Grier, H.E., Shamberger, R.C., Korf, B., and Morton, C.C. (1991) Consistent cytogenetic aberrations in hepatoblastoma: a common pathway of genetic alterations in embryonal liver and skeletal muscle malignancies? *Genes Chromosom. Cancer* **3,** 37–43.

339. Bonnici, A.V., Birjandi, F., Spencer, J.D., Fox, S.P., and Berry, A.C. (1992) Chromosomal abnormalities in Dupuytren's contracture and carpel tunnel syndrome. *J. Hand Surg.* **17B,** 349–355.

340. Mugneret, F., Sidaner, I., Favre, B., Manone, L., Maynadie, M., Caillot, D., and Solary, E. (1995) der(16)t(1;16)(q10;p10) in multiple myeloma: a new non-random abnormality that is frequently associated with Burkitt's type translocations. *Leukemia* **9,** 277–281.

341. Mugneret, F., Dastugue, N., Favre, B., Sidaner, I., Huguet-Rigal, F., Solary, E., and Salles, B. (1995) der(16)t(1;16)(q11;q11) in myelodysplastic syndromes: a new non-random abnormality characterized by cytogenetic and fluorescence *in situ* hybridization studies. *Br. J. Haematol.* **90,** 119–124.

342. Mrózek, K. and Bloomfield, C.D. (1998) Der(16)t(1;16) is a secondary chromosome aberration in at least eighteen different types of human cancer. *Genes Chromosom. Cancer* **23,** 78–80.

343. Mrózek, K., Arthur, D.C., Karakousis, C.P., Koduru, P.R.K., Le Beau, M., Pettenati, M.J., Tantravahi, R., Mrózek, E., Perez-Mesa, C., Rao, U.N.M., Frankel, S.R., Davey, F.R., and Bloomfield, C.D. (1995) der(16)t(1;16) is a nonrandom secondary chromosome aberration in many types of human neoplasia, including myxoid liposarcoma, rhabdomyosarcoma and Philadelphia chromosome-positive acute lymphoblastic leukemia. *Int. J. Oncol.* **6,** 531–538.

344. Atkin, N.B. (1986) Chromosome 1 aberrations in cancer. *Cancer Genet. Cytogenet.* **21,** 279–285.

345. Teixeira, M.R., Pandis, N., Bardi, G., Andersen, J.A., Mandahl, N., Mitelman, F., and Heim, S. (1994) Cytogenetic analysis of multifocal breast carcinomas: Detection of karyotypically unrelated clones as well as clonal similarities between tumor foci. *Br. J. Cancer* **70,** 922–927.

346. Pandis, N., Bardi, G., and Heim, S. (1994) Interrelationship between methodological choices and conceptual models in solid tumor cytogenetics. *Cancer Genet. Cytogenet.* **76,** 77–84.

347. Meltzer, P.S. and Trent, J.M. (1998) Chromosome rearrangements in human solid tumors, in *The Genetic Basis of Human Cancer* (Vogelstein, B., Kinzler, K.W., eds.), McGraw-Hill, New York, pp. 143–160.

348. Ladanyi, M. (1995) The emerging molecular genetics of sarcoma translocations. *Diagn. Mol. Pathol.* **4,** 162–173.

349. Jeon, I.-S., Davis, J.N., Braun, B.S., Sublett, J.E., Roussel, M.F., Denny, C.T., and Shapiro, D.N. (1995) A variant Ewing's sarcoma translocation (7;22) fuses the *EWS* gene to the ETS gene *ETV1*. *Oncogene* **10,** 1229–1234.

350. Kaneko, Y., Yoshida, K., Handa, M., Toyoda, Y., Nishihira, H., Tanaka, Y., Sasaki, Y., Ishida, S., Higashino, F., and Fujinaga, K. (1996) Fusion of an *ETS*-family gene *EIAF,* to *EWS* by t(17;22)(q12;q22) chromosome translocation in an undifferentiated sarcoma of infancy. *Genes Chromosom. Cancer* **15,** 115–121.

351. Delattre, O., Zucman, J., Melot, T., Garau, X.S., Zucker, J.M., Lenoir, G.M., Abmros, P.F., Sheer, D., Turc-Carel, C., Triche, T.J., Aurias, A., and Thomas, G. (1994) The Ewing

family of tumors—a subgroup of small-round-cell tumors defined by specific chimeric transcripts. *N. Eng. J. Med.* **331,** 294–299.

352. Panagopoulos, I., Höglund, M., Mertens, F., Mandahl, N., Mitelman, F., and Åman, P. (1996) Fusion of the *EWS* and *CHOP* genes in myxoid liposarcoma. *Oncogene* **12,** 489–494.

353. Grosschedl, R., Giese, K., and Pagel, J. (1994) MHG domain proteins: architectural elements in the assembly of nucleoprotein structures. *Trends Genet.* **10,** 94–100.

354. Ashar, H.R., Schoenberg Fejzo, M., Tkachenko, A., Zhou, X., Fletcher, J.A., Weremowicz, S., Morton, C.C., and Chada, K. (1995) Disruption of the architectural factor HMGI-C: DNA-binding AT hook motifs fused in lipomas to distinct transcriptional regulatory domains. *Cell* **82,** 57–65.

355. Hess, J.L. (1998) Chromosome translocations in benign tumors. The HMGI proteins. *Am. J. Clin. Pathol.* **109,** 251–261.

356. Sakurai, M. and Sandberg, A.A. (1973) Prognosis in acute myeloblastic leukemia: chromosomal correlation. *Blood* **41,** 93–104.

357. Schapers, R.F.M., Smeets, A.W.G.B., Pauwels, R.P.E., Van den Brandt, P.A., and Bosman, F.T. (1993) Cytogenetic analysis in transitional cell carcinoma of the bladder. *Br. J. Urol.* **72,** 887–892.

358. Ganju, V., Jenkins, R.B., O'Fallon J.R., Scheithauer, B.W., Ransom, D.T., Katzmann, J.A., and Kimmel, D.W. (1994) Prognostic factors in gliomas. A multivariate analysis of clinical, pathologic, flow cytometric, cytogenetic, and molecular markers. *Cancer* **74,** 920–927.

359. Pejovic, T., Himmelmann, A., Heim, S., Mandahl, N., Flodérus, U.-M., Furgyik, S., Elmfors, B., Helm, G., Willén, H., and Mitelman, F. (1992) Prognostic impact of chromosome aberrations in ovarian cancer. *Br. J. Cancer* **65,** 282–286.

360. Lundgren, R., Heim, S., Mandahl, N., Anderson, H., and Mitelman, F. (1992) Chromosome abnormalities are associated with unfavorable outcome in prostatic cancer patients. *J. Urol.* **147,** 784–788.

361. Bardi, G., Johansson, B., Pandis, N., Bak-Jensen, E., Örndal, C., Heim, S., Mandahl, N., Andrén-Sandberg, Å., and Mitelman, F. (1993) Cytogenetic aberrations in colorectal adenocarcinomas and their correlation with clinicopathologic features. *Cancer* **71,** 306–314.

362. Elfving, P., Mandahl, N., Lundgren, R., Limon, J., Bak-Jensen, E., Fernö, M., Ohlsson, H., and Mitelman, F. (1997) Prognostic implications of cytogenetic findings in kidney cancer. *Br. J. Urol.* **80,** 698–706.

363. Åkervall, J.A., Jin, Y., Wennerberg, J.P., Zätterström, U.K., Kjellén, E., Mertens, F., Willén, R., Mandahl, N., Heim, S., and Mitelman, F. (1995) Chromosomal abnormalities involving 11q13 are associated with poor prognosis in patients with squamous cell carcinoma of the head and neck. *Cancer* **76,** 853–859.

364. Mitelman, F., Johansson, B., Mandahl, N., and Mertens, F. (1997) Clinical significance of cytogenetic findings in solid tumors. *Cancer Genet. Cytogenet.* **95,** 1–8.

365. Trent, J.M., Meyskens, F.L., Salmon, S.E., Ryschon, K., Leong, S.P.L., Davis, J.R., and McGee, D.L. (1990) Relation of cytogenetic abnormalities and clinical outcome in metastatic melanoma. *N. Engl. J. Med.* **322,** 1508–1511.

366. Choong, P.F.M., Mandahl, N., Mertens, F., Willén, H., Alvegård, T., Kreicbergs, A., Mitelman, F., and Rydholm, A. (1996) 19p+ marker chromosome correlates with relapse in malignant fibrous histiocytoma. *Genes Chromosom. Cancer* **16,** 88–93.

367. Brodeur, G.M. (1995) Molecular basis for heterogeneity in human neuroblastomas. *Eur. J. Cancer* **31A,** 505–510.

368. Schwab, M., Praml, C., and Amler, L.C. (1997) Genomic instability in 1p and human malignancies. *Genes Chromosom. Cancer* **16,** 211–229.

369. Christiansen, H., Sahin, K., Berthold, F., Hero, B., Terpe, H.J., and Lampert, F. (1995)

Comparison of DNA aneuploidy, chromosome 1 abnormalities, *MYCN* amplification and CD44 expression as prognostic factors in neuroblastoma. *Eur. J. Cancer* **31A,** 541–544.

370. Hayashi, Y., Kanda, N., Inaba, T., Hanada, R., Nagahara, N., Muchi, H., and Yamamoto, K. (1989) Cytogenetic findings and prognosis in neuroblastoma with emphasis on marker chromosome 1. *Cancer* **63,** 126–132.

371. Dewald, G.W., Schad, C.R., Lilla, V.C., and Jalal, S.M. (1993) Frequency and photographs of HGM11 chromosome anomalies in bone marrow samples of 3,996 patients with malignant hematologic neoplasms. *Cancer Genet. Cytogenet.* **68,** 60–69.

IV
Beyond Chromosomes

Editor's Foreword to Section IV

The progress that has been made in the clinical analysis of chromosomes in the four decades since Tjio and Levan is impressive. The resolution possible with the light microscope has become so good that to go any further almost requires analysis of the DNA itself, and indeed the distinction between classic cytogenetics and molecular analysis is no longer as discrete as it once was.

In addition, the information provided by the cytogenetics laboratory has become so complex that it often requires the expertise of a specialist who can assist both patient and health care provider with its interpretation. This is the role traditionally played by the genetic counselor.

In this section, we explore the continuum that is cytogenetics, molecular genetics, and molecular cytogenetics, as well as the practical utilization of data via genetic counseling.

14

Fragile X

From Cytogenetics to Molecular Genetics

Patricia N. Howard-Peebles, Ph.D.

FRAGILE SITES IN HUMANS

The first fragile site identified in man was on chromosome 9q, as described by Dekaban in 1965 *(1)*, but the term "fragile site" was first used in 1970 when Magenis et al. described a site on 16q in a large Oregon family *(2)*. Fragile sites continued as an active area of cytogenetic research during the late 1970s and most of the 1980s, stimulated by a link between the fragile site at Xq27 and X-linked mental retardation, the discovery that fragile site expression was directly related to the tissue culture conditions used for cell preparations *(3)*, and a possible relationship between fragile sites and cancer/cancer cytogenetics. The application of molecular techniques to fragile sites began in the early 1990s with the discovery of a new mutation mechanism for fragile Xq27, which resulted in the identification of a new type of human disease.

Definition and Classification

In 1979, Sutherland defined a fragile site as a specific point on a chromosome that appears as a nonstaining gap, usually on both arms or chromatids *(4)*. In a family, this site is always in the same location, is inherited as a Mendelian codominant, and results in chromosome fragility under appropriate tissue culturing conditions. Over time, this definition came to include nonrandom points (gaps or breaks) caused by exposing chromosomes to specific tissue culture conditions or chemical agents ("induction") *(5)*. Over 100 human fragile sites have been described, and they can be classified into two major categories, rare or common. The rare fragile sites can vary from being present in only one reported family to a frequency of 1 in 40 chromosomes, as with the fragile site at 16q22 in the German population *(6)*, and the three fragile sites that have been found only in the Japanese population *(7)*. In contrast, common fragile sites are thought to be present on every human chromosome. Because spontaneous expression of fragile sites is rare as well as inconsistent, they are grouped by the induction system needed for expression. **Table 1** summarizes the classification of all fragile sites by induction system with examples.

From: *The Principles of Clinical Cytogenetics*
Edited by: S. Gersen and M. Keagle © Humana Press Inc., Totowa, NJ

Table 1
Classification of Human Fragile Sites

Type	Group no.	Induction system	n^a	Examples	(Gene symbol)[b]
Rare	1	Folate-sensitive	21	Xq27.3	(*FRAXA*)
	2	Distamycin-A-inducible		11q23.3	(*FRA11B*)
		Subgroup a	2	16q22.1	(*FRA16B*)
		Subgroup b[c]	3	11p15.1	(*FRA11I*)
	3	Bromodeoxyuridine-requiring	2	10q25.2	(*FRA10B*)
Common	4	Aphidicolin-inducible	75	3p14.2	(*FRA3B*)
				Xq27.2	(*FRAXD*)
	5	5-Azacytidine-inducible	4	19q13	(*FRA19A*)
	6	Bromodeoxyuridine-inducible	7	5p13	(*FRA5A*)
				13q21	(*FRA13B*)

Adapted from Howard-Peebles (8).
[a]n = number of reported sites (9).
[b]Gene symbols as assigned by Human Gene Mapping 8 (10).
[c]Fragile sites identified in Japanese only (7).

Significance

Clinical significance has been established for two fragile sites: FRAXA (Xq27.3) and FRAXE (Xq28). Both are rare, folate-sensitive fragile sites. FRAXA is the fragile X (fraX) that is associated with the fragile X syndrome, the most common form of familial mental retardation. FRAXE is associated with a mild form of X-linked mental retardation.

Two other fragile sites are candidates for a role in oncogenesis: FRA3B(3p14.2) and FRA11B(11q23.3). The former is a common site that may act as a tumor suppressor *(11)*. FRA11B, a rare folate-sensitive site, is located within the CBL2 proto-oncogene and has provided the first direct evidence of in vivo chromosome breakage at a fragile site *(12)*.

Cytogenetics of Fragile Sites

The majority of cytogenetic studies have been performed on phytohemagglutinin (PHA)-stimulated lymphocytes, owing to induction problems in other tissue types. Tissue culture conditions and modifications for induction of all groups of fragile sites are detailed by Sutherland *(5,13)*. Scoring of fragile sites involves looking for points of discontinuous staining, which are more easily seen in solid stained (nonbanded) chromosomes. However, G-banding is required for correct site identification, making the sites more difficult to locate *(5)*. One folate-sensitive fragile site (FRA12D) has only been identified in individuals having another folate-sensitive site *(14)*. Individuals with two rare fragile sites, for which there is no known explanation, have been reported multiple times.

The folate-sensitive fragile sites can be induced by multiple methods, as summarized in **Table 2**. Media systems deprived of folic acid and thymidine (Method 1) require serum supplements of ≤5%, pH maintained at ≥7.3 and extra precautions to

Table 2
Methods for Inducing Folate-Sensitive Fragile Sites

Number	Method	Additive[a]
1	Thymidine and folic acid deprivation	—
2	Inhibiting folate metabolism	Aminopterin
		Methotrexate
		Trimethoprin
3	Inhibiting thymidylate synthetase	Fluorodeoxyuridine (FUdR)
		Fluorodeoxycytidine (FCdR)
4	Excess thymidine	300–600 mg/L
5	Combination—nos. 3 and 4 *(15)*	FUdR + thymidine (300–600 mg/L)

[a]All additives required last 24 h of culturing time.

Table 3
Inhibitors of Folate-Sensitive Fragile Site Expression

Folic acid	Folinic acid
Thymidine	Methionine (lack of)
Bromodeoxyuridine (BrdU)	Bromodeoxycytidine (BrdC)

ensure sterility of cultures. Method 4 has an advantage of being somewhat less cyto-toxic to the cells during culturing *(16)*. All methods require maintaining the cultures for at least 72 h and, under some culture conditions, 96 h gives a higher number of cells that express the fragile site. Some medium components and additives inhibit expression (**Table 3**), which likely explains why some media completely inhibit fragile site expression. Numerous physical and chemical factors affect the ultimate level of expression of these sites, including the pH of the culture medium, storage and transportation of the specimen, the age of the blood sample when cultures are initiated, medium composition and serum concentration, length of culture time, harvesting procedures, and cell density *(4,5,13,17)*. Sutherland and Hecht provide extensive details and documentation of requirements for expression *(13)*. Because of the stringent requirements for folate-sensitive fragile site expression, basic guidelines were developed to ensure quality clinical testing for fraX *(18,19)*; see also Chapter 6.

Although most of the above work was done on lymphocyte cultures, fraX expression is possible in other tissue types. In general, inducers are required for expression and the level of expression is reduced, which made prenatal diagnosis difficult with cytogenetic technology.

GENETICS OF FRAGILE X SYNDROME, PRIOR TO THE AVAILABILITY OF MOLECULAR ANALYSIS

X-Linked Mental Retardation

In 1938, Penrose noted a higher incidence of mental retardation (MR) in males, and reports of families with only affected males *(20)*. These observations were compatible with X-linked inheritance, and numerous reports appeared in the literature *(21)*. Based

Table 4
Fragile Sites on X Chromosome

Gene symbol	Location	Type	Subtype
FRAXA	Xq27.3	Rare	Folate-sensitive
FRAXB	Xp22.31	Common	Aphidicolin-inducible
FRAXC	Xq22.1	Common	Aphidicolin-inducible
FRAXD	Xq27.2	Common	Aphidicolin-inducible
FRAXE	Xq28	Rare	Folate-sensitive
FRAXF	Xq28	Rare	Folate-sensitive

on this early work, a clinically nonspecific X-linked MR disorder was delineated and called Renpenning's syndrome, Martin–Bell syndrome, or nonspecific X-linked MR. In 1959, Lubs described the first family with cytogenetic expression of the "marker X" (which became the fragile X), and the heterogeneity of this nonspecific X-linked MR disorder became apparent *(22)*. Numerous disorders have been delineated from this original subgroup of MR males and, in a continuing effort, a total of 105 X-linked disorders involving MR have been described *(23)*. The fraX subgroup was unique, as there was a diagnostic laboratory test; the name Martin–Bell syndrome was attached when this family, first described in 1943, was shown to be positive for fraX *(24)*. However, the popular name for this disorder became fragile X syndrome.

Inheritance of fraX

It became apparent soon after the cytogenetic test became available that the inheritance and penetrance of fragile X syndrome was unlike that of any previously described X-linked disorder, although it came closest to an X-linked dominant with reduced penetrance. It was determined that some males who inherited the fraX were clinically normal, but passed the disorder to their normal daughters and frequently had affected grandchildren. The term transmitting male (TM) was coined to describe such unaffected carrier males. These TMs were thought to be the missing 20% of affected males described by Sherman et al. from 206 fraX families *(25,26)*. Their observation, that the mothers of TMs are much less likely to have affected offspring than are the TM's unaffected daughters, became known as the Sherman paradox. Other unusual features of fragile X syndrome are that TMs have fewer mentally retarded daughters than do unaffected carrier females, affected females occur more frequently (about one of three) than in other X-linked disorders, and affected females have more affected offspring that do unaffected carrier females.

Cytogenetic Expression of fraX

The fraX site is located in band Xq27.3, one of 6 fragile sites located on the X chromosome (**Table 4**). It can be identified by both conventional solid stained and banded preparations (**Fig. 1**). Banded preparations are required, as other fragile sites and lesions can mimic fraX *(13,17,27)*. Three other fragile sites have been found in bands Xq27-28; FRAXD *(28)*, FRAXE *(29)*, and FRAXF *(30)*. The latter two sites (**Table 4**) cannot be cytogenetically distinguished from fraX. The standard ISCN

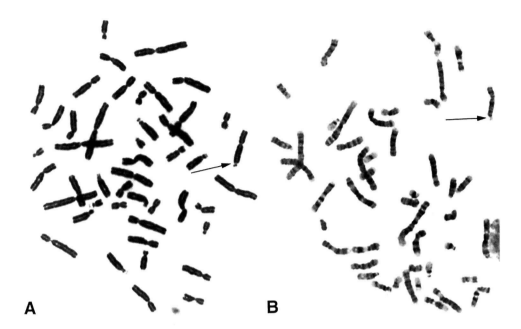

Fig. 1. Appearance of FRAXA. (**A**) Conventional stain (Giemsa) and (**B**) GTG-banded. The arrow ($\rightarrow$) indicates the location of the fraX site.

nomenclature (see Chapter 3) to cytogenetically designate fraX is 46,Y,fra(X)(q27.3) for an affected male and 46,X,fra(X)(q27.3) for an expressing female *(31)*.

FraX is not a chromosome abnormality. It is a chromosomal "marker" that allowed us to diagnose fragile X syndrome in most cases while better techniques were being developed. Thus, the table developed for chromosome mosaicism *(32)* does not apply.

Cytogenetic Expression in Affected Males and Carrier Females

In affected males, fraX expression varied from less than 4% to 50% of cells, with the low-expressing males comprising a minority of the diagnosed cases. However, this group is the origin of the false-negative males diagnosable with molecular techniques. Why fraX does not express in >50% of metaphases is still not known. Cytogenetic testing of carrier (heterozygous) females was even more problematic. Among obligate carriers, only about 50% tested positive, and clearly about one third of these carriers were affected to some degree. In general, fraX expression was easier to demonstrate (although lower than in males) in affected females than in those with normal intelligence. Guidelines were established for interpretation of these data *(18,19)*.

Prenatal Diagnosis

Prenatal testing was available on an experimental basis beginning in 1981. Testing was done on fetal blood, amniocytes, or chorionic villus cells with varying degrees of success. False-negative males were reported with all three tissue types. In the United States, amniocentesis was the major procedure, while chorionic villus sampling (CVS) was the standard in Europe and Australasia. England had the major experience with fetal blood sampling. Worldwide experience with prenatal diagnosis by cytogenetic analysis or cytogenetic analysis plus DNA polymorphism analysis (see discussion later)

Table 5
Classification of Trinucleotide Repeat Diseases

Class	n	Repeat	Position of repeat	Examples (locus)
1	5	CCG	5' Untranslated region	Fragile X syndrome (*FMR1*)
				FRAXE syndrome (*FMR2*)
2	1	CTC	3' Untranslated region	Myotonic dystrophy (DM)
3	6	CAG	Inside coding region	Huntington disease (HD)
				Spinocerebellar ataxia type 1 (SCA1)
				Kennedy disease (*AR*)
4	1	GAA	In first intron	Friedreich ataxia (*FRDA*)

Adapted from Howard-Peebles *(8)*.

exceeded over 400 cases. The "state of the art" was summarized at the Fourth International Workshop on Fragile X and X-linked Mental Retardation *(33)*.

MOLECULAR ASPECTS OF FRAGILE X SYNDROME

Analysis Using Linked Polymorphisms

From the mid-1980s through 1991, molecular (DNA) analysis using linked polymorphisms was used in confirmed fraX families to help with prenatal diagnosis and carrier status. Although the gene for fragile X syndrome had not been identified, its relative location on a linear map of the distal X long arm was known. Using genes and polymorphisms on both sides of fraX allowed molecular geneticists to track fraX chromosomes through families. The risks of inheriting the fraX chromosome were expressed as probabilities. Success with the method depended on the distance between the tested polymorphism/gene and the fragile X syndrome gene, size of the family, and which polymorphism/genes were informative. Regardless of these limitations, the combination of cytogenetic and linkage analysis allowed many families to get more reliable results than with cytogenetic analysis alone.

Trinucleotide Repeats

In 1991, the mechanism causing the fragile X syndrome mutation was first identified *(34–36)*. The mutation results from the expansion of a trinucleotide repeat located in or near an expressed sequence; see Chapter 2 for a discussion of nucleotides. This observation was followed by an intensive search for other trinucleotide, or triplet, repeats associated with genes causing inherited disorders. To date, there are 13 triplet repeats associated with 11 diseases.

Triplet repeat disorders are grouped according to the specific trinucleotide sequence. **Table 5** summarizes the four classes presently known *(37–39)*. One characteristic of these disorders, each generation showing an earlier onset and increasing severity, is known as *anticipation.* All the disorders are either X-linked or autosomal dominant except Friedreich ataxia, which is autosomal recessive.

The CGG triplet repeats (Class 1) are located at folate-sensitive fragile sites and their characteristics are summarized in **Table 6** *(9,40)*. When the number of trinucle-

Table 6
Characteristics of the Cloned Folate-Sensitive Fragile Sites

			Copy number		
Symbol	*Location*	*Disease*	*Normal*	*Premutation*	*Full mutation*
FRAXA	Xq27.3	Fragile X syndrome	6–54	60–200	230 to > 1000
FRAXE	Xq28	Fragile XE syndrome	6–25	?50–200	200 to > 800
FRAXF	Xq28	None	6–29	?	300 to 1000
FRA16A	16p13.1	None	16–50	?50–200	?1000–2000
FRA11B	11q23.3	Offspring predisposed to Jacobsen syndrome	11	85–100	100–1000

Adapted from Howard-Peebles *(8)*.

otide repeats exceeds a specific number, the resulting "CpG island" near the region becomes hypermethylated, resulting in cytogenetic expression of the fragile site. This is called a full mutation. In the case of fraX, methylation prevents production of the gene product (protein).

Based on the triplet repeat size in FRAXA and FRAXE, an individual can be classified as *normal, premutation,* or *full mutation.* An individual with *normal* repeat size is characterized by length stability and normal intelligence, whereas an individual with a *premutation* repeat size shows instability of the repeat length from generation to generation, but normal intelligence. In contrast, *full mutation* individuals have massive, unstable somatic changes resulting in fragile X syndrome. The values of these repeat lengths for fraX are listed in **Table 6.**

The Fragile X Gene—FMR1

The fragile X mental retardation-1 (*FMR1*) gene was identified in mid-1991. It has 17 exons (coding regions) containing about 38 kilobases of the genome *(41)*. The CGG triplet repeat is in the 5′ untranslated region of exon 1, about 250 basepairs downstream of a CpG island. The most common repeat size in normal X chromosomes is 29 or 30, but ranges from 6 to 54. Not unexpectedly, direct mutations such as deletions and point mutations have been found in *FMR1*. Prior et al. *(42)* reported a case of germline mosaicism, an important issue when counseling deletion families. These are rare, with fewer than 20 reported through 1996 *(43)*.

In patients expressing fraX full mutations, the upstream CpG island is methylated. This methylation inactivates the *FMR1* locus and suppresses transcription of mRNA, which eliminates protein production. The protein (FMRP) is an RNA-binding protein *(44,45);* this type of protein usually has regulatory functions. FMRP is primarily found in the cytoplasm of the cell and interacts with as much as 4% of mRNA from human brain *(45)*. The direct mutations described above confirm that the loss of FMRP results in the phenotype of fragile X syndrome.

The instability of the trinucleotide repeat is the important factor that differentiates fraX families from normal families. As noted in **Table 6**, X chromosomes having 50–59 repeats are "between" normal and premutation chromosomes. The overlap of normal and premutation repeat size is called the "gray zone" and varies between 40 and 60 repeats *(46).* These "gray zone" alleles tell us that length is not the only factor determining triplet repeat expansion. The likely mechanism of repeat expansion is slippage during replication of the DNA strands. Stability is thought to be provided by AGG triplets scattered within the CGG repeat region, with most normal alleles of 30 repeats having two AGG triplets *(35).* Thus, instability and further expansion may result from loss of these anchor triplets resulting in a longer pure CGG sequence. A premutation repeat size of 52 has been reported in a fragile X family *(36),* but the smallest repeat observed to expand to a full mutation in the next generation is 56 *(46).* A "gray zone" allele of 50–59 in non-fraX families has a low empirical risk for expansion *(47),* and prenatal testing is available. Several groups have developed tables for risk of a premutation in a mother expanding to a full mutation in her offspring. A recent summary of three groups totaling 532 meioses confirmed that premutations of ≥90 almost always expand to full mutations *(47).*

Genes homologous to *FMR1* have been found in yeast, nematodes, and many other vertebrates, but not in the fruit fly *(35).* The mouse homologue *(Fmr1)* has been cloned and sequenced. Compared to *FMR1, Fmr1* has 95% sequence identity with the DNA sequence, and 97% protein sequence identity with FMRP *(48).*

An animal model (knockout mouse) of fragile X has been developed *(49).* This strain resembles the human condition having no mRNA, but has an interruption of exon 5 instead of a trinucleotide repeat expansion *(50).* Transgenic mice with unstable triplet repeats (CTG and CAG) have recently been described by three groups *(51).* Thus, it may be possible to produce mice with the CGG triplet.

Two autosomal genes, *FXR1* at 3q28 *(52)* and *FXR2* at 17p13.1 *(53),* show significant homology to *FMR1,* and proteins produced by all three genes interact with each other and with ribosomes. The FXR1 and FXR2 proteins are normal in fraX males without FMRP; no mutations of these two genes have been identified. [Note: FXR1 has a pseudogene located on chromosome 12q13 *(54).*]

The term *mosaicism* (or mosaic when referring to the individual) is used in two ways in fragile X syndrome *(55).* One is methylation mosaicism, which is expressed as the percentage of full mutation molecules that are methylated. The other is full mutation/premutation mosaicism, where both types of molecules are found with molecular analysis techniques. Both have been proposed as factors that can result in a milder phenotype, but this is unpredictable owing to tissue-specific differences.

CLINICAL ASPECTS OF FRAGILE X SYNDROME

Physical Phenotype

In males, the classic features of fragile X syndrome are X-linked mental retardation, macroorchidism, and minor dysmorphic facial features including a long, oblong face with a large mandible and large and/or prominent ears. At least 80% of affected males have one or more of these features, but expression varies with age. Other frequent features are a high-arched palate, hyperextensible finger joints, velvetlike skin, and flat feet. A small subgroup of males with a "Prader–Willi-like" phenotype has been described by Fryns et al. *(56).* Heterozygous females express these same features of fragile X syndrome, with manifestations being more common in females with mental disability than in those of normal intelligence. Hull and Hagerman showed the same relationship for full mutation vs premutation females *(57)*.

Behavioral Phenotype

The behavior of fragile X males, especially in childhood, is more consistent and diagnostic than the physical features. They are typically hyperactive and speech delayed. Other complicating features can include irritability, hypotonia, perseveration in speech and behavior, and autistic-like features such as hand flapping, hand biting, and poor eye contact. Social anxiety and avoidance are prominent features of fragile X syndrome in both sexes.

Premutation carriers of both sexes scored significantly lower on two psychometric tests than did their same sex relatives who had a normal *FMR1* gene *(58)*. Transmitting males likely form a spectrum from normal to mildly affected *(59)*. In the case of adult premutation carrier females, available data suggest they are not different from normal controls on measures of psychiatric diagnoses or symptoms, neuropsychologic function, or self-rated personality profiles *(60,61)*.

Recently, Hagerman reviewed in detail the physical and behavioral phenotype of fragile X syndrome *(62, see also 43)*. The variability of expression makes clinical diagnosis difficult. Therefore, fragile X syndrome should be considered in the differential diagnosis of all mentally retarded individuals.

Cognitive Phenotype

In males, preliminary evidence suggests there are specific deficits in arithmetic, sequential motor planning, short-term auditory memory, and spatial skills. The IQ appears to decrease with age, although the reason for this longitudinal decline is unclear *(63)*. Most adult fraX males are moderately retarded. IQ is not correlated with the size of the triplet repeat. However, it does appear to be correlated with the mosaic status of the male. Affected males with both somatic full mutation and premutation size repeats or those who are methylation mosaics have higher IQs than the affected males who are nonmosaic or fully methylated. On occasion, such mosaic males will test in the normal/low normal range *(64)*.

In females, cognitive studies consistently indicate specific weaknesses on short-term auditory memory, arithmetic, and visuospatial tasks. Full mutation females have mean IQs in the low average range (IQ = 80–91), whereas premutation females have mean IQs at or above the population mean (IQ = 98–117). Attempts to correlate IQ with triplet repeat size and/or proportion of inactive fraX chromosomes have produced conflicting results. Thus, it remains impossible to predict the IQ of a full mutation female. Preliminary data suggest an IQ decrease over time in young fraX females *(65)*.

Other Clinical Aspects

A recent review explores the neurologic and pathologic findings in fragile X syndrome *(43)*. Medical follow-up, pharmacotherapy, treatment of emotional and behavioral problems, and intervention approaches for fragile X syndrome have also been reviewed *(66)*.

Reproductive Issues

One of the yet unsolved questions is when in development the expansion from premutation to full mutation occurs. Expansion could occur during oögenesis (meiotic) or after fertilization (mitotic). Reyniers et al. *(67)* showed that full mutation or mosaic full/premutation males produce only premutation sperm, and premutation daughters, as repeat expansion occurs only in females *(68)*. Testicular selection against full mutation sperm is unlikely, as male Frm1 mice show fertility *(49)*. These data support a model of expansion only in somatic cells, and protection of the premutation in the germline cells. However, Malter et al. showed in full mutation fetuses that only full mutation alleles (in the unmethylated state) were found in oöcytes from intact ovaries or in immature testes from 13-week fetuses, but both full and premutation alleles were found in the germ cells of a 17-week male fetus *(69)*. They hypothesize that the full mutation contracts in the fetal testes, with subsequent selection for the premutation sperm. In females, the expansion could occur during maternal oögenesis or very early in embryogenesis, prior to general methylation. The answer requires analysis of oöcytes from premutation females.

There is increasing evidence *(70–73)* that premutation carrier females are at increased risk to develop premature ovarian failure (POF), that is, early menopause (onset of menopause before 40 years of age). This is one of the factors that can limit the usefulness of preimplantation genetic testing (PGT) as a reproductive option for carrier females *(74)*. The objective of fraX PGT is to utilize only those embryos that receive the normal X chromosome from the mother. Donor egg, where available, is another reproductive option that allows carrier females, even those with POF, to have unaffected children *(75)*.

CURRENT GENETIC ASPECTS OF FRAGILE X SYNDROME

Incidence

Many approaches to determining the prevalence of fragile X syndrome in populations were used prior to availability of molecular diagnosis. A prevalence figure of about 1 in 1000 males was estimated from these studies. However, numerous population studies reported during this time failed to identify the expected number of individuals based on this prevalence figure. Turner et al. reevaluated the males from two studies using molecular methods, and a revised prevalence figure of about 1 in 4000 males was determined *(76)*. The false-positives were due to utilizing low positive cytogenetic percentages (≤5%) and/or identifying the wrong fragile site (see earlier).

Fragile X syndrome has been found in all ethnic groups studied, and has been identified in individuals with cytogenetic abnormalities such as XXY, XXX, XYY, and +21, as well as in those with other genetic disorders such as neurofibromatosis. These cases are likely coincidental as a result of the frequency of both disorders in human populations.

The premutation carrier rate in females is as yet unclear and likely will vary among populations. The largest survey of 10,624 anonymous women in Quebec City who were tested for premutations yielded a carrier rate (55⁺ repeats) of 1 in 259 *(77)*. Screening of 474 pregnant women and 271 egg donors with no family history of MR or learning disability identified three premutation carriers (60$^+$ repeats) for a carrier rate of 1 in 248 *(78)*. An earlier study had found one premutation carrier (75 repeats) in 561 females sampled from families with genetic disorders without MR *(79)*. More data are necessary to determine population dynamics of fraX *(77,80)* including the reported founder effect *(81)*.

Molecular Rules of Inheritance

DNA analysis of fraX can detect all stages of the triplet repeat expansion. Reduced penetrance, the Sherman paradox, and other unusual characteristics of fragile X syndrome were explained by the silent premutation state. The rules of inheritance *(43)* as we now understand them include:

1. Every affected individual has a carrier mother with observable expansions. No new mutation has gone directly from normal to full. Full mutation males do not pass a full mutation to their daughters.
2. Carrier females may have a full mutation or a premutation. Affected females have full mutations and unaffected females may be premutations or nonpenetrant full mutations. As a result, a female with a full mutation has an obligate carrier mother, but a female with a premutation could have received that X chromosome from either parent.
3. The risk that a female carrier will have a child with a full mutation is directly related to the size of her expansion. A repeat size of 90 appears to be the point of significance, as nearly all premutations ≥90 become full mutations in subsequent offspring *(47,68)*.
4. Premutations appear to be inherited silently for many generations. No family has been found in which the normal allele to premutation allele has been documented. Thus, many present families may have the same ancestral premutation but cannot be traced reliably. Using polymorphism analysis, Smits et al. showed one family with five living fraX males who share an X chromosome to be related through their last common ancestry six or more generations in the past *(82)*.

DIAGNOSTIC LABORATORY TESTING FOR FRAGILE X SYNDROME

Cytogenetic Testing

From 1977 to 1992, the standard laboratory test for diagnosis of fragile X syndrome was cytogenetic scoring for expression of the fraX in metaphase cells. Compared to routine karyotype analysis, fraX testing had many technical difficulties as well as biologic limitations. One significant advantage was that the test was combined with routine chromosome analysis, and as a result other chromosome abnormalities could be diagnosed as well. In a developmentally delayed population negative for Down syndrome, there is a higher chance of finding a chromosome abnormality than of diagnosing fragile X syndrome.

Reevaluation of Negative Results

Cytogenetic testing appears to have been over 90% effective in diagnosing males with fragile X syndrome *(55,77)*. However, premutations carriers usually do not express cytogenetically at all, and full mutation females frequently have lower expression levels than affected males. Thus, all potential carriers in fraX families who tested negative by the cytogenetic test should be retested with DNA methodology. The same applies to individuals with a strong fragile X syndrome phenotype.

Reevaluation of Positive Results

It is now apparent that the false-positive rate for cytogenetic testing was significant in both affected and carrier individuals *(76,77)*. The other three fragile sites in the Xq27–28 region, one common site (FRAXD) and two rare sites (FRAXE and FRAXF), were the major contributors, as the rare sites cannot be cytogenetically distinguished from FRAXA. However, technical and interpretative problems in the laboratory were also factors. Any family with a cytogenetic diagnosis of fragile X syndrome should have one family member (affected or obligate carrier) tested with DNA methodology, especially prior to carrier testing in other family members via DNA technology. Normal females who were defined as carriers based on low-level cytogenetic expression should be retested as well *(77)*.

Molecular Testing

By the time DNA-based diagnosis of fraX became available, the problems with cytogenetic testing had become apparent *(83)*. First, fraX expression was variable (between one and 50%), with females usually having fewer positive cells than males, and carriers often tested negative. Second, the presence of the other three fragile sites on Xq reduced the reliability of cytogenetic scoring. Lastly, lower expression in cell types other than lymphocytes compromised prenatal diagnosis. DNA-based testing solved all these problems and usually costs less as well. **Thus, cytogenetic fraX testing should be retired, because it is less accurate and more expensive.**

The objective of all DNA-based methods for fraX is to identify a piece of DNA containing the CGG repeat and determine its length by electrophoresis in order to classify it as normal, premutation, or full mutation.

DNA-Based Methods

The two methods utilized for DNA testing for fraX are Southern blot, with or without methylation, and polymerase chain reaction (PCR). PCR is more sensitive for premutations or carrier testing, and the results are usually expressed as a total repeat number. Southern blots are better for full mutations and, if double digestion is utilized, the methylation status can be determined. The results are expressed as Δ kb (delta kb defined as the difference between the patient and a normal reference). Detailed descriptions of these techniques and illustrations are provided by Maddalena et al. *(43)*.

Protein/mRNA-Based Diagnosis

Monoclonal antibodies against FMRP have been used with success to diagnose affected males and some affected females *(84)*. It cannot be used for premutation testing, but is more rapid than DNA-based testing. It has been successfully utilized for prenatal diagnosis *(85)*.

FRAXE Syndrome

Cytogenetically, FRAXE was described in 1992 *(29)*. The gene (*FMR2*) is located 600 kb distal to FRAXA, and the repeat sizes in normal, premutation, and full mutation individuals are similar (**Table 6**). FRAXE expansion can decrease or increase in both males and females, and two deletions have been identified (86). The phenotype of FRAXE syndrome appears to be mild MR (IQ = 60–80); however, the collection of cognitive and behavioral data from FRAXE families may further define this phenotype from that of FRAXA *(87).* FRAXE expansions appear to be quite rare in human populations *(88),* and, although available, DNA analysis for the FRAXE expansion is not widely utilized.

INDICATIONS FOR PRENATAL DIAGNOSIS AND CARRIER TESTING

Carrier Testing

Women who have affected children are obligate carriers. Determining DNA status for these women is indicated if future pregnancies are planned. Other family members who could share an X chromosome with an obligate carrier are at risk and should be referred for counseling and possible testing.

Carrier testing could be elected by any individual whether he or she has a positive family history or not, especially because the frequency of premutation carriers in human populations appears to be high. Family members whose carrier status was determined by DNA linkage should be tested to confirm the result. Likewise, DNA testing is recommended for low-expressing family members who are diagnosed cytogenetically.

Prenatal Diagnosis

Prenatal DNA testing is indicated in families where the mother is a known carrier of a premutation/full mutation CGG repeat. This is the only situation in which the offspring is at risk to inherit a full mutation. Specimens from either amniocentesis or chorionic villus sampling can be used to determine the allele size of the fetus. Timing and availability are issues that help determine the procedure selected.

CVS is done early in pregnancy and, if sufficient tissue is obtained, testing can be performed on uncultured cells. In CVS tissue, full mutations are not always methylated, but interpretation can be based on the size of the allele, not its methylation pattern. Maternal cell contamination, if present, can be seen by fetal to maternal comparison. This is most evident when using double digests and the CVS specimen is not methylated *(89).*

Interpretation of results of testing is usually unremarkable, except in the case of full mutation females. The severity of the disorder cannot be predicted in an individual female, but is based on the risk probabilities developed in family studies of such females.

Genetic Counseling

Genetic counseling is a vital part of a multidisciplinary approach to helping families adjust to and cope with the stresses of fragile X syndrome and its impact on the family. [See the excellent review by Cronister *(90).*] Genetic counseling covers a multitude of areas such as diagnosis, prognosis, recurrence risks, family planning options, management, psychosocial issues, to name a few. It provides the family with educational and emotional support so they can adjust to and cope with present as well as future circumstances. Genetic counseling is covered in Chapter 17.

REFERENCES*

1. Dekaban, A. (1965) Persisting clone of cells with an abnormal chromosome in a woman previously irradiated. *J. Nucl. Med.*, **6,** 740–746.

*Literature search ended February, 1997.

2. Magenis, R.E., Hecht, F., and Lovrien, E.W. (1970) Heritable fragile site on chromosome 16: probable localization of haptoglobin locus in man. *Science* **170,** 85–87.

3. Sutherland, G.R. (1977) Fragile sites on human chromosomes: demonstration of their dependence on the type of tissue culture medium. *Science* **197,** 265–266.

4. Sutherland, G.R. (1979) Heritable fragile sites on human chromosomes. I. Factors affecting expression in lymphocyte culture. *Am. J. Med. Genet.* **31,** 125–135.

5. Sutherland, G.R. (1991) Chromosomal fragile sites. *Genet. Analysis Techn. Appl.* **8,** 161–166.

6. Schmid, M., Feichtinger, W., Jessberger, A., Kohler, J., and Lange, R. (1986) The fragile site (16)(q22). I. Induction by AT-specific DNA-ligands and population frequency. *Hum. Genet.* **74,** 67–73.

7. Hori, T., Takahashi, E., and Murata, M. (1988) Nature of distamycin A-inducible fragile sites. *Cancer Genet. Cytogenet.* **34,** 189–194.

8. Howard-Peebles, P.N. (1997) Fragile sites and trinucleotide repeats. *Appl. Cytogenet.* **23,** 1–6.

9. Sutherland, G.R. and Richards, R.I. (1995) The molecular basis of fragile sites in human chromosomes. *Curr. Opinion Genet. Dev.* **5,** 323–327.

10. McAlpine, P.J., Shows, T.B., Miller, R.L., and Pakstis, A.J. (1985) The 1985 catalog of mapped genes and report of the nomenclature committee. *Cytogenet. Cell Genet.* **40,** 8–66.

11. Pennisi, E. (1996) New gene forges link between fragile site and many cancers. *Science* **272,** 649.

12. Jones, C., Penny, L., Mattina, T., Yu, S., Baker, E., Voullaire, L., Langdon, W.Y., Sutherland, G.R., Richards, R.I., and Tunnacliffe, A. (1995) Association of a chromosome deletion syndrome with a fragile site within the proto-oncogene *CBL2*. *Nature* **376,** 145–149.

13. Sutherland, G.R. and Hecht, F. (1985) *Fragile Sites on Human Chromosomes.* Oxford University Press, New York (*Oxford Monographs on Medial Genetics,* No. 13).

14. Sutherland, G.R. and Baker, E. (1993) Unusual behaviour of a human autosome having two rare folate sensitive fragile sites. *Ann. Genet.* **36,** 159–162.

15. Howard-Peebles, P.N. (1991) Letter to the editor. Fragile X expression: use of a double induction system. *Am. J. Med. Genet.* **38,** 445–446.

16. Sutherland, G.R., Baker, E., and Fratini, A. (1985) Excess thymidine induces folate sensitive fragile sites. *Am. J. Med. Genet.* **22,** 433–443.

17. Howard-Peebles, P.N. (1983) Conditions affecting fragile X chromosome structure *in vitro*. In *Cytogenetics of the Mammalian X Chromosome, Part B: X Chromosome Anomalies and Their Clinical Manifestations* (Sandberg, A.A., ed.), Alan R. Liss, New York, pp. 431–443.

18. Jacky, P.B., Ahuja, Y.R., Anyane-Yeboa, K., Breg, W.R., Carpenter, N.J., Froster-Iskenius, U.G., Fryns, J-P., Glover, T.W., Gustavson, K.-H, Hoegerman, S.F., Holmgren G., Howard-Peebles, P.N., Jenkins, E.C., Krawczun, M.S., Neri G., Pettigrew, A., Schaap, T., Schonberg, S.A., Shapiro, L.R., Spinner, N., Steinbach, P., Vianna-Morgante, A.M., Watson, M.S., and Wilmot, P.L. (1991) Guidelines for the preparation and analysis of the fragile X chromosome in lymphocytes. *Am. J. Med. Genet.* **38,** 400–403.

19. Knutsen, T., Bixenman, H., Lawce, H., and Martin, P. (1989) Chromosome analysis guidelines. Preliminary report. *Karyogram* **15,** 131–135.

20. Penrose, L.S. (1938) A clinical and genetic study of 1,280 cases of mental defect. *Special Report Series No. 229.* Medical Research Council, London.

21. Howard-Peebles, P.N. (1982) Non-specific X-linked mental retardation: background, types, diagnosis and prevalence. *J. Ment. Defic. Res.* **26,** 205–213.

22. Lubs, H.A. (1969) A marker X chromosome. *Am. J. Hum. Genet.* **21,** 231–244.

23. Lubs, H.A., Chiurazzi, P., Arena, J., Schwartz, C., Tranebjerg, L., and Neri, G. (1996) XLMR genes: update 1996. *Am. J. Med. Genet.* **64,** 147–157.

24. Martin, J.P. and Bell, J. (1943) A pedigree of mental defect showing sex-linkage. *J. Neurol. Psychiatry* **6,** 154–157.

25. Sherman, S.L., Morton, N.E., Jacobs, P.A., and Turner, G. (1984) The marker (X) syndrome: a cytogenetic and genetic analysis. *Annu. Hum. Genet.* **48**, 21-37.

26. Sherman, S.L., Jacobs, P.A., Morton, N.E., Froster-Iskenius, U., Howard-Peebles, P.N., Nielsen, K.B., Partington, M.W., Sutherland, G.R., Turner, G., and Watson, M. (1985) Further segregation analysis of the fragile X syndrome with special reference to transmitting males. *Hum. Genet.* **69**, 289-299, Erratum: *Hum. Genet.* **71**, 184-186.

27. Steinbach, P., Barbi, G., and Boller, T. (1982) On the frequency of telomeric chromosomal changes induced by culture conditions suitable for fragile X expression. *Hum. Genet.* **61**, 160-162.

28. Sutherland, G.R. and Baker, E. (1990) The common fragile site in band q27 of the human X chromosome is not coincident with the fragile X. *Clin. Genet.* **37**, 167-172.

29. Sutherland, G.R. and Baker, E. (1992) Characterization of a new rare fragile site easily confused with the fragile X. *Hum. Mol. Genet.* **1**, 111-113.

30. Hirst, M.C., Barnicoat, A., Flynn, G., Wang, Q., Daker, M., Buckle, V.J., Davies, K.E., and Bobrow, M. (1993) The identification of a third fragile site, FRAXF, in Xq27-q28 distal to both FRAXA and FRAXE. *Hum. Mol. Genet.* **2**, 197-200.

31. Mitelman, F., ed. (1995) *ISCN(1995): An International System for Human Cytogenetic Nomenclature.* S. Karger, Basel.

32. Hook, E.B. (1977) Exclusion of chromosomal mosaicism: table of 90%, 95% and 99% confidence limits and comments on use. *Am. J. Hum. Genet.* **29**, 94-97.

33. Brown, W.T., Jenkins, E., Neri, G., Lubs, H., Shapiro, L.R., Davies, K.E., Sherman, S., Hagerman, R., and Laird, C. (1991) Conference report: fourth international workshop on the fragile X and X-linked mental retardation. *Am. J. Med. Genet.* **38**, 158-172.

34. Oberlé, I., Rousseau, F., Heitz, D., Kretz, C., Devys, D., Hanauer, A., Boué, J., Bertheas, M.F., and Mandel, J.-L. (1991) Instability of a 550-base pair DNA segment and abnormal methylation in fragile X syndrome. *Science* **252**, 1097-1102.

35. Verkerk, J.M.H., Pieretti, M., Sutcliffe, J.S., Fu, Y.-H., Kuhl, D.P.A., Pizzuti, A., Reiner, O., Richards, S., Victoria, M.F., Zhang, F., Eussen, B.E., van Ommen, G.-J.B., Blonden, L.A.J., Riggins, G.J., Chastain, J.L., Kunst, C.B., Galjaard, H., Caskey, C.T., Nelson, D.L., Oostra, B.A., and Warren, S.T. (1991) Identification of a gene (*FMR-1*) containing a CGG repeat coincident with a breakpoint cluster region exhibiting length variation in fragile X syndrome. *Cell* **65**, 905-914.

36. Yu, S., Pritchard, M., Kremer, E., Lynch, M., Nancarrow, J., Baker, E., Holman, K., Mulley, J.C., Warren, S.T., Schlessinger, D., Sutherland, G.R., and Richards, R.I. (1991) Fragile X genotype characterized by an unstable region of DNA. *Science* **252**, 1179-1181.

37. Hummerich, H. and Lehrach, H. (1995) Trinucleotide repeat expansion and human disease. *Electrophoresis* **16**, 1698-1704.

38. Campuzano, V., Montermini, L., Moltò, M.D., Pianese, L., Cossée, M., Cavalcanti, F., Monros, E., Rodius, F., Duclos, F., Monticelli, A., Zara, F., Cañizares, J., Koutnikova, H., Bidichandani, S.I., Gellera, C., Brice, A., Trouillas, P., De Michele, G., Filla, A., De Frutos, R., Palau, F., Patel, P.I., Di Donato, S., Mandel, J.-L., Cocozza, S., Koenig, M., and Pandolfo, M. (1996) Friedreich's ataxia: autosomal recessive disease caused by an intronic GAA triplet repeat expansion. *Science* **271**, 1423-1427.

39. Zoghbi, H.Y. (1996) The expanding world of ataxins. *Nat. Genet.* **14**, 237-238.

40. Ashley, C.T. and Warren, S.T. (1995) Trinucleotide repeat expansion and human disease. *Annu. Rev. Genet.* **29**, 703-728.

41. Eichler, E.E., Richards, S., Gibbs, R.A., and Nelson, D.L. (1993) Fine structure of the human FMR1 gene. *Hum. Mol. Genet.* **2**, 1147-1153.

42. Prior, T.W., Papp, A.C., Snyder, P.J., Sedra, M.S., Guida, M., and Enrile, B.G. (1995) Germline mosaicism at the fragile X locus. *Am. J. Med. Genet.* **55**, 384-386.

43. Maddalena, A., Schneider, N.R., and Howard-Peebles, P.N. (1997) Fragile X syndrome. In *The Molecular and Genetic Basis of Neurological Disease,* 2nd edition (Rosenberg,

R.N., Prusiner, S.B., DiMauro, S., and Barchi, R.L., eds.), Butterworths, Stoneham, MA, pp. 81–99.

44. Siomi, H., Siomi, M.C., Nussbaum, R.L., and Dreyfuss, G. (1993) The protein product of the fragile X gene, *FMR1*, has characteristics of an RNA-binding protein. *Cell* **74**, 291–298.

45. Ashley, C.T., Wilkinson, K.D., Reines, D., and Warren, S.T. (1993) *FMR1* protein: conserved RNP family domains and selective RNA binding. *Science* **262**, 563–566.

46. Eichler, E.E., Holden, J.J.A., Popovich, B.W., Reiss, A.L., Snow, K., Thibodeau, S.N., Richards, C.S., Ward, P.A., and Nelson, D.L. (1994) Length of uninterrupted CGG repeats determines instability in the *FMR1* gene. *Nat. Genet.* **8**, 88–94.

47. Nolin, S.L., Lewis, F.A. III, Ye, L.L., Houck, G.E. Jr., Glicksman, A.E., Limprasert, P., Li, S.Y., Zhong, N., Ashley, A.E., Feingold, E., Sherman, S.L., and Brown, W.T. (1996) Familial transmission of the FMR1 CGG repeat. *Am. J. Hum. Genet.* **59**, 1252–1261.

48. Ashley, C.T., Sutcliffe, J.S., Kunst, C.B., Leiner, H.A., Eichler, E.E., Nelson, D.L., and Warren, S.T. (1993) Human and murine *FMR-1*: alternative splicing and translocational initiation downstream of the CGG-repeat. *Nat. Genet.* **4**, 244–251.

49. Bakker, C.E., Verheij, C., Willemsen, R., van der Helm, R., Oerlemans, F., Vermey, M., Bygrave, A., Hoogeveen, A.T., Oostra B.A., Reyniers, E., De Boulle, K., D'Hooge, R., Cras, P., van Velzen, D., Nagels, G., Martin, J.-J., De Deyn, P.P., Darby, J.K., and Willems, P.J. (The Dutch-Belgian Fragile X Consortium) (1994) *Fmr1* knockout mice: a model to study fragile X mental retardation. *Cell* **78**, 23–33.

50. Oostra, B.A. (1996) FMR1 protein studies and animal models for fragile X syndrome. In *Fragile X Syndrome: Diagnosis, Treatment, and Research*, 2nd edition (Hagerman, R.J. and Cronister, A., eds.), The Johns Hopkins University Press, Baltimore, pp. 193–209.

51. Korneluk, R.G. and Narang, M.A. (1997) Anticipating anticipation. *Nat. Genet.* **15**, 119.

52. Siomi, M.C., Siomi, H., Sauer, W.H., Srinivasan, S., Nussbaum, R.L., and Dreyfuss, G. (1994) FXR1, an autosomal homologue of the fragile X mental retardation gene. *EMBO J.* **14**, 2401–2408.

53. Zhang, Y., O'Connor, J.P., Siomi, M.C., Srinivasan, S., Dutra, A., Nussbaum, R.L., and Dreyfuss, G. (1995) The fragile X mental retardation syndrome protein interacts with novel homologs FXR1 and FXR2. *EMBO J.* **14**, 5358–5366.

54. Coy, J.F., Sedlacek, Z., Bächner, D. Hameister, H., Joos, S., Lichter, P., Delius, H., and Poustka, A. (1995) Highly conserved 3′ UTR and expression pattern of FXR1 points to a divergent gene regulation of FXR1 and FMR1. *Hum. Mol. Genet.* **4**, 2209–2218.

55. Rousseau, F., Heitz, D., Biancalana, V., Blumenfeld, K., Kretz, C., Boué, J., Tommerup, N., Van Der Hagen, C., DeLozier-Blanchet, C., Croquette, M.-F, Gilgenkrantz, S., Jalbert, P., Voelckel, M.-A., Oberlé, I., and Mendel, J.-L. (1991) Direct diagnosis by DNA analysis of the fragile X syndrome of mental retardation. *N. Engl. J. Med.* **325**, 1673–1681.

56. Fryns, J.P., Haspeslagh, M., Dereymaeker, A.M., Volcke, P., and Van den Berghe, H. (1987) A peculiar subphenotype in the fra(X) syndrome: extreme obesity, short stature, stubby hands and feet, diffuse hyperpigmentation. Further evidence of disturbed hypothalamic function in the fra(X) syndrome? *Clin. Genet.* **32**, 388–392.

57. Hull, C. and Hagerman, R.J. (1993) A study of the physical, behavioral, and medical phenotype, including anthropometric measures of females with fragile X syndrome. *AJDC* **147**, 1236–1241.

58. Loesch, D.Z., Hay, D.A., and Mulley, J. (1994) Transmitting males and carrier females in fragile X-revisited. *Am. J. Med. Genet.* **51**, 392–399.

59. Hagerman, R.J., Staley, L.W., O'Conner, R., Lugenbeel, K., Nelson, D., McLean, S.D., and Taylor, A. (1996) Learning-disabled males with a fragile X CGG expansion in the upper premutation size range. *Pediatrics* **97**, 122–126 (pp. 8–12).

60. Sobesky, W.E., Pennington, B.F., Porter, D., Hull, C.E., and Hagerman, R.J. (1994) Emotional and neurocognitive deficits in fragile X. *Am. J. Med. Genet.* **51**, 378–385.

61. Reiss, A.L., Freund, L., Abrams, M.T., Boehm, C., and Kazazian, H. (1993) Neuro-behavioral effects of the fragile X premutation in adult women: a controlled study. *Am. J. Hum. Genet.* **52,** 884–894.

62. Hagerman, R.J. (1996) Physical and behavioral phenotype. In *Fragile X Syndrome: Diagnosis, Treatment, and Research,* 2nd edition (Hagerman, R.J. and Cronister, A., eds.), The Johns Hopkins University Press, Baltimore, pp. 3–87.

63. Bennetto, L. and Pennington, B.F. (1996) The neuropsychology of fragile X syndrome. In *Fragile X Syndrome: Diagnosis, Treatment, and Research,* 2nd edition (Hagerman, R.J. and Cronister, A., eds.), The Johns Hopkins University Press, Baltimore, pp. 210–248.

64. De Vries, B.B.A., Jansen, C.C.A.M., Duits, A.A., Verheij, C., Willemsen, R., Van Hemel, J.O., Van den Ouweland, A.M.W., Niermeijer, M.F., Oostra, B.A., and Halley, D.J.J. (1996) Variable FMR1 gene methylation of large expansions leads to variable phenotype in three males from one fragile X family. *J. Med. Genet.* **33,** 1007–1010.

65. Fisch, G.S., Simensen, R., Arinami, T., Borghgraef, M., and Fryns, J.-P. (1994) Longitudinal changes in IQ among fragile X females: a preliminary multicenter analysis. *Am. J. Med. Genet.* **51,** 353–357.

66. Hagerman, R.J. and Cronister, A., eds. (1996) *Fragile X Syndrome: Diagnosis, Treatment, and Research,* 2nd edition. The Johns Hopkins University Press, Baltimore.

67. Reyniers, E., Vits, L., De Boulle, K., Van Roy, B., Van Velzen, D., de Graaff, E., Verkerk, A.J.M.H., Jorens, H.Z.J., Darby, J.K., Oostra, B., and Willems, P.J. (1993) The full mutation in the FMR-1 gene of male fragile X patients is absent in their sperm. *Nat. Genet.* **4,** 143–146.

68. Fu, Y.H., Kuhl, D.P.A., Pizzuti, A., Pieretti, M., Sutcliffe, J.S., Richards, S., Verkerk, A.J.M.H., Holden, J.J.A., Fenwick, R.G. Jr., Warren, S.T., Oostra, B.A., Nelson, D.L., and Caskey, C.T. (1991) Variation of the CGG repeat at the fragile X site results in genetic instability: resolution of the Sherman paradox. *Cell* **67,** 1047–1058.

69. Malter, H.E., Iber, J.C., Willemsen, R., de Graaff, E., Tarleton, J.C., Leisti, J., Warren, S.T., and Oostra, B.A. (1997) Characterization of the full fragile X syndrome mutation in fetal gametes. *Nat. Genet.* **15,** 165–169.

70. Turner, G., Robinson, H., Wake, S., and Martin, N. (1994) Dizygous twinning and premature menopause in fragile X syndrome (letter). *Lancet* **344,** 1500.

71. Schwartz, C.E., Dean, J., Howard-Peebles, P.N., Bugge, M., Mikkelsen, M., Tommerup, N., Hull, C., Hagerman, R., Holden, J.J.A., and Stevenson, R.E. (1994) Obstetrical and gynecological complications in fragile X carriers: a multicenter study. *Am. J. Med. Genet.* **51,** 400–402.

72. Conway, G.S., Hettiarachchi, S., Murray, A., and Jacobs, P.A. (1995) Fragile X premutations in familial premature ovarian failure. *Lancet* **346,** 309–310.

73. Partington, M.W., Moore, D.Y., and Turner, G.M. (1996) Confirmation of early menopause in fragile X carriers. *Am. J. Med. Genet.* **64,** 370–372.

74. Black, S.H., Levinson, G., Harton, G.L., Palmer, F.T., Sisson, M.E., Schoener, C., Nance, C., Fugger, E.F., and Fields, R.A. (1995) Preimplantation genetic testing (PGT) for fragile X (fraX)(abstract). *Am. J. Med. Genet.* **57,** A31.

75. Howard-Peebles, P.N. (1996) Successful pregnancy in a fragile X carrier by donor egg (letter). *Am. J. Med. Genet.* **64,** 377.

76. Turner, G., Webb, T., Wake, S., and Robinson, H. (1996) Prevalence of fragile X syndrome. *Am. J. Med. Genet.* **64,** 196–197.

77. Rousseau, F., Rouillard, P., Morel, M.-L., Khandjian, E.W., and Morgan, K. (1993) Prevalence of carriers of premutation-size alleles of the *FMR1* gene—and implications for the population genetics of the fragile X syndrome. *Am. J. Human Genet.* **57,** 1006–1018.

78. Spence, W.C., Black, S.H., Fallon, L., Maddalena, A., Cummings, E., Menapace-Drew, G., Bick, D.P., Levinson, G., Schulman, J.D., and Howard-Peebles, P.N. (1996) Molecular fragile X screening in normal populations. *Am. J. Med. Genet.* **64,** 181–183.

79. Reiss, A.L., Kazazian, H.H. Jr., Krebs, C.M., McAughan, A., Boehm, C.D., Abrams, M.T., and Nelson, D.L. (1994) Frequency and stability of the fragile X premutation. *Hum. Mol. Genet.* **3,** 393–398.

80. Sherman, S.L. (1995) Invited Editorial: the high prevalence of fragile X premutation carrier females: is this frequency unique to the French Canadian population? *Am. J. Hum. Genet.* **57,** 991–993.

81. Richards, R.I., Holman, K., Friend, K., Kremer, E., Hillen, D., Staples, A., Brown, W.T., Goonewardena, P., Tarleton, J., Schwartz, C., and Sutherland, G.R. (1992) Evidence of founder chromosomes in fragile X syndrome. *Nat. Genet.* **1,** 257–260.

82. Smits, A.P.T., Dreesen, J.C.F.M., Post, J.G., Smeets, D.F.C.M., de Die-Smulders, C., Spaans-van der Bijl, T., Govaerts, L.C.P., Warren, S.T., Oostra, B.A., and van Oost, B.A. (1993) The fragile X syndrome: no evidence for any recent mutations. *J. Med. Genet.* **30,** 94–96.

83. Howard-Peebles, P.N. (1994) Fragile X. In *The Cytogenetic Symposia, 1994* (Kaplan, B.L. and Dale, K.S., eds.), The Association of Cytogenetic Technologists, Burbank, CA, pp. 14-1–14-6.

84. Willemsen, R., Mohkamsing, S., De Vries, B., Devys, D., van den Ouweland, A., Mandel, J.-L., Galjaard, H., and Oostra, B. (1995) Rapid antibody test for fragile X syndrome. *Lancet* **345,** 1147.

85. Willemsen, R., Osterwijk, J.C., Los, F.J., Galjaard, H., and Oostra, B.A. (1996) Prenatal diagnosis of fragile X syndrome. *Lancet* **348,** 967–968.

86. Gedeon, A.K., Keinanen, M., Ades, L.C., Kaariainen, H., Gecz, J., Baker, E., Sutherland, G.R., and Mulley, J.C. (1995) Overlapping submicroscopic deletions in Xq28 in two unrelated boys with developmental disorders: identification of a gene near FRAXE. *Am. J. Hum. Genet.* **56,** 907–914.

87. Abrams, M.T., Doheny, K.F., Mazzocco, M.M.M., Knight, S.J.L., Baumgardner, T.L., Freund, L.S., Davies, K.E., and Reiss, A.L. (1997) Cognitive, behavioral, and neuroanatomical assessment of two unrelated male children expressing FRAXE. *Am. J. Med. Genet. (Neuropsych. Genet.)* **74,** 73–81.

88. Allingham-Hawkins, D.J. and Ray, P.N. (1995) FRAXE expansion is not a common etiological factor among developmentally delayed males. *Am. J. Hum. Genet.* **56,** 72–76.

89. Maddalena, A., Hicks, B.D., Spence, W.C., Levinson, G., and Howard-Peebles, P.N. (1994) Prenatal diagnosis in known fragile X carriers. *Am. J. Med. Genet.* **51,** 490–496.

90. Cronister, A. (1996) Genetic counseling. In *Fragile X Syndrome: Diagnosis, Treatment, and Research,* 2nd edition (Hagerman, R.J. and Cronister, A., eds.), The Johns Hopkins University Press, Baltimore, pp. 251–282.

15

Fluorescence *In Situ* Hybridization

Jan K. Blancato, Ph.D.

INTRODUCTION

Within the past decade, the cytogenetics laboratory has witnessed a major advance in diagnostic and prognostic capability with the advent of molecular cytogenetics. The product of a combination of cytogenetics and molecular biology, the technique of fluorescence *in situ* hybridization (FISH), has increased the resolution and application of traditional cytogenetics *(1)*. The availability of quality-controlled DNA probes from commercial sources has expedited the clinical use and acceptance of these tests, although most applications of this technology are still considered investigational. FISH is a technique that allows DNA sequences to be detected on metaphase chromosomes, in interphase nuclei, in a tissue section, or in a blastomere or gamete. This technique uses DNA probes that hybridize to entire chromosomes or single unique sequence genes, and serves as a powerful adjunct to classic cytogenetics. The applications of FISH include ploidy analysis, translocation and structural breakpoint analysis, microdeletion resolution, and gene mapping.

The steps of a FISH protocol are similar to those of a Southern blot hybridization but in an *in situ* experiment; the DNA is not extracted and run in a gel, but studied in its original place in the nucleus or on the chromosome. The DNA and surrounding material is first fixed. The specimen is then treated with heat and formamide to denature the double-stranded DNA, rendering it single-stranded. The target DNA is now available for binding, under appropriate conditions, to a similarly denatured and single-stranded DNA probe with a complementary sequence. The probe and target DNA then hybridize or anneal to each other in a duplex, based upon complementary base pairing. The probe DNA is tagged with a hapten such as biotin or digoxigenin. Following the hybridization, the detection of the hapten can be accomplished through use of an antibody tagged with a fluorescent dye, or the probe may be directly labeled with a fluorescent dye. Fluorescence microscopy then allows the visualization of the hybridized probe on the target material. This process is depicted in **Fig. 1**.

In situ hybridization (ISH) with radionuclide-labeled probes has been used by a select number of laboratories for more than 20 years *(2)*. Autoradiography, however, requires long exposure periods, not acceptable for clinical applications. In the mid-1980s, the concept of using biotin-labeled DNA probes was put into practice and fluorescence detection of α-satellite (centromere) sequences was made practical for the cytogenetics laboratory. Fluorescence has the advantage of providing dual- or triple-

From: *The Principles of Clinical Cytogenetics*
Edited by: S. Gersen and M. Keagle © Humana Press Inc., Totowa, NJ

FLUORESCENCE *IN SITU* HYBRIDIZATION (FISH)
TECHNICAL STEPS

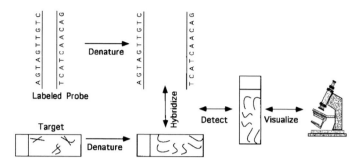

Fig. 1. Diagram of the steps of a FISH experiment. The double-stranded DNA on the microscope slide is denatured (made single-stranded) and hybridized to a labeled probe, which has also been denatured. Nonhybridized probe is washed way, and the region of study is visualized with fluorescence microscopy.

target detection through the use of different fluorescent dyes in the same experiment, and the signal intensity with fluorescence is greater than with *in situ* hybridization performed with immunochemical stains such as horseradish peroxidase. Versatile labeling systems, streamlined protocols, and vastly improved fluorescence microscopes have enabled most cytogenetics laboratories and some pathology laboratories to perform FISH tests as a part of their repertoire.

This chapter focuses only on *in situ* hybridization with DNA probes, which involves copy number analysis and investigation of structural rearrangements. The primary topic will be FISH, although nonfluorescent ISH applications are discussed when relevant. *In situ* hybridization experiments that investigate aspects of RNA and gene expression require the use of gene amplification techniques or the increased sensitivity of radionuclides, and are not discussed here. A review of these techniques can be found for various applications *(3–6)*.

FISH IN THE CYTOGENETICS LABORATORY

The Probes

There are three major categories of DNA sequences used as probes used for clinical FISH studies: satellite sequence probes, whole chromosome probes, and unique sequence probes. Examples of the use of each of these follow later in the chapter. It should be pointed out that, in addition to many commercially available probes, many disorders lend themselves to FISH analysis on a case-by-case basis with probes developed by and obtained from academic/research laboratories. With the exception of a few commercially available probes, most are not approved for use in the United States by the FDA, and for this reason laboratories must report clinical applications of FISH as investigational.

Satellite Sequence Probes

Satellite sequences are polymorphic, repetitive DNA sequences that are present in the genome but do not code for gene products. Different individuals have variations in

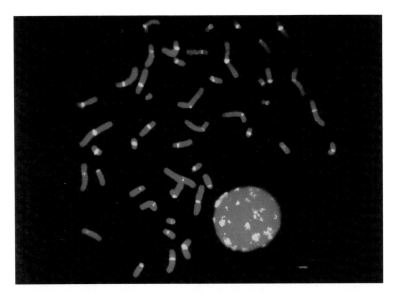

Fig. 2. A normal human metaphase spread is hybridized with a probe for total human centromere (Oncor, Inc.). The probe (bright yellow) is localized at the centromeres of all of the chromosomes (orange-red) and also hybridized to the adjacent interphase nucleus.

the number of copies of these DNAs. Alpha-satellite DNA is a 171-base-pair (bp) DNA monomer that is tandemly repeated *n* times. This entire block of tandemly repeated DNA is then copied *n* number of times in a higher order repeat at the centromere of each chromosome *(7,8)*. The majority of this DNA is identical in all human chromosomes (see **Fig. 2**), but 2–3% is variable to the degree that the centromeres of most individual chromosomes can be distinguished and probes unique to those chromosomes can be produced *(9)*. The exception is the shared homology that exists between the centromeres of chromosomes 13 and 21 and also between those of chromosomes 14 and 22, which render these pairs of chromosomes indistinguishable through centromeric FISH studies. Other satellite DNAs include β-satellite, which is a 68-bp unit that repeats in the same fashion as α-satellite DNA and is found at the tip of the acrocentric chromosomes (pairs 13, 14, 15, 21, and 22) *(10)*; classic satellite I DNA, which is an AATGG repeat found on chromosomes 1, 9, 16, and Y *(11)*; and the telomeric repeat found at the ends of both arms of all chromosomes, which is conserved over species and is composed of a TTAGGG that repeats *n* number of times *(12)*.

Repeat sequence probes are useful for determining the number of specific chromosomes present (ploidy determination) and can be used on both interphase and metaphase cells. Simultaneous visualization of different chromosomes can be accomplished with up to three different satellite probes through the use of different colored labels or label mixing. These probes are robust because the targets are large and repeated many times, which allows the hybridization to take place rapidly and produce a very large signal. See **Fig. 3**.

Whole Chromosome Paints

Whole chromosome probes, also known as whole chromosome "paints" (WCPs) are composed of numerous unique and repetitive sequences, each derived from one entire

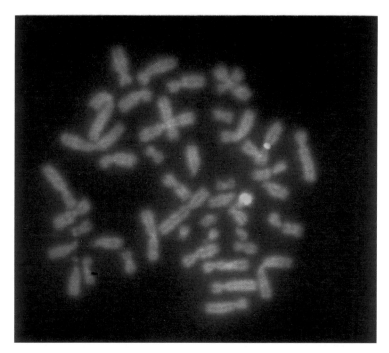

Fig. 3. Dual-labeled satellite probes are used to observe the Texas Red labeled X α-satellite and the FITC labeled Y classic satellite (green) in this DAPI counterstained blue metaphase preparation. (Image provided by Oncor, Inc.)

chromosome *(13)*. They can be produced using somatic cell hybrids, via flow sorting of the specific chromosome, or by microdissection of specific chromosomes with subsequent PCR amplification of the DNA *(14,15)*. Whole chromosome probes are also called painting probes because of the painted appearance of the metaphase chromosomes after hybridization (see **Fig. 4**). These probes are designed for use on metaphase chromosome preparations only; their use in interphase results in splotchy, undefined signals because the interphase chromatin to which they hybridize is decondensed, as opposed to the compact, condensed state of the metaphase chromosome.

Whole chromosome paints for all human chromosomes are used to determine the composition of marker chromosomes; to detect subtle or cryptic translocations; to confirm the presence of deletions, duplications, and insertions; and in the analysis of complex rearrangements involving multiple chromosomes *(16)*.

Chromosome arm-specific probes represent a subset of WCPs, made from the individual short and long arms of each chromosome. These are used for similar purposes, but allow the focus to be narrowed to one chromosome arm.

As the facility of probe development increases, laboratories are devising probe systems that allow investigation of specific abnormalities. These probes may be composed of yeast artificial chromosomes (YACs); contiguous cosmids; or chromosome arms, bands, or sub-bands that have been isolated through microdissection. Depending on the size of the probe and the specific application, these can be used on interphase and/or metaphase preparations. The simplest use of such probes is in screening cells for the presence of monosomy or trisomy for a specific chromosome arm or region,

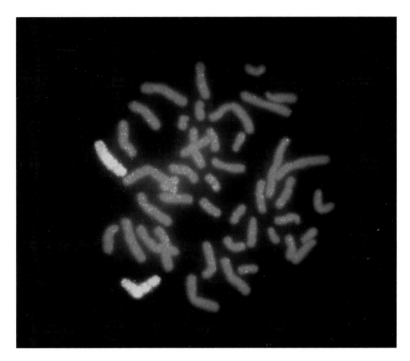

Fig. 4. Chromosome painting. A whole chromosome probe for chromosome 7 (Oncor, Inc.) is used to identify both homologues (yellow). The probe is detected with a fluorescein labeled antibody, while the chromosomes are counterstained with propidium iodide. These probes are suitable for analysis of metaphase chromosomes.

often the case in tumor cells. This may be related to loss of heterozygosity or deletion of a dominant tumor suppressor gene.

Another example is the use of probes customized for the 11q23 region associated with the *MLL (ALL–1/HRX)* gene *(17)*. Rearrangements involving this gene are associated with acute monoblastic leukemia and acute lymphoblastic leukemia, and may involve subtle translocations that can be detected in metaphase cells or interphase nuclei via FISH with YAC probes to this region *(18)*.

A FISH probe for the region 5q31 composed of 15 individual nonoverlapping probes is used by some laboratories for analysis of del(5q) where the band 5q31 appears to be a critical region and is thought to contain a tumor suppressor gene involved in malignant myeloid neoplasms. The pericentric inversion of chromosome 16 with occasional deletion of a region of 16p is associated with acute myeloid leukemia. These cytogenetic abnormalities can be detected using FISH probes made from cosmid contigs and YACS *(19)*.

Unique Sequence Probes

Unique sequence probes have target regions that are not repeated in the genome and may code for genes. The various sorts of FISH unique sequence probes currently used in clinical cytogenetics laboratories are aimed at identification of microdeletion syndromes; oncogenes such as n-*myc*, c-*myc*, and *her-2-neu*; and unique sequences in subtelomeric regions.

Subtelomeric probes can be produced from the chromosomal regions proximal to the telomere and contain unique sequences that are specific for each chromosome. Unique sequences in close proximity to the ends of chromosomes are used for studies of cryptic translocations and for gene mapping *(20)*.

The Labels

Although some success has been demonstrated with *in situ* hybridization using a light microscopy to detect horseradish peroxidase and other immunocytochemical reagents, probes are typically visualized with fluorescence microscopy. Fluorochromes or fluorophores such as fluorescein, rhodamine, or coumarin are molecules that can be bound to organic matter and are used to visualize the target DNA by absorbing light of one wavelength and reradiating it at another.

In general, only α-satellite probes can be easily detected without the use of fluorescent labels. Unique sequence *in situ* hybridization probes cannot be easily resolved when using light microscopy. However, there are cases in which light microscopic detection becomes desirable, such as when autofluorescent material is analyzed. This occurs when specimens are of certain tissue types or when plastic embedding reagents or coverslips that have fluorescent capacity have been used. This internal fluorescence interferes with the discernment of the label. Also, some pathology specimens or research studies in which adjunct experiments involve the use of antibodies detected with light microscopy are best coupled with transmitted light *in situ* hybridization. Finally, some laboratories exclusively use centromeric probes with transmitted light ISH systems, because updating the microscope system to fluorescence is not possible owing to budget constraints.

Two methods of probe labeling, direct and indirect, are available. In the direct method the reporter molecule, usually a fluorescent dye, is bound directly to the nucleic acid probe. The bound probe and target can then be visualized with fluorescence microscopy. Probes can be directly labeled in different ways, but this typically involves chemically binding a fluorescent tag into a nucleoside triphosphate, and then using polymerases and other enzymes to replace some of the probe's nucleosides with labeled ones *(16)*.

Indirect labeling involves similarly tagging the probe, but in this case with an intermediate molecule that can then be coupled to the fluorescent dye after hybridization, usually be employing an antigen–antibody or similar reaction. Two labels commonly used for this purpose are biotin and digoxigenin. Biotin is popular because it binds to avidin or streptavidin with high affinity, and either of these can easily be modified to incorporate a fluorochrome. Antibodies to digoxigenin are also available for detection of this hapten, and either system can also be used in a nonfluorescent fashion.

One of the advantages to FISH is the fact that different probes can be labeled with different colors. Two or three distinguishable colors can be used concurrently to study various target regions of interest. Dual- and triple-color labeling and detection schemes are gaining popularity because of this flexibility in labeling reagents.

Visualization of the surrounding chromatin or nuclear material is accomplished through use of counterstains. This allows the microscopist to scan for cells under low power and have a frame of reference for analysis of the DNA sequences of interest. The common counterstains used in FISH studies are propidium iodide and diamidino-2-

phenylindole (DAPI). Both are DNA intercalators, and fluoresce under similar wavelengths as the commonly used fluorochromes Fluorescein, Texas Red, Rhodamine, Spectrum Orange,* and Spectrum Green.* Either can be used. A general rule of thumb is that when working primarily with a yellow dye such as fluorescein, it is better to use the orange propidium iodide dye as a counterstain. When using a red fluorochrome such as Texas Red or Rhodamine (or in a dual-label study, which uses a red and green/yellow dye) the blue DAPI counterstain is best, as it may be difficult to see a red dye on an orange background. The use of DAPI as a counterstain provides an additional advantage, because DAPI produces a banding pattern that facilitates identification of chromosomes.

When performing studies with fluorochromes, an antibleaching agent is commonly used to preserve the signal during storage and microscopy and photography. Quenching (fading) of fluorochromes upon excitation is a photochemical process; light-induced damage in the presence of oxygen or non-oxygen-generated radicals is a source of photochemical fluorochrome destruction. Mounting media containing diphenylene diamine or other similar agents, which act as radical scavengers and/or antioxidants, prevent quenching without altering experimental results.

The Specimen

The stability of DNA allows FISH to be performed on most types of clinical specimens. Potentially, any cell with a nucleus can be examined using FISH techniques. Situations in which DNA is degraded, such as in tissues from spontaneous abortions or stillbirths, are the most challenging. FISH has been performed on specimens ranging from archival pathology specimens to bladder washed epithelial cells, uncultured amniocytes, or routine cytogenetic preparations. The specimen used depends upon the clinical application needed.

CLINICAL APPLICATIONS OF FISH

One of the benefits of FISH is its flexibility in providing clinical information that is otherwise difficult or impossible to obtain, or that can take longer to obtain with standard test methods. This technology can provide such assistance in microdeletion analysis, identification of marker chromosomes, characterization of structural rearrangements, analysis of gene rearrangements associated with neoplasia, ploidy analysis for both prenatal and tumor diagnoses, monitoring of unlike gender bone marrow transplants, preimplantation analysis, and gene amplification studies. In addition, innovative applications of FISH continue to be developed.

Microdeletion Syndromes

There are an ever-increasing number of syndromes characterized by deletions that may or may not be visible with high-resolution karyotyping. These are referred to as microdeletion syndromes. High-resolution chromosome analysis can sometimes detect these deletions, but in most cases they are either too small to be seen microscopically or involve a cryptic translocation. FISH, on the other hand, can resolve submicroscopic deletions to a lower limit of approximately 3 Mb *(21),* and is rapidly becoming the method of choice for diagnosis of these disorders. Interest in this category of genetic

*Spectrum Orange and Spectrum Green are trademarks of Vysis, Inc.

Table 1
Microdeletion Syndromes and Percentage of Cases Resolved with FISH

Microdeletion syndrome	Chromosomal location	Probe	Percentage detected with FISH
Prader–Willi	(15q11-13)	*SNRPN*	70%[63]
Angelman	(15q11-1 3)	*D15S12*	70%[63]
Miller–Dieker	(17p13.3)	*LIS-1*	90%[63]
Isolated lissencephaly sequence	(17p13.3)	*LIS-1*	33%[63]
DeGeorge	(22q11 .2)	*D22S75(N25)*	85%[64]
Velocardiofacial (VCFS)	(22q11.2)	*D22S75(N25)*	53%[64]
Williams	(7 q11.23)	*LIM* kinase *Elastin* gene	97%[65]
Smith–Magenis	(17p11.2)	*D17S258*	95%[66]
Rubinstein Taybi	(16p13.3)	Cosmids *RT1* and N2	25%[67]
X-Linked icthyosis	(Xp22.3)	Steroid sulfatase gene	85%[68]
Kallmann	Xp22.3	*KAL* gene	Unknown[68]
X-Linked ocular albinism	Xp22.3	*DXS1140*	Unknown[68]
Wolf–Hirschorn	4p16.3	*WHSCR* 165 Kb	presumably >95%[69]

diseases has grown over the past 10 years as research has focused on the critical region or smallest possible deletion that results in these rather complex phenotypes. Genetic theory involving uniparental disomy and genomic imprinting (see Chapter 8) has evolved in large part through study of the chromosomal region responsible for Prader–Willi and Angelman syndromes. Of significance is the fact that these microdeletion disorders often can now be quite easily diagnosed with the use of FISH probes for the loci that are deleted in affected individuals.

The most useful probes for these analyses are cosmid clones that range from 30 to 50 kb in size, or are cosmid contigs, collections of contiguous cosmid DNAs, that span the region of interest up to 80–100 kb. **Table 1** shows a list of syndromes detectable by such probes, the chromosomal location of the microdeletion, and the percentage of cases that can be resolved using FISH. See Chapter 9, Table 1, for additional clinical information regarding deletion syndromes.

The usual FISH probe for a microdeletion syndrome is a cocktail, or probe mixture, that includes a DNA sequence for the critical region plus an internal marker (control) used to confirm the identity of the target chromosome. These probes are designed for use on metaphase preparations. A normal cell would therefore show four signals when probing for an autosomal deletion syndrome: two for the control markers and two for the critical region. An abnormal metaphase spread produces three signals, as the normal chromosome generates both a control signal and one for the locus being tested, whereas on the chromosome with a deletion, only the control signal is seen (see **Fig. 6**).

Prader–Willi and *Angelman syndromes* are frequently caused by deletions of chromosome 15 at band q12. Most of these are about four Mb in size, with rare individuals exhibiting smaller deletions. Prader–Willi syndrome results from loss (or functional loss) of a paternal allele, while Angelman syndrome involves a missing maternal allele. The smallest deletion reported in a patient who manifests the Prader–Willi phenotype is one that involves the small ribonucleoprotein N gene (*SNRPN*) *(22)*.

Seventy percent of patients with Prader-Willi syndrome will exhibit a deletion, while 25% are due to maternal uniparental disomy of chromosome 15 (see Chapter 8), and 5% result from mutations on the maternal chromosome 15. Angelman syndrome can be diagnosed via FISH studies of the γ-aminobutyric acid (GABA) region, and is caused by a *de novo* maternal deletion of 15q11–13 in 70% of cases. A small number, 2–3%, are caused by paternal uniparental disomy of the chromosome 15 (see Chapter 8). The remaining 25% of cases may be familial and are likely to be the result of mutations expressed exclusively from the maternal 15. In one family, a mutation in the *UBE3A/ E6–AP* ubiquitin–protein ligase gene has been shown to be associated with Angelman syndrome *(23)*. A combination of probes for *SNRPN, GABRB3, D15S10,* and *D15S11* is often used to identify deletions in Prader–Willi/Angelman syndrome patients, as all of these loci are not always deleted in positive patients.

In DiGeorge syndrome, the gene *DGCR5* has been implicated as a regional transcriptional controller of distal genes *(24)*, although this is not considered to be a single-gene disorder. Through FISH, a deletion of region q11.2 of chromosome 22 can be detected in 90% of cases. Velocardiofacial syndrome, which has some clinical overlap with DiGeorge syndrome, has recently been associated with a deletion of the clathrin heavy chain gene in the 22q11.2 region *(25)*, and can also be diagnosed with a DiGeorge region probe.

The Smith–Magenis syndrome is a microdeletion disorder that has been associated with a deletion of chromosome 17 at band p11.2. The phenotype includes abnormal growth and development, hoarse voice, brachycephaly, midface hypoplasia, hyperactivity, and self-destructive behavior. A gene for a microfibril-associated glycoprotein is one gene deleted in most Smith–Magenis patients. This gene and its protein product are being studied to investigate its role in normal development *(26)*, and FISH probes for diagnosing deletions are available. FISH can also be used to diagnose deletions or duplications of 17p11.2 that result in two other neurologic disorders *(27,28)*. The duplication disorder results in the autosomal dominant Charcot–Marie–Tooth disease, which is characterized by peripheral neuropathy with nerve conductance abnormalities as the result of a duplication of the gene for peripheral myelin, a 22-kD protein. The deletion of a copy of the same gene results in *H*ereditary *N*europathy with liability to *P*ressure *P*alsy (HNPP). Although metaphase analysis for the duplication is not possible because a doublet signal cannot be resolved in the more highly condensed chromatin present in chromosomes, FISH studies on interphase cells have been used to successfully diagnose this disorder.

Many of the microdeletion syndromes do not appear to be contiguous gene deletion syndromes as was originally believed; the deletion of a single gene is responsible for the phenotypic effects in many. However, Williams syndrome is caused by the deletion of at least two genes, the *elastin* gene and the *LIM–kinase* gene *(29,30)* (see **Fig. 5**). Affected individuals with deletions only in the *elastin* gene have supravalvular aortic stenosis, but hypercalcaemia, mental retardation, and growth deficiency are not present.

The contiguous gene syndrome WAGR (*W*ilms tumor, *A*niridia, *G*enitourinary malformations, and mental *R*etardation) is a rare disorder. It is caused by the deletion of

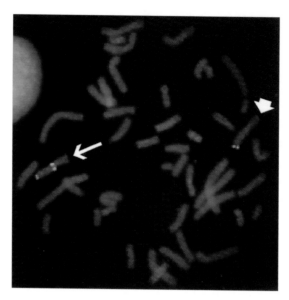

Fig. 5. Microdeletion analysis. The normal chromosome 7 (long arrow) shows yellow signals at both the telomeric control region and the Williams syndrome region at 7q11.23. The abnormal chromosome (short arrow) shows only the control signal, indicating a deletion at the Williams syndrome locus. (Photo courtesy of Dr. Gordon DeWald, Mayo Clinic. Probes from Oncor, Inc.)

two genes on chromosome 11p13, the gene *WT-1*, and the *aniridia* gene. Analyses for detection include FISH, as well as pulsed field gel electrophoresis and cytogenetic analysis *(31)*.

In the del(Xp) syndrome ("Xp–"), the entire band Xp21 may be deleted, which would certainly be analyzed with standard cytogenetics. However, this contiguous gene syndrome often causes patients to present with numerous X-linked Mendelian disorders as a result of much smaller deletions in the X short arm. For example, an affected boy may present with Duchenne muscular dystrophy, glycerol kinase deficiency, and adrenal hypoplasia *(32)*. A deletion of a restricted area of one of the genes on Xp could result in Kallman syndrome or steroid sulfatase deficiency *(33)*. FISH can be used as a diagnostic tool if any of these disorders are suspected. Deletions in female carriers can be determined through FISH if an index case is informative with the same FISH probe.

The del(4p) and del(5p) syndromes ("4p–" and "5p–") are not categorized as microdeletion syndromes because of the large size of the deletions usually present. The 4p16.3 region encompasses the locus for Wolf–Hirschhorn syndrome, a multiple anomaly syndrome with growth and mental retardation, microcephaly, cardiac septal defects, prominent glabella, and cleft lip or palate *(34,35)*. The smallest deletion of 4p16.3 that resulted in a phenotype consistent with Wolf–Hirschorn syndrome was as large as 2.2 Mb *(36)* and could be resolved with conventional cytogenetics, but FISH analysis is sometimes helpful in cases involving translocations. The del(5p) syndrome includes the locus for cri du chat ("cat cry") syndrome and a locus for laryngeal hoarseness or isolated catlike cry feature which is associated with a better prognosis *(37)*.

It is also possible that there are specific point mutations that can cause phenocopies of some of these microdeletion syndromes. Some disorders have been mapped to loci other than those commonly observed, which suggests that genetic heterogeneity also exists for these syndromes. Therefore, although the use of FISH for confirmation of diagnosis of these disorders is quite efficient, it is prudent to order a conventional karyotype first.

Because many of these disorders cannot be visualized with standard cytogenetics, the determination of possible carrier status in a parent, in the form of a cryptic translocation, is equally difficult. FISH with the appropriate probes to parental metaphase preparations can be used to diagnose such subtle rearrangements.

Structural Rearrangements

Most translocations, deletions, and other chromosome abnormalities can be readily diagnosed with standard cytogenetic analysis. However, some complex rearrangements, particularly in cancer cells, cannot easily be interpreted or cannot be interpreted at all with standard methods. FISH analysis with whole chromosome paints can usually provide the data needed to elucidate such complex changes. Whole chromosome paints are also a useful augmentation to standard chromosome analysis (frequently in a confirmatory role for improved patient care/counseling) for the demonstration or interpretation of deletions, insertions, inversions, or visible but subtle translocations. Multicolor analysis can be particularly useful here, and 24-color FISH (see later) can be used for patients with phenotypic abnormalities or for individuals experiencing reproductive difficulties to diagnose subtle rearrangements that cannot be appreciated with routine chromosome analysis.

Gene Amplification Analysis

FISH analysis using other unique sequences probes is targeted at genes that, when amplified, provide information concerning the diagnosis and prognosis of specific cancers. These include the genes for *her-2-neu, p53,* n-*myc* and c-*myc*. Amplification of *her-2-neu* is a prognostic indicator in node-negative breast cancer (see **Fig. 6**), while *p53* may be deleted in a subset of B-CLL patients. The gene n-*myc* is amplified in neuroblastoma and c-*myc* amplification is common in breast carcinoma *(38)*. For the most part, these probes are used on interphase cells in touch print preparations or paraffin-embedded tissues, and FISH in these cases is used to enhance the information from other biologic marker studies.

Gene Rearrangement (Translocation) Analysis

Translocation probes are composed of unique sequences known to be involved in fusion gene events that have been associated with leukemias and lymphomas. The identification of such a translocation can be used as both a diagnostic and a prognostic indicator and is also considered useful in patient follow-up studies. The two genes involved in the translocation event are typically located on different chromosomes and can therefore be easily visualized in metaphase studies. The power of FISH in these cases results from the fact that it performed on interphase nuclei, facilitating the analy-

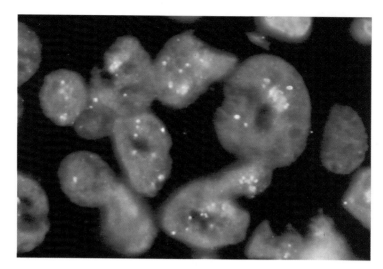

Fig. 6. An increase in copy number of the *her-2-neu (ERBB2)* oncogene on chromosome 17, shown in green, is seen in these formalin-fixed, paraffin-embedded breast cells from an infiltrating ductal carcinoma. The two red signals represent a chromosome 17 α-satellite (centromere) probe, while the nuclei are counterstained blue with DAPI. An increase in copy number of the oncogene is a manifestation of gene amplification and suggests poor prognosis. (Courtesy of Roman Giraldez, Oncor/Appligene, Heidelberg, Germany.)

sis of many more cells and providing detail concerning percentages of cells that are positive or negative for the rearrangement. Interphase FISH also does not require that the sample actually produce analyzable metaphases, often a problem with certain specimen types.

The probes are labeled with different colored fluorochromes, which results in a third fusion color when the translocation brings them into close proximity. Positive and negative controls should be used for these studies, as chromosomes can show false-positive "fusion" signals merely from being close to each other in the interphase nucleus. The most common analysis performed with translocation probes involves the *bcr/abl* gene rearrangement, which occur in the cells of almost all patients with chronic myelogenous leukemia (CML) *(39,40)*, and also in acute myeloid leukemia (AML) and acute lymphocytic leukemia (ALL). Different breakpoints in *bcr* can result in different fusion signals (see **Fig. 7**). Translocation probes are also used to visualize the promyelocytic leukemia (PML)/retinoic acid receptor alpha (RARα) translocation associated with acute promyelocytic leukemia (APL) *(41)*. More complex FISH systems can also be designed for detection of split signals where a single-color labeled DNA probe shows a large signal in the normal chromosome and two smaller signals when mutation such as deletions or translocations take place in disease states.

Bone marrow transplant follow-up studies in cases with mixed sex donor–recipient pairs can be accomplished using FISH probes to α-satellite sequences in the X and Y chromosomes. Using dual-color detection, these probes can be used to determine the proportions of male and female cells in a bone marrow or peripheral blood sample. Samples for evaluation can be scored at various time intervals following bone marrow transplantation, and low levels of chimerism can be detected using this methodology

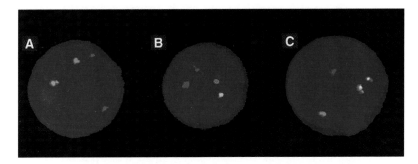

Fig. 7. Interphase FISH analysis of *bcr/abl* gene rearrangements with probes designed to differentiate between major and minor breakpoint regions. **A**: A cell without a *bcr/abl* rearrangement, showing two *abl* signals (red) and two *bcr* signals (green). **B**: Major (M-*bcr/abl*) gene rearrangement (seen in CML, AML, and ALL), with one native *bcr* signal (green), one native *abl* (large red) signal, one residual *abl* (small red) signal, and one *bcr/abl* fusion signal (yellow). **C**: Minor (m-*bcr/abl*) gene rearrangement, associated with ALL. Note the native *bcr* and *abl* signals, plus two *bcr/abl* fusion signals. (Images of *bcr/abl* ES™ probes provided by Vysis, Inc.)

because of the high sensitivity of the test and the large numbers of cells that can easily be scored.

Solid Tumor Analysis

Pathology specimens of tumor tissues that are nonmitotic can be analyzed via interphase FISH. The most commonly available specimens are formalin-fixed, paraffin-embedded tissue sections. FISH studies on these sections have the advantage of analysis of chromosomal information such as aneuploidy or amplification while retaining the architectural integrity of the tissue being studied. Numerical aberrations such as trisomy 12 (+12) in ovarian tumors, and +7 and –9 in bladder tumors can be detected with α-satellite probes on embedded material that has been specially treated *(42)*. Similarly, tumor-specific amplifications such as *her-2-neu* amplification in breast cancer and n-*myc* amplification in neuroblastomas can be recognized using unique sequence probes. It is important to properly control studies on tissue sections because of the difficulty in scoring hybridization signals in three-dimensional nuclei.

Some solid tumor FISH studies are performed on touch print preparations of the tumor tissue or on disaggregated nuclei from tissues. The touch print is performed by touching a section of unfixed tissue directly on a slide that has been coated with an adhesive, resulting in single-cell impressions. The preparation is then fixed using standard methods. This approach retains some of the advantages of the tissue section, while enabling a more straightforward FISH scoring of individual cells. Disaggregation allows analysis of difficult samples and has been used both in tumor studies and in placental tissues.

Prenatal Sample Analysis

Prenatal diagnosis is an area of cytogenetic investigation that requires rapid and accurate laboratory data. The advantage that FISH presents to studies of prenatal specimens is the rapid detection of selected chromosomal abnormalities, typically the ploidy

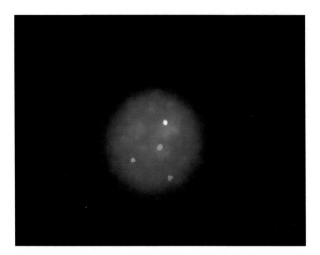

Fig. 8. Prenatal ploidy detection for chromosomes 18 (green), X (red), and Y (yellow) in an uncultured amniotic fluid cell. The pattern indicates two chromosomes 18 and an XY sex chromosome complement. (Photo provided by Dr. Avirachan Tharapel. Probes from Oncor, Inc.)

status of chromosomes 13, 18, 21, X, and Y. Standard turnaround time for cytogenetic analysis of amniotic fluid samples averages 6–10 days because of the need to culture the cells. Interphase analyses on amniocytes, chorionic villus cells, blood smears from PUBS (percutaneous umbilical blood sampling) specimens, and fetal cells from maternal blood samples have all been successful *(43–45)*. If a particular abnormality is suspected because of a family history of a translocation, an abnormal ultrasonography result, or other clinical data, the cells can be studied directly via FISH, with no culture time required.

Standard FISH has the limitation of being incapable of evaluating all that can be revealed when examining the entire chromosome complement via standard chromosome analysis, and so it continues to be important to evaluate the entire karyotype of a fetus following amniocentesis or chorionic villus sampling.

Amniotic Fluid

In a standard amnio FISH screening scenario, α-satellite probes for chromosomes 18, X, and Y and unique sequence probes for the long arms of chromosomes 13 and 21 are used. The centromeres of chromosomes 13 and 21 cannot be distinguished with α-satellite probes, and hence unique sequence, nonoverlapping cosmid contigs derived from these chromosomes are used for prenatal studies. This creates some difficulty, because such unique sequence probes produce smaller hybridization signals, and longer hybridization times are required. However, some systems utilize no α-satellite probes at all (see **Figs. 8 and 9**).

With the capability of using two fluorochromes to simultaneously visualize three colors per interphase nucleus (red, green, and red + green = yellow), the α-satellite probes can be hybridized to one cell suspension drop and the unique sequence probes on another. Alternatively, each autosome can be studied separately, along with a dual-color X and Y system.

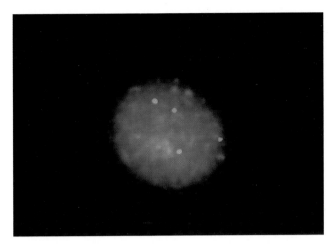

Fig. 9. Prenatal ploidy detection in a CVS cell demonstrating two copies each of chromosomes 13 (red) and 21 (green). (Photo provided by Dr. Avirachan Tharapel. Probes from Oncor, Inc.)

Uncultured amniocytes for prenatal diagnosis are typically obtained from pregnancies ranging from 13 to 18 weeks of gestation, although there are clinical reasons for employing FISH on both earlier and later samples. The successful application of interphase FISH with a probe set for chromosomes 13, 18, 21, X, and Y has been reported *(46)*, although there are some technical difficulties involved in working with amniocytes. The typical amniotic fluid sample contains many dead cells, and those that attach and grow in culture are used for standard chromosome analysis. The process of selecting the appropriate cells for FISH is important, as the denatured DNA in nonviable cells will not provide an appropriate hybridization target for the FISH probe, while maternal blood cells present in the specimen can hybridize and produce erroneous results. As laboratories continue to develop specimen preparation techniques that select for viable cells and protocols to eliminate the risk of maternal contamination, FISH studies are becoming more successful. The trend of earlier amniocentesis performed at many centers also results in a larger percentage of viable cells, which improves FISH results.

It must be remembered, however, that at this point in time prenatal ploidy detection via FISH is still considered to be an adjunct to standard chromosome analysis. Several prominent medical organizations have drafted position statements that specify that no irreversible therapeutic action be taken on the basis of prenatal FISH analysis alone.

CVS

A recent study using this strategy on uncultured mesenchymal chorionic villus cells examined more than 2900 samples with FISH and standard cytogenetic techniques *(47)*. The samples were categorized as normal or abnormal according to karyotype. Of the 103 abnormal CVS samples, 59 (57%) had abnormalities that the FISH assay of targeting chromosomes 13, 18, 21, X, and Y could identify. This type of prenatal FISH program can be considered an alternative to the more labor-intensive "direct" CVS preparation performed by some laboratories as the first analysis of such samples. However, it must be repeated that not all chromosome abnormalities will be identified with such a strategy (see **Figs. 8 and 9**).

PUBS

Perinatologists sometimes perform PUBS in high-risk pregnancies. This technique enables the physician to obtain fetal blood for various analytical needs. The fetal blood can be used to prepare a smear, and FISH for a specific suspected chromosome abnormality can be performed. Similarly, the experimental procedure of obtaining fetal nucleated blood cells from maternal blood cells employs FISH for interphase analysis *(48)*.

Preimplantation Studies

Preimplantation genetics is a research method that analyzes the genetic composition of an embryo by examining a single blastomere at the eight-cell stage. Genetic analysis of these embryos for disorders such as Lesch–Nyhan disease, cystic fibrosis, Duchenne muscular dystrophy, and others has been accomplished using gene amplification techniques *(49)*. FISH studies can also be performed on this cell if one parent is known to carry a chromosomal rearrangement, or to establish the gender of the embryo in families with sex-linked disorders. For those blastomeres that are found to be normal (or female, in the case of an X-linked disorder) after PCR or FISH studies, the remainders of the multicellular embryos can be transferred to the mother's uterus with a catheter.

A few groups have successfully devised FISH protocols for examination of the haploid genome of human sperm. This requires special pretreatment because of the tightly coiled and packed chromatin in the differentiated sperm cell *(50)*. Studies performed on these cells investigate effects of radiation and other environmental exposures on the gamete.

Marker Chromosome Analysis

An abnormal chromosome whose origin cannot be determined with standard cytogenetic analysis is called a marker chromosome (see Chapter 8). These are also referred to as extra structurally abnormal chromosomes (ESACs) or supernumerary chromosomes. Some markers are small, some are satellited or bisatellited, some are rings, and some involve the large complex rearrangements seen in tumor cells. Although it can always be clinically helpful to identify the origin of a marker chromosome, those that are found in prenatal samples or in newborns are particularly important to characterize, as the chromosomal make-up of such a marker can be of prognostic significance. FISH has become the method of choice for this analysis. Through the use of probes specific for certain chromosomes or chromosomal regions and/or whole-chromosome probes, the chromosomal origin of most markers can be deduced *(51)* (see **Fig. 10**).

EMERGING FISH TECHNOLOGIES

A number of research laboratories have been working on experimental FISH-oriented procedures that would be useful in answering some basic genetic questions. Although these are not currently available in most clinical cytogenetics laboratories, the technology is rapidly becoming available to specialized centers and this is likely to increase with time. Some of these procedures are particularly useful when no aberration is provided via standard analysis about which chromosome or gene may be in-

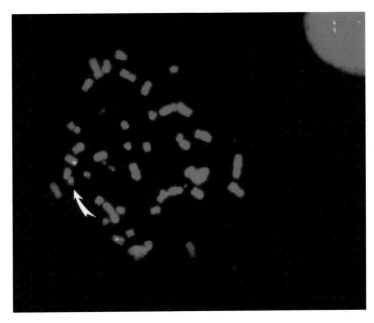

Fig. 10. Marker chromosome analysis. An α-satellite probe for chromosome 12 (yellow signals) hybridizes to the centromeres of both chromosomes 12, as well as to the centromere of the small marker chromosome present (*arrow*), indicating its origin from chromosome 12. In this case, FISH was used to confirm a diagnosis of Pallister-Killian syndrome [47, +i(12p)]. (Probes from Oncor, Inc.)

volved in a mutation event. In typical clinical FISH studies, the FISH probe is chosen because of phenotypic or genotypic clues provided by either the patient or the karyotype. When this type of specific direction is not available, such as in the case of an uncharacterized solid tumor or a blood sample with a marker chromosome, one or more of these tests may prove useful.

Comparative Genomic Hybridization (CGH)

This technique uses a reverse hybridization strategy in which the genetic material to be investigated, usually tumor tissue DNA, is used as the probe. This DNA is labeled with one color (for example FITC, a green fluorochrome) and an equal amount of normal male DNA is labeled with a different color (such as with the red dye rhodamine). The two DNA probes are then hybridized simultaneously to a metaphase preparation of normal male chromosomes. Where the tumor contains normal chromosomes, the two probes will be present in equal amounts, and the two dyes will either combine to produce a third color (yellow in this case) or will "cancel each other out" and be removed by computer. Where the tumor contains an excess of chromosomal material, such as additional chromosomes or chromosomal regions, or the gene amplification represented by double minutes or HSRs (homogeneously staining regions), the color will be green. Regions of the genome deleted in the tumor will result in the excess normal DNA probe producing a red signal. The ratio of the intensity of each fluorochrome is measured along each chromosome, usually through use of a specialized digital image analysis system with appropriate software. Comparative genomic hybridization (CGH) has

CGH: TECHNICAL STEPS

**1- Prepare genomic DNA
from the blood of a normal
control male and a tissue to
be tested.**

2- Label both differentially.

**3- Mix both together with
Cot - 1 DNA.**

4- Denature the DNA.

**5- Hybridize to a normal
male metaphase spread.**

**6-Detect with 2 different
fluorochromes and counterstain with DAPI.**

7-Acquire images.

**8- Using computer software,
obtain a ratio image and
analyze it.**

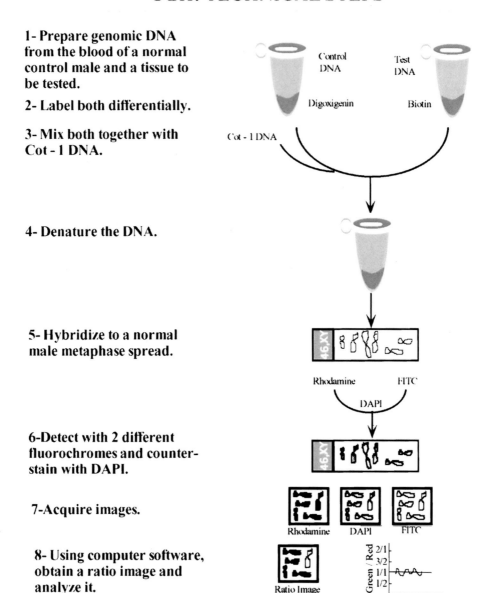

Fig. 11. The technical steps of a comparative genomic hybridization (CGH) experiment.
(Diagram courtesy of Dr. Bassem Hassad, Georgetown University.)

been instrumental in elucidating chromosome regions of interest in a number of solid
tumors, where little data were previously available *(52,53)*. The fact that actual chromosome preparations from the study tissue are not necessary is one benefit of CGH in
tumor studies, because it can often be technically difficult to grow cells or analyze the
complex karyotypes typical of cancer. CGH can also be helpful in identifying small
marker chromosomes (see **Figs. 11 and 12**).

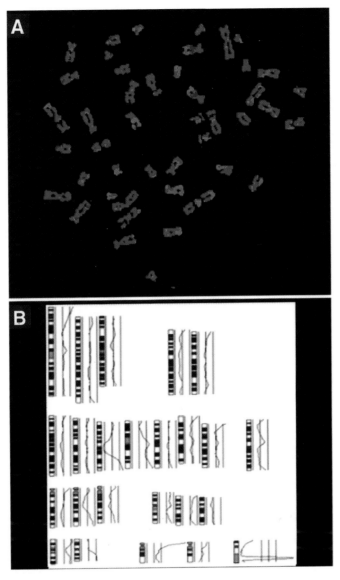

Fig. 12. CGH analysis of DNA prepared from a prostate cancer tumor. **A:** Ratio image via hybridization to a normal metaphase cell. Regions of balanced copy number are blue, whereas red indicates loss of sequences in the tumor genome and green indicates gain of sequences. Average ratio profiles are shown in **B.** The three vertical lines on the right of the chromosome ideograms represent different values of the fluorescence intensities comparing the tumor and reference genomes. This ratio profile is calculated as the mean of at least five metaphases. (courtesy of Dr. Bassem Haddad.)

24-Color FISH

FISH with multicolor probes for simultaneous visualization is highly desirable in certain situations, but the actual limit of distinguishable separate fluorochromes in standard FISH work is, at this time, three colors by using two probes (A, B, and A+B or 2^n-1). Using a third probe with a different label can produce seven colors (A, B, C,

A+B, A+C, B+C, and A+B+C or $2^3-1 = 7$). Using this approach, the reader is invited to demonstrate that labeling DNA probes with five separate fluorochromes can result in 31 combinations *(54),* more than the 24 needed to represent each human chromosome. Using Fourier spectroscopy, CCD-imaging, and computer-generated pseudo color, each chromosome pair is given a different color.

For this multicolor karyotyping, the target chromosomes are from the subject; therefore the study material must produce a reasonable metaphase preparation. The FISH probe is a cocktail of all the human chromosomes that can be individually distinguished as separate colors with special hardware and dedicated software. This analysis is capable of evaluating structural anomalies involving all chromosomes in one FISH experiment and is uniquely powerful. Its uses appear to be best suited to solid tumor cytogenetics, marker identification, and elucidation of complex or cryptic rearrangements. Because each chromosome is represented by a unique color, rearrangements (whether visible or not with standard cytogenetic analysis) and marker chromosomes can be readily characterized (see **Figs. 13 and 14**).

MicroFISH

MicroFISH or FISH following microdissection of probe DNA is another form of reverse *in situ* hybridization. Microdissection of DNA from cytogenetic preparations is performed to obtain DNA to produce a FISH probe to a whole chromosome, or arm of a chromosome, a specific band, or a marker. In microFISH studies, the DNA of interest is scraped from a number of metaphase spreads with a glass needle from a micromanipulator while the process is visualized through the microscope *(55).* The scraped DNA is then amplified to sufficient quantity with specialized polymerase chain reaction (PCR) techniques with a degenerate oligonucleotide primer system. The DNA can then be labeled and used as a probe in a FISH experiment. This has also been shown to be an effective method for FISH probe production. If the microdissected probe is derived from an unknown region such as an HSR, double minute, or other marker, the probe will bind to its chromosome region of origin when hybridized to a normal male metaphase chromosome spread. For example, if the DNA from a microdissected double minute involves the *p53* gene, the microFISH probe will hybridize to chromosome 17, the region of the genome where *p53* maps. The utility of microFISH for this purpose has been documented *(56,57).* Confirmatory experiments using DAPI fluorescence banding or cohybridization with a control probe can assist in defining the region of hybridization.

Although CGH can also be used for this purpose, it should be noted that although microFISH requires metaphase spreads where CGH does not, microFISH is more sensitive in areas of the genome with large number of repeats, such as the centromere. Also, CGH can provide data concerning additional copies of genetic material, but not necessarily the composition of one particular piece, such as a marker, HSR, or double minute.

Primed **In Situ** Labeling (PRINS)

The PRINS technique is a derivative of an *in situ* PCR experiment in which the chromosomal sequences of interest from a sample on a glass microscope slide are

SKY: TECHNICAL STEPS

1- Prepare individual chromosomal libraries from each of the 24 different human chromosomes.

2- Label each chromosomal library using 5 different fluorochromes in unique combinations.

Example:
Spectrum Green™ (A)
Cy3 (B)
Texas Red (C)
Cy5 (D)
Cy5.5 (E)

3- Prepare a probe mixture of all 24 different labeled libraries + Cot-1 DNA.

4- Denature the DNA.

5- Hybridize to metaphase spreads.

6- Acquire images.

*The emitted light is visualized through a triple band pass filter, sent through an interferometer, and imaged through a CCD camera.
*The interferogram that is generated for each pixel is analyzed by Fourier transformation, a process that makes it possible to define the spectrum of the light.

7- Classify the image.

*The measured spectra can be converted to display colors or to classification colors.
*Chromosomes with similar spectra have similar colors.

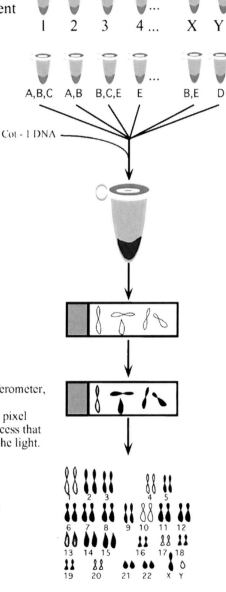

Fig. 13. Diagram showing the steps involved in one type of combinatorial multicolor FISH procedure, spectral karyotyping (SKY™). (Courtesy of Dr. Bassem Haddad.)

amplified via a PCR reaction. The reaction incorporates fluorochrome-labeled nucleotides, which, when amplified with specific oligonucleotide primers and subsequent primer extension by Taq polymerase *(58,59)*, allows for the direct detection on the chromosome with fluorescence microscopy (see **Fig. 15**).

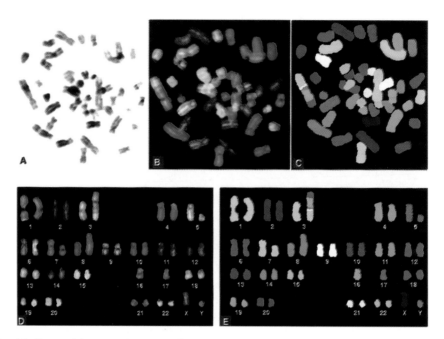

Fig. 14. Spectral karyotyping™ analysis of a bone marrow sample with complex chromosome abnormalities. **A**: DAPI banded image. **B**: Original color metaphase. **C**: Classified 24-color metaphase. **D**: Color karyotype. **E**: 24-color karyotype. Note the derivative chromosome 3, which can be seen to contain segments of five different chromosomes. Other abnormal chromosomes are also readily identified. (Courtesy of V. R. Babu, Ph.D. and V. Sukova, Quest Diagnostics, Inc.)

PRINS experiments can be completed rapidly and have been used most effectively with α-satellite sequences in both interphase and metaphase cells. With PRINS it is possible to distinguish between the α-satellite sequences of chromosomes 13 and 21 because of the specificity of the primers used in the PCR reaction, whereas standard FISH with α-satellite probes does not. In fact, large target probes on the q arms of chromosomes 13 and 21 are recommended for detection of aneuploidies involving these chromosomes, and both require overnight hybridization.

PRINS is a relatively simple, inexpensive technique relative to the other technologies presented here. It is similar to standard FISH in that the investigator must identify the probe or primer of interest, but because of constraints of the PCR reaction only one sequence can be amplified or probed at a time.

FICTION

Fluorescence immunophenotyping and interphase cytogenetics as a tool for investigation of neoplasms (FICTION) is a technique that combines immunophenotyping or cell identification with antibodies to cell surface markers and FISH with centromeric probes on the same cells from paraffin-embedded and frozen tissue sections *(60)*. Interphase cell types from many different tissue sources are amenable to investigation through use of this technique. The studies to date using FICTION have employed the use of antibodies to various cell lineages such as granulocytes (CD15), monocytes

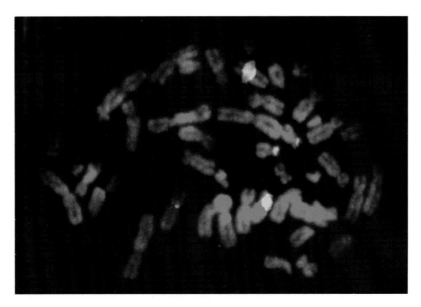

Fig. 15. Primed random *in situ* hybridization (PRINS). Metaphase chromosomes are subjected to PRINS with α-satellite oligonucleotides specific for chromosomes X, 11, and 17. Bright yellow fluorescein staining is seen at the centromeres of these chromosomes. See text for details. (Courtesy of Drs. Steen Kolvraa and Lars Bolund, Aarhus University, Sweden.)

(CD14), T lymphocytes (CD3), B lymphocytes (CD 20), and the involvement of trisomy 8 and monosomy 7 in myelodysplastic syndromes *(61)*.

The demonstration of the "Philadelphia" translocation through use of dual-color FISH in paraffin-embedded sections of bone marrow from patients with chronic myelogenous leukemia has also been combined with immunochemical detection of membrane-bound or cytoplasmic antigens. These studies demonstrated fusion of *bcr* and *abl* in megakaryocytes with Factor VIII and cells positive for CD 45. These types of experiments may help to provide information about specific chromosome aberrations and the cell lines in which they originate *(62)*.

Use of DNA probes labeled with fluorescent dyes has expanded the scope of analysis in the clinical cytogenetics laboratory. The range of uses for molecular cytogenetic techniques includes hematologic malignancy, bone marrow transplantation, prenatal diagnosis, diagnosis of microdeletion syndromes, elucidation of complex and cryptic rearrangements, determination of the origin of marker chromosomes, and analysis of solid tumors from archival pathology specimens. In most cases, it remains important to first evaluate the standard G-banded karyotype before choosing a suitable FISH application. The FISH analysis then can serve as an adjunct test, a follow-up screening test, or a method for rapid evaluation of interphase cells. The new techniques, such as comparative genomic hybridization and 24-color FISH, provide a means to evaluate complex chromosomal changes in solid tumors and other cancers, which previously could not be completely analyzed. Data derived from these studies may add to our knowledge of specific chromosomal changes in tumors, which may assist in the analysis of tumor progression and also in the development of future targeted applications of FISH.

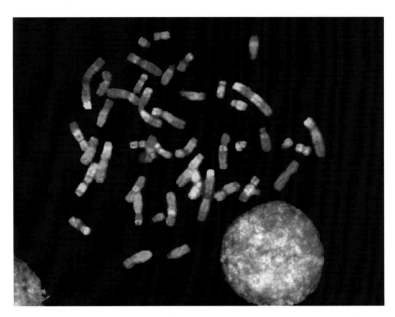

Fig. 16. Cross-species color banding (Rx FISH). Gibbon chromosomes are used as the source of DNA for multicolor FISH probes. Evolutionary divergence between species creates the observed banding pattern. See text for details. (Photo provided by Applied Imaging.)

Rx FISH (Color Banding)

Cross-species color banding is a specialized application of combinatorial-labeled FISH that provides a colored banding pattern on human metaphase chromosomes. This technique utilizes a set of subregional DNA probes to distinguish each chromosome in a single metaphase *(70)*. The chromosome-specific painting probe set used for color banding is derived by flow sorting individual gibbon chromosomes, through which subsequent combinatorial labeling uses three fluorochromes to generate seven colors (see **Fig. 16**).

Probes from primate species other than human can be used in the identification of homologous human chromosomes, because only conserved (unchanged) sequences hybridize. Repetitive DNA has diverged significantly through evolution, which inhibits cross-hybridization and eliminates the need for blocking to prevent background. Relative to the human, the gibbon has a highly rearranged genome, with at least 31 translocations and numerous intrachromosomal rearrangements *(71)*. Approximately 80 differently colored syntenic blocks have been observed in the human karyotype at this time.

Color banding serves as an adjunct to standard G-banding in the analysis of complex cases with both gross and cryptic chromosomal rearrangements, or of samples with poor chromosome morphology, as is often the case with cancer specimens. Although not yet in widespread use, it will be particularly helpful in resolving intrachromosomal rearrangements such as duplications, deletions, insertions, and inversions (see Chapter 9).

ACKNOWLEDGMENT

Many thanks to Dr. Frank Greenberg for his reading of this manuscript and his helpful suggestions.

REFERENCES

1. Pinkel, D., Gray, J., Trask, B., van den Engh, G., Fuscoe, J., and van Dekken, H. (1986) Cytogenetic analysis by in situ hybridization with fluorescently labeled nucleic acid probes. *Cold Spring Harbor Symp. Quant. Biol.* **51,** 151–157.
2. Gall, J. and Pardue, M.L. (1969) Molecular hybridization of radioactive DNA to the DNA of cytological preparations. *Proc. Natl. Acad. Sci. USA* **63,** 378.
3. Angerer, L. and Angerer, R. (1993) In situ hybridization to cellular RNA with radiolabeled probes. In *In Situ Hybridization: A Practical Approach* (D.G. Wilkinson, ed.), IRL Press, Oxford, England, pp. 15–32.
4. Hockfield, S., Carlson, S., Evans, C., Levitt, P., Pintar, J., and Silberstein, L. (1993) *Selected Methods for Antibody and Nucleic Acid Probes.* Cold Spring Harbor Press, Cold Spring Harbor, New York.
5. Nuovo, G.J. (1994) *PCR In Situ Hybridization.* Raven Press, New York.
6. Isaacson, S., Asher, D., Gajdusek, C., and Gibbs, C. (1994) Detection of RNA viruses in archival brain tissue by *in situ* RT–PCR amplification and labeled probe hybridization. *Cell Vision* **1,** 25–28.
7. Jabs, E.W., Wolf, S.F., and Migeon, B.R. (1984) Characterization of a cloned DNA sequence that is present at centromeres of all human autosomes and the X chromosome and shows polymorphic variation. *Proc. Natl. Acad. Sci. USA* **81,** 4882–4888.
8. Waye, J.S. and Willard, H.F. (1985) Chromosome specific alpha satellite DNA: Nucleotide sequence analysis of the 2.0 kilobase repeat from the human X chromosome. *Nucl. Acids Res.* **12,** 2731–2734.
9. Aleixandre, C., Miller, D., Mitchell, A., Warburton, D., Gersen, S., Disteche, C., and Miller, O.J. (1987) p82H identifies sequences at every human centromere. *Hum. Genet.* **77,** 46–50.
10. Waye, J.S. and Willard, H. (1989) Human beta satellite DNA: genomic organization and sequence definition of a class of highly repetitive tandem DNA. *Proc. Natl. Acad. Sci. USA* **86,** 6250–6254.
11. Nakahori, Y., Mitani, K., Yamada, M,., and Nakagome, Y. (1986) A human Y chromosome specific repeated DNA family (DYZ1) consists of a tandem array of pentanucleotides. *Nucl. Acids Res.* **14,** 7569–7580.
12. Moyzis, R.K. (1991) The human telomere. *Sci. Am.* **265,** 48–55.
13. Cremer, T., Lichter, P., Borden, J., Ward, D.C., and Manuelidis, L. (1988) Detection of chromosome aberrations in metaphase and interphase tumor cells by in situ hybridization using chromosome-specific library probes. *Hum. Genet.* **80,** 235–246.
14. Lichter, P., Ledbetter, S.A., Ledbetter, D.H., and Ward, D.C. (1990) Fluorescence in situ hybridization with ALU and L1 polymerase chain reaction probes for rapid characterization of human chromosomes in hybrid cell lines. *Proc. Natl. Acad. Sci. USA* **85,** 9138–9142.
15. Guan, X.Y., Meltzer, P., and Trent, J. (1994) Rapid generation of whole chromosome painting probes (WCPs) by chromosome microdissection. *Genomics* **22,** 101–107.
16. Mundy, C.R., Cunningham, M.W., and Read, C.A. (1991) Nucleic acid labeling and detection. In *Essentials of Molecular Biology, A Practical Approach*, Vol. II (T.A. Brown, ed.), IRL Press, Oxford, England, pp. 57–82.
17. Talmant, P., Berger, R., Robillard, N., Mechineau-Lacroix, F., and Garand, R. (1996)

Childhood B-cell acute lymphoblastic leukemia with FAB-L1 morphology and a t(9;11) translocation involving the MLL gene. *Hematol. Cell Ther.* **38,** 265–268.

18. Martinez-Climent, J., Thirman, M., Espinosa, R., LeBeau, M.M., and Rowley, J.D. (1995) Detection of 11q23/MLL rearrangements in infant leukemias with florescence in situ hybridization and molecular analysis. *Leukemia* **9,** 1299–1304.

19. Marlton, P., Claxton, D., Liu, P., Estey, E., Beran, M., LeBeau, M., Testa, J., Collins, F., Rowley, J.D., and Siciliano, M.J. (1995) Molecular characterization of 16p deletions associated with inversion 16 defines the critical fusion for leukemogenesis. *Blood* **85,** 772–779.

20. National Institutes of Health and Institute of Molecular Medicine Collaboration. (1996) A complete set of human telomeric probes and their clinical application. *Nat. Genet.* **14,** 86–90.

21. Ledbetter, D.H. and Ballabio, A. (1995) Molecular cytogenetics of contiguous gene syndromes: mechanisms and consequences of gene dosage imbalances. In *The Metabolic and Molecular Bases of Inherited Disease,* 7th edition (Scriver, C.R., Beaudet, A., Sly, W., and Valle, D. eds.), McGraw Hill, New York, pp. 811–839.

22. Reed, M.L. and Leff, S. (1994) Maternal imprinting of human SNRPN, a gene deleted in Prader–Willi syndrome. *Nat. Genet.* **6,** 163–167.

23. Kishino, T., Lalande, M., and Wagstaff, J. (1997) UBE3A/E6-AP mutations cause Angelman syndrome. *Nat. Genet.* **15,** 70–73.

24. Budarf, M., Collins, J., Gong, W., Roe, B., Wang, Z., Bailey, L.C., Sellinger, B., Michaud, D., Driscoll, D., and Emanuel, B. (1995) Cloning a balanced translocation associated with DiGeorge syndrome and identification of a disrupted candidate gene. *Nat. Genet.* **13,** 269–278.

25. Crifasi, P., Michels, V., Driscoll, D.J., Jalal, S., and Dewald, G. (1995) DNA fluorescent probes for diagnosis of velocardiofacial and related syndromes. *Mayo Clin. Proc.* **70,** 1148–1153.

26. Zhao, A., Lee, C.C., Jiralerspong, S., Juyal, R., Lu, F., Baldidi, A., Greenberg, F., Caskey, T., and Patel, P. (1995) The gene for a human microfibril-associated glycoprotein is commonly deleted in Smith–Magenis syndrome patients. *Hum. Genet.* **44,** 589–597.

27. Pentao, L., Wise, C.A., Chinault, A., Patel, P.I., and Lupski, J.R. (1992) Charcot–Marie–Tooth Type IA duplication appears to arise from recombination at repeat sequences flanking the 1.2 Mb monomer unit. *Nat. Genet.* **2,** 292.

28. Chance, P. Alderson, M., Leppig, K., Lensch, M., Matsunami, N., Smith, B., Swanson, P., Odelberg, S., Disteche, C., and Bird, T. (1993) DNA deletions associated with hereditary neuropathy with liability to pressure palsies. *Cell* **72,** 143.

29. Nickerson, E., Greenberg, F., Keating, M., McCaskill, C., and Shaffer, L. (1995) Deletions of the elastin gene at 7q11.23 occur in ≅90% of patients with William's syndrome. *Am. J. Hum. Genet.* **56,** 1156–1161.

30. Tassabehji, M., Metcalfe, K., Fergusson, W., Carette, M., Dore, J., Donnai, D., and Read, A. (1996) LIM-kinase deleted in Williams syndrome. *Nat. Genet.* **13,** 272–273.

31. Dreschler, M., Meijers-Heijboer, S., Schneider, S., Schurich, B., Grond-Ginsbach, C., Tariverdian, G., Blankenagel, A., Kaps, Schroeder-Kurth, M., and Royer-Pokora, B. (1994) Molecular analysis of aniridia patients for deletions involving the Wilm's tumor gene. *Hum. Genet.* **94,** 331–338.

32. McCabe, E., Towbin, J.A., van den Engh, G., and Trask, B. (1992) X p21 Contiguous gene syndromes: deletion quantitation with bivariate flow karyotyping allows mapping of patient breakpoints. *Am. J. Hum. Genet.* **51,** 1277.

33. Ballabio, A., Baldini, A., Carrozzo, R., Andria, G., Bick, D., Campbell, L., Hamel, B., Ferguson-Smith, M., Gimelli, G., Fraccaro, M., Maraschio, P., Zuffardi, O., Guilio, S., and Camerino, G. (1989) Contiguous gene syndromes due to deletions in the short arm of the human X chromosome. *Proc. Natl. Acad. Sci. USA* **86,** 10,001.

34. Rief, O., Winkelman, B., and Epplen, J. (1994) Toward the complete genomic map and molecular pathology of human chromosome 4. *Hum. Genet.* **94,** 1–18.
35. Estabrooks, L., Rao, K., Driscoll, D., Crandall, B., Dean, J., Ikonen, E., Korf, B., and Aylsworth, A. (1995) Preliminary phenotypic map of chromosome 4p16 based on 4p deletions. *Am. J. Med. Genet.* **57,** 581–586.
36. Gandelman, K.-Y., Gibson, L., Stephen-Meyn, M., and Yang-Feng, T. (1992) Molecular definition of the smallest region of overlap in Wolf–Hirschhorn syndrome. *Am. J. Hum. Genet.* **51,** 571–578.
37. Gersh, M., Goodart, S.A., Pasztor, L., Harris, D., Weiss, L., and Overhaus, J. (1995) Evidence for a distinct region causing a cat-like cry in patients with 5p deletions. *Am. J. Hum. Genet.* **56,** 1404–1410.
38. Slamon, D., Clark, G., Wong, S., Levin, W., Ullrich, A., and McGuire, W. (1987) Human breast cancer: correlation of relapse and survival with amplification of HER-2/neu oncogene. *Science* **244,** 707–712.
39. Tkachuk, D.C., Westbrook, C.A., Andreef, M., Donlon, T.A., Cleary, M.L., Suryanarayan, K., Homge, M., Redner, A., Gray, J., and Pinkel, D. (1990) Detection of bcr–abl fusion in chronic myelogenous leukemia by in situ hybridization. *Science* **250,** 559–562.
40. Dewald, G., Schad, C., Christensen, E., Tiede, A., Zinmeister, A., Spurbeck, J., Thibodeau, S., and Jalal, S. (1993) The application of fluorescence in situ hybridization to detect Mbcr/abl fusion in variant ph chromosomes in CML and ALL. *Cancer Genet. Cytogenet.* **71,** 7–14.
41. Schad, C., Hanson, C., Pairatta, E., Casper, J., Jalal, S., and Dewald, G. (1994) Efficacy of fluorescence in situ hybridization for detecting PML/RARA gene fusion in treated and untreated acute promyelocytic leukemia. *Mayo Clinic Proc.* **69,** 1047–1053.
42. Wolman, S. (1995) Application of fluorescent in situ hybridization (FISH) to genetic analysis of human tumors. In *Pathology Annuals,* (Rosen/Fechner, eds.) Appleton and Lange, Stamford, CT, pp. 227–243.
43. Kuo, W.L., Tenjin, H., Segraves, R., Pinkel, D., Golbus, M., and Gray, J. (1991) Detection of aneuploidy involving chromosomes 13, 18, or 21 by fluorescence in situ hybridization (FISH) to interphase and metaphase amniocytes. *Am. J. Hum. Genet.* **49,** 112–119.
44. Ward, B., Gersen, S., Carelli, M., McGuire, N., Dackowitz, W., Weinstein, M., and Sandlin, C. (1993) Rapid prenatal diagnosis of chromosomal aneuploidies by fluorescence in situ hybridization: clinical experience with 4,500 cases. *Am. J. Hum. Genet.* **52,** 854–865.
45. Schwartz, S. (1993) Efficacy and applicability of interphase fluorescence in situ hybridization for prenatal diagnosis. *Am. J. Hum. Genet.* **52,** 851–853.
46. Klinger, K., Landes, G., Shook, D., Harvey, R., Lopez, L., Locke, P., Lerner, T., Osathanondh, R., Leverone, B., Houseal, T., Pavelka, K., and Dackowski, W. (1992) Rapid detection of chromosomal aneuploidies in uncultured amniocytes using fluorescence in situ hybridization (FISH). *Am. J. Hum. Genet.* **51,** 55–65.
47. Bryndorf, T., Christensen, B., Vad, M., Parner, J., Carelli, M., Ward, B., Klinger, K., Bang, J., and Philip, J. (1996) Prenatal detection of chromosome aneuploidies in uncultured chorionic villus samples by FISH. *Am. J. Hum. Genet.* **59,** 918–926.
48. Zheng, Y., Craigo, S., Price, C., and Bianchi, D. (1995) Demonstration of spontaneously dividing male fetal cells in maternal blood by negative cell sorting and FISH. *Prenat. Diagn.* **15,** 573–578.
49. Griffin, D., Handyside, A., Penketh, R., Winston, R., and Delhanty, J. (1991) Fluorescent in-situ hybridization to interphase nuclei of human preimplantation embryos with X and Y chromosome specific probes. *Hum. Reprod.* **6,** 101–105.
50. Coonen, E., Pieters, M., Dumoulin, J., Myer, H., Evers, J., Ramaekers, F., and Geraedts, J. (1991) Non-isotopic in situ hybridization as a method for nondisjunction studies in human spermatazoa. *Mol. Reprod. Dev.* **28,** 18–22.

51. Blennow, E., Anneren, G., Bui, T-H, Berggren, E., Asadi, E., and Nordenskjold, M. (1993) Characterization of supernumerary ring marker chromosomes by fluorescence situ hybridization. *Am. J. Hum. Genet.* **53,** 433–442.

52. Kallioniemi, A., Kallioniemi, O.P., Suder, D., Rutovitz, D., Gray, J., Waldman, F., and Pinkel, D. (1992) Comparative genome hybridization for molecular cytogenetic analysis of solid tumors. *Science* **258,** 818–821.

53. Kallioniemi, A., Kallioniemi, O.P., Piper, J., Tanner, M., Stokke, T., Chen, L., Smith, H.S., Pinkel, D., Gray, J.W., and Waldman, F.M. (1994) Detection and mapping of amplified DNA sequences in breast cancer by comparative genome hybridization. *Proc. Natl. Acad. Sci. USA* **91,** 2156–2160.

54. Schrock, E., duManoir, S., Veldman, T., Schoell, B., Weinberg, J., Ferguson-Smith, M., Ning, Y., Ledbetter, D.H., Bar-Am, I., Soenksen, D., Garini, Y., and Ried, T. (1996) Multicolor spectral karyotyping of human chromosomes. *Science* **273,** 494–497.

55. Su, Y., Trent, J., Guan, X.Y., and Meltzer, P. (1994) Direct isolation of genes encoded within a homogenously staining region by chromosome microdissection. *Proc. Natl. Acad. Sci. USA* **91,** 9121–9125.

56. Muller-Navia, J., Nebel, A., and Schleiermacher, E. (1995) Complete and precise characterization of marker chromosomes by application of microdissection in prenatal diagnosis. *Hum. Genet.* **96,** 661–667.

57. Guan, X.Y., Meltzer, P., Dalton, W., and Trent, J. (1994) Identification of cryptic sites of DNA sequence amplification in human breast cancer by chromosome microdissection. *Nat. Genet.* **8,** 155–160.

58. Koch, J., Hindkjaer, J., Mogensen, J., Kolvraa, S., and Bolund, L. (1991) An improved method for chromosome specific labeling of alpha satellite DNA in situ using denatured double-stranded DNA probes as primers in a primed in situ labeling (PRINS) procedure. *Genet. Anal. Tech. Appl.* **8,** 171–178.

59. Pellestor, F., Girardet, A., Andreo, B., and Charlieu, J. (1994) A polymorphic alpha satellite sequence specific for human chromosome 13 detected by oligonucleotide primed in situ labeling (PRINS). *Hum. Genet.* **94,** 346–348.

60. Weber-Matthiesen, K., Pressi, S., Schlegelburger, B., and Grote, W. (1993) Combined immunophenotyping and interphase cytogenetics on cryostat sections by the new FICTION method. *Leukemia* **7,** 646–649.

61. Soenen, V., Fenaux, P., Flactif, M., Lepelley, P.M., Lai, J.L., Cosson, A., and Preudhomme, C. (1995) Combined iommunophenotyping and in situ hybridization (FICTION)—a rapid method to study cell lineage involvement in myelodysplastic disorder. *Br. J. Haemotol.* **90,** 701–706.

62. Nolte, M., Werner, M., Ewing, M., Von Wasielewski, R., Wilkens, L., and Georgii, A. (1995) Demonstration of the Philadelphia translocation by fluorescence situ hybridization in paraffin sections and identification of aberrant cells by a combined FISH/immunophenotyping approach. *Histopathology* **26,** 433–437.

63. Ledbetter, D. and Ballabio, A. (1995) Molecular cytogenetics of contiguous gene syndromes. In *The Metabolic Bases of Inherited Disease,* 7th edition (Scriver, C., Beaudet, A., Sly, W., and Valle, D., eds.), McGraw Hill, New York, pp. 822–825.

64. Crifasi, P.A., Michels, V.V., Driscoll, D.J., Jala, S.M., and Dewald, G.W. (1995) DNA fluorescent probes for diagnosis of velocardiofacial and related syndromes. *Mayo Clin. Proc.* **70,** 1148–1153.

65. Nickerson, E., Greenberg, F., Keating, M., McCaskill, C., and Shaffer, L.G. (2995) Deletions of the elastin gene at 7q11.23 occur in ~90% of patients with Williams syndrome. *Am. J. Hum. Genet.* **56,** 1156–1161.

66. Zhao, Z., Lee, C., Jiralerspong-Juyal, R., Lu, F., Baldini, A., Grenberg, F., Caskey, T., and Patel, P. (1995) The gene for a human microfibrill-associated glycoprotein in Smith-Magenis syndrome patients. *Hum. Mol. Genet.* **4,** 589–597.

67. Bruening, M., Dauwerse, H., Fugazza, G., Saris, J., Spruit, L., Wijnen, H., Tommerup, N., vander Hagen C.B., Imaizuni, K., Kuroki, Y., van der Boogaard, M-J, de Pater, J., Mariman, E., Hamel, B., Himmelbauer-Frischauf, A.M., Stallings, R., Beverstock, G., van Ommen, G.-B., and Hennekam, R. (1993) Rubinstein Taybi syndrome caused by submicroscopic deletions within 16p13.3. *Am. J. Hum. Genet.* **52,** 249–254.

68. Muroya, K., Ogata, T., Masuo, N., Nagai, T., Franco, B., Ballabio, A., Rappold, G., Sakura, N., and Fukushioma, Y. (1996) Mental retardation in a boy with an interstitial deletion of Xp22.3 involving STS, KAL1, and OA1: implication for the MRX locus. *Am. J. Med. Genet.* **64,** 583–587.

69. Wright, T., Ricks, D., Denison, K., Abmayr, S., Cotter, P., Hirschorn-Keinanen, M., McDonald-McGinn, D., Somer, M., Spinner, N., Yang-Feng, T., Zackai, E., and Altherr, M. (1997) A transcript map of the newly defined 165 kb Wolf–Hirschorn syndrome critical region. *Hum. Mol. Genet.* **6,** 317–324.

70. Muller, S., Rocchi, M., Ferguson-Smith, M., and Wienberg, J. (1997) Toward a multicolor chromosome bar code for the entire human karyotype by fluorescence in situ hybridization. *Hum. Genet.* **100,** 271–278.

71. Weinberg, J. and Stanyon, R. (1995) Chromosome painting in mammals as an approach to comparative cytogenetics. *Curr. Opinion Genet. Dev.* **5,** 792–797.

16

Genomic Imprinting and Uniparental Disomy

Jin-Chen C. Wang, M.D.

INTRODUCTION

Genomic imprinting refers to the process of differential modification and expression of parental alleles. As a result, the same gene may function differently depending on whether it is maternally or paternally derived. This concept is contrary to that of the traditional Mendelian inheritance in which genetic information contributed by either parent is assumed to be equivalent.

The term "Imprinting" was coined by Crouse *(1)* to describe the modification and the selective elimination of paternal X chromosomes from somatic and germline cells of the fly *Sciara*, in which the "imprint" a chromosome bears is determined only by the sex of the parent through which the chromosome has been inherited. It has since been used in many other species, including humans *(2)*.

Evidence for the existence of genomic imprinting is manifold. Initial experimental approaches include studies in mouse embryos using nuclear-transplantation techniques *(3–7)*. These experiments involve the removal and reintroduction of pronuclei into zygotes, thus creating embryos that have either only the maternal or only the paternal genome. In parthenogenetic eggs, that is, eggs that contain two maternal pronuclei and no paternal pronucleus, fetal development is relatively good but extraembryonic tissue development is poor. In contrast, in androgenetic eggs, that is, eggs containing two paternal pronuclei and no maternal pronucleus, the development of extraembryonic tissue is good but fetal development is poor. In either case, the embryos fail to reach term. Thus, both maternal and paternal genomes are required for normal development, and it appears that, at least in mice, the maternal genome is essential for embryogenesis while the paternal genome is essential for placental development.

The human equivalents to these observations in mice are the ovarian teratoma, the complete hydatidiform mole, and the two types of triploidy—digynic triploidy and diandric partial hydatidiform mole (see Chapter 8). Ovarian teratoma is an embryonal tumor that contains tissues derived predominantly from ectodermal, but also mesodermal and endodermal germ layers. The ovarian teratoma has been shown to be parthenogenetic and contains two sets of the maternal genome and no paternal genome *(8)*. The complete mole, on the other hand, is androgenetic and contains two sets of the paternal genome and no maternal genome *(9,10)*. Studies of the parental origin of the extra haploid set of chromosomes in triploids reveal that this is maternal (digynic triploidy) when severe intrauterine growth retardation and abnormally small placentas are seen,

From: *The Principles of Clinical Cytogenetics*
Edited by: S. Gersen and M. Keagle © Humana Press Inc., Totowa, NJ

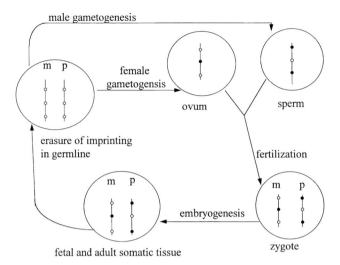

Fig. 1. Diagrammatic representation of the imprinting process. Open circles represent undermethylated genes, filled circles represent hypermethylated genes. m = maternally derived chromosome, p = paternally derived chromosome. See text for details.

while it is paternal (diandric triploidy) in partial hydatidiform moles, in which the placenta is abnormally large *(11–13)*. Intercross experiments in mice between either Robertsonian or reciprocal translocation carriers further demonstrate that maternal duplication/paternal deficiency or maternal deficiency/paternal duplication of certain mouse chromosomes or regions of chromosomes results in different phenotypic abnormalities *(14)*.

Observations of X chromosome inactivation in different species and different tissues provide further evidence of imprinting. While inactivation of the X chromosome in females of placental mammals is in general random in somatic cells *(15)*, studies in interspecies crosses between marsupials reveal that the paternally derived X chromosome is preferentially inactivated in female kangaroo somatic tissues *(16)*. In extra-embryonic tissues, the paternally derived X chromosome is preferentially inactivated in both mice *(17–19)* and humans *(20,21)*. Direct evidence that genomic imprinting exists in humans is provided by the observation of a variety of human conditions or diseases such as Prader–Willi syndrome (PWS) and Angelman syndrome (AS) and certain types of cancer. These are discussed in detail later.

Not all chromosomes or all regions of one chromosome are involved in genomic imprinting. This is true for both mice *(14,22)* and humans *(23)*.

MECHANISM

Imprinting is a phenomenon that is reversible from generation to generation. The process must therefore involve the establishment of the imprint during gametogenesis, the maintenance of the imprint through embryogenesis and in adult somatic tissues, and then the erasure of the imprint in the germline *(24,25)* (see **Fig. 1**). Thus, stable and differential modification of chromatin is required. Differential methylation of the cytosine residues of DNA on maternal and paternal chromosomes appears, at least in part, to fulfill this role.

DNA methylation is involved in human X chromosome inactivation. Using 5-azacytidine, which causes hypomethylation of DNA, Mohandas et al. were able to achieve reactivation of an inactive human X chromosome *(26).* Yen et al. showed that the human *HPRT* (hypoxanthine phosphoribosyltransferase) gene is hypomethylated on the active X chromosome relative to the inactive X *(27).* Furthermore, DNA methylation has been shown, in experiments involving gene insertion into mouse L cells, to render these sequences insensitive to both DNase I and restriction endonucleases, by directing DNA into an inactive supranucleosome structure *(28).* These observations suggest that DNA methylation may exert its effect on gene transcription by altering interactions between DNA and nuclear proteins.

The involvement of methylation in the initiation and/or maintenance of genomic imprinting has been examined extensively. Experiments with transgenic mice, in which a foreign gene is inserted into the mouse genome by microinjection, have demonstrated that some transgenes show different states of methylation specific to the parent of origin, and that the methylation pattern changes from generation to generation depending upon the sex of the parent transmitting the transgene *(29–31).* In most cases, a paternally inherited transgene is less methylated than one that is maternally inherited. In a study of transgene-bearing elements of the Rous sarcoma virus (RSV) and a fused c-*myc* gene, the paternally inherited transgene is undermethylated in all tissues and is expressed only in the heart *(31).* This observation suggests that methylation status alone does not determine the expression of a transgene, and that undermethylation may be necessary, but not sufficient, for gene expression. In this same study, the somatic organs of a male animal with a maternally inherited transgene exhibited a methylated transgene pattern, but in the testes the transgene was undermethylated, suggesting that the maternally derived methylation pattern is eliminated in the testes of male offspring during gametogenesis.

The role of DNA methylation in genomic imprinting is further demonstrated by observations made in three imprinted endogenous genes in mice: insulin-like growth factor 2 (*Igf2*), *H19* (these two genes are closely linked on mouse chromosome 7), and the Igf2 receptor gene (*Igf2r*, on mouse chromosome 17).

Studies of mouse *H19* showed that it is subject to transcriptional regulation by genomic imprinting, with the maternal allele expressed and the paternal allele silent *(32).* By comparing CpG methylation and nuclease sensitivity of chromatin in mouse embryos, Ferguson-Smith et al. *(33)* showed that hypermethylation and chromatin compaction in the region of the *H19* promoter are associated with repression of the paternally inherited copy of the gene.

Studies of the mouse *Igf2* gene showed that, contrary to *H19,* the paternal allele is expressed in embryos, whereas the maternal allele is silent, but both alleles are transcriptionally active in the choroid plexus and leptomeninges *(34).* Therefore, imprinting may also be tissue specific. In addition, studies using mouse embryos with maternal duplication and paternal deficiency of the region of chromosome 7 that encompasses *Igf2* showed that the chromatin of the 5′ region of the repressed maternal *Igf2* allele is potentially active for transcription, that is, it is hypomethylated and contains DNase I hypersensitive sites *(35).* Therefore, the parental-origin-specific difference in methylation of the *Igf2* gene may not reside at the promoters, or there may be other unidentified modifications of the promoters.

Studies of the mouse *Igf2r* gene indicated that the maternal allele is expressed and the paternal allele is silent *(36)*. The parental-origin-specific difference in methylation for this gene has been demonstrated in two distinct CpG islands *(37)*. Here, while the promoter is methylated on the inactive paternal allele, an intronic CpG island is methylated only on the expressed maternal allele, suggesting that methylation of the latter site is necessary for expression of the *Igf2r* gene.

In vivo evidence that a direct correlation is present between DNA methylation and gene activity is provided by the finding that the normally silent mouse paternal *H19* allele is activated in DNA methyltransferase-deficient embryos *(38)*. Thus, these studies demonstrate that although DNA methylation plays a critical role in genomic imprinting, the process is much more complex than simply inactivating a gene by methylation.

In humans, the methylation patterns of the parental alleles have been determined for several imprinted loci on chromosome 15 at bands 15q11–q13. These include the *ZNF127*/DN34 gene (D15S9) studied in PWS and AS patients *(39)* and in complete hydatidiform moles *(40)*, the small nuclear ribonucleoprotein polypeptide N *(SNRPN)* gene *(41,42)*, and the DNA sequence PW71 (D15S63) *(43)*. Distinct differences in methylation of the parental alleles are observed in all instances. This is also true for some of the other known imprinted genes in humans: *H19* (maternal allele active) *(44)* and *IGF2* (paternal allele active) *(45,46)*, both located on the short arm of chromosome 11 at band 11p15. In the case of *IGF2*, although it is the paternal allele that is active, the maternal allele is hypomethylated whereas the paternal allele is methylated at the 5′ portion of exon 9, again demonstrating the complexity of the involvement of methylation in genomic imprinting. Unlike that in mice, the human *IGF2R* gene is not imprinted *(47)*.

A difference in DNA replication timing of maternal and paternal alleles of imprinted genes has also been observed *(48–51)*. Cell-cycle replication timing has been shown to correlate with gene activity: genes that are expressed generally replicate earlier *(52,53)*. Furthermore, most genes on homologous chromosomes replicate synchronously *(54)*. This is not the case for imprinted genes. Using fluorescence *in situ* hybridization in interphase nuclei and scoring for the stage of the two alleles in S phase, Kitsberg et al. *(48)* showed that the imprinted genes *H19, Igf2, Igf2r*, and *Snrpn* in mice and their corresponding genes in humans all replicate asynchronously, with the paternal allele replicating early. Studies of genes in the 15q11–q13 region in humans demonstrated that most show a paternal-early/maternal-late pattern, with some exhibiting the opposite pattern *(49,50)*. Therefore it appears that imprinted genes are embedded in DNA domains with differential replication patterns, which may provide a structural imprint for parental identity *(49)*.

Thus, the process of genomic imprinting is very complex. It may involve an interaction between DNA methylation, chromatin compaction, DNA replication timing, and potentially other mechanisms *(55)*.

GENOMIC IMPRINTING AND HUMAN DISEASES

Genomic imprinting provides an explanation for the observation that the transmission of certain genetic diseases cannot be explained by traditional Mendelian inheritance, but that rather the phenotype depends upon whether the gene involved is maternally or paternally inherited. Conversely, the existence of such diseases provides evidence that genomic imprinting occurs in man. Human conditions that fall into this category include certain deletion/duplication syndromes, a number of cancers, and many situations arising from uniparental disomy.

Chromosome Deletion/Duplication Syndromes

Prader–Willi Syndrome/Angelman Syndrome

The best-studied examples of genomic imprinting in human disease are the Prader–Willi and Angelman syndromes. These are clinically distinct disorders; both map to chromosome 15q11–q13 region *(56–58)* but they involve different genes *(59,60)*. The etiologies for these disorders include (a) the absence of a parent-specific contribution of this region due to either deletion *(61–64)* or uniparental disomy (UPD) *(65–69)*, (b) mutations in the imprinting process *(70–74)*, and (c) possibly mutations within the gene *(60,75)*.

The clinical phenotype of PWS has been well characterized *(76)*. In brief, it includes hypotonia during infancy, obesity, hyperphagia, hypogonadism, characteristic facies, small hands and feet, hypopigmentation, and mental deficiency. Approximately 70% of cases have a deletion of 15q11–q13 on the paternally derived chromosome 15 *(61)*. Twenty-five percent are due to maternal uniparental disomy for chromosome 15 *(65,68)*, and 2% or so as a result of an abnormality of the imprinting process, causing a maternal methylation imprint on the paternal chromosome 15 *(72,73)*. The *SNRPN* gene remains a candidate gene for PWS *(59)*.

The clinical phenotype of AS patients is distinct from that of PWS *(77)*. In brief, it includes microcephaly, ataxia, characteristic gait, inappropriate laughter, seizures, severe mental retardation, and hypopigmentation. Approximately 70% of AS patients have a deletion of 15q11–q13 on the maternally derived chromosome 15 *(62–64)*. From 2% to 5% are due to paternal uniparental disomy for chromosome 15 *(66,67,69)*, approximately 5% as a result of an abnormality of the imprinting process, causing a paternal methylation imprint on the maternal chromosome 15 *(70–72,74)*, and 20% as a result of a mutation within the AS gene *(60,75)*. Recently, the gene for E6-associated protein (E6-AP) ubiquitin-protein ligase (*UBE3A*) (maternal allele active) has been shown to be a candidate gene for AS *(60,75)*. The imprinting of *UBE3A* appears to be tissue specific, being restricted to brain *(78–80)*.

In both PWS and AS patients with abnormalities of the imprinting process, Buiting et al. identified inherited microdeletions in the 15q11–q13 region *(81)*. They proposed that these deletions probably affect a single genetic element that they called an "imprinting center." Mutations of the imprinting center can be transmitted silently through the germline of one sex, but appear to block the resetting of the imprint in the germline of the opposite sex. Thus, it appears that there may be a central element (the "imprinting center") that controls imprint switching in the germline.

These observations in PWS and AS indicate that the PWS gene is active only on the paternal chromosome 15 and the AS gene is active only on the maternal chromosome 15. These two syndromes serve as classic examples of genomic imprinting in humans.

Beckwith–Wiedemann Syndrome

Beckwith–Wiedemann syndrome (BWS) is an overgrowth disorder associated with neonatal hypoglycemia, abdominal wall defects, macroglossia, visceromegaly, gigantism, midface hypoplasia, and a predisposition to embryonal tumors including Wilms tumor, rhabdomyosarcoma, and hepatoblastoma *(82,83)* (see next section). Most cases are sporadic. Paternal UPD for the p15.5 region of chromosome 11 has been seen in approximately 20% of sporadic cases *(84,85)*. In some patients, cytogenetic abnormalities involving 11p15 have been observed. These include duplication of the paternal 11p15 region as a result of either a *de novo* rearrangement or a familial translocation/inversion *(86,87)*, and maternally inherited balanced rearrangements involving 11p15 *(87,88)*. In familial cases, the segregation appears to be autosomal dominant with incomplete penetrance *(83)*. Furthermore, penetrance appears to be more complete with maternal inheritance, that is, there is an excess of transmitting females *(89,90)*. These parent-of-origin-dependent transmissions indicate that genomic imprinting plays a role in BWS.

Linkage studies confirm that BWS maps to 11p15 *(91,92)*. By studying UPD for various segments of chromosome 11, Catchpoole et al. *(85)* refined the critical region to a 25 cM interval between segments D11S861 and D11S2071. This region contains the imprinted genes *H19* (maternal allele active), *IGF2* (paternal allele active), and *p57KIP2* (maternal allele active) *(93,94)*. In some BWS patients who inherited an 11p15 allele from both parents, an altered pattern of allelic methylation of *H19* and *IGF2* has been reported *(85,95)*. In these patients, a paternal imprint pattern is seen on the maternal allele, which results in the nonexpression of *H19*, while *IGF2* continues to be expressed from both parental alleles. This observation suggests an "imprinting center" mutation, as in PWS/AS, which prevents the resetting of imprinting in the maternal germline and explains the observation that the affected individuals are usually born to carrier mothers in familial cases. The same explanation can be applied to the observation that inherited balanced rearrangements involving 11p15 causing BWS are usually maternal in origin: a disruption/mutation of the "imprinting center" has occurred in the process of rearrangement.

p57KIP2 is a negative regulator of cell proliferation; its overexpression arrests cells in G_1. Recently, *p57KIP2* mutations have been described in two BWS patients, in one of them the mutation was transmitted from the phenotypically normal mother *(96)*. These observations suggest that the *p57KIP2* gene and the *IGF2* gene may interact and both, and conceivably other genes, may be involved in the pathogenesis of the BWS phenotype.

Cancer

Paraganglioma

A type of nonchildhood tumor, paraganglioma of the head and neck (glomus tumor) has been mapped to chromosome 11 in the region q23→qter by linkage analysis *(97)*. Inheritance of familial glomus tumors is autosomal dominant with both males and females affected. However, transmission is almost exclusively through the father *(97–99)*. Only male gene carriers will have affected offspring. The disease is not observed in the offspring of affected females until subsequent generations, when transmission of the gene through a male carrier has occurred. These observations can be explained by genomic imprinting. The gene is inactivated during oögenesis and is reactivated only during spermatogenesis.

Wilms Tumor/Rhabdomyosarcoma

In a number of embryonal tumors, loss of heterozygosity (LOH) of a specific parental allele has been observed. In all cases studied, the maternal allele is preferentially lost. This suggests that duplication of some paternal alleles results in enhanced cell proliferation, whereas duplication of certain maternal alleles may inhibit cell proliferation.

In Wilms tumor and rhabdomyosarcoma, LOH involves chromosome 11 *(100,102)*. LOH does not involve markers for 11p13, the proposed Wilms tumor locus, but only markers on 11p15.5 *(101)*. Known imprinted genes in the 11p15.5 region include *H19*, *IGF2*, and *p57^{KIP2}* (see earlier). The expression of *p57^{KIP2}* is reduced in Wilms tumor *(94)*. In addition, by using several overlapping subchromosomal transferable fragments from 11p15 distinct from *H19* and *IGF2*, Koi et al. *(103)* were able to obtain in vitro growth arrest of rhabdomyosarcoma cells. These observations suggest that *p57^{KIP2}*, which is normally active on the maternal allele only, may be a candidate for a tumor suppressor gene. Loss of the active *p57^{KIP2}* allele on the maternal chromosome results in tumor development. Besides LOH, another possible mechanism, known as loss of imprinting (LOI), has been proposed. Ogawa et al. *(104)* reported biallelic *IGF2* RNA synthesis in 4 of 30 Wilms tumors they studied. Thus, "relaxation" of *IGF2* gene imprinting on the maternal allele has occurred, resulting in its expression. This would be equivalent to having two copies of an active *IGF2* gene, as would occur with a paternal duplication or with paternal UPD. Recently, a similar biallelic expression of *IGF2* has been reported in 30% of breast cancer patients studied *(105)*. Disruption of the imprinting mechanism, that is, LOI, may therefore also play a role in tumorigenesis. A third possible mechanism has also been proposed in a proportion of Wilms tumor patients. In some patients, LOI was observed in both the Wilms tumor tissue and the normal adjacent kidney tissue, but *IGF2* expression was significantly higher in tumor tissue. The overexpression in tumor tissue was accompanied by activation of all four *IGF2* promoters *(106)*. These studies indicate that although genomic imprinting plays an important role in tumorigenesis, a single mechanism does not account for all cases.

Retinoblastoma/Osteosarcoma

In retinoblastoma and osteosarcoma, loss of both functional copies of the retinoblastoma gene *(Rb)* on chromosome 13 at band q14 has been observed *(107)*. In familial cases, a mutation in one of the alleles is present in the germline. The *de novo* mutations in the germline occur preferentially in the paternal chromosome *(108,109)*, consistent with the general observation that new germline mutations arise predominantly during spermatogenesis. In sporadic, nonfamilial tumors, loss of function of both alleles occurs somatically. In sporadic osteosarcomas, the initial mutation occurs preferentially on the paternal chromosome 13 *(110)*, suggesting that genomic imprinting may be involved. Data are less clear in sporadic retinoblastomas. No predilection in the parental origin of the somatic allele loss was noted in some studies *(109,111)*; but a preferential loss of the maternal allele, which implies a preferential initial somatic mutation on the paternal allele, was reported in one study *(112)*. Thus, the role of genomic imprinting in retinoblastoma is unclear at this time.

Neuroblastoma

In neuroblastoma, deletions of chromosome 1p and amplification of the N-*MYC* gene on chromosome 2 are frequently seen *(113)*. Preferential amplification of the paternal N-*MYC* allele in neuroblastoma tumor tissues has been reported *(114)*; loss of parental 1p alleles was found to be random in this same study. On the other hand, preferential loss of maternal alleles in band p36 of chromosome 1 was reported in another study of neuroblastomas with a single copy of the N-*MYC* gene *(115)*. It appears that different genomic imprinting processes are involved in this neuronal tumor.

UNIPARENTAL DISOMY

The term uniparental disomy (UPD) was introduced by Engel in 1980 *(116)*. It describes a phenomenon in which both homologues (or segments) of a chromosome pair are derived from a single parent. An example of the latter is the paternal UPD for 11p15 in BWS discussed previously. Discussion here is restricted to disomies for whole chromosomes, of which there are two types. Uniparental *iso*disomy describes a situation in which both copies of a chromosome are not only derived from one parent, but also represent the same homologue (i.e., two copies of the same exact chromosome). Uniparental *hetero*disomy refers to both of one parent's homologues being represented. The type of UPD present is not always readily apparent, and it should be noted that, because of the recombination that takes place during meiosis, UPD along the length of an involved chromosome pair can be iso- for certain loci and hetero- for others.

UPD for an entire chromosome can occur as a result of gamete complementation, as suggested by Engel *(116)*. Because aneuploidy is relatively frequent in gametes, the chance union of two gametes, one hypo-, the other hyper-haploid for the same chromosome, will result in a diploid zygote with UPD for that chromosome. Structural rearrangements, such as Robertsonian or reciprocal translocations, increase the chance of meiotic malsegregation and thus may predispose to UPD. This is best illustrated by the case reported by Wang et al. *(117),* in which UPD for chromosome 14 was observed in a child with a paternal (13;14) Robertsonian translocation and a maternal (1;14) reciprocal translocation (see **Fig. 2**). Studies in animals also support this concept. Maternal or paternal disomies are readily produced in mice with intercrosses between either Robertsonian or reciprocal translocation carriers *(14)*.

Another mechanism for the occurrence of UPD is by "trisomy rescue" *(118)*. The vast majority of trisomic conceptuses are nonviable; they may survive to term only if one of the trisomic chromosomes is lost postzygotically. In one third of these cases, such loss will result in UPD in the now disomic cells (see **Fig. 3**). Mosaicism in such conceptuses is often observed, sometimes confined to the placenta (see Chapter 11). A third possible mechanism is by duplication of the single chromosome in monosomic conceptuses, thus leading to UPD *(119)*.

Two mechanisms contribute to the phenotypic effects of UPD. Unmasking of a recessive gene can occur as a result of uniparental isodisomy, in which the disomic chromosomes are homozygous. This was illustrated initially in an individual with cystic fibrosis who had maternal uniparental isodisomy for chromosome 7 *(119)*, and later in many other patients with recessive disorders and UPD (see later). The second mechanism is the effect caused by imprinted genes on the involved chromosome. This is best illustrated by PWS/AS patients who have no deletion of 15q11.2, but rather have UPD, as discussed previously. In cases where UPD arises as a result of "trisomy rescue," the presence of a mosaic trisomic cell line in the placenta and/or fetus may also modify the phenotype.

Twenty-five different maternal and paternal UPDs for different chromosomes were reviewed in a 1995 publication *(23)*, and at least four others have been described since. Some provide clear evidence for imprinting and some seem to suggest no such effect,

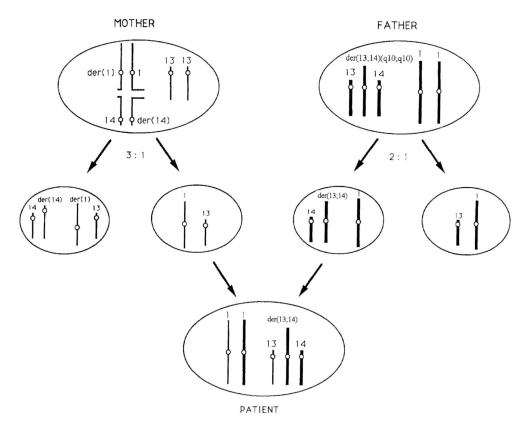

Fig. 2. An example of paternal UPD formation by gamete complementation. Malsegregation involving chromosome 14 occurred in both parents as the result of structural rearrangements [mother: reciprocal translocation t(1;14)(q32;q32); father: Robertsonian translocation der(13;14)(q10;q10)]. The patient inherited both chromosomes 14 from the father and neither from the mother. Segregation is normal for chromosome 13 in the mother and for chromosome 1 in the father.

whereas others will require accumulation of additional data before their status in this regard can be determined.

upd(1)mat

A single case of maternal UPD for chromosome 1 has been reported *(119a)*. The proband had lethal autosomal recessive Herlitz type junctional epidermolysis bullosa as a result of homozygosity for a nonsense mutation in the *LAMB3* gene on chromosome 1. The mother was a heterozygous carrier for the mutation and the father had two normal *LAMB3* alleles. The patient died at 2 months of age. Autopsy was not performed but weight and length were reportedly normal and no overt dysmorphisms or malformations were noted. It appears that maternal UPD for chromosome 1 may not have an imprinting effect.

upd(1)pat

Two cases of paternal UPD for chromosome 1 were reported recently in abstracts *(119b,119c)*. A 7-year-old boy presented with pycnodysostosis as a result of a homozy-

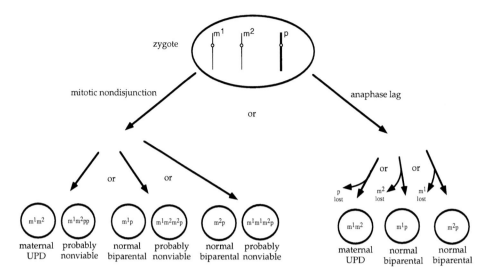

Fig. 3. A diagrammatic representation of maternal UPD formation by "trisomy rescue." A trisomic zygote resulting from maternal meiosis I nondisjunction is depicted here. Loss of one of the trisomic chromosomes through either mitotic nondisjunction or anaphase lag results in euploidy. Uniparental disomy occurs in one third of these cases. m^1 and m^2 = maternally derived chromosomes. p = paternally derived chromosome.

gous mutation of the cathepsin K gene for which the father was a heterozygote and the mother was normal. The child was otherwise developmentally normal. The second patient was a 43-year-old female with short stature, ptosis, micro/retrognathia, scoliosis, hearing loss, myopathy, and infertility. She has isochromosomes for the short arm and long arm of chromosome 1 [i(1)(p10),i(1)(q10)]. It was not clear whether the abnormal phenotype in this woman resulted from an imprinting effect or from homozygosity for recessive alleles. These observations provide no clear evidence for an imprinting effect of paternal UPD 1.

upd(2)mat

Maternal UPD for chromosome 2 has been reported in at least five cases *(120–124)*. Three cases were associated with confined placental mosaicism (CPM) for trisomy 2, and two cases resulted from *de novo* isochromosome formation of the short arm and long arm of chromosome 2 [i(2)(p10),i(2)(q10); see Chapters 3 and 9]. In one case with the isochromosomes, no phenotypic abnormalities were present *(122)*. A common phenotype was observed in the other four cases. These include intrauterine growth retardation (IUGR), oligohydramnios, pulmonary hypoplasia, hypospadias (in two patients), and normal development in the three surviving patients at ages 8 months, 31 months, and 8 years, respectively. IUGR, oligohydramnios, and pulmonary hypoplasia can be explained by placental dysfunction as a result of trisomy 2 mosaicism. However, these same features were also present in the second case with the isochromosomes *(124)*. These observations suggest a possible, although not certain, imprinting effect of maternal UPD 2.

upd(4)mat

A single case of maternal UPD for chromosome 4 as a result of isochromosome formation of the short arm and long arm of chromosome 4 [i(4)(p10),i(4)(q10)] has been reported *(125)*. Cytogenetic studies were performed because of multiple early miscarriages. The patient was otherwise phenotypically normal. There is no clear evidence to date that maternal UPD for chromosome 4 confers an imprinting effect.

upd(5)pat

Paternal UPD for chromosome 5 was reported in a child with autosomal recessive spinal muscular atrophy *(126)*. The child had no other developmental abnormalities. Spinal muscular atrophy in this case can be explained by the paternal transmission of two copies of the defective gene. Paternal UPD 5 is unlikely to have an imprinting effect.

upd(6)mat

Maternal uniparental isodisomy for chromosome 6 was identified in a renal transplant patient in the process of HLA typing *(127)*. There is no evidence for an imprinting effect.

upd(6)pat

At least six cases of paternal uniparental isodisomy for chromosome 6 have been reported *(128–132)*. Three of these had transient neonatal diabetes associated with very low birth weight *(130,132)*, and one had pancreatic β-cell aplasia and autosomal recessive methylmalonic acidemia and died at 16 days of age *(129)*. These findings suggest the possibility of an imprinted gene on chromosome 6; the maternal allele may be functionally active during fetal development of pancreatic tissue.

upd(7)mat

At least 11 patients with maternal UPD for chromosome 7 have been reported *(23,133,134)*. This was the first documented UPD in humans, identified in two individuals with cystic fibrosis and short stature *(119,135)*. Approximately 10% of patients with Silver–Russell syndrome, in which intrauterine and persistent postnatal growth retardation is a major feature, are noted to have maternal UPD 7 *(134,136)*. In the unusual case reported by Eggerding et al. *(137)*, UPD resulted from the presence of an isochromosome 7p that was paternal in origin and an isochromosome 7q that was maternal in origin. This female child also had growth retardation at 27 months. These observations suggest that an imprinting effect of chromosome 7 exists and that the lack of paternal gene(s) on 7q may cause growth retardation.

upd(7)pat

One case of paternal UPD for the entire chromosome 7 has been reported *(138)*. This patient, who also had recessive congenital chloride wasting diarrhea, had normal growth and development. In addition, one patient had paternal isodisomy 7p and maternal isodisomy 7q as described previously *(137)*. The growth retardation in that patient was considered to be a result of maternal isodisomy for 7q. Thus, it appears unlikely that paternal UPD 7 confers an imprinting effect.

upd(8)pat

A single case of paternal uniparental isodisomy for chromosome 8 has been reported *(139)*. This $5^1/_2$-year-old girl had normal development and lipoprotein lipase (LPL) deficiency due to a mutation of the *LPL* gene. The patient was ascertained due to a diagnosis of chylomicronemia. The father was a heterozygous carrier for the same mutation. It appears that normal development can occur in paternal UPD 8 and that an imprinting effect of this UPD may not exist.

upd(9)mat

Three cases of maternal UPD for chromosome 9 have been reported *(140,141)*. Two patients had recessive cartilage hair hypoplasia, a disorder that maps to the short arm of chromosome 9. The third case was a fetus associated with CPM for trisomy 9. Pathologic examination of the abortus was not possible. The available data indicate that maternal UPD 9 may not have an imprinting effect.

upd(10)mat

A single case of prenatally diagnosed maternal UPD for chromosome 10 associated with CPM has been reported *(142)*. The infant was phenotypically and developmentally normal at 8 months of age. There is no evidence to date that this UPD confers an imprinting effect.

upd(11)pat

Paternal UPD for the entire chromosome 11 has been reported in two cases *(143,144)*. One patient had hemihypertrophy, congenital adrenal carcinoma, and Wilms tumor. The second had associated CPM for trisomy 11 and intrauterine death occurred between 19 and 20 weeks' gestation. This fetus had growth retardation, aberrant intestinal rotation, and hypospadias. In addition, many cases of paternal segmental UPD for distal 11p associated with BWS have been observed (see earlier). The existence of an imprinting effect due to paternal UPD 11 is clear.

upd(13)mat

At least two cases of maternal UPD for chromosome 13 have been reported *(145,146)*. In both cases, a normal phenotype was associated with the presence of an isochromosome for the long arm of chromosome 13. These indicate that an imprinting effect due to maternal UPD 13 is very unlikely.

upd(13)pat

A single case of paternal UPD for chromosome 13 was reported in the mother of one of the maternal UPD 13 patients described previously *(145)*. This phenotypically normal individual presumably received this isochromosome 13q from her father, who was not available for study, but DNA polymorphism studies in her mother revealed the absence of maternal chromosome 13 alleles in this patient *(147)*. This observation suggests that paternal UPD 13 may not have imprinting effect.

upd(14)mat

Maternal UPD for chromosome 14 has been reported in at least 10 cases, some in abstracts only *(148,149–151)*. A distinct clinical phenotype appears to be present and consists of mild to moderate motor and/or mental developmental delay, short stature, and precocious puberty. Less frequent features include hydrocephalus, small hands, hyperextensible joints, scoliosis, and recurrent otitis media. Evidence for an imprinting effect due to maternal UPD 14 appears clear.

upd(14)pat

Four cases of paternal UPD for chromosome 14 have been reported *(117,152–154)*. A similar phenotype is present in these patients and includes polyhydramnios, low birth weight, blepharophimosis/short palpebral fissures, protruding philtrum, small ears, small thorax, abnormal ribs, simian creases, and joint contractures. Severe mental retardation was seen in the only patient who was beyond 20 months of age at the time of reporting *(117)*. These observations indicate that an imprinting effect due to paternal UPD 14 does exist.

Human chromosome 14 has significant homology to mouse chromosomes 12 and 14 *(155)*. Mouse chromosome 12 is imprinted and both maternal and paternal disomy cause early embryonic death *(156)*. Thus, the observation of imprinting effects for both maternal and paternal UPD 14 in humans is not unexpected.

upd(15)mat

More than 100 cases of maternal UPD for chromosome 15 have been reported in association with PWS *(65,68,157,158)*. Many had associated trisomy 15 mosaicism, which was confined to the placenta in most cases. Comparison of the phenotype of PWS patients with different etiologies has shown that advanced maternal age was present in mothers of patients with maternal UPD, whereas a higher frequency of hypopigmentation is seen in patients due to deletion of paternal 15q11–q13 *(157,159)*. Advanced maternal age can be expected in UPDs that result from "trisomy rescue," as advanced maternal age is associated with meiotic nondisjunction. Hypopigmentation results from mutation/deletion of the *P* gene (mouse homologue pink-eyed dilution *p* gene) located at 15q11–q13 *(160–162)*. The human *P* gene is not imprinted and both copies are functional in UPD patients. Hypopigmentation is therefore more prominent in PWS patients due to deletion. Differences in other clinical features between these two groups are less clear cut. Although there may not be a significant difference in the overall severity, female UPD patients were found to be less severely affected than female deletion patients in one study *(157)*, and UPD patients were found to be less likely to have "typical" facial appearance than deletion patients in another study *(158)*.

upd(15)pat

More than 10 cases of paternal UPD for chromosome 15 associated with AS have been reported *(67,69,163,164)*. AS patients with paternal UPD may have a milder phenotype than those with a maternal deletion of 15q11–q13 *(69,163,164)*, although reports showing no difference in clinical severity in these two groups of patients are also available *(164)*. One possible mechanism for the milder phenotype in UPD patients may be the presence of many nonimprinted genes in the 15q11–q13 region in these patients, whereas these are absent in deletion patients. Alternatively, as proposed by Bottani et al. *(69)*, it may be due to the "leaky" expression of the imprinted paternal genes, where two copies of the allele will result in an expression higher than in deletion cases, in which only one imprinted paternal allele is present.

Both maternal and paternal UPD 15 clearly confer imprinting effects.

upd(16)mat

More than 10 cases of maternal UPD for chromosome 16 have been described *(165–170)*, and potentially many more cases are not reported (Hsu et al., *Am J Med Genet*, in press). Again, associated trisomy 16 mosaicism, usually confined to the placenta, is present in most cases. IUGR is a frequent finding, but this may result from the presence of trisomy 16 cells in the placenta *(165)*. Development has been normal in all cases, the oldest reported at 4 years of age *(168)*. Imperforate anus has been reported in two cases *(165,166)* and congenital cardiac abnormalities observed in two cases with atrioventricular (A–V) canal defect in one *(167)* and atrial septal defect (ASD) and ventricular septal defect (VSD) in another *(169)*. A clinical phenotype of maternal UPD 16 has not been clearly defined; the possibility of the presence of an undetected trisomy 16 cell line complicates the comparison among reported cases. In a patient the author is currently following, in whom trisomy 16 cells have been extensively excluded in lymphocytes and fibroblasts, mild facial dysmorphism (slightly upslanted palpebral fissures, almond shaped eyes, broad nasal root, upturned nares, long philtrum, thin upper lip, prominent ears, and triangular face) is present (Wang et al., *Am J Med Genet*, in press). The facial features appear similar to those in a patient reported by Woo et al. *(170)*. Although not yet certain, the existence of an imprinting effect due to maternal UPD 16 is a distinct possibility.

upd(16)pat

A single case of paternal UPD for chromosome 16 was reported in an abstract *(171)*. A fetus presenting with hydrops at 20 weeks of gestation was found to have α-thalassemia, for which the father was a heterozygous carrier. No other pathologic findings were available. Without additional cases, it is unknown whether paternal UPD 16 confers imprinting effects.

upd(20)pat

A single case of paternal UPD for chromosome 20 was reported in an abstract *(172)*. UPD in this case resulted from a structurally abnormal chromosome 20 derived from a terminal rearrangement that joined two chromosomes 20 at band p13. DNA polymorphism studies indicated that the two chromosomes 20 in this terminal rearrangement were derived from one paternal chromosome, thereby representing paternal isodisomy. The patient had multiple anomalies including anotia, microcephaly, congenital heart disease, and Hirschsprung disease. However, this case was complicated by the presence of trisomy 20 cells in skin and the possibility of deletion of genes at the terminal rearrangement site. Therefore, although an imprinting effect is possible for paternal UPD 20, a definitive conclusion cannot be drawn without further case reports.

upd(21)mat

Maternal UPD for chromosome 21 has been reported in one patient *(173)*. The phenotypically normal woman had a balanced *de novo* (21;21) Robertsonian translocation ascertained through the birth of a Down syndrome child. Although maternal UPD 21 has been reported in early abortus specimens (174), it has not been possible to clearly attribute embryonal death to UPD. Therefore, maternal UPD 21 may be considered at this time to have no imprinting effect.

upd(21)pat

Two cases of paternal UPD for chromosome 21 have been reported *(149,175)*. In both cases, UPD resulted from *de novo* formation of a Robertsonian translocation. Both individuals were phenotypically normal. Paternal UPD 21 does not appear to have an imprinting effect.

upd(22)mat

Maternal UPD for chromosome 22 not associated with mosaic trisomy 22 has been reported in three cases *(176–178)*. All three phenotypically normal individuals were ascertained via history of multiple spontaneous abortions and found to have balanced (22;22) Robertsonian translocations. It appears clear that maternal UPD 22 does not have an imprinting effect.

upd(22)pat

A single case of paternal UPD for chromosome 22 was reported in an abstract *(179)*. It was observed in a phenotypically normal individual with a balanced (22;22) Robertsonian translocation. Paternal UPD 22 is not likely to have imprinting effect.

upd(X)mat

Maternal UPD for the two X chromosomes in females has been reported in three cases *(180,181)*. The first two cases were detected by screening a normal population of 117 individuals. The third patient had Duchenne muscular dystrophy due to homozygosity of a maternally inherited deletion of exon 50 of the dystrophin gene. These observations indicate that maternal UPD for the X chromosome does not have an imprinting effect.

upd(X)pat

A single case of paternal UPD for the two X chromosomes in the 46,XX cell line of a 14-year-old girl with 45,X/46,XX mosaicism has been reported *(182)*. This patient had impaired gonadal function and short stature. The presence of a 45,X cell line makes it difficult to determine if the observed clinical features in this patient can be attributed to paternal UPD for the X chromosome. Therefore, it is unknown at this time if paternal UPD X has an imprinting effect.

upd(XY)pat

A single case of paternal contribution of both the X and Y chromosomes in a male patient was reported in an abstract *(183)*. This patient was ascertained because he had hemophilia A which was transmitted from his father. No abnormalities other than hemophilia were described. Paternal UPD for XY may therefore not have an imprinting effect.

In summary, of 47 possible maternal and paternal UPDs for whole chromosomes in humans, 29 have been reported. Among them, six clearly have imprinting effects (7mat, 11pat, 14mat, 14pat, 15mat, 15pat); three may have imprinting effects (2mat, 6pat, 20pat); and seventeen are unlikely to have imprinting effects [1mat, 1pat, 4mat, 5pat, 6mat, 7pat, 8pat, 9mat, 10mat, 13mat, 13pat, 21mat, 21pat, 22mat, 22pat, Xmat,

(XY)pat]. In addition, 16mat probably does have an imprinting effect, whereas the status is not known for 16pat and Xpat at this time. A better understanding of the effects of UPD will be possible as more data are accumulated.

ACKNOWLEDGMENT

I thank Jo Ann Rieger for assistance in preparation of the manuscript.

REFERENCES

1. Crouse, H.V. (1960) The controlling element in sex chromosome behaviour in Sciara. *Genetics* **45,** 1429–1443.
2. Hall, J.G. (1990) Genomic Imprinting: Review and Relevance to Human Diseases. *Am. J. Hum. Genet.* **46,** 857–873.
3. Hoppe, P.C. and Illmensee, K. (1977) Microsurgically produced homozygous–diploid uniparental mice. *Proc. Natl. Acad. Sci. USA* **74,** 5657–5661.
4. McGrath, J. and Solter, D. (1984) Completion of mouse embryogenesis requires both the maternal and paternal genomes. *Cell* **37,** 179–183.
5. Surani, M.A.H., Barton, S.C., and Norris, M.L. (1984) Development of reconstituted mouse eggs suggests imprinting of the genome during gametogenesis. *Nature* **308,** 548–550.
6. Barton, S.C., Surani, M.A.H., and Norris, M.L. (1984) Role of paternal and maternal genomes in mouse development. *Nature* **311,** 374–376.
7. Surani, M.A.H., Barton, S.C., and Norris, M.L. (1986) Nuclear transplantation in the mouse: heritable differences between parental genomes after activation of the embryonic genome. *Cell* **45,** 127–136.
8. Linder, D., McCaw, B.K., and Hecht, F. (1975) Parthenogenic origin of benign ovarian teratomas. *N. Engl. J. Med.* **292,** 63–66.
9. Kajii, T. and Ohama, K. (1977) Androgenetic origin of hydatidiform mole. *Nature* **268,** 633–634.
10. Lawler, S.D., Povey, S., Fisher, R.A., and Pickthal, V.J. (1982) Genetic studies on hydatiform moles. II. The origin of complete moles. *Ann. Hum. Genet.* **46,** 209–222.
11. McFadden, D.E. and Kalousek, D.K. (1991) Two different phenotypes of fetuses with chromosomal triploidy: correlation with parental origin of the extra haploid set. *Am. J. Med. Genet.* **38,** 535–538.
12. Jacobs, P.A., Szulman, A.E., Funkhouser, J., Matsuura, J.S., and Wilson, C.C. (1982) Human triploidy: relationship between parental origin of the additional haploid complement and development of partial hydatidiform mole. *Ann. Hum. Genet.* **46,** 223–231.
13. McFadden, D.E., Kwong, L.C., Yam, I.Y., and Langlois, S. (1993) Parental origin of triploidy in human fetuses: evidence for genomic imprinting. *Hum. Genet.* **92,** 465–469.
14. Cattanach, B.M. (1986) Parental origin effects in mice. *J. Embryol. Exp. Morphol.* **97** (Suppl), 137–150.
15. Lyon, M.F. (1988) The William Allan Memorial award address: X-chromosome inactivation and the location and expression of X-linked genes. *Am. J. Hum. Genet.* **42,** 8–16.
16. Sharman, G.B. (1971) Late DNA replication in the paternally derived X chromosome of female kangaroos. *Nature* **230,** 231–232.
17. Takagi, N. and Sasaki, M. (1975) Preferential inactivation of the paternally derived X chromosome in the extraembryonic membranes of the mouse. *Nature* **256,** 640–642.
18. West, J.D., Freis, W.I., Chapman, V.M., and Papaioannou, V.E. (1977) Preferential expression of the maternally derived X chromosome in the mouse yolk sac. *Cell* **12,** 873–882.
19. Harper, M.I., Fosten, M., and Monk, M. (1982) Preferential paternal X inactivation in extraembryonic tissues of early mouse embryos. *J. Embryol. Exp. Morphol.* **67,** 127–138.

20. Harrison, K.B. (1989) X-chromosome inactivation in the human cytotrophoblast. *Cytogenet. Cell Genet.* **52**, 37–41.
21. Goto, T., Wright, E., and Monk, M. (1997) Paternal X-chromosome inactivation in human trophoblastic cells. *Mol. Hum. Reprod.* **3**, 77–80.
22. Cattanach, B.M. and Krik, M. (1985) Differential activity of maternally and paternally derived chromosome regions in mice. *Nature* **315**, 496–498.
23. Ledbetter, D.H. and Engel, E. (1995) Uniparental disomy in humans: development of an imprinting map and its implications for prenatal diagnosis. *Hum. Mol. Genet.* **4**, 1757–1764.
24. Monk, M. (1988) Genomic imprinting. *Genes Dev.* **2**, 921–925.
25. Razin, A. and Cedar, H. (1994) DNA methylation and genomic imprinting. *Cell* **77**, 473–476.
26. Mohandas, T., Sparkes, R.S., and Shapiro, L.J. (1981) Reactivation of an inactive human X-chromosome: evidence for inactivation by DNA methylation. *Science* **211**, 393–396.
27. Yen, P.H., Patel, P., Chinault, A.C., Mohandas, T., and Shapiro, L.J. (1984) Differential methylation of hypoxanthine phosphoribosyltransferase genes on active and inactive human X chromosomes. *Proc. Natl. Acad. Sci. USA* **81**, 1759–1763.
28. Keshet, I., Lieman-Hurwitz, J., and Cedar, H. (1986) DNA methylation affects the formation of active chromatin. *Cell* **44**, 535–543.
29. Reik, W., Collick, A., Norris, M.L., Barton, S.C., and Surani, M.A. (1987) Genomic imprinting determines methylation of parental alleles in transgenic mice. *Nature* **328**, 248–251.
30. Sapienza, C., Peterson, A.C., Rossant, J., and Balling, R. (1987) Degree of methylation of transgenes is dependent on gamete of origin. *Nature* **328**, 251–254.
31. Swain, J.L., Stewart, T.A., and Leder, P. (1987) Parental legacy determines methylation and expression of an autosomal transgene: a molecular mechanism for parental imprinting. *Cell* **50**, 719–727.
32. Bartolomei, M., Zemel, S., and Tilghman, S.M. (1991) Parental imprinting of the mouse H19 gene. *Nature* **351**, 153–155.
33. Ferguson-Smith, A.C., Sasaki, H., Cattanach, B.M., and Surani, M.A. (1993) Parental-origin-specific epigenetic modification of the mouse *H19* gene. *Nature* **362**, 751–755.
34. DeChiara, T.M., Robertson, E.J., and Efstratiadis, A. (1991) Parental imprinting of the mouse insulin-like growth factor II gene. *Cell* **64**, 849–859.
35. Sasaki, H., Jones, P.A., Chaillet, J.R., Ferguson-Smith, A.C., Barton, S.C., Reik, W., and Surani, M.A. (1992) Parental imprinting: potentially active chromatin of the repressed maternal allele of the mouse insulin-like growth factor II (*Igf2*) gene. *Genes Dev.* **6**, 1843–1856.
36. Barlow, D.P., Stöger, R., Hermann, B.G., Saito, K., and Schweifer, N. (1991) The mouse insulin-like growth factor type-2 receptor is imprinted and closely linked to the *Tme* locus. *Nature* **349**, 84–87.
37. Stöger, R., Kubicka, P., Liu, C.G., Kafri, T., Razin, A., Cedar, H., and Barlow, D.P. (1993) Maternal-specific methylation of the imprinted mouse *Igf2r* locus identifies the expressed locus as carrying the imprinting signal. *Cell* **73**, 61–71.
38. Li, E., Beard, C., and Jaenisch, R. (1993) Role for DNA methylation in genomic imprinting. *Nature* **366**, 362–365.
39. Driscoll, D.J., Waters, M.F., Williams, C.A., Zori, R.T., Glenn, C.C., Avidano, K.M., and Nicholls, R.D. (1992) A DNA methylation imprint, determined by the sex of the parent, distinguishes the Angelman and Prader–Willi syndrome. *Genomics* **13**, 917–924.
40. Mowery-Rushton, P.A., Driscoll, D.J., Nicholls, R.D., Locker, J., and Surti, U. (1996) DNA methylation patterns in human tissues of uniparental origin using a zinc-finger gene (*ZNF127*) from the Angelman/Prader–Willi region. *Am. J. Med. Genet.* **61**, 140–146.

41. Glenn, C.C., Porter, K.A., Jong, M.T., Nicholls, R.D., and Driscoll, D.J. (1993) Functional imprinting and epigenetic modification of the human *SNRPN* gene. *Hum. Mol. Genet.* **2,** 2001–2005.

42. Glenn, C.C., Saitoh, S., Jong, M.T.C., Filbrandt, M.M., Surti, U., Driscoll, D.J., and Nicholls, R.D. (1996) Gene Structure, DNA methylation, and imprinted expression of the human *SNRPN* gene. *Am. J. Hum. Genet.* **58,** 335–346.

43. Dittrich, B., Buiting, K., Gross, S., and Horsthemke, B. (1993) Characterization of a methylation imprint in the Prader–Willi syndrome chromosome region. *Hum. Mol. Genet.* **2,** 1995–1999.

44. Zhang, Y., Shields, T., Crenshaw, T., Hao, Y., Moulton, T., and Tycko, B. (1993) Imprinting of human H19: allele-specific CpG methylation, loss of the active allele in Wilms tumor, and potential for somatic allele switching. *Am. J. Hum. Genet.* **53,** 113–124.

45. Schneid, H., Seurin, D., Vazquez, M-P., Gourmelen, M., Cabrol, S., and Bouc, Y.L. (1993) Parental allele specific methylation of the human insulin-like growth factor II gene and Beckwith–Wiedemann syndrome. *J. Med. Genet.* **30,** 353–362.

46. Ohlsson, R., Nyström, A., Pfeifer-Ohlsson, S., Töhönen, V., Hedborg, F., Schofield, P., Flam, F. and Ekström, T.J. (1993) *IGF2* is parentally imprinted during human embryogenesis and in the Beckwith–Wiedemann syndrome. *Nat. Genet.* **4,** 94–97.

47. Kalscheuer, V.M., Mariman, E.C., Schepens, M.T., Rehder, H., and Ropers, H-H. (1993) The insulin-like growth factor type-2 receptor gene is imprinted in the mouse but not in humans. *Nat. Genet.* **5,** 74–78.

48. Kitsberg, D., Selig, S., Brandeis, M., Simon, I., Keshet, I., Driscoll, D.J., Nicholls, R.D., and Cedar, H. (1993) Allele-specific replication timing of imprinted gene regions. *Nature* **364,** 459–463.

49. Knoll, J.H.M., Cheng, S-D., and Lalande, M. (1994) Allele specificity of DNA replication timing in the Angelman/Prader–Willi syndrome imprinted chromosomal region. *Nat. Genet.* **6,** 41–46.

50. LaSalle, J.M. and Lalande, M. (1995) Domain organization of allele-specific replication within the *GABRB3* gene cluster requires a biparental 15q11–13 contribution. *Nat. Genet.* **9,** 386–394.

51. White, L.M., Rogan, P.K., Nicholls, R.D., Wu, B-L, Korf, B., and Knoll, J.H.M. (1996) Allele-specific replication of 15q11-q13 loci: a diagnostic test for detection of uniparental disomy. *Am. J. Hum. Genet.* **59,** 423–430.

52. Goldman, M.A., Holmquist, G.P., Gray, M.C., Caston, L.A., and Nag, A. (1984) Replication timing of genes and middle repetitive sequences. *Science* **224,** 686–692.

53. Dhar, V., Skoultchi, A.I., and Schildkraut, C.L. (1989) Activation and repression of a beta-globin gene in cell hybrids is accompanied by a shift in its temporal replication. *Mol. Cell Biol.* **9,** 3524–3532.

54. Selig, S., Okumura, K., Ward, D.C., and Cedar, H. (1992) Delineation of DNA replication time zones by fluorescence *in situ* hybridization. *EMBO J.* **11,** 1217–1225.

55. Nicholls, R.D. (1994) New insights reveal complex mechanisms involved in genomic imprinting. *Am. J. Hum. Genet.* **54,** 733–740.

56. Ledbetter, D.H., Riccardi, V.M., Airhart, S.D., Strobel, R.J., Keenan, B.S., and Crawford, J.D. (1981) Deletions of chromosome 15 as a cause of the Prader–Willi syndrome. *N. Engl. J. Med.* **304,** 325–329.

57. Ledbetter, D.H., Mascarello, J.T., Riccardi, V.M., Harper, V.D., Airhart, S.D., and Strobel, R.J. (1982) Chromosome 15 abnormalities and the Prader–Willi syndrome: a follow-up report of 40 cases. *Am. J. Hum. Genet.* **34,** 278–285.

58. Magenis, R.E., Brown, M.G., Lacy, D.A., Budden, S., and LaFranchi, S. (1987) Is Angelman syndrome an alternate result of del(15)(q11q13)? *Am. J. Med. Genet.* **28,** 829–838.

59. Reed, M.L. and Leff, S.E. (1994) Maternal imprinting of human *SNRPN*, a gene deleted in Prader–Willi syndrome. *Nat. Genet.* **6,** 163–167.

60. Matsuura, T., Sutcliffe, J.S., Fang, P., Galjaard, R-J., Jiang, Y-H., Benton, C.S., Rommens, J.M., and Beaudet, A.L. (1997) *De novo* truncating mutations in E6-AP ubiquitin–protein ligase gene (*UBE3A*) in Angelman syndrome. *Nat. Genet.* **15**, 74–77.

61. Butler, M.G. and Palmer, C.G. (1983) Parental origin of chromosome 15 deletion in Prader–Willi syndrome (letter). *Lancet* **i**, 1285–1286.

62. Knoll, J.H., Nicholls, R.D., Magenis, R.E., Graham, J.M. Jr., Lalande, M., and Latt, S.A. (1989) Angelman and Prader–Willi syndromes share a common chromosome 15 deletion but differ in parental origin of the deletion. *Am. J. Med. Genet.* **32**, 285–290.

63. Magenis, R.E., Toth-Fejel, S., Allen, L.H., Black, M., Brown, M.G., Budden, S., Cohen, R., Friedman, J.M., Kalousek, D., Zonana, J., Lacy, D., LaFranchi, S., Lahr, M., Macfarlane, J., and Williams, C.P.S. (1990) Comparison of the 15q deletions in Prader–Willi and Angelman syndromes: specific regions, extent of deletions, parental origin, and clinical consequences. *Am. J. Med. Genet.* **35**, 333–349.

64. Williams, C.A., Zori, R.T., Stone, J.W., Gray, B.A., Cantu., E.S., and Ostrer, H. (1990) Maternal origin of 15q11–13 deletions in Angelman syndrome suggests a rule for genomic imprinting. *Am. J. Med. Genet.* **35**, 350–353.

65. Nicholls, R.D., Knoll, J.H.M., Butler, M.G., Karam, S., and Lalande, M. (1989) Genetic imprinting suggested by maternal heterodisomy in non-deletion Prader–Willi syndrome. *Nature* **342**, 281–285.

66. Knoll, J.H.M., Glatt, K.A., Nicholls, R.D., Malcolm, S., and Lalande, M. (1991) Chromosome 15 uniparental disomy is not frequent in Angelman syndrome. *Am. J. Hum. Genet.* **48**, 16–21.

67. Malcolm, S., Clayton-Smith, J., Nicols, M., Robb, S., Webb, T., Armour, J.A., Jeffreys, A.J., and Pembrey, M.E. (1991) Uniparental paternal disomy in Angelman's syndrome. *Lancet* **337**, 694–697.

68. Mascari, M.J., Gottlieb, W., Rogan, P.K., Butler, M.G., Waller, D.A., Armour, A.L., Jeffreys, A.J., Ladda, R.L., and Nicholls, R.D. (1992) The frequency of uniparental disomy in Prader–Willi syndrome: implications for molecular diagnosis. *N. Engl. J. Med.* **326**, 1599–1607.

69. Bottani, A., Robinson, W.P., DeLozier-Blanchet, C.D., Engel, E., Morris, M.A., Schmitt, B., Thun-Hohenstein, L., and Schinzel, A. (1994) Angelman syndrome due to paternal uniparental disomy of chromosome 15: a milder phenotype? *Am. J. Med. Genet.* **51**, 35–40.

70. Wagstaff, J., Knoll, J.H.M., Glatt, K.A., Shugart, Y.Y., Sommer, A., and Lalande, M. (1992) Maternal but not paternal transmission of 15q11–13-linked nondeletion Angelman syndrome leads to phenotypic expression. *Nat. Genet.* **1**, 291–294.

71. Glenn, C.C., Nicholls, R.D., Robinson, W.P., Saitoh, S., Niikawa, N., Schinzel, A., Horsthemke, B., and Driscoll, D.J. (1993) Modification of 15q11–q13 DNA methylation imprints in unique Angelman and Prader–Willi patients. *Hum. Mol. Genet.* **2**, 1377–1382.

72. Reis, A., Dittrich, B., Greger, V., Buiting, K., Lalande, M., Gillessen-Kaesbach, G., Anvret, M., and Horsthemke, B. (1994) Imprinting mutations suggested by abnormal DNA methylation patterns in familial Angelman and Prader–Willi syndromes. *Am. J. Hum. Genet.* **54**, 741–747.

73. Sutcliffe, J.S., Nakao, M., Christian, S., Orstavik, K.H., Tommerup, N., Ledbetter, D.H., and Beaudet, A.L. (1994) Deletions of a differentially methylated CpG island at the SNRPN gene define a putative imprinting control region. *Nat. Genet.* **8**, 52–58.

74. Bürger, J., Buiting, K., Dittrich, B., Groß, S., Lich, C., Sperling, K., Horsthemke, B., and Reis, A. (1997) *Am. J. Hum. Genet.* **61**, 88–93.

75. Kishino, T., Lalande, M., and Wagstaff, J. (1997) *UBE3A*/E6-AP mutations cause Angelman syndrome. *Nat. Genet.* **15**, 70–73.

76. Holm, V.A., Cassidy, S.B., Butler, M.G., Hanchett, J.M., Greenswag, L.R., Whitman,

B.Y., and Greenberg, F. (1993) Prader–Willi syndrome: consensus diagnostic criteria. *Pediatrics* **91**, 398–402.

77. Angelman, H. (1965) "Puppet" children: a report on three cases. *Dev. Med. Child. Neurol.* **7**, 681–688.

78. Nakao, M., Sutcliffe, J.S., Durtschi, B., Mutirangura, A., Ledbetter, D.H., and Beaudet, A.L. (1994) Imprinting analysis of three genes in the Prader–Willi/Angelman region: SNRPN, E6-associated protein, and PAR-2 (D15S225E). *Hum. Mol. Genet.* **3**, 309–315.

79. Vu, T.H. and Hoffman, A.R. (1997) Imprinting of the Angelman syndrome gene, *UBE3A*, is restricted to brain. *Nat. Genet.* **17**, 12–13.

80. Rougeulle, C., Glatt, H., and Lalande, M. (1997) The Angelman syndrome candidate gene, *UBE3A/E6-AP*, is imprinted in brain. *Nat. Genet.* **17**, 14–15.

81. Buiting, K., Saitoh, S., Gross, S., Dittrich, B., Schwartz, S., Nicholls, R.D., and Horsthemke, B. (1995) Inherited microdeletions in the Angelman and Prader–Willi syndromes define an imprinting centre on human chromosome 15. *Nat. Genet.* **9**, 395–400.

82. Beckwith, J.B. (1969) Macroglossia, omphalocele, adrenal cytomegaly, gigantism, and hyperplastic visceromegaly. *Birth Defects* **5**, 188.

83. Pettenati, M.J., Haines, J.L., Higgins, R.R., Wappner, R.S., Palmer, C.G., and Weaver, D.D. (1986) Wiedemann–Beckwith syndrome: presentation of clinical and cytogenetic data on 22 new cases and review of the literature. *Hum. Genet.* **74**, 143–154.

84. Henry, I., Bonaiti-Pellie, C., Chehensse, V., Beldjord, C., Schwartz, C., Utermann, G., and Junien, C. (1991) Uniparental paternal disomy in a genetic cancer-predisposing syndrome. *Nature* **351**, 665–667.

85. Catchpoole, D., Lam, W.W.K., Valler, D., Temple, I.K., Joyce, J.A., Reik, W., Schofield, P.N., and Maher, E.R. (1997) Epigenetic modification and uniparental inheritance of H19 in Beckwith–Wiedemann syndrome. *J. Med. Genet.* **34**, 353–359.

86. Brown, K.W., Gardner, A., Williams, J.C., Mott, M.G., McDermott, A., and Maitland, N.J. (1992) Paternal origin of 11p15 duplications in the Beckwith–Wiedemann syndrome. A new case and review of the literature. *Cancer Genet. Cytogenet.* **58**, 66–70.

87. Weksberg, R., Teshima, I., Williams, B.R., Greenberg, C.R., Pueschel, S.M., Chernos, J.E., Fowlow, S.B., Hoyme, E., Anderson, I.J., Whiteman, D.A., Fisher, N., and Squire, J. (1993) Molecular characterization of cytogenetic alterations associated with the Beckwith–Wiedemann syndrome (BWS) phenotype refines the localization and suggests the gene for BWS is imprinted. *Hum. Mol. Genet.* **2**, 549–556.

88. Tommerup, N., Brandt, C.A., Pedersen, S., Bolund, L., and Kamper, J. (1993) Sex dependent transmission of Beckwith–Wiedemann syndrome associated with a reciprocal translocation t(9;11)(p11.2;p15.5). *J. Med. Genet.* **30**, 958–961.

89. Moutou, C., Junien, C., Henry, I., and Bonaiti-Pellie, C. (1992) Beckwith–Wiedemann syndrome: a demonstration of the mechanisms responsible for the excess of transmitting females. *J. Med. Genet.* **29**, 217–220.

90. Viljoen, D. and Ramesar, R. (1992) Evidence for paternal imprinting in familial Beckwith–Wiedemann syndrome. *J. Med. Genet.* **29**, 221–225.

91. Ping, A.J., Reeve, A.E., Law, D.J., Young, M.R., Boehnke, M., and Feinberg, A.P. (1989) Genetic linkage of Beckwith–Wiedemann syndrome to 11p15. *Am. J. Hum. Genet.* **44**, 720–723.

92. Koufos, A., Grundy, P., Morgan, K., Aleck, K.A., Hadro, T., Lampkin, B.C., Kalbakji, A., and Cavenee, W.K. (1989) Familial Wiedemann–Beckwith syndrome and a second Wilms tumor locus both map to 11p15.5. *Am. J. Hum. Genet.* **44**, 711–719.

93. Matsuoka, S., Thompson, J.S., Edwards, M.C., Bartletta, J.M., Grundy, P., Kalikin, L.M., Harper, J.W., Elledge, S.J., and Feinberg, A.P. (1996) Imprinting of the gene encoding a human cyclin-dependent kinase inhibitor, $p57^{KIP2}$, on chromosome 11p15. *Proc. Natl. Acad. Sci. USA* **93**, 3026–3030.

94. Hatada, I., Inazawa, J., Abe, T., Nakayama, M., Kaneko, Y., Jinno, Y., Niikawa, N.,

Ohashi, H., Fukushima, Y., Iida, K., Yutani, C., Takahashi, S., Chiba, Y., Ohishi, S., and Mukai, T. (1996) Genomic imprinting of human *p57^KIP2* and its reduced expression in Wilms' tumors. *Hum. Mol. Genet.* **5,** 783–788.

95. Reik, W., Brown, K.W., Schneid, H., Le Bouc, Y., Bickmore, W., and Maher, E.R. (1995) Imprinting mutations in the Beckwith–Wiedemann syndrome suggested by altered imprinting pattern in the *IGF2–H19* domain. *Hum. Mol. Genet.* **4,** 2379–2385.

96. Hatada, I., Hirofumi, O., Fukushima, Y., Kaneko, Y., Inoue, M., Komoto, Y., Okada, A., Ohishi, S., Nabetani, A., Morisaki, H., Nakayama, M., Niikawa, N., and Mukai, T. (1996) An imprinted gene p57^KIP2 is mutated in Beckwith–Wiedemann syndrome. *Nat. Genet.* **14,** 171–173.

97. Heutink, P., van der Mey, A.G.L., Sandkuijl, L.A., van Gils, A.P.G., Bardoel, A., Breedveld, G.J., van Vliet, M., van Ommen, G-J.B., Cornelisse, C.J., Oostra, B.A., Weber, J.L., and Deville, P. (1992) A gene subject to genomic imprinting and responsible for hereditary paragangliomas maps to chromosome 11q23-qter. *Hum. Mol. Genet.* **1,** 7–10.

98. van der Mey, A.G., Maaswinkel-Mooy, P.D., Cornelisse, C.J., Schmidt, P.H., and van de Kamp, J.J. (1989) Genomic imprinting in hereditary glomus tumours: evidence for new genetic theory. *Lancet* **ii,** 1291–1294.

99. van Gils, A.P., van der Mey, A.G., Hoogma, R.P., Sankuijl, L.A., Maaswinkel-Mooy, P.D., Falke, T.H., and Pauwels, E.K. (1992) MRI screening of kindred at risk of developing paragangliomas: support for genomic imprinting in hereditary glomus tumours. *Br. J. Cancer* **65,** 903–907.

100. Schroeder, W.T., Chao, L-Y., Dao, D.D., Strong, L.C., Pathak, S., Riccardi, V., Lewis, W.H., and Saunders, G.F. (1987) Nonrandom loss of maternal chromosome 11 alleles in Wilms tumors. *Am. J. Hum. Genet.* **40,** 413–420.

101. Mannens, M., Slater, R.M., Heyting, C., Bliek, J., de Kraker, J., Coad, N., de Pagter-Holthuizen, P., and Pearson, P.L. (1988) Molecular nature of genetic changes resulting in loss of heterozygosity of chromosome 11 in Wilms' tumors. *Hum. Genet.* **81,** 41–48.

102. Scrable, H., Cavenee, W., Ghavimi, F., Lovell, M., Morgan, K., and Sapienza, C. (1989) A model for embryonal rhabdomyosarcoma tumorigenesis that involves genome imprinting. *Proc. Natl. Acad. Sci. USA* **86,** 7480–7484.

103. Koi, M., Johnson, L.A., Kalikin, L.M., Little, P.F., Nakamura, Y., and Feinberg, A.P. (1993) Tumor cell growth arrest caused by subchromosomal transferable DNA fragments from chromosome 11. *Science* **260,** 361–364.

104. Ogawa, O., Eccles, M.R., Szeto, J., McNoe, L.A., Yun, K., Maw, M.A., Smith, P.J., and Reeve, A.E. (1993) Relaxation of insulin-like growth factor II gene imprinting implicated in Wilms' tumour. *Nature* **362,** 749–751.

105. Wu, H.K., Squire, J.A., Catzavelos, C.G., and Weksberg, R. (1997) Relaxation of imprinting of human insulin-like growth factor II gene, *IGF2,* in sporadic breast carcinomas. *Biochem. Biophys. Res. Commun.* **235,** 123–129.

106. Wang, W.H., Duan, J.X., Vu, T.H., and Hoffman, A.R. (1996) Increased expression of the insulin-like growth factor-II gene in Wilms' tumor is not dependent on loss of genomic imprinting or loss of heterozygosity. *J. Biol. Chem.* **271,** 27863–27870.

107. Friend, S.H., Bernards, R., Rogelj, S., Weinberg, R.A., Rapaport, J.M., Albert, D.M., and Dryja, T.P. (1986) A human DNA segment with properties of the gene that predisposes to retinoblastoma and osteosarcoma. *Nature* **323,** 643–646.

108. Ejima, Y., Sasaki, M.S., Kaneko, A., and Tanooka, H. (1988) Types, rates, origin and expressivity of chromosome mutations involving 13q14 in retinoblastoma patients. *Hum. Genet.* **79,** 118–123.

109. Dryja, T.P., Mukai, S., Petersen, R., Rapaport, J.M., Walton, D., and Yandell, D.W. (1989) Parental origin of mutations of the retinoblastoma gene. *Nature* **339,** 556–558.

110. Toguchida, J., Ishizaki, K., Sasaki, M.S., Nakamura, Y., Ikenaga, M., Kato, M., Sugimot,

M., Kotoura, Y., and Yamamuro, T. (1989) Preferential mutation of paternally derived RB gene as the initial event in sporadic osteosarcoma. *Nature* **338,** 156–158.

111. Kato, M.V., Ishizaki, K., Shimizu, T., Ejima, Y., Tanooka, H., Takayama, J., Kaneko, A., Toguchida, J., and Sasakio, M.S. (1994) Parental origin of germ-like and somatic mutations in the retinoblastoma gene. *Hum. Genet.* **94,** 31–38.

112. Leach, R.J., Magewu, A.N., Buckley, J.D., Benedict, W.F., Rother, C., Murphree, A.L., Griegel, S., Rajewsky, M.F., and Jones, P.A. (1990) Preferential retention of paternal alleles in human retinoblastoma: evidence for genomic imprinting. *Cell Growth Differ.* **1,** 401–406.

113. Mitelman, F. (1994) *Catalog of Chromosome Aberrations in Cancer,* 5th edition. Wiley-Liss, New York.

114. Cheng, J.M., Hiemstra, J.L., Schneider, S.S., Naumova, A., Cheung, N-K.V., Cohn, S.L., Diller, L., Sapienza, C., and Brodeur, G.M. (1993) Preferential amplification of the paternal allele of the n-myc gene in human neuroblastomas. *Nat. Genet.* **4,** 191–194.

115. Caron, H., Peter, M., van Sluis, P., Speleman, F., de Kraker, J., Laureys, G., Michon, J., Brugieres, L., Vo"ute, P.A., Westerveld, A., Slater, R., DeLattre, O., and Versteeg, R. (1995) Evidence for two tumour suppressor loci on chromosomal bands 1p35-36 involved in neuroblastoma: one probably imprinted, another associated with n-myc amplification. *Hum. Mol. Genet.* **4,** 535–539.

116. Engel, E. (1980) A new genetic concept: uniparental disomy and its potential effect, isodisomy. *Am. J. Med. Genet.* **6,** 137–143.

117. Wang, J-C.C., Passage, M.B., Yen, P.H., Shapiro, L.J., and Mohandas, T.K. (1991) Uniparental heterodisomy for chromosome 14 in a phenotypically abnormal familial balanced 13/14 Robertsonian translocation carrier. *Am. J. Hum. Genet.* **48,** 1069–1074.

118. Robinson, W.P., Wagstaff, J., Bernasconi, F., Baccichette, C., Artifoni, L., Franzoni, E., Suslak, L., Shih, L-Y., Aviv, H., and Schinzel, A.A. (1993) Uniparental disomy explains the occurrence of the Angelman or Prader–Willi syndrome in patients with an additional small inv dup(15) chromosome. *J. Med. Genet.* **30,** 756–760.

119. Spence, J.E., Perciaccante, R.G., Greig, G.M., Williard, H.F., Ledbetter, D.H., Hejtmancik, J.F., Pollack, M.S., O'Brien, W.E., and Beaudet, A.L. (1988) Uniparental disomy as a mechanism for human genetic disease. *Am. J. Hum. Genet.* **42,** 217–226.

119a. Pulkkinen, L., Bullrich, F., Czarnecki, P., Weiss, L., and Uitto, J. (1997) Maternal uniparental disomy of chromosome 1 with reduction to homozygosity of the LAMB3 locus in a patient with Herlitz junctional epidermolysis bullosa. *Am. J. Hum. Genet.* **61,** 611–619.

119b. Gelb, B.D., Willner, J.P., Verloes, A., Herens, C., and Desnick, R.J. (1997) Mutation analysis of pycnodysostosis reveals uniparental disomy of chromosome 1. *Am. J. Hum. Genet.* **61** (Suppl), A39.

119c. Chen, H., Young, R., Nandi, K., Mu, X., Fan, R., Miao, S., Batte, L., Howell, N., Prouty, L., Ursin, S., Gonzalez, J., and Yanamandra, K. (1977) Short stature, ptosis, micro/retrognathia, myopathy, and sterility in a female patient with 46,XX,i(1p),i(1q). *Am. J. Hum. Genet.* **61** (Suppl), A120.

120. Harrison, K., Eisenger, K., Anyane-Yeboa, K., and Brown, S. (1995) Maternal uniparental disomy of chromosome 2 in a baby with trisomy 2 mosaicism in amniotic fluid culture. *Am. J. Med. Genet.* **58,** 147–151.

121. Webb, A.L., Sturgiss, S., Warwicker, P., Robson, S.C., Goodship, J.A., and Wolstenholme, J. (1996) Maternal uniparental disomy for chromosome 2 in association with confined placental mosaicism for trisomy 2 and severe intrauterine growth retardation. *Prenat. Diagn.* **16,** 958–962.

122. Bernasconi, F., Karagüzel, A., Celep, F., Keser, I., Lüleci, G., Dutly, F., and Schinzel, A.A. (1996) Normal phenotype with maternal isodisomy in a female with two isochromosomes: i(2p) and i(2q). *Am. J. Hum. Genet.* **59,** 1114–1118.

123. Hansen, W.F., Bernard, L.E., Langlois, S., Rao, K.W., Chescheir, N.C., Aylsworth, A.S., Smith, D.I., Robinson, W.P., Barrett, I.J., and Kalousek, D.K. (1997) Maternal uniparental disomy of chromosome 2 and confined placental mosaicism for trisomy 2 in a fetus with intrauterine growth restriction, hypospadias, and oligohydramnios. *Prenat. Diagn.* **17**, 443–450.

124. Shaffer, L.G., McCaskill, C., Egli, C.A., Baker, J.C., and Johnston, K.M. (1997) Is there an abnormal phenotype associated with maternal isodisomy for chromosome 2 in the presence of two isochromosomes? *Am. J. Hum. Genet.* **61**, 461–462.

125. Lindenbaum, R.H., Woods, C.G., Norbury, C.G., Povey, S., and Rysiecki, G. (1991) An individual with maternal disomy of chromosome 4 and iso (4p), iso 4(q). *Am. J. Hum. Genet.* **49** (Suppl), A285.

126. Brzustowicz, L.M., Allitto, B.A., Matseoane, D., Theve, R., Michaud, L., Chatkupt, S., Sugarman, E., Penchaszadeh, G.K., Suslak, L., Koenigsberger, M.R., Gilliam, T.C., and Handelin, B.L. (1994) Paternal isodisomy for chromosome 5 in a child with spinal muscular atrophy. *Am. J. Hum. Genet.* **54**, 482–488.

127. van den Berg-Loonen, E.M., Savelkoul, P., van Hooff, H., van Eede, P., Riesewijk, A., and Geraedts, J. (1996) Uniparental maternal disomy 6 in a renal transplant patient. *Hum. Immunol.* **45**, 46–51.

128. Welch, T.R., Beischel, L.S., Choi, E., Balakrishnan, K., and Bishof, N.A. (1990) Uniparental isodisomy 6 associated with deficiency of the fourth component of complement. *J. Clin. Invest.* **86**, 675–678.

129. Abramowicz, M.J., Andrien, M., Dupont, E., Dorchy, H., Parma, J., Duprez, L., Ledley, F.D., Courtens, W., and Vamos, E. (1994) Isodisomy of chromosome 6 in a newborn with methylmalonic acidaemia and agenesis of the pancreatic beta cells causing diabetes mellitus. *J. Clin. Invest.* **94**, 418–421.

130. Temple, I.K., James, R.S., Crolla, J.A., Sitch, F.L., Jacobs, P.A., Howell, W.M., Betts, P., Baum, J.D., and Shield, J.P.H. (1995) An imprinted gene(s) for diabetes? *Nat. Genet.* **9**, 110–112.

131. Bittencourt, M.C., Morris, M.A., Chabod, J., Gos, A., Lamy, B., Fellmann, F., Antonarakis, S.E., Plouvier, E., Herve, P., and Tiberghien, P. (1997) Fortuitous detection of uniparental isodisomy of chromosome 6. *J. Med. Genet.* **34**, 77–78.

132. Whiteford, M.L., Narendra, A., White, M.P., Cooke, A., Wilkinson, A.G., Robertson, K.J., and Tolmie, J.L. (1997) Paternal uniparental disomy for chromosome 6 causes transient neonatal diabetes. *J. Med. Genet.* **34**, 167–168.

133. Langlois, S., Yong, S.L., Wilson, R.D., Kwong, L.C., and Kalousek, D.K. (1995) Prenatal and postnatal growth failure associated with maternal heterodisomy for chromosome 7. *J. Med. Genet.* **32**, 871–875.

134. Preece, M.A., Price, S.M., Davies, V., Clough, L., Stanier, P., Trembath, R.C., and Moore, G.E. (1997) Maternal uniparental disomy 7 in Silver–Russell syndrome. *J. Med. Genet.* **34**, 6–9.

135. Voss, R., Ben-Simon, E., Avital, A., Godfrey, S., Zlotogora, J., Dagan, J., Tikochinski, Y., and Hillel, J. (1989) Isodisomy of chromosome 7 in a patient with cystic fibrosis: could uniparental disomy be common in humans? *Am. J. Hum. Genet.* **45**, 373–380.

136. Kotzot, D., Schmitt, S., Bernasconi, F., Robinson, W.P., Lurie, J.W., Hyina, H., Mehes, M., Hamel, B.C.J., Otten, B.J., Hergersberg, M., Werder, E., Schoenle, E., and Schinzel, A. (1995) Uniparental disomy 7 in Silver–Russell syndrome and primordial growth retardation. *Hum. Mol. Genet.* **4**, 583–587.

137. Eggerding, F.A., Schonberg, S.A., Chehab, F.F., Norton, M.E., Cox, V.A., and Epstein, C.J. (1994) Uniparental isodisomy for paternal 7p and maternal 7q in a child with growth retardation. *Am. J. Hum. Genet.* **55**, 253–265.

138. Höglund, P., Holmberg, C., de la Chapelle, A., and Kere, J. (1994) Paternal isodisomy

for chromosome 7 is compatible with normal growth and development in a patient with congenital chloride diarrhea. *Am. J. Hum. Genet.* **55**, 747–752.

139. Benlian, P., Foubert, L., Gagné, E., Bernard, L., De Gennes, J.L., Langlois, S., Robinson, W., and Hayden, M. (1996) Complete paternal isodisomy for chromosome 8 unmasked by lipoprotein lipase deficiency. *Am. J. Hum. Genet.* **59**, 431–436.

140. Sulisalo, T., Francomano, C.A., Sistonen, P., Maher, J.F., McKusick, V.A., de la Chapelle, A., and Kaitila, I. (1994) High-resolution genetic mapping of the cartilage-hair hypoplasia (CHH) gene in Amish and Finnish families. *Genomics* **20**, 347–353.

141. Wilkinson, T.A., James, R.S., Crolla, J.A., Cockwell, A.E., Campbell, P.L., and Temple, I.K. (1996) A case of maternal uniparental disomy of chromosome 9 in association with confined placental mosaicism for trisomy 9. *Prenat. Diagn.* **16**, 371–374.

142. Jones, C., Booth, C., Rita, D., Jazmines, L., Spiro, R., McCulloch, B., McCaskill, C., and Shaffer, L.G. (1995) Identification of a case of maternal uniparental disomy of chromosome 10 associated with confined placental mosaicism. *Prenat. Diagn.* **15**, 843–848.

143. Grundy, P., Telzerow, P., Paterson, M.C., Habier, D., Berman, B., Li, F., and Garber, J. (1991) Chromosome 11 uniparental isodisomy predisposing to embryonal neoplasms (Letter). *Lancet* **338**, 1079–1080.

144. Webb, A., Beard, J., Wright, C., Robson, S., Wolstenholme, J., and Goodship, J. (1995) A case of paternal uniparental disomy for chromosome 11. *Prenat. Diagn.* **15**, 773–777.

145. Slater, H., Shaw, J.H., Dawson, G., Bankier, A., and Forrest, S.M. (1994) Maternal uniparental disomy of chromosome 13 in a phenotypically normal child. *J. Med. Genet.* **31**, 644–646.

146. Staffard, R., Krueger, S., James, R.S., and Schwartz, S. (1995) Uniparental isodisomy 13 in a normal female due to transmission of a maternal t(13q13q). *Am. J. Med. Genet.* **57**, 14–18.

147. Slater, H., Shaw, J.H., Bankier, A., Forrest, S.M., and Dawson, G. (1995) UPD 13: no indication of maternal or paternal imprinting of genes on chromosome 13. *J. Med. Genet.* **32**, 493.

148. Healey, S., Powell, F., Battersby, M., Chenevix-Trench, G., and McGill, J. (1994) Distinct phenotype in maternal uniparental disomy of chromosome 14. *Am. J. Med. Genet.* **51**, 147–149.

149. Robinson, W.P., Bernasconi, F., Basaran, S., Yüksel-Apak, M., Neri, G., Serville, F., Balicek, P., Haluza, R., Farah, L.M.S., Lüleci, G., and Schinzel, A.A. (1994) A somatic origin of homologous Robertsonian translocations and isochromosomes. *Am. J. Hum. Genet.* **54**, 290–302.

150. Sirchia, S.M., De Andreis, C., Pariani, S., Grimoldi, M.G., Molinari, A., Buscaglia, M., and Simoni, G. (1994) Chromosome 14 maternal uniparental disomy in the euploid cell line of a fetus with mosaic 46,XX/47,XX,+14 karyotype. *Hum. Genet.* **94**, 355–358.

151. Coviello, D.A., Panucci, E., Mantero, M.M., Perfumo, C., Guelfi, M., Borrone, C., and Dagna-Bricarelli, F. (1996) Maternal uniparental disomy for chromosome 14. *Acta Genet. Med. Gemellol (Roma)* **45**, 169–172.

152. Papenhausen, P.R., Mueller, O.T., Johnson, V.P., Sutcliffe, M., Diamond, T.M., and Kousseff, B.G. (1995) Uniparental isodisomy of chromosome 14 in two cases: an abnormal child and a normal adult. *Am. J. Med. Genet.* **59**, 271–275.

153. Walter, C.A., Shaffer, L.G., Kaye, C.I., Huff, R.W., Ghidoni, P.D., McCaskill, C., McFarland, M.B., and Moore, C.M. (1996) Short-limb dwarfism and hypertrophic cardiomyopathy in a patient with paternal isodisomy 14: 45,XY,idic(14)(p11). *Am. J. Med. Genet.* **65**, 259–265.

154. Cotter, P.D., Kaffe, S., McCurdy, L.D., Jhaveri, M., Willner, J.P., and Hirschhorn, K. (1997) Paternal uniparental disomy for chromosome 14: a case report and review. *Am. J. Med. Genet.* **70**, 74–79.

155. Cox, D.W., Gedde-Dahyl, T., Menon, A.G., Nygaard, T.G., Tomlinson, I.M., Peters, J., St. George-Hyslop, P.H., Walter, M.A., and Edwards, J.H. (1995) Report of the second international workshop on human chromosome 14 mapping 1994. *Cytogenet. Cell Genet.* **69**, 159–174.

156. Cattanach, B.M., Barr, J., and Jones, J. (1995) Use of chromosome rearrangements for investigations into imprinting in the mouse. In *Genomic Imprinting, Causes and Consequences* (Ohlsson, R., Hall, K. and Ritzen, M., eds.), Cambridge University Press, Cambridge, pp. 327–341.

157. Mitchell, J., Schinzel, A., Langlois, S., Gillessen-Kaesbach, G., Schuffenhauer, S., Michaelis, R., Abeliovich, D., Lerer, I., Christian, S., Guitart, M., McFadden, D.E., and Robinson, W.P. (1996) Comparison of phenotype in uniparental disomy and deletion Prader–Willi syndrome: sex specific differences. *Am. J. Med. Genet.* **65**, 133–136.

158. Cassidy, S.B., Forsythe, M., Heeger, S., Nicholls, R.D., Schork, N., Benn, P., and Schwartz, S. (1997) Comparison of phenotype between patients with Prader–Willi syndrome due to deletion 15q and uniparental disomy15. *Am. J. Med. Genet.* **68**, 433–440.

159. Gillessen-Kaesbach, G., Robinson, W., Lohmann, D., Kaya-Westerloh, S., Passarge E., and Horsthemke, B. (1995) Genotype-phenotype correlation in a series of 167 deletion and nondeletion patients with Prader–Willi syndrome. *Hum. Genet.* **96**, 638–643.

160. Gardner, J.M., Nakatsu, Y., Gondo, Y., Lee, S., Lyon, M.F., King, R.A., and Brilliant, M.H. (1992) The mouse pink-eyed dilution gene: association with human Prader–Willi and Angelman syndromes. *Science* **257**, 1121–1124.

161. Rinchik, E.M., Bultman, S.J., Horsthemke, B., Lee, S.T., Strunk, K.M., Spritz, R.A., Avidano, K.M., Jong, M.T., and Nicholls, R.D. (1993) A gene for the mouse pink-eyed dilution locus and for human type II oculocutaneous albinism. *Nature* **361**, 72–76.

162. Lee, S-T., Nicholls, R.D., Phil, D., Bundey, S., Laxova, R., Musarella, M., and Spritz, R.A. (1994) Mutations of the *P* gene in oculocutaneous albinism, ocular albinism, and Prader–Willi syndrome plus albinism. *N. Engl. J. Med.* **330**, 529–534.

163. Smith, A., Marks, R., Haan, E., Dixon, J., and Trent, R.J. (1997) Clinical features in four patients with Angelman syndrome resulting from paternal uniparental disomy. *J. Med. Genet.* **34**, 426–429.

164. Prasad, C. and Wagstaff, J. (1997) Genotype and phenotype in Angelman syndrome caused by paternal UPD 15. Genotype and phenotype in Angelman syndrome caused by paternal UPD 15. *Am. J. Med. Genet.* **70**, 328–329.

165. Kalousek, D.K., Langlois, S., Barrett, I., Yam, I., Wilson, D.R., Howard-Peebles, P.N., Johnson, M.P., and Giorgiutti, E. (1993) Uniparental disomy for chromosome 16 in humans. *Am. J. Hum. Genet.* **52**, 8–16.

166. Vaughan, J., Zehra, A., Bower, S., Bennett, P., Chard, T., and Moore, G. (1994) Human maternal uniparental disomy for chromosome 16 and fetal development. *Prenat. Diagn.* **14**, 751–756.

167. Whiteford, M.L., Coutts, J., Al-Roomi, L., Mather, A., Lowther, G., Cooke, A., Vaughan, J.I., Moore, G.E., and Tolmie J.L. (1995) Uniparental isodisomy for chromosome 16 in a growth-retarded infant with congenital heart disease. *Prenat. Diagn.* **15**, 579–584.

168. Schneider, A.S., Bischoff, F.Z., McCaskill, C., Coady, M.L., Stopfer, J.E., and Shaffer, L.G. (1996) Comprehensive 4-year follow-up on a case of maternal heterodisomy for chromosome 16. *Am. J. Med. Genet.* **66**, 204–208.

169. O'Riordan, S., Greenough, A., Moore, G.E., Bennett, P., and Nicolaides, K.H. (1996) Case report: uniparental disomy 16 in association with congenital heart disease. *Prenat. Diagn.* **16**, 963–965.

170. Woo, V., Bridge, P.J., and Bamforth, J.S. (1997) Maternal uniparental heterodisomy for chromosome 16: case report. *Am. J. Med. Genet.* **70**, 387–390.

171. Ngo, K.Y., Lee, J., Dixon, B., Liu, D., and Jones, O.W. (1993) Paternal uniparental

isodisomy in a hydrops fetalis α-thalassemia fetus. *Am. J. Hum. Genet.* **53** (Suppl), A1207.

172. Spinner, N.B., Rand, E., Bucan, M., Jirik, F., Gogolin-Ewens, C., Riethman, H.C., McDonald-McGinn, D.M., and Zackai, E.H. (1994) Paternal uniparental isodisomy for human chromosome 20 and absence of external ears. *Am. J. Hum. Genet.* **55** (suppl), A118.

173. Creau-Goldberg, N., Gegonne, A., Delabar, J., Cochet, C., Cabanis, M.O., Stehelin, D., Turleau, C., and de Grouchy, J. (1987) Maternal origin of a de novo balanced t(21q21q) identified by ets-2 polymorphism. *Hum. Genet.* **76**, 396–398.

174. Henderson, D.J., Sherman, L.S., Loughna, S.C., Bennett, P.R., and Moore, G.E. (1994) Early embryonic failure associated with uniparental disomy for human chromosome 21. *Hum. Mol. Genet.* **3**, 1373–1376.

175. Blouin, J-L., Avramopoulos, D., Pangalos, C., and Antonarakis, S.E. (1993) Normal phenotype with paternal uniparental isodisomy for chromosome 21. *Am. J. Hum. Genet.* **53**, 1074–1078.

176. Palmer, C.G., Schwartz, S., and Hodes, M.D. (1980) Transmission of a balanced homologous t(22q;22q) translocation from mother to normal daughter. *Clin. Genet.* **17**, 418–422.

177. Kirkels, V.G., Hustinx, T.W., and Scheres, J.M. (1980) Habitual abortion and transloca-tion (22q;22q): unexpected transmission from a mother to her phenotypically normal daughter. *Clin. Genet.* **18**, 456–461.

178. Schinzel, A.A., Basaran, S., Bernasconi, F., Karaman, B., Yüksel-Apak, M., and Robinson, W.P. (1994) Maternal uniparental disomy 22 has no impact on the phenotype. *Am. J. Hum. Genet.* **54**, 21–24.

179. Miny, P., Koopers, B., Rogadanova, N., Schulte-Vallenun, M., Horst, J., and Dwornizak, B. (1995) *European Society of Human Genetics 17th Annual Meeting.* H-76 (Abstract).

180. Avivi, L., Korenstein, A., Braier-Goldstein, O., Goldman, B., and Ravia, Y. (1992) Uni-parental disomy of sex chromosome in man. *Am. J. Hum. Genet.* **51** (Suppl), A11.

181. Quan, F., Janas, J., Toth-Fejel, S., Johnson, D.B., Wolford, J.K., and Popovich, B.W. (1997) Uniparental disomy of the entire X chromosome in a female with Duchenne mus-cular dystrophy. *Am. J. Hum. Genet.* **60**, 160–165.

182. Schinzel, A.A., Robinson, W.P., Binkert, F., Torresani, T., and Werder, E.A. (1993) Exclu-sively paternal X chromosomes in a girl with short stature. *Hum. Genet.* **92**, 175–178.

183. Vidaud, D., Vidaud, M., Plassa, F., Gazengel, C., Noel, B., and Goossens, M. (1989) Father-to-son transmission of hemophilia A due to uniparental disomy. *Am. J. Hum. Genet.* **45** (Suppl), A226.

17
Genetic Counseling

Yael Furman, M.S.

INTRODUCTION

Genetic counseling is a communication process that deals with the human problems associated with the occurrence, or the risk of occurrence, of a genetic disorder in a family. This process involves an attempt by one or more appropriately trained persons to help the individual or family: (a) comprehend the medical facts, including the diagnosis, the probable course of the disorder, and the available management; (b) appreciate the way heredity contributes to the disorder, and the risk of recurrence in specified relatives; (c) understand the options for dealing with the risk of occurrence; (d) choose the course of action that seems appropriate to them in view of their risk and the family goals and act in accordance with that decision; and (e) make the best possible adjustment to the disorder in an affected family member and/or to the risk of recurrence of that disorder *(1)*.

This definition of genetic counseling was accepted in 1975 by the American Society of Human Genetics. However, the foundation of genetic counseling lies in the eugenics movement. In 1883 the English scientist Francis Galton introduced the term "eugenic" ("well born" in Greek) which refers to improvement of the human population over generations by giving "the more suitable races or strains of blood a better chance of prevailing speedily over the less suitable" *(2)*. Galton and his colleagues promoted the idea of an improved human species by selective breeding and founded the eugenics movement. During the 1920s and 1930s the movement was widely supported in the United States, Canada, and Britain and some eugenics laws were enforced such as sterilization of mentally retarded individuals. Some of these laws remained in effect until the Second World War.

The Nazi party adopted the eugenics idea, specifically by promoting sterilization of the "unfit." This raised opposition by both the public and the scientists who initially supported the eugenics concept, and as a result eugenics laws were reversed in the United States and Great Britain.

The eugenics movement changed accordingly. It adopted the view that society needs the reproductive contribution of all competent people, with the common belief that valuable characteristics were to be found in most social groups. The reformed eugenics movement was medically oriented and enhanced the theory of preventative medicine.

Between 1930 and 1945 human genetics was a new discipline populated by a small group of enthusiastic pioneers. By the mid-1940s, human genetics became a multi-

From: *The Principles of Clinical Cytogenetics*
Edited by: S. Gersen and M. Keagle © Humana Press Inc., Totowa, NJ

disciplined science including genetics, medicine, statistics, physiology, biochemistry, demography, and psychology *(3)*.

During that time, genetic evaluations were mostly provided by physicians with an interest in genetics or Ph.D. geneticists with an interest in medicine. It became apparent to them that in addition to giving accurate genetic information and risk assessment, there was a need to deal with the psychological and emotional impact of this information.

To describe the process of providing counseling and support together with genetic information, Sheldon Reed introduced the term "genetic counseling" in 1947. In contrast to the goal of the eugenics movement to serve the interests of the society, Reed felt that genetic counseling should serve the interest of families *(4)*.

The American Society of Human Genetics was formed by a group of geneticists in 1950, and 4 years later the *American Journal of Human Genetics* was established.

Subsequent dramatic progress in understanding genetic diseases and in developing and improving technology to diagnose genetic conditions increased the need for professionals who could deliver the new knowledge and options to the people who needed it.

In response to this need, in 1969 the first program in genetic counseling was established at Sarah Lawrence College in New York, and in the early 1970s a new category of medical professional emerged—the genetic counselor with a Master's degree.

Since then, with the increasing demand for genetic counselors, several other programs have been established *(5)*. Today, genetic counselors can be M.D.s, Ph.D.s, nurses, social workers, and other health care professionals who have been trained in medical genetics and counseling, and most are graduates of one of these programs. The genetic counselor's training involves a combination of genetics, medicine, laboratory work, counseling, social work, and ethical analysis.

The National Society of Genetic Counselor was formed in 1979 and it provides a network of genetic counselors throughout the country. The American Board of Genetic Counseling (ABGC) was formed in 1993 and certifies training programs that grant Master's degrees in genetic counseling. Graduates of these programs must then pass a certifying examination administered by this Board. Genetic counselors practice in a variety of settings, including hospitals, universities, laboratories, private offices, research units, and state and federal government offices.

INDICATIONS FOR REFERRAL FOR GENETIC COUNSELING

When should a patient and his or her family be referred to a genetic counselor? There are several appropriate situations:

An existing condition

- The patient, a previous child, or a family member has or is suspected of having a genetic condition based on clinical evaluation and laboratory test results and seeks diagnosis or confirmation of diagnosis.
- The patient, a previous child, or a family member is known to have a congenital abnormality and seeks information regarding this condition. (A congenital abnormality is a condition that is present at birth, and may be of genetic or environmental etiology, or a combination of both).
- The patient is a known carrier of:
  ❖ A balanced chromosomal rearrangement. An individual may have the normal amount of chromosomal material; however, it may be rearranged. As long as the rearrangement is balanced, the individual is not expected to have any clinical symp-

toms (see Chapter 9). Individuals with balanced chromosomal rearrangements have a higher risk of producing gametes that will be chromosomally unbalanced and result in spontaneous abortions and/or the birth of a child with an unbalanced karyotype. This will most likely have clinical implications (mental and/or physical disabilities).

❖ An autosomal dominant (AD) condition, which is caused by one mutated gene. The expressivity and severity of these conditions vary within families, as is the case with neurofibromatosis (NF), for example. Sometimes gene carriers (heterozygotes) may not have any apparent symptoms. Some dominant conditions exhibit reduced penetrance (for example, split-hand deformity). Penetrance is an all-or-none expression of a mutant gene. If a condition is expressed in fewer than 100% of individuals who carry the mutated gene, it is said to have reduced penetrance.

A carrier of a dominant disease gene has a 50% chance of transmitting the mutated gene to each conception. Both sexes are equally at risk. Isolated cases of an autosomal dominant condition in a family may represent a new mutation.

❖ An autosomal recessive condition, which is caused by two mutated genes on two homologous chromosomes. Carriers of only one mutated gene (heterozygotes) are usually not clinically symptomatic. When both parents are carriers there is a 1 in 4 (25%) risk of having an affected child. Both sexes are equally at risk. A sibling of a carrier has 50% risk of being a carrier. An unaffected sibling of an affected individual has 67% risk of being a carrier. Carriers of autosomal recessive disease genes are usually identified after the birth of an affected child, a family history of the condition, or general population screening. There are some recessive genetic conditions that are more prevalent among specific ethnic groups, such as Tay–Sachs disease among Ashkenazic Jews or sickle cell disease among African–Americans.

❖ An X-linked condition, in which the mutated gene is located on the X chromosome. Most X-linked conditions are recessive. Therefore, mostly males are affected, because they have only one X chromosome, and females are usually carriers. A female can be affected if her father is affected and her mother is a carrier, such as in the case of hemophilia. Females who are carriers of an X-linked recessive disease gene have a 50% risk with each conception of transmitting the mutant gene to their offspring. Half of the sons will be affected and half of the daughters will be carriers. There is no male-to-male transmission. In some X-linked recessive conditions, females may be mildly symptomatic. Females are considered obligate carriers if they have more than one first-degree affected male relative. An isolated case of an X-linked condition may represent a new mutation, such as in Duchenne muscular dystrophy.

X-Linked dominant conditions are very rare, and may be expressed in both males and females. There is no male to male transmission. Some X-linked dominant conditions are lethal in males, for example, incontinentia pigmenti.

❖ A polygenic/multifactorial condition, in which there is a genetic component but it does not follow any of the known Mendelian patterns, as is the case with open neural tube defects (ONTD), for example. It is assumed that a variety of genes and/or environmental factors attribute to the condition. Recurrence risks for siblings are lower than Mendelian risks. The recurrence risk is usually based on population studies and is influenced by the number of affected members of a family, relationship to affected members, severity, and sex.

• The patient has a history of reproductive failure:

❖ Infertility. The inability to conceive may result from different causes. One is having a sex chromosome abnormality such as 47,XXY, 46,XY/47,XXY, and 47,XYY in males and Turner syndrome (45,X) in females *(6)*. See Chapter 10. Also, some chromosomal structural rearrangements have been associated with infertility. Therefore, chromosome analysis should always be offered to couples who are experiencing infertility. Although infertility that is due to a chromosome abnormality cannot be restored, the diagnosis of such an abnormality will provide an explanation, and other reproductive options such as sperm or egg donation can be discussed with the couple *(7)*.

❖ Recurrent spontaneous abortions. There are various possible causes for having two or more spontaneous miscarriages, including genetic, endocrinologic, immunologic, and obstetric or gynecologic factors. The genetic etiology for multiple spontaneous pregnancy loss includes an unbalanced chromosome rearrangement, which may be the result of one parent being a carrier for a balanced chromosome rearrangement *(8)*. In approximately 5–10% of couples with multiple pregnancy losses, the cause is a balanced translocation or other chromosomal rearrangement in one member. The unbalanced karyotypes that can result are associated with an increased pregnancy loss rate and an increased risk for chromosomally abnormal liveborn offspring.

❖ Chromosomally abnormal products of conception (POC). About 20% of all recognized pregnancies end in miscarriage, mostly in the first trimester. About half of these are chromosomally abnormal. The most common chromosome abnormalities seen in spontaneous miscarriages are trisomy 16 and monosomy X. The diagnosis of an abnormal karyotype in a POC may provide a chromosomal basis for the pregnancy loss. It may also help in clarifying the risk for future miscarriages or for the birth of a chromosomally abnormal child. In those cases where the POC karyotype revealed a structural rearrangement, parental chromosome analysis should be offered. After the spontaneous abortion of a chromosomally abnormal fetus when parental chromosomes are normal, the recurrence risk for subsequent pregnancies is not significantly higher than the age related risk for delivering a baby with a chromosome abnormality *(9)*. Chromosome analysis of POCs is not always successful, usually due to the fetal demise that frequently precedes the pregnancy loss.

Prenatal genetic counseling

• A maternal illness during pregnancy: IDDM, seizure disorder, etc.
• Maternal exposure during pregnancy: viral exposure (rubella, varicella), environmental exposure (X-ray), chemical exposure (medications).
• Advanced maternal age (also referred to as increased maternal age). The risk of having a child with a chromosome abnormality increases with the mother's age *(10)* (see **Table 1**; see also Chapter 11, **Table 2**). Age 35 was initially used as a cutoff, because by age 35, the statistical chance that a chromosomal defect may be present in the baby was considered to be equal to or greater than the risk of amniocentesis. The possible risks of amniocentesis are of course decreasing with improvements in technology and the ever-increasing experience of the medical community with the procedure.
• Abnormal maternal serum screen. Initially, an elevated maternal serum α-fetoprotein (AFP), measured at about the 16th week of pregnancy, was used as a marker to identify patients at risk of having a fetus with an open neural tube defect. Later, a low maternal serum AFP level was associated with an increased risk of fetal Down syndrome. The addition of two other markers, human chorionic gonadotropin (hCG) and unconjugated Estriol (uE3), have increased the sensitivity for identifying patients at risk for Down syndrome or other chromosome abnormalities, specifically trisomy 18. A screening test only indicates the likelihood of an abnormality; when a screening test is abnormal, a diagnostic test, amniocentesis, is typically recommended *(11)*.
• Abnormal ultrasound examination. For decades, ultrasound (and more recently high-resolution ultrasound) has been a diagnostic tool for anatomical and structural abnor-

Table 1
Risks for Chromosome Abnormalities at Term by Maternal Age

Maternal Age at Term	Risk for Trisomy 21 (29)[a]	Risk for Any Chromosome Abnormality (30)[a,b]
15	1:1578	1:454
16	1:1572	1:475
17	1:1565	1:499
18	1:1556	1:525
19	1:1544	1:555
20	1:1528	1:525
21	1:1507	1:525
22	1:1481	1:499
23	1:1447	1:499
24	1:1404	1:475
25	1:1351	1:475
26	1:1286	1:475
27	1:1208	1:454
28	1:1119	1:434
29	1:1018	1:416
30	1:909	1:384
31	1:796	1:384
32	1:683	1:322
33	1:574	1:285
34	1:474	1:243
35	1:384	1:178
36	1:307	1:148
37	1:242	1:122
38	1:189	1:104
39	1:146	1:80
40	1:112	1:62
41	1:85	1:48
42	1:65	1:38
43	1:49	1:30
44	1:37	1:23
45	1:28	1:18
46	1:21	1:14
47	1:15	1:10
48	1:11	1:8
49	1:8	1:6
50	1:6	data not available

[a]risks based on maternal age at term. Term risks do not include chromosomally abnormal fetuses spontaneously lost before term.
[b]includes risk for trisomy 21. Does not include 47,XXX.

malities such as open neural tube, abdominal wall, and heart or limb reduction defects or choroid plexus cyst, cystic hygroma, or more subtle abnormalities. After such findings, amniocentesis may be indicated, as some sonographic observations have been associated with chromosomal abnormalities: increased nuchal thickness and short femur lengths with Down syndrome, choroid plexus cysts, and club foot with trisomy 18.

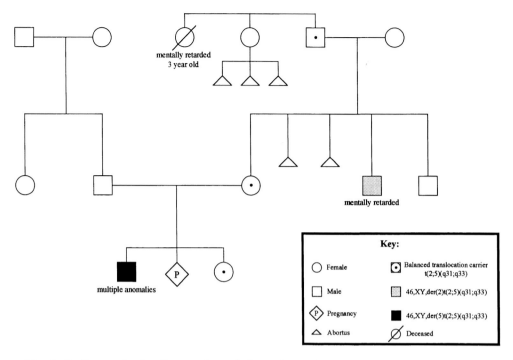

Fig. 1. Pedigree of a family carrying a balanced translocation involving the long arms of chromosomes 2 and 5. See key for interpretation of symbols.

WHAT TO EXPECT FROM A GENETIC COUNSELING SESSION

A genetic referral differs from other medical referrals because it raises questions and anxiety concerning not only the patient, but also other family members. Therefore, it is important to clarify with the patient the reason for the referral and identify the patient's concerns and expectations from the session. Sometimes, the family, and not just the person affected or potentially affected with a genetic disease, is truly the "patient."

A correct diagnosis is a prerequisite for genetic evaluation but, unfortunately, making a diagnosis is not always a simple task. It is often time consuming and may require various tests. Sometimes a diagnosis cannot be made. An inability to pinpoint a clear diagnosis may stimulate additional questions and concerns.

Review of family history, medical history, relevant lab test results, and medical records of the affected individual/s are crucial to confirm or make a diagnosis. If information is not sufficient, a referral for special evaluations and testing may be indicated to establish a diagnosis.

Family history is recorded and analyzed by way of a *pedigree*. This involves a detailed interview of the patient (and often of other family members present during the session) concerning as much of the extended family as possible, and utilizes a standardized system of symbols to represent each family member and his or her status concerning relevant medical and/or genetic issues *(12)*. An example is shown in **Fig. 1**.

Based on all of this information, the counselor will discuss the recurrence risk with the patient and/or family members.

Risk perception and interpretation of the results may differ for different people. The patient's expectations, previous experience with the condition in question, and his or her own morals and beliefs may influence these. The counselor must be sensitive to these issues when discussing the outcome of the risk involved and the options available. Often there is a need to suggest additional testing such as chromosome analysis or carrier testing by DNA analysis or biochemical methods to clarify the risk.

Prenatal testing by amniocentesis is usually done at about 16–18 weeks' gestation, although samples may be obtained from as early as 10 weeks essentially until term. From a sample of amniotic fluid, amniocytes are cultured and fetal chromosomes are analyzed. The level of AFP in the amniotic fluid is also measured for possible diagnosis of open neural tube and abdominal wall defects. The risk of a miscarriage following this procedure has always been quoted at about 0.5%, but as indicated above is frequently much lower.

Prenatal diagnosis by chorionic villus sampling (CVS) is usually done at 9–12 weeks of gestation. A small sample of chorionic villi is obtained, then cultured and analyzed. The cytotrophoblasts present in the sample are actively dividing at this gestational age, and can be directly prepared for analysis without culture, one of the initially attractive features of the CVS procedure. Although either the direct or cultured methods can be used, most laboratories will examine cells using both methods whenever possible. The risk of a miscarriage following CVS is about 1–2%.

If there is a need for additional tests, some can be done directly on the amniocytes or chorionic villi and some can be done on amniotic fluid, whereas others require cultured cells. For a detailed description and discussion of these procedures, see Chapter 11.

Prenatal diagnosis is an optional procedure. For some patients, the decision to have such a procedure may be a difficult one. The counselor may bring up some points that will help the patient through this decision-making process: Does the risk involved with the procedure outweigh the risk of having an affected child? If the procedure is not done, what will the anxiety be like until the baby is born? If the procedure is done and reveals abnormal results, what are the options? If termination of the pregnancy is chosen, what will be the impact of a second trimester termination, what is the recurrence risk, and what is the time period before they can try again? If continuation of the pregnancy is chosen, knowing the results may change the management of the pregnancy and delivery. Advanced knowledge allows the family to make the necessary psychological and economical adjustments, and the medical staff will be prepared to provide treatment, if available, at the birth of the child. Discussing these issues and exploring the patient's feelings toward them may help the patient and family better understand the situation and make a decision that will be best for them.

Once testing has been completed, the counselor should communicate the results to the patient/family. If a patient is identified as a carrier of a chromosomal translocation or other genetic condition, the genetic counselor explains the risk for siblings and discusses further testing if a pregnancy is or will be involved.

The identification of a phenotypically normal individual as a carrier may bring up additional counseling issues. The patient may feel stigmatized, suffer reduced self-esteem, or be burdened by guilt as he or she feels "responsible" for the increased risk.

If a pregnancy is determined to be abnormal ("affected"), the genetic counselor explains the prognosis and available options. Some genetic conditions are not compatible with life, or involve severe mental and physical retardation. There are conditions in which the severity varies and cannot be clarified by prenatal testing (e.g., cystic fibrosis, Gaucher disease), and some conditions, such as phenylketonuria (PKU), can be treated.

Receiving an abnormal result following prenatal diagnosis is always unexpected, shocking, and upsetting to the parents. The explanation of the results should include a discussion of the accuracy and reliability of the test, the clinical presentation of the condition (including photographs when possible), the likelihood that the baby will die during pregnancy or shortly after birth, the requirements of caring for a baby with such a condition, and the impact on family dynamics.

Should a couple decide to terminate a pregnancy, they should be given information regarding the procedure and its implications. Termination of a wanted pregnancy has an enormous emotional impact. Sometimes the genetic counselor can help deal with these emotions, while at other times the counselor may find it necessary to refer the patients to a support group. Regardless of the individual dynamic, the counselor must always remain nondirective concerning this difficult decision. This is not always easy, as the counselor is often asked "What would you do if you were in this position?"

When a couple decides to continue pregnancy with a known abnormality, they should be given information regarding possible treatment and early stimulation programs, and should be referred to a support group dedicated to individuals with the same condition.

CHROMOSOME ABNORMALITY DETECTED BY PRENATAL DIAGNOSIS—REFERRALS AND RISK ASSESSMENT

In some settings, counseling is a routine part of prenatal diagnosis, and occurs before the diagnostic procedure itself. In this way, the counselor (or, in some cases, the physician or midwife) can explain the risks and limitations of the test, and prepare the patient for the potential for an abnormal outcome. However, a patient or family is often referred to a genetic counselor only after a genetic condition, usually a chromosome abnormality, has been diagnosed. In this scenario, patient and counselor have never met, and the important relationship between them must evolve during an emotionally trying time. The nature of such a counseling session depends upon the abnormality present.

Autosomal Trisomy

Trisomies 21, 13, and 18 are among the most common prenatally diagnosed chromosome abnormalities (see Chapter 8). Trisomy 21 (Down syndrome) results in mental retardation and variable clinical severity. Individuals with Down syndrome have characteristic facial features and may also have other birth defects and long term medical problems. These include congenital heart defects, gastrointestinal abnormalities, immunologic problems, thyroid dysfunction, and visual and hearing disorders, as well as an increased risk of developing leukemia. The level of disability cannot be predicted from the chromosome diagnosis, even if mosaicism involving a normal cell line (e.g., 47,XY,+21/46,XY) is diagnosed. Children with Down syndrome learn to walk and talk. However, they usually accomplish their developmental milestones at a slower rate than do normal children. They always require special education and can benefit from early infant stimulation programs *(13)*.

Trisomy 13 results in multiple congenital anomalies and severe mental retardation. There are major defects of the brain, heart, and gastrointestinal and urinary tracts associated with this condition. There may also be cleft lip and/or palate, eye abnormalities, and polydactyly. The infants tend to be of low birth weight and fail to thrive. Survival beyond the first year of life is unusual and those who survive are severely retarded *(14)*.

Trisomy 18 is a syndrome of severe mental retardation. There are also multiple congenital anomalies involving several organ systems, including the brain, heart, gastrointestinal system, and kidneys. Infants with trisomy 18 also have recognizable facial features. The majority of infants diagnosed with trisomy 18 do not survive beyond the first year of life *(15)*.

The genetic counselor is usually involved with the parents during the decision-making process. After termination of a pregnancy, if this option is chosen, the counselor may be involved with the grief counseling and/or may refer the patients to a support group made up of people who have had similar experiences.

After the prenatal diagnosis of a trisomy, the recurrence risk for future pregnancies is about 0.8% when the mother is under 30 years of age. For women who are over 30 years old, the risk is based on age *(6)*. Prenatal diagnosis should be recommended for all future pregnancies.

Sex Chromosome Abnormality

Individuals with Klinefelter syndrome (47,XXY) have a characteristic body habitus that includes tall stature, lack of normal secondary sexual development, and gynecomastia. Virtually all nonmosaics are infertile. Intelligence levels are variable and will fall in the normal to low normal range. Mild to moderate mental retardation is seen in a small percentage of patients. Learning disabilities, language delay, and behavior problems are not uncommon. IQ tend to be 15–20 points lower than in normal siblings. Exogenous testosterone therapy will deepen the voice, promote growth of facial hair, and prevent breast development *(16)*.

Individuals with Turner syndrome are females with a characteristic body habitus and short stature. The average adult height ranges from 4 ft 7 in to 5 ft (140–152 cm). Other features of Turner syndrome include a short neck that appears to be webbed, occasional heart defects, and kidney abnormalities. Puberty is delayed and most of these women are infertile. Because most individuals with Turner syndrome do not produce adequate amounts of female hormones, estrogen replacement therapy is recommended in early adolescence to induce the onset of puberty. Growth hormone replacement therapy may also be an option. Girls with Turner syndrome may be at risk for learning disabilities, particularly abstract skills. For the most part IQ is normal; however, psychological evaluation is recommended before formal school entry *(17)*.

Other commonly found sex chromosome abnormalities include 47,XXX and 47,XYY. The clinical features of these sex chromosome abnormalities are not as striking as those mentioned above. Sex chromosome abnormalities are covered in detail in Chapter 10.

The decision-making process involving a pregnancy that has been diagnosed with a sex chromosome abnormality is much more difficult than with autosomal abnormalities because the clinical picture is less clear, particularly concerning mental development, and these conditions are rarely life-threatening. This is particularly true with such ill-defined entities as 47,XXX and 47,XYY. The genetic counselor should offer moral support and clarifications regarding possible phenotypic effects of the condition in question. If the pregnancy is continued, the parents should be referred to the appropriate support groups and educational consultants.

In the above-mentioned examples, the chromosome analysis is definitive and the conditions diagnosed are well recognized. But this is not always the case. Occasionally, there are situations in which the diagnosis is not clear-cut, as described in the following sections.

Chromosomal Mosaicism

Chromosomal mosaicism is defined as the presence of more than one population of cells, in an individual, tissue or culture, that differ in their chromosomal constitution. Mosaicism presents unique challenges to the genetic counselor.

Mosaicism Finding Following Amniocentesis

About 1–2% of amniocenteses result in reports of mosaicism. However, only a small percentage of these cases (0.1–0.4%) reflect true mosaicism *(18–21)*. Most involve pseudomosaicism, which represents culture artifact and does not reflect the fetal karyotype. True mosaicism may actually represent the presence of more than one cell line in the fetus, or may be due to a cell line (or cell lines) present only in the placenta or other extraembryonic tissues, and therefore also may not reflect the fetal karyotype. Thus the distinction between true and pseudomosaicism is not always clear or certain. Because of this, several categories of mosaicism have been defined, based on the number of abnormal cells, the number of clones, and the number of cultures in which mosaicism is detected *(22)*.

Level 1: A single abnormal cell. All such instances represent pseudomosaicism, and would be ignored by most cytogeneticists. If mentioned in a laboratory report, the parents should be reassured that this result does not reflect the baby's chromosomal constitution.

Level 2: Two or more cells with the same chromosome abnormality in a single colony. More than 80% of those are pseudomosaicism, as a true mixed colony essentially demonstrates that the observed chromosomal difference arose in culture. However, because of the small but finite possibility that what appears to be a mixed colony might actually be two intermingled colonies, a clear-cut distinction may be difficult and the parents should be informed that this type of result might still represent a true low-level mosaicism. Based on the specific chromosome(s) involved, a review of the literature may provide more information regarding the outcome of pregnancies with the same chromosomal mosaicism.

Level 3: Two or more cells with the same chromosome abnormality seen in two or more colonies. More than 50% of these are true mosaicism. If all abnormal colonies arose in a single culture, this finding more than likely represents culture artifact. If the same abnormality is observed in colonies that originated in separate cultures (and assuming correct laboratory practices that ensure this), the result is almost always representative of true mosaicism. However, as discussed previously, this may involve nonfetal tissues. About half of the cases described in the literature involve an autosome, about 40% involve a sex chromosome, and the remainder involve an extra structurally abnormal chromosome (marker chromosome). The ultimate phenotypic effects of true mosaicism cannot be predicted, as the proportion of different chromosome complements may vary from tissue to tissue.

Because some chromosome abnormalities have been associated with sonographic findings, high-resolution ultrasound may be indicated. A repeat amniocentesis may also be offered. However, this cannot always clarify the question of true vs pseudomosaicism. If the result of the second amnio is normal, it does not invalidate the results of the first amniocentesis (but will lower the risk). However, if the second amniocentesis reveals the same mosaicism as the first, it supports the finding of true mosaicism, but again does not prove that the mosaicism does not involve extrafetal tissue. Fetal blood sampling can be offered to look at the karyotype of a different fetal tissue, but again, a normal result does not eradicate the original finding, although it does increase the likelihood of mosaicism of extraembryonic origin.

PUBS (percutaneous umbilical blood sampling) is a method to obtain fetal blood for diagnostic purposes. The procedure is usually done after the 16th week of pregnancy. Under ultrasound guidance, the umbilical cord is punctured and a blood sample is drawn. The risk involved is about 1–2%. Chromosome analysis as well as other biochemical and DNA analyses can be performed. Analysis to determine whether maternal blood is present in the sample should be performed as well.

If a pregnancy that was diagnosed with true mosaicism is terminated, chromosome analysis of different fetal tissues should be recommended. If the pregnancy goes to term, chromosome analysis of a cord or neonatal blood sample should be performed.

AUTOSOMAL MOSAICISM

Most normal/autosomal trisomy mosaics detected in amniotic fluid represent pseudomosaicism; however, most mosaicisms for trisomies 13, 18, and 21 represent true mosaicism associated with phenotypic abnormalities *(19)*.

SEX CHROMOSOME MOSAICISM

Patients with 45,X/46,XX mosaicism can express a range of clinical features, varying from normal female maturation to full expression of a Turner syndrome phenotype. About 95% of fetuses with 45,X/46,XY mosaicism will have normal male genitalia, even though there is a wide range of phenotypic abnormalities ranging from females with Turner syndrome to ambiguous genitalia to males with hypospadias, undescended testes, rudimentary phallus, and scrotal fusion. There is about 27% risk for abnormal gonadal histology. Ultrasound examination to determine the phenotypic sex may be reassuring for the parents. Individuals with 45,X/46,XY mosaicism are at risk for gonodoblastoma and should be followed closely for signs of the development of gonadal tumors *(23)*. True 46,XX/46,XY chimerism is extremely rare. XX/XY mixoploidy is usually pseudomosaicism resulting from maternal cell contamination (MCC), with subsequent growth in culture of maternal cells, in a 46,XY pregnancy. Maternal cell contamination cannot be detected with a 46,XX pregnancy, and therefore the estimate of its occurrence is obtained by measuring the observed frequency in XY fetuses and multiplying by a factor of 2.

Mosaicism Following CVS

As discussed previously and in Chapter 11, chromosome analysis can be performed directly on cytotrophoblasts from the chorionic villi with results available in 24–48 h. Cells are also cultured and analyzed and results are available in 7–10 days. Mosaicism is seen in about 1% of direct preparations. Most of the time these results do not reflect the fetus' karyotype. If the results of the cultured cells are normal, then mosaicism seen in the direct preparation is considered confined placental mosaicism *(24)*. If mosaicism is seen in both direct and cultured analyses, ultrasound and amniocentesis are recommended.

Structural Rearrangements

Structural chromosome rearrangements are a result of chromosome breakage in which chromosomal material is reapportioned in some way. These breaks can occur anywhere in any chromosome, although some chromosomal regions (so-called "hot spots") are involved in rearrangements more often than others are. Chromosomal rearrangements can be present in all cells, or in mosaic fashion. They can be balanced, when all chromosomal material is present but rearranged, or unbalanced, when there is extra or missing chromosomal material. For details, see Chapter 9.

A structural chromosome rearrangement is detected in about 1 in every 1000 prenatal cytogenetic diagnoses. Parental chromosome analysis should be done as soon as possible after such a finding is reported. If the rearrangement appears to be balanced and is found to be familial because one of the parents carries the same apparently balanced rearrangement, the family can be reassured that there is no increased risk for abnormalities in the fetus as a result of this rearrangement. However, recently there has been a concern that, even with a parental balanced rearrangement, there appears to be about a 1% risk for unpredictable abnormalities in the fetus. Possible sources of this risk are: (a) that the fetal rearrangement is actually not identical to that found in the parent and is in fact unbalanced, but this fact is beyond the resolution of routine cytogenetics; (b) uniparental disomy exists, resulting from trisomy due to the parental translocation which was subsequently "rescued" by loss of the homologue from the other parent; and (c) a recessive allele on the normal homologue has been exposed *(6)*.

If a rearrangement is unbalanced, there is a significant risk for mental and/or physical abnormalities. Unbalanced structural rearrangements typically result in partial monosomy (when a segment of a chromosome is missing) and/or partial trisomy (when a segment of a chromosome is duplicated). When amniocentesis or CVS detects an unbalanced rearrangement, parental chromosomes should be studied in an attempt to determine its origin. If one of the parents is a carrier of a balanced rearrangement, the recurrence risk can be clarified. Other family members should be offered testing as well.

When a structural rearrangement is not familial, it is considered to be *de novo* (new). With a *de novo* apparently balanced structural rearrangement, interpretation of the findings is more difficult. The rearrangement could be a reciprocal translocation, the result of breakage of nonhomologous chromosomes with a reciprocal exchange of the broken segments. In *de novo* apparently balanced reciprocal translocations, there is about 6% (3% over the risk in the general population) for mental retardation and/or phenotypic abnormalities *(25)*. Ultrasound can be offered to diagnose any apparent anatomical abnormalities. However, it can not detect mental deficiencies. With a *de novo* Robertsonian translocation (the result of fusion of two acrocentric chromosomes at or near the centromere with inconsequential loss of the short arms) there is about 3.7% risk for abnormalities (less than 1% over the general population risk) *(25)*. An inversion is the result of a segment of a chromosome that was broken and reinserted in reverse orientation. With *de novo* inversions there is about a 9.4% risk for abnormalities (6–7% risk over the general population risk) *(25)* (see Chapter 9).

If a fetal chromosome rearrangement is not found in either of the parents, the possibility of non-paternity might also be considered, since this can dramatically change the associated risk. If the genetic counselor has reason to believe that the man who has been tested may not be the biological father of the pregnancy, the patient should be contacted, separately from her partner, and the possibility of non-paternity should be raised. If such possibility exists, a blood sample for chromosome analysis should be collected from the potential biological father.

When a pregnancy is conceived via sperm or egg donor and a fetal chromosome rearrangement is detected, a blood sample from the donor should be obtained if possible. Medical records might also be requested, as such donors often have chromosome analysis performed as part of a standard work-up. However, individuals with known rearrangements are now typically considered ineligible as sperm or egg donors.

Marker or Supernumerary Chromosomes

Markers may be present in addition to the normal chromosome complement and are often in mosaic fashion. These represent both numerical and structural abnormalities. Because the identification of these markers may be difficult, the clinical interpretation may be difficult as well. With a *de novo* supernumerary chromosome, there is about a 13% risk for abnormalities. The risk does not seem to be different for mosaic or nonmosaic cases *(25)*.

GENETIC COUNSELING BY AGE GROUP

The role of the genetic counselor varies according to the age of the affected individual. If the patient is a newborn or young child, it is the parents who will receive counseling. If the patient is an adolescent (often the case with sex chromosome abnormalities), the parents may wish for their child to be present during counseling. If the affected individual is an adult, a spouse or significant other may attend, and extended family members may become involved if the patient(s) choose to share information with them.

Neonatal

Dysmorphic features and congenital anomalies observed or suspected at birth are indications for cytogenetic studies. Sometimes, additional tests such as biochemical studies, X-ray films, or ultrasound examination are indicated. Chromosome analysis as a follow-up to prenatal studies, whether normal or abnormal, is often performed for reassurance or confirmation and closure for the parents. This is particularly valuable after chorionic villus sampling suggests confined placental mosaicism (see Chapter 11).

Pediatric

Dysmorphic features, physical disabilities, medical problems, speech and development delay, and behavioral problems are indications for cytogenetic studies. Increased spontaneous chromosome breakage may be diagnostic for some conditions, such as Bloom syndrome. Although rare, sex chromosome abnormalities may also be suspected in young children.

Adolescent

Delayed puberty, amenorrhea, tall/short stature, and developmental delay are indications for cytogenetic studies, specifically for sex chromosome abnormalities.

Giving parents the news that their newborn baby, child, or adolescent is diagnosed or is suspected to have a genetic condition or birth defect is one of the most painful tasks of genetic counseling.

Once a diagnosis is established, or if a diagnosis cannot be made, the counselor will communicate these results to the parents and provide information regarding additional testing, prognosis, available treatment, educational resources, the impact of the diagnosis on the family, and available support groups. If the condition is genetic, the recurrence risk and the significance for other family members should be explained along with an offer of additional testing if available and appropriate.

Adults

Adults may be referred for genetic evaluation for various reasons:

1. Reproductive failure (see indications for genetic counseling).
2. Late onset genetic conditions. In some nonchromosomal genetic conditions, symptoms appear later in life, often after the reproductive years. These conditions are usually autosomal dominant and therefore impart a significant risk (50%) for offspring. Affected individuals may seek genetic counseling to clarify the risk for their children and to learn about available testing.
3. Cancer. Our understanding of the genetics of cancer and the association between cancer and chromosomal or molecular abnormalities has grown tremendously in the last decade and continues to expand. Advances in molecular techniques have also enabled researchers to uncover genes that may predispose patients to different types of cancer.

    Some cancers are associated with specific chromosomal translocations (see Chapter 13). When these are suspected, bone marrow chromosome analysis may be a helpful diagnostic tool. For example:
    - t(9;22)(q34;q11) is found in about 90–95% of patients with chronic myelogenous leukemia (CML), in 10–20% of patients with acute lymphocytic leukemia (ALL), and in 3% of patients with acute myeloid leukemia (AML) *(26,27)*.
    - t(8;14)(q24;q32) is found in about 80% of patients with Burkitt's lymphoma *(28)*.
    - t(11;14)(q13;q32) is found in about 10–30% of patients with chronic lymphocytic leukemia.

For a detailed discussion of such rearrangements, refer to Chapter 13.

Occasionally, when chromosome analysis is performed on cancer patients, nonspecific and apparently balanced chromosomal rearrangements will be found. In these cases, additional testing is usually suggested in order to determine whether such rearrangements are constitutional (present since birth and probably unrelated to the cancer) or acquired (clinically significant concerning the cancer). If constitutional, the patient should be counseled regarding about the significance of these findings for reproduction (there is an increased risk for spontaneous miscarriages and/or chromosomally abnormal children), either for themselves or other family members who are of reproductive age. Informing and testing of other family members should be encouraged.

SUMMARY

Since the 1975 definition of genetic counseling, there have been significant changes in the field of human genetics and genetic counseling.

The awareness of the general population of genetic conditions and the availability of testing and genetic counseling has increased the number of individuals who seek genetic services. As a result, accessibility to genetic services, the number of providers, and the quality of services have improved.

The diversity of the population that seeks genetic counseling has enhanced the counselor's knowledge and abilities to meet the special needs of different individuals.

Technological advancements and an increased understanding of the nature of genetic conditions, particularly in the field of molecular genetics, have significantly increased the availability of testing for various genetic conditions. A recent challenge for genetic counselors has been the development of predictive testing for adult onset disorders, including breast cancer and Alzheimer's disease. DNA-based testing can modify the relative probability that an individual will develop a genetic disease later in life, yet for some diseases there is no cure or treatment. For many people, the benefits of such testing therefore remain questionable.

Even though genetic counseling can raise complex ethical questions that often do not have clear and simple answers, it is an integral part of the ever-developing field of medical genetics. As new genetic information and knowledge become available, new ethical, medical, social, and legal problems arise. Genetic counselors are available to meet these challenges and help families cope with these issues.

REFERENCES

1. Ad Hoc Committee on Genetic Counseling (1975) Genetic Counseling. *Am. J. Hum. Genet.* **27,** 240–242.
2. Pearson, K. (1914–1930) *The LIfe, Letters and Labours of Francis Galton.* Cambridge University Press.
3. Kevles, D. (1985) *In the Name of Eugenics.* Alfred A. Knoff, New York.
4. Reed, S.C. (1980) In the beginning. In *Counseling in Medical Genetics,* 3rd edition. John Wiley, New York.
5. Fine, B., Baker, D.L., Fiddler, M.B., and ABGC Consensus Development Consortium (1996) Practice-Based Competencies for Accreditation of and Training in Graduate Programs in Genetic Counseling. *J. Genet. Counsel.* **5,** 113–121.
6. Gardner, R.J.M. and Sutherland, G.R. (1996) *Chromosome Abnormalities and Genetic Counseling. Second edition.* Oxford University Press.
7. Saenger, P. (1996) Turner's Syndrome. *N. Engl. J. Med.* **335,** 1749–1754.
8. Gadow, E.C., Lippold, S., Otano, L., Serafin, E., Scarpati, R., and Matayoshi, T. (1991) Chromosome rearrangements among couples with pregnancy losses and other adverse reproductive outcomes. *Am. J. Med. Genet.* **41,** 279–281.

9. Warburton, D., Kline, J., Stein, Z., Hutzler, M., Chin, A., and Hassold, T. (1987) Does the karyotype of a spontaneous abortion predict the karyotype of a subsequent abortion?— evidence from 273 women with two karyotyped spontaneous abortions. *Am. J. Hum. Genet.* **41,** 465–483.

10. Hook, E.B. (1994) Down's syndrome epidemiology and biochemical screening. In *Screening for Down's Syndrome* (Grudzinskas, J.G., Chard, T., Chapman, M., and Cuckle, H. eds.) Cambridge University Press, Cambridge, pp. 1–18.

11. American College of Obstetricians and Gynecologists (1996) Maternal Serum Screening. *ACOG Educ. Bull. No. 228.*

12. Bennett, R.L., Steinhaus, K.A., Uhrich, S.B., O'Sullivan, C.K., Resta, R.G., Lochner-Doyle, D., Markel, D.S., Vincent, V., and Hamanishi, J. (1995) Recommendations for standardized human pedigree nomenclature. *Am. J. Hum. Genet.* **56,** 745–752.

13. Down, J.L.H. (1866) Observations on an ethnic classification of idiots. *Clin. Lect. Rep. Lond. Hosp.* **3,** 259.

14. Patau, K., Smith, D.W., Therman, E., and Inhorn, S.L. (1960) Multiple congenital anomaly caused by an extra chromosome. *Lancet* **i,** 790–793.

15. Edwards, J.H., Harnden, D.G., Cameron, A.H., Cross, V.M., and Wolff, O.H. (1960) A new trisomic syndrome. *Lancet* **i,** 711–713.

16. Klinefelter, H.F. Jr., Reifenstein, E.C. Jr., and Albright, F. (1942) Syndrome characterized by gynecomastia, aspermatogenesis without aleydigism, and increased secretion of follicle-stimulating hormone (gynecomastia). *J. Clin. Endocrinol. Metab.* **2,** 615.

17. Turner, H.H. (1938) A syndrome of infantilism, congenital webbed neck, and cubitus valgus. *Endocrinology* **28,** 566–574.

18. Welborn, J.L. and Lewis, J.P. (1990) Analysis of mosaic states in amniotic fluid using the *in situ* colony technique. *Clin. Genet.* **38,** 14–20.

19. Hsu, L.Y.F. and Perlis, T.E. (1984) United States survey on chromosome mosaicism and pseudomosaicism in prenatal diagnosis. *Prenat. Diagn.* **4,** 97–130.

20. Moertel, C.A., Stupca, P.J., and Dewald, G.W. (1992) Pseudomosaicism, true mosaicism and maternal cell contamination in amniotic fluid processed with in situ culture and robotic harvesting. *Prenat. Diagn.* **12,** 671–683.

21. Hsu, L.Y.F., Kaffe, S., Jenkins, E.C., Alonso, L., Benn, P.A., David, K., Hirschhorn, K., Lieber, E., Shanske, A., Shapiro, L., Schutta, E., and Warburton, D. (1992) Proposed guidelines for diagnosis of chromosome mosaicism in amniocytes based on data derived from chromosome mosaicism and pseudomosaicism studies. *Prenat. Diagn.* **12,** 555–573.

22. Worton, R.G. and Stern, R. (1984) A Canadian collaborative study of mosaicism in amniotic fluid cell cultures. *Prenat. Diagn.* **4,** 131–144.

23. Chang, H.J., Clark, R.B., and Bachman, H. (1990) The phenotype of 45,X/46,XY mosaicism: an analysis of 92 prenatally diagnosed cases. *Am. J. Hum. Genet.* **46,** 156–167.

24. Kalousek, D.K., Howard-Peebles, P.N., Olson, S.B., Barrett, I.J., Dorfman, A., Black, S.H., Schulman, J.D., and Wilson, R.D. (1991) Confirmation of CVS mosaicism in term placentae and high frequency of intrauterine growth retardation: Association with confined placental mosaicism. *Prenat. Diagn.* **11,** 743–750.

25. Warburton, D. (1991) De novo balanced chromosome rearrangements and extra marker chromosomes identified at prenatal diagnosis: clinical significance and distribution of breakpoints. *Am. J. Hum. Genet.* **49,** 995–1013.

26. Rowley, J.D. (1990) The Philadelphia chromosome translocation: a paradigm for understanding leukemia. *Cancer* **65,** 2178–2184.

27. Heim, S. and Mitelman, F. (1995) *Cancer Cytogenetics,* 2nd edition. Wiley-Liss, New York.

28. Croce, C.M. and Nowel, P.C. (1985) Molecular basis of human B cell neoplasias. *Blood* **65,** 1–7.

29. Cuckle, H.A., Wald, N.J and Thompson, S.C. (1987) Estimating a woman's risk of having a pregnancy associated with Down's Syndrome using her age and serum alpha-fetoprotein level. *Br. J. Obstet. Gynaecol.* **94**:387.
30. Hook, E.B. (1981) Rates of chromosomal abnormalities at different maternal ages. *Obstet. Gynaecol.* **58**(3), 282–285.

---

*Page references to figures or tables are set in *italic type*.